T0297506

Fundamentals of University Mathematics
Third edition

Related titles:

Mathematical analysis and proof
Second edition
(ISBN 978-1-904275-40-4)
This fundamental and straightforward text addresses a weakness observed among present-day students, namely a lack of familiarity with formal proof. Beginning with the idea of mathematical proof and the need for it, associated technical and logical skills are developed with care and then brought to bear on the core material of analysis in such a lucid presentation that the development reads naturally and in a straightforward progression. Retaining the core text, the second edition has additional worked examples which users have indicated a need for, in addition to more emphasis on how analysis can be used to tell the accuracy of the approximations to the quantities of interest which arise in analytical limits.

Mathematical methods for mathematicians, physical scientists and engineers
(ISBN 978-1-904275-10-7)
This practical introduction encapsulates the entire content of teaching material for UK honours degree courses in mathematics, physics, chemistry and engineering, and is also appropriate for post-graduate study. It imparts the necessary mathematics for use of the techniques, with subject-related worked examples throughout. The text is supported by challenging Problem-Exercises (and Answers) to test student comprehension. Index notation used in the text simplifies manipulations in the sections on vectors and tensors. Partial differential equations are discussed, and special functions introduced as solutions. The book will serve for postgraduate reference worldwide, with variation for the USA.

Mathematics teaching practice: Guide for university and college lecturers
(ISBN 978-1-898563-79-2)
Clarifying the distinction between mathematical research and mathematics education, this book offers hundreds of suggestions for making small and medium sized changes for lectures, tutorials, task design, or problem solving. Here is guidance and inspiration for effective mathematics teaching in a modern technological environment, directed to teachers who are unhappy with results or experience, or those now in teacher training or new to the profession. Commencing with a range of student behaviours and attitudes that have struck and amazed tutors and lecturers, Professor Mason offers a wealth of partial diagnoses, followed by specific advice and suggestions for remedial actions.

Details of these and other Woodhead Publishing mathematics books can be obtained by:

- visiting our web site at www.woodheadpublishing.com
- contacting Customer Services (e-mail: sales@woodheadpublishing.com; fax: +44 (0) 1223 893694; tel.: +44 (0) 1223 891358 ext. 130; address: Woodhead Publishing Limited, Abington Hall, Granta Park, Great Abington, Cambridge CB21 6AH, UK)

If you would like to receive information on forthcoming titles, please send your address details to: Francis Dodds (address, tel. and fax as above; e-mail: francis.dodds@ woodheadpublishing.com). Please confirm which subject areas you are interested in.

Fundamentals of University Mathematics

Third edition

Colin McGregor, Jonathan Nimmo and
Wilson Stothers

Department of Mathematics, University of Glasgow

WOODHEAD
PUBLISHING

Oxford   Cambridge   Philadelphia   New Delhi

Published by Woodhead Publishing Limited, Abington Hall, Granta Park, Great Abington, Cambridge CB21 6AH, UK
www.woodheadpublishing.com

Woodhead Publishing, 525 South 4th Street #241, Philadelphia, PA 19147, USA

Woodhead Publishing India Private Limited, G-2, Vardaan House, 7/28 Ansari Road, Daryaganj, New Delhi – 110002, India
www.woodheadpublishingindia.com

First edition 1994, Albion Publishing Limited
(since renamed Horwood Publishing Limited)
Second edition 2000, Horwood Publishing Limited, reprinted in 2007
Third edition 2010, Woodhead Publishing Limited

© C. M. McGregor, J. J. C. Nimmo and W. W. Stothers, 2010
The authors have asserted their moral rights.

This book contains information obtained from authentic and highly regarded sources. Reprinted material is quoted with permission, and sources are indicated. Reasonable efforts have been made to publish reliable data and information, but the authors and the publisher cannot assume responsibility for the validity of all materials. Neither the authors nor the publisher, nor anyone else associated with this publication, shall be liable for any loss, damage or liability directly or indirectly caused or alleged to be caused by this book.

Neither this book nor any part may be reproduced or transmitted in any form or by any means, electronic or mechanical, including photocopying, microfilming and recording, or by any information storage or retrieval system, without permission in writing from Woodhead Publishing Limited.

The consent of Woodhead Publishing Limited does not extend to copying for general distribution, for promotion, for creating new works, or for resale. Specific permission must be obtained in writing from Woodhead Publishing Limited for such copying.

Trademark notice: Product or corporate names may be trademarks or registered trademarks, and are used only for identification and explanation, without intent to infringe.

British Library Cataloguing in Publication Data
A catalogue record for this book is available from the British Library.

ISBN 978-0-85709-223-6 (print)
ISBN 978-0-85709-224-3 (online)

The publisher's policy is to use permanent paper from mills that operate a sustainable forestry policy, and which has been manufactured from pulp which is processed using acid-free and elemental chlorine-free practices. Furthermore, the publisher ensures that the text paper and cover board used have met acceptable environmental accreditation standards.

Table of contents

Preface to the Third Edition

This book is based on lectures given by the authors to first year students at the University of Glasgow. We record, with sadness, the passing of Wilson Stothers who died during the preparation of the third edition.

The object of the book is to provide, in a single volume, a unified treatment of first year topics fundamental to university mathematics. Full mathematical rigour has not been attempted but the general approach is suitable for students aiming for an honours degree in mathematics. Also, students of computer science, physics and statistics will find the basic mathematics they need. The book covers clearly and comprehensively much of the material that other books tend to assume. This is particularly valuable now that less mathematics is taught in schools. Indeed, the book might be used as a supplementary text for pre-university work.

The contents are set out in nineteen chapters with seven appendices. Worked examples are provided on all topics covered, and each chapter concludes with a comprehensive selection of exercises to which answers are given. For students seeking further challenges, a selection of problems intersperses the text, ranging from the straightforward to the very hard. For these, complete solutions are provided.

In Glasgow, the book was originally used to complement approximately 100 one-hour lectures, divided into two courses running concurrently. One, dealing mainly with topics in algebra and geometry, made use of Chapters 1, 3–6 and 10–13. The other course was essentially calculus and was covered by Chapters 2, 7–9 and 14–19. The rigorous treatment of limits and continuity, introduced in Appendix C, was not covered in lectures but provided a summer reading opportunity for intending honours students. Over the years since the book was first written the system of teaching in Glasgow has changed from traditional to modular. The structure and contents of the book, however, remain as appropriate and relevant now as they did then.

The third edition contains a number of changes and additions, mainly in Chapter 18. The formal but unorthodox definition of sequence limit introduced in the first edition and maintained in the second edition has been abandoned in favour of a more informal approach akin to that used for function limits in Chapter 7. To compensate, the section on sequence limits in Appendix C has been extended. In earlier editions, only the comparison test and the alternating series test were used to investigate series convergence. In the third edition the ratio test is introduced together with its limit version. The limit version of the comparison test is also included and a new appendix provides help in deciding which test for convergence to try. More generally, there are more worked examples, more exercises with answers, more problems with solutions, and many of the diagrams have been improved.

The book splits naturally into the following parts:

- Chapters 1–6 cover number systems, sets and functions. Several techniques of

ix

proof are developed—proof by counter-example, proof by contradiction and proof by induction. Complex numbers are introduced and are used to extend results on trigonometry and factorisation.

- Chapters 7–10 are concerned with various aspects of differential calculus. This part of the book begins with an informal treatment of limits and continuity, a topic vital to all branches of calculus. A rigorous approach to this material is given in Appendix C. Differentiation is introduced and developed in Chapters 8 and 9. Chapter 10 is concerned with curve sketching.

- Chapter 11 is about matrices and their application to solving systems of linear equations. The method of finding matrix inverses by elementary row operations is included.

- Chapters 12 and 13 are about vectors and their applications to three dimensional geometry. Topics include the section formula, lines and planes, and vector products.

- Chapters 14–16 are concerned with integral calculus. Fundamental concepts are established and used to introduce the logarithmic and exponential functions. Methods of integration and applications to length, area and volume complete this part.

- Chapter 17 introduces ordinary differential equations of first and second order. Methods are described for finding both general solutions, and particular solutions satisfying initial or boundary conditions.

- Chapters 18 and 19, while different from each other in character, are both concerned with sequences and their limits. Series and tests for convergence, Taylor's theorem and Maclaurin series are discussed in Chapter 18, while in Chapter 19 numerical methods are applied to finding roots of equations, evaluating integrals and solving first order ordinary differential equations.

- Appendices A–G contain answers to the exercises, solutions to problems, a rigorous treatment of limits and continuity, trigonometric formulae and tables of integrals, a 'route map' of tests for convergence, and a list of Maclaurin series.

We gratefully acknowledge the readers whose vigilance has enabled us to avoid transferring errors from earlier editions.

To Woodhead Publishing Limited we offer our sincere thanks for allowing us the opportunity to complete this third edition and for the help they have provided.

The manuscript was typeset by the authors using LaTeX. The figures were drawn using METAPOST, making use of the `mfpic` macro package.

University of Glasgow   C M McGregor
August 2010   J J C Nimmo

Notation

Sets

Here S and T are sets, $\mathcal{P}$ is a statement.

$x \in S$	x belongs to S, x is member of S
$x \notin S$	x does not belong to S
$\{a, b, c, \ldots\}$	the set whose elements are a, b, c, ...
$\{x : \mathcal{P}\}$	the set of elements x for which $\mathcal{P}$ holds
$\{x \in S : \mathcal{P}\}$	the set of elements x of S for which $\mathcal{P}$ holds
$S \subseteq T$	S is a subset of T or $S = T$
$S \subset T$	S is a subset of T and $S \neq T$
$S \cap T = \{x : x \in S \text{ and } x \in T\}$	the intersection of S and T
$S \cup T = \{x : x \in S \text{ or } x \in T\}$	the union of S and T
$S - T = \{x \in S : x \notin T\}$	the relative difference of S and T
S'	the complement of S, the set of elements in a universal set which are not in S
$S \times T = \{(x, y) : x \in S \text{ and } y \in T\}$	the Cartesian product of S and T, the set of ordered pairs (x, y) where x belongs to S and y belongs to T

$$S^2 = S \times S = \{(x, y) : x \in S \text{ and } y \in S\} \quad \text{and, for } n \in \{1, 2, 3, \ldots\},$$
$$S^n = \underbrace{S \times S \times \cdots \times S}_{n \text{ factors}} = \{(x_1, x_2, \ldots, x_n) : x_i \in S \text{ for } i = 1, 2, \ldots, n\}$$

Logic

Here $\mathcal{P}$ and $\mathcal{Q}$ are statements.

$\mathcal{P} \Rightarrow \mathcal{Q}$	$\mathcal{P}$ implies $\mathcal{Q}$, if $\mathcal{P}$ then $\mathcal{Q}$
$\mathcal{P} \Leftarrow \mathcal{Q}$	$\mathcal{P}$ is implied by $\mathcal{Q}$, $\mathcal{P}$ if $\mathcal{Q}$
$\mathcal{P} \Leftrightarrow \mathcal{Q}$	$\mathcal{P}$ is equivalent to $\mathcal{Q}$, $\mathcal{P}$ if and only if $\mathcal{Q}$

Geometry

Here A, B and C are points, $\mathcal{L}$ and $\mathcal{M}$ are straight lines.

$A(x, y)$	A has coordinates (x, y)
$A(x, y, z)$	A has coordinates (x, y, z)
$|AB|$	the distance between A and B
$\triangle ABC$	triangle ABC
$\angle ABC$	angle ABC
$\mathcal{L} \parallel \mathcal{M}$	$\mathcal{L}$ is parallel to $\mathcal{M}$
$\mathcal{L} \perp \mathcal{M}$	$\mathcal{L}$ is perpendicular to $\mathcal{M}$

Greek Alphabet

A	α	alpha	N	ν	nu
B	β	beta	Ξ	ξ	xi
Γ	γ	gamma	O	o	omicron
Δ	δ	delta	Π	π	pi
E	ε	epsilon	P	ρ	rho
Z	ζ	zeta	Σ	σ	sigma
H	η	eta	T	τ	tau
Θ	θ	theta	Υ	υ	upsilon
I	ι	iota	Φ	ϕ	phi
K	κ	kappa	X	χ	chi
Λ	λ	lambda	Ψ	ψ	psi
M	μ	mu	Ω	ω	omega

Chapter 1

Preliminaries

All of the mathematics in this book is based on *real numbers*. Historically, mathematics began with the set of *natural numbers* (or *positive integers*) on which we have the operations of *addition* and *multiplication*. Much later, this set was extended by adding *zero* and the *negative integers*, thus giving the set of *integers*. In this set, subtraction is always possible, but *division* is only possible in certain cases. For example, we can divide 20 by 5, but not by 6. Mathematicians were thus led to introduce the *rational numbers* (or *fractions*), so division was always possible (except by zero). Eventually it was realised that yet more 'numbers' were needed to do even quite simple calculations. For example, the diagonal of a unit square has length $\sqrt{2}$, but $\sqrt{2}$ cannot be written as a fraction (see Section 1.8). To resolve this problem, mathematicians introduced *irrational numbers*, thus arriving finally at the *set of real numbers*.

In this first chapter, we recall results about the set of real numbers and some subsets which are important in their own right. The reader should be familiar with most of the material in Sections 1.1 to 1.6, but should read these sections carefully since they introduce notation which will be used in the remainder of the book. The final sections may well be new. Section 1.7 looks in detail at the idea of divisibility in the set of integers. Section 1.8 considers the numbers which cannot be written as fractions—the irrational numbers. These sections also introduce the idea of *proof by contradiction*, a technique much used throughout mathematics.

1.1   Number Systems

Informally, a *number system* consists of a set $\mathbb{F}$ of numbers which is *closed* under operations of *addition* and *multiplication*, *i.e.* if $a \in \mathbb{F}$ and $b \in \mathbb{F}$ then $a + b \in \mathbb{F}$ and $ab \in \mathbb{F}$. The number systems introduced in this section are all subsets of the real numbers and should be familiar to the reader. Indeed, it is assumed that the reader has a working knowledge of these systems. They are:

$\mathbb{N}$, the set of *positive integers* or *natural numbers*

$$\mathbb{N} = \{1, 2, 3, \ldots\};$$

$\mathbb{Z}$, the set of *integers*

$$\mathbb{Z} = \{\ldots, -2, -1, 0, 1, 2, \ldots\};$$

$\mathbb{Q}$, the set of *rational numbers* or *fractions*

$$\mathbb{Q} = \{k/n : k \in \mathbb{Z}, n \in \mathbb{N}\};$$

$\mathbb{R}$, the set of *real numbers*, represented by the points on an infinite straight line.

We shall further assume that the reader is familiar with the language of arithmetic, including such terms as *reciprocal* and *quotient*.

Example 1.1.1 Assuming that $\mathbb{N}$ and $\mathbb{Z}$ are number systems, prove that $\mathbb{Q}$ is a number system. Show, also, that if $q \in \mathbb{Q}$ with $q \neq 0$ then $1/q \in \mathbb{Q}$.

Solution Let $p, q \in \mathbb{Q}$. Then $p = h/m$ and $q = k/n$ where $h, k \in \mathbb{Z}$ and $m, n \in \mathbb{N}$. So

$$p + q = \frac{nh + mk}{mn} \qquad \text{and} \qquad pq = \frac{hk}{mn}.$$

Since $\mathbb{N}$ and $\mathbb{Z}$ are closed under addition and multiplication and $\mathbb{N} \subseteq \mathbb{Z}$ we have $nh + mk \in \mathbb{Z}$, $hk \in \mathbb{Z}$ and $mn \in \mathbb{N}$. Hence $p + q \in \mathbb{Q}$ and $pq \in \mathbb{Q}$ as required.

Now let $q \in \mathbb{Q}$ with $q \neq 0$. Then $q = k/n$ where $k \in \mathbb{Z}$, $k \neq 0$ and $n \in \mathbb{N}$. Hence $1/q = n/k = (-n)/(-k) \in \mathbb{Q}$ since n and $-n$ are in $\mathbb{Z}$ and k *or* $-k$ is in $\mathbb{N}$. $\square$

Problem 1.1.2 Let $\mathbb{D} = \{1, 3, 5, \ldots\}$ and $\mathbb{E} = \{2, 4, 6, \ldots\}$. Is either $\mathbb{D}$ or $\mathbb{E}$ a number system? Can you find a number system $\mathbb{F}$ such that $\mathbb{Z} \subset \mathbb{F} \subset \mathbb{Q}$?

The reader is probably aware that

$$\mathbb{N} \subset \mathbb{Z} \subset \mathbb{Q} \subset \mathbb{R}.$$

As number systems, the four sets differ in a more fundamental way—in terms of the solutions of equations. The equation $x + 5 = 3$, which is expressed in terms of the number system $\mathbb{N}$, has no solution in $\mathbb{N}$. It *does* have a solution, namely $x = -2$, in the larger system $\mathbb{Z}$. Similarly, the equation $2x - 3 = 0$, which is expressed in terms of the system $\mathbb{Z}$ has no solution in $\mathbb{Z}$ but *does* have a solution, $x = 3/2$, in $\mathbb{Q}$. We shall see later that $x^2 = 2$, an equation in $\mathbb{Q}$, has no solution in $\mathbb{Q}$. It has a solution in $\mathbb{R}$. Even in $\mathbb{R}$ there are simple equations with no solution. For example, $x^2 = -1$. There is a yet larger number system $\mathbb{C}$, the set of *complex numbers*, in which all such equations have solutions. We shall deal with $\mathbb{C}$ in Chapter 6.

In addition to the *arithmetic* in the number systems $\mathbb{N}$, $\mathbb{Z}$, $\mathbb{Q}$ and $\mathbb{R}$, there is also the idea of *order*. For example, $1 < 2$ and $\sqrt{2} > 0$. This is *not* a feature of every number system. There is, for example, no sensible meaning of order for the number system $\mathbb{C}$.

A real number x can be *positive*, written $x > 0$, *negative*, written $x < 0$, or zero. We call x *non-negative* and write $x \geq 0$ if $x > 0$ or $x = 0$, and we call x *non-positive* and write $x \leq 0$ if $x < 0$ or $x = 0$.

When we write $x > 0$, $x \leq 0$, *etc.* without specifying to which number system x belongs, it will be assumed that $x \in \mathbb{R}$.

We shall see more of order in Section 1.6.

As has been mentioned, each real number can be regarded as a point on an infinite straight line called the *real line*. Figure 1.1.1 illustrates the way in which each of the isolated numbers -2, 1 and 3 is marked on the line by •. The arrowhead indicates the positive side of 0 and the positive direction on the line.

Figure 1.1.1

When we wish to indicate more complicated sets of points we use a thick line to indicate a continuous set of points and ∘ for a point which is specifically excluded. Thus, in Figure 1.1.2, we have represented the set $S = \{x \in \mathbb{R} : -2 \le x < 0\} \cup \{1, 3\}$.

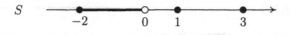

Figure 1.1.2

1.2   Intervals

Definitions 1.2.1 The connected subsets of $\mathbb{R}$, those represented on the real line by a continuous set of points, are called *intervals*. In the different types listed below, we let $a, b \in \mathbb{R}$ with $a < b$. Their real line representations are shown in Figure 1.2.1. They are:

$\emptyset$	the *empty set*,
$\{a\}$	a *singleton*,
$(a, b) = \{x \in \mathbb{R} : a < x < b\}$	a *bounded open* interval,
$[a, b] = \{x \in \mathbb{R} : a \le x \le b\}$	a *bounded closed* interval,
$\begin{aligned}(a, b] &= \{x \in \mathbb{R} : a < x \le b\} \\ [a, b) &= \{x \in \mathbb{R} : a \le x < b\}\end{aligned}\Bigg\}$	*bounded half-open* intervals,
$\begin{aligned}(a, \infty) &= \{x \in \mathbb{R} : x > a\} \\ (-\infty, a) &= \{x \in \mathbb{R} : x < a\}\end{aligned}\Bigg\}$	*semi-infinite open* intervals,
$\begin{aligned}[a, \infty) &= \{x \in \mathbb{R} : x \ge a\} \\ (-\infty, a] &= \{x \in \mathbb{R} : x \le a\}\end{aligned}\Bigg\}$	*semi-infinite closed* intervals,
$(-\infty, \infty) = \mathbb{R}$	the set of real numbers.

We shall refer to the empty set and singletons as *trivial* intervals and all other types as *non-trivial*.

Remark The symbols ∞ and $-\infty$ which are read as 'infinity' and 'minus infinity', respectively, are *not* real numbers and are not included in any interval. Sometimes $+\infty$, read as 'plus infinity', is used instead of ∞.

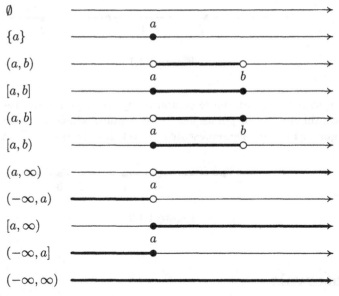

Figure 1.2.1

Example 1.2.2 Let $A = (-2, 3)$, $B = (-\infty, -5] \cup [-2, 5)$ and $C = (4, \infty)$. Express $A' \cap (B - C)$ as a union of intervals.

Solution Figure 1.2.2 shows the steps taken to calculate this expression. It shows that

$$A' \cap (B - C) = (-\infty, -5] \cup \{-2\} \cup [3, 4].$$ □

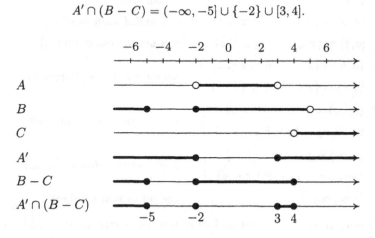

Figure 1.2.2

Frequent use is made, in mathematics, of sets of real numbers 'near' a given number. We make this idea precise as follows.

Definitions 1.2.3 Let $c \in \mathbb{R}$. A *neighbourhood* of c is any set of the form $(c - \delta, c + \delta)$ where $\delta > 0$, *i.e.* any bounded open interval centred on c. A *punctured neighbourhood* of c is any set of the form $N - \{c\}$ where N is a neighbourhood of c, *i.e.* a set of the form $(c - \delta, c) \cup (c, c + \delta)$ where $\delta > 0$. See Figure 1.2.3.

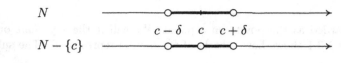

Figure 1.2.3

A *left neighbourhood* of c is any set of the form $(c - \delta, c]$ where $\delta > 0$ and a *punctured left neighbourhood* of c is any set of the form $L - \{c\}$ where L is a left neighbourhood of c. A *right neighbourhood* R and a *punctured right neighbourhood* $R - \{c\}$ are defined in the obvious way. See Figure 1.2.4.

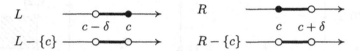

Figure 1.2.4

Notes 1.2.4 (1) If M and N are both neighbourhoods [both punctured neighbourhoods, *etc.*] of $c \in \mathbb{R}$ then $M \cap N$ is also a neighbourhood [punctured neighbourhood, *etc.*] of c.

(2) If $c \in (a, b)$ then (a, b) contains a neighbourhood of c. See Figure 1.2.5. More generally, if $c \in (a, b) \subseteq A$ then A contains a neighbourhood of c.

Figure 1.2.5

(3) Let $c \in \mathbb{R}$ and let $\delta > 0$. Let $N = (c - \delta, c + \delta)$ and $I = (-\delta, \delta)$. Then N and I are neighbourhoods of c and 0, respectively, and, if $x = c + h$,

$$x \in N \iff h \in I \qquad \text{and} \qquad x \in N - \{c\} \iff h \in I - \{0\}.$$

See Figure 1.2.6.

1.3 The Plane

The elements of the set

$$\mathbb{R}^2 = \{(x, y) \colon x, y \in \mathbb{R}\}$$

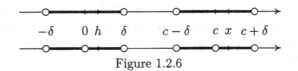

Figure 1.2.6

are often regarded as the points on a plane. We call it the *x,y-plane* or simply *the plane*. Figure 1.3.1 shows how subsets of $\mathbb{R}^2$ may be represented. The subsets shown are

$$S_1 = \{(1,0), (0,2), (-1,1)\},$$
$$S_2 = \{(x,y) \in \mathbb{R}^2 : x + y = 1, \ 0 < x \le 1\},$$
$$S_3 = [1,3) \times [1,2).$$

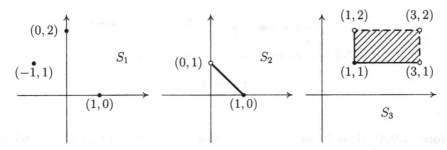

Figure 1.3.1

 Hazard When $x, y \in \mathbb{R}$ with $x < y$, the notation (x, y) can mean both an open interval and a point in the plane. The context usually makes the meaning clear.

1.4   Modulus

Definition 1.4.1 For $x \in \mathbb{R}$ the *modulus* or *absolute value* of x, denoted by $|x|$, is defined by

$$|x| = \begin{cases} x & \text{if } x \ge 0, \\ -x & \text{if } x < 0. \end{cases}$$

For example,

$$|2| = 2, \qquad |0| = 0, \qquad |-5/4| = 5/4, \qquad |3 - 5| = |5 - 3| = 2.$$

The curves $y = x$ and $y = |x|$ are shown in Figure 1.4.1.

More generally, Figure 1.4.2 shows the relationship between curves $y = f(x)$ and $y = |f(x)|$. The two curves coincide where $f(x) \ge 0$ and they are mirror images of each other in the x-axis where $f(x) < 0$. For more on this see Section 2.2.

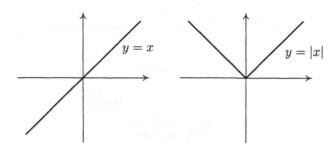

Figure 1.4.1

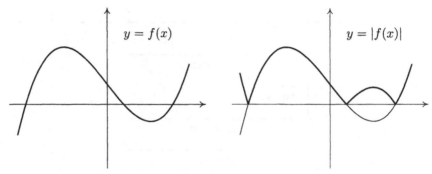

Figure 1.4.2

For $x \in \mathbb{R}$, the number $|x|$ can be thought of as the distance on the real line from 0 to x. So x is the point on the line reached by starting at 0 and taking a *step* of length $|x|$, in the positive direction if $x > 0$, or in the negative direction if $x < 0$.

More generally, we can think of x or 'adding' x as a step of length $|x|$ in the appropriate direction starting at any point on the line. Then, for $x, y \in \mathbb{R}$, $x + y$ is represented by a step x followed by a step y. Since $x = y + (x - y)$ it follows that $|x - y|$ is the length of the step from x to y, *i.e.* $|x - y|$ is the distance from x to y on the real line.

These ideas are illustrated in Figure 1.4.3.

Example 1.4.2 Express each of the sets

$$A = \{x \in \mathbb{R} : |x + 1| \le 3\}, \qquad B = \{x \in \mathbb{R} : |x - 8| < |x|\},$$

as an interval.

Solution We use the geometrical interpretation of modulus. Since $|x+1| = |x-(-1)|$, the set A consists of those x in $\mathbb{R}$ whose distance from -1 is less than or equal to 3, *i.e.* $A = [-4, 2]$. The set B consists of those x in $\mathbb{R}$ nearer to 8 than to 0, *i.e.* $B = (4, \infty)$. See Figure 1.4.4. □

Properties 1.4.3 The following properties of modulus follow directly from the definition. Let $x, y \in \mathbb{R}$.

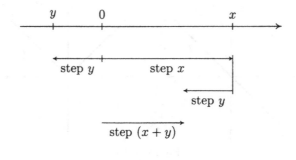

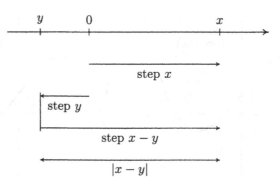

Figure 1.4.3

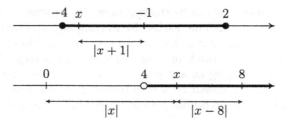

Figure 1.4.4

(1) $|x| \geq 0$ and $|x| = 0 \Leftrightarrow x = 0$.

(2) $|-x| = |x|$.

(3) $|x|^2 = x^2$.

(4) $|x - y| = 0 \Leftrightarrow x = y$.

Lemma 1.4.4 *Let $x \in \mathbb{R}$ and $a \geq 0$. Then*

$$|x| \leq a \Leftrightarrow -a \leq x \leq a \Leftrightarrow x \leq a \text{ and } -x \leq a.$$

Proof Using the definition of modulus, we have

$$|x| \le a \Leftrightarrow [x \ge 0 \text{ and } x \le a] \text{ or } [x < 0 \text{ and } -x \le a]$$

$$\Leftrightarrow [0 \le x \le a] \text{ or } [-a \le x < 0]$$

$$\Leftrightarrow -a \le x \le a$$

$$\Leftrightarrow x \le a \text{ and } -x \le a,$$

the last equivalence being immediate. ☐

Theorem 1.4.5 (The Triangle Inequality) *Let $n \in \mathbb{N}$ and suppose that $x_1, x_2, \ldots, x_n \in \mathbb{R}$. Then*

$$|x_1 + x_2 + \cdots + x_n| \le |x_1| + |x_2| + \cdots + |x_n|.$$

Proof For $i = 1, 2, \ldots, n$, it is obvious that $|x_i| \le |x_i|$ so, by Lemma 1.4.4, $-|x_i| \le x_i \le |x_i|$. Then adding gives

$$-(|x_1| + |x_2| + \cdots + |x_n|) \le x_1 + x_2 + \cdots + x_n \le |x_1| + |x_2| + \cdots + |x_n|.$$

This is justified in Theorem 1.6.5 (2).
   Applying Lemma 1.4.4 again gives

$$|x_1 + x_2 + \cdots + x_n| \le |x_1| + |x_2| + \cdots + |x_n|.$$ ☐

Corollary 1.4.6 *Let $x, y \in \mathbb{R}$. Then*

(1) $|x \pm y| \le |x| + |y|$ *and* (2) $\big||x| - |y|\big| \le |x \pm y|.$

Proof (1) This is just a special case of Theorem 1.4.5 together with the observation that $|x - y| = |x + (-y)| \le |x| + |-y| = |x| + |y|$.
   (2) We have

$$|x| = |y + (x - y)| \le |y| + |x - y|$$

so that

$$|x| - |y| \le |x - y| \quad \text{and} \quad -(|x| - |y|) = |y| - |x| \le |y - x| = |x - y|.$$

Thus

$$-|x - y| \le |x| - |y| \le |x - y|$$

and hence, by Lemma 1.4.4, $\big||x| - |y|\big| \le |x - y|$. Replacing y with $-y$ completes the proof. ☐

Theorem 1.4.7 *Let $n \in \mathbb{N}$ and $x_1, x_2, \ldots, x_n \in \mathbb{R}$. Then*

$$|x_1 x_2 \ldots x_n| = |x_1||x_2| \ldots |x_n|.$$

Proof We shall prove the special case that, for $x, y \in \mathbb{R}$, $|xy| = |x||y|$. The general case is dealt with in Example 1.5.12.

Case 1. Let $x, y \geq 0$. Then $xy \geq 0$ and $|xy| = xy = |x||y|$.

Case 2. Let $x \geq 0$ and $y < 0$. Then $xy \leq 0$ and $|xy| = -xy = x(-y) = |x||y|$.

Case 3. Let $x < 0$ and $y \geq 0$. Then $xy \leq 0$ and $|xy| = -xy = (-x)y = |x||y|$. [Since $ab = ba$ for $a, b \in \mathbb{R}$ this is essentially the same as Case 2.]

Case 4. Let $x, y < 0$. Then $xy > 0$ and $|xy| = xy = (-x)(-y) = |x||y|$.

Thus, in all four cases, $|xy| = |x||y|$. □

Remark The above proof of Theorem 1.4.7 is an example of *proof by cases*. It is important to ensure that the cases considered cover all possibilities.

Corollary 1.4.8 *Let $x, y \in \mathbb{R}$ with $y \neq 0$. Then $\left|\dfrac{x}{y}\right| = \dfrac{|x|}{|y|}$.*

Proof Using Theorem 1.4.7, $|x/y||y| = |(x/y)y| = |x|$ and the result follows immediately. □

Definitions 1.4.9 For $x, y \in \mathbb{R}$, we define

$$\min\{x, y\} = \begin{cases} x & \text{if } x \leq y, \\ y & \text{if } x > y \end{cases} \quad \text{and} \quad \max\{x, y\} = \begin{cases} x & \text{if } x \geq y, \\ y & \text{if } x < y. \end{cases}$$

More general definitions of min and max are given in Section 2.4.

Example 1.4.10 Show that, for $x, y \in \mathbb{R}$,

   (1) $\min\{x, y\} = \frac{1}{2}[(x + y) - |x - y|]$,

   (2) $\max\{x, y\} = \frac{1}{2}[(x + y) + |x - y|]$.

Deduce that, for $x \in \mathbb{R}$, $|x| = \max\{x, -x\}$.

Solution We shall leave (1) to the reader and establish (2).

Case 1. Let $x \geq y$. Then $x - y \geq 0$ and

$$\tfrac{1}{2}[(x + y) + |x - y|] = \tfrac{1}{2}[(x + y) + (x - y)] = x = \max\{x, y\}.$$

Case 2. Let $x < y$. Then $x - y < 0$ and

$$\tfrac{1}{2}[(x + y) + |x - y|] = \tfrac{1}{2}[(x + y) - (x - y)] = y = \max\{x, y\}.$$

Thus, in both cases equation (2) holds.

Now let $x \in \mathbb{R}$. Then, using (2),

$$\max\{x, -x\} = \tfrac{1}{2}[x + (-x) + |x - (-x)|] = \tfrac{1}{2}|2x| = |x|$$

as required. □

1.5 Rational Powers

Our aim in this section is to define, as far as is possible, the *rational powers* of real numbers so that they obey certain *index laws*. More general *real powers* must wait until Chapter 15. For $q \in \mathbb{Q}$ and $x \in \mathbb{R}$, the qth *power* of x, when it exists, is denoted by x^q. We proceed in four stages.

Stage 1. For $n \in \mathbb{N}$ and x in *any* number system, we define

$$x^n = \underbrace{xx \ldots x}_{n \text{ factors}}.$$

For example,

$$3^4 = 81, \qquad (-2)^5 = -32, \qquad (-5/4)^2 = 25/16, \qquad 0^n = 0 \ \ (n \in \mathbb{N}).$$

The general shape of the curve $y = x^n$ is shown in Figure 1.5.1.

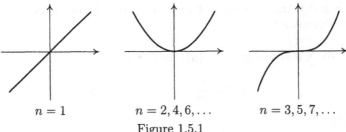

$$n = 1 \qquad\qquad n = 2, 4, 6, \ldots \qquad\qquad n = 3, 5, 7, \ldots$$

Figure 1.5.1

Properties 1.5.1 These properties of positive integer powers follow directly from the definition. Let $n \in \mathbb{N}$ and $x \in \mathbb{R}$.

(1) $x = 0 \ \Leftrightarrow \ x^n = 0$.

(2) $x > 0 \ \Rightarrow \ x^n > 0$.

(3) $x < 0 \ \Rightarrow \ \begin{cases} x^n < 0 & \text{if } n \text{ is odd,} \\ x^n > 0 & \text{if } n \text{ is even.} \end{cases}$

(4) $(-x)^n = \begin{cases} -x^n & \text{if } n \text{ is odd,} \\ x^n & \text{if } n \text{ is even.} \end{cases}$

Lemma 1.5.2 *Let $x, y \in \mathbb{R}$ and let $n \in \mathbb{N}$. Then*

$$x^n - y^n = (x - y)(x^{n-1} + x^{n-2}y + x^{n-3}y^2 + \cdots + x^2y^{n-3} + xy^{n-2} + y^{n-1})$$

and, if n is odd,

$$x^n + y^n = (x + y)(x^{n-1} - x^{n-2}y + x^{n-3}y^2 - \cdots + x^2y^{n-3} - xy^{n-2} + y^{n-1}).$$

In particular,

$$x^2 - y^2 = (x - y)(x + y),$$
$$x^3 - y^3 = (x - y)(x^2 + xy + y^2),$$
$$x^3 + y^3 = (x + y)(x^2 - xy + y^2).$$

Proof When the right-hand sides of the equations are multiplied out, the terms cancel in pairs except for x^n and $\pm y^n$. $\square$

Lemma 1.5.3 *Let $x^n = y^n$ where $n \in \mathbb{N}$ and either $x, y \geq 0$ or $x, y \leq 0$. Then $x = y$.*

Proof If $x = 0$, Property 1.5.1 (1) gives $x^n = 0$, so that $y^n = 0$, and then $y = 0$, so that $x = y(= 0)$. If $y = 0$ the proof is similar.

Let $A = x^{n-1} + x^{n-2}y + x^{n-3}y^2 + \cdots + x^2y^{n-3} + xy^{n-2} + y^{n-1}$. Then, by Lemma 1.5.2,

$$(x - y)\,A = x^n - y^n = 0.$$

If $x, y > 0$ then Property 1.5.1 (2) leads to $A > 0$ so we must have $x - y = 0$, *i.e.* $x = y$. If $x, y < 0$ then $-x, -y > 0$ and Property 1.5.1 (4) gives $(-x)^n = (-y)^n$ so that $-x = -y$, *i.e.* $x = y$. $\square$

Stage 2. For $n \in \mathbb{N}$ and $x \in \mathbb{R} - \{0\}$, we define

$$x^{-n} = \frac{1}{x^n} \qquad \text{and} \qquad x^0 = 1.$$

For example,

$$2^{-3} = 1/8, \qquad (-1/3)^{-2} = 9, \qquad (4^{-1})^{-1} = 4, \qquad 0^0 \text{ is undefined.}$$

We have now defined all sensible integer powers of real numbers.

Lemma 1.5.4 (Index Laws for Integer Powers) *Let $h, k \in \mathbb{Z}$ and $x, y \in \mathbb{R}$. Then, whenever both sides of the equation are defined,*

$$(1)\ \ x^h x^k = x^{h+k}, \qquad (2)\ \ (x^h)^k = x^{hk}, \qquad (3)\ \ x^h y^h = (xy)^h.$$

Proof This is left to the reader. For each index law, various cases must be considered. For example, let $h \in \mathbb{N}$ and $-k = n \in \mathbb{N}$ with $h < n$. Then, for $x \neq 0$,

$$x^h x^k = \frac{x^h}{x^n} = \frac{\overbrace{xx \ldots x}^{h \text{ factors}}}{\underbrace{xx \ldots x}_{n \text{ factors}}} = \frac{1}{\underbrace{xx \ldots x}_{n-h \text{ factors}}} = \frac{1}{x^{n-h}} = x^{h-n} = x^{h+k}$$

which proves (1) in this case. $\square$

Stage 3. For $n \in \mathbb{N}$ and $x \geq 0$ we define

$$x^{1/n} = u \quad \text{where} \quad u \geq 0 \quad \text{and} \quad u^n = x.$$

and for $n \in \mathbb{N}$, n odd, and $x < 0$ we define

$$x^{1/n} = v \quad \text{where} \quad (v < 0 \text{ and}) \quad v^n = x.$$

When $n = 1$, these definitions agree with the Stage 1 definition. When $n \geq 2$, the existence of u and v is an *assumption* but their uniqueness follows from Lemma 1.5.3.

The number $x^{1/n}$ is called the nth *root* of x. Traditionally, $x^{1/2}$ is called the *square root* and $x^{1/3}$ the *cube* root. The nth root of x is also written $\sqrt[n]{x}$ or, in the case $n = 2$, $\sqrt{x}$. For example,

$$4^{1/2} = 2, \quad \sqrt{1/9} = 1/3, \quad (-1/8)^{1/3} = -1/2, \quad \sqrt[4]{81} = 3, \quad 0^{1/n} = 0 \ \ (n \in \mathbb{N}).$$

 Hazard Remember that $x^{1/2}$ is always non-negative. Thus, for example, $4^{1/2}$ and $\sqrt{4}$ mean 2; they *never* mean -2.

Lemma 1.5.5 *Let $n \in \mathbb{N}$ and $x \in \mathbb{R}$ such that $x^{1/n}$ is defined. Then*

$$(x^{1/n})^n = x = (x^n)^{1/n}.$$

Proof Since, by definition, $x^{1/n} = w$ where $w^n = x$, it follows immediately that $(x^{1/n})^n = w^n = x$. This proves the first equation.

When $x \geq 0$, $x^n \geq 0$ and $(x^n)^{1/n} = u$ where $u \geq 0$ and $u^n = x^n$. Then Lemma 1.5.3 gives $u = x$. When $x < 0$ and $x^{1/n}$ exists, n must be odd so that $x^n < 0$ and $(x^n)^{1/n} = v$ where $v < 0$ and $v^n = x^n$. Then Lemma 1.5.3 gives $v = x$. This proves the second equation. □

Example 1.5.6 Solve the equation

$$\sqrt{x+4} + \sqrt{3x+1} = \sqrt{12x-11}. \tag{1.5.1}$$

Solution Squaring both sides of equation (1.5.1) gives

$$(x+4) + 2\sqrt{x+4}\sqrt{3x+1} + (3x+1) = 12x - 11,$$

i.e.

$$2\sqrt{x+4}\sqrt{3x+1} = 8x - 16,$$

i.e.

$$\sqrt{x+4}\sqrt{3x+1} = 4(x-2). \tag{1.5.2}$$

Then squaring both sides of equation (1.5.2) gives

$$(x+4)(3x+1) = 16(x-2)^2,$$

i.e.

$$13x^2 - 77x + 60 = 0,$$

i.e.

$$(x - 5)(13x - 12) = 0,$$

i.e.

$$x = 5 \quad \text{or} \quad x = 12/13$$

Substituting these values of x in equation (1.5.1) we find that only $x = 5$ satisfies the original equation and is therefore its only solution. $\qquad\square$

Hazard The check by substitution at the end of the last example was essential. By squaring both sides of an equation we create a new equation which may have more solutions than the original. For example, $x = -2$ implies that $x^2 = 4$ which has solutions $x = -2$ *and* $x = 2$. In Example 1.5.6 it was squaring equation (1.5.2) which caused the problem.

Lemma 1.5.7 *Let $h, k \in \mathbb{Z}$ and $m, n \in \mathbb{N}$ with $h/m = k/n$ and let $x > 0$. Then*

$$(x^{1/m})^h = (x^{1/n})^k.$$

Proof Let $a = (x^{1/m})^h$ and $b = (x^{1/n})^k$. Then, using Lemmas 1.5.4 and 1.5.5,

$$a^{mn} = ((x^{1/m})^h)^{mn} = (x^{1/m})^{hmn} = ((x^{1/m})^m)^{hn} = x^{hn}.$$

Similarly, $b^{mn} = x^{mk}$. Since $hn = mk$, it follows that $a^{mn} = b^{mn}$. Also, $a, b > 0$. Hence, Lemma 1.5.3 applies to give $a = b$. $\qquad\square$

Stage 4. For $k \in \mathbb{Z}$, $n \in \mathbb{N}$ and $x > 0$ we define

$$x^{k/n} = (x^{1/n})^k.$$

When $k = 1$, $k = n$ or $k = -n$, this definition agrees with the definitions of Stages 1 and 2. Lemma 1.5.7 guarantees that this definition does not depend on the fractional representation of k/n when $x > 0$.

The definition also applies when $x = 0$ provided $k \neq 0$ and when $x < 0$ provided k and n have no common factor and n is odd. For example,

$$8^{5/3} = 32, \qquad 9^{-3/2} = 1/27, \qquad (-1/27)^{2/3} = 1/9, \qquad 0^r = 0 \ \ (r \in \mathbb{Q} - \{0\}).$$

In the definition we could stipulate throughout that k and n have no common factors, thereby avoiding the need for Lemma 1.5.7. However, it is convenient to have the less restrictive definition for $x > 0$.

We have now defined all sensible rational powers of real numbers.

Note 1.5.8 Let $q \in \mathbb{Q}$ with $q = k/n$ where $k \in \mathbb{Z}$, $n \in \mathbb{N}$ and k and n have no common factor. Then, in Stages 1 to 4, we have defined x^q for $x \in D(q)$ where

$$D(q) = \begin{cases} \mathbb{R} & \text{if } q > 0 \text{ and } n \text{ is odd,} \\ \mathbb{R} - \{0\} & \text{if } q \leq 0 \text{ and } n \text{ is odd,} \\ [0, \infty) & \text{if } q > 0 \text{ and } n \text{ is even,} \\ (0, \infty) & \text{if } q < 0 \text{ and } n \text{ is even.} \end{cases}$$

Theorem 1.5.9 (Index Laws for Rational Powers) *Let $p, q \in \mathbb{Q}$ and $x, y \in \mathbb{R}$. Then, whenever x^p and x^q are defined,*

$$(1) \quad x^p x^q = x^{p+q} \qquad and \qquad (2) \quad (x^p)^q = x^{pq},$$

and whenever x^p and y^p are defined,

$$(3) \quad x^p y^p = (xy)^p.$$

In particular, (1), (2) and (3) hold for $p, q \in \mathbb{Q}$ and $x, y > 0$.

Proof (For the case $x, y > 0$) Let $p = h/m$ and $q = k/n$ where $p, q \in \mathbb{Z}$ and $m, n \in \mathbb{N}$.

(1) We have

$$x^p x^q = x^{h/m} x^{k/n} = x^{hn/mn} x^{mk/mn} = (x^{1/mn})^{hn} (x^{1/mn})^{mk}$$
$$= (x^{1/mn})^{hn+mk} \quad \text{(using Lemma 1.5.4 (1))}$$
$$= x^{(hn+mk)/mn} = x^{(h/m)+(k/n)} = x^{p+q}.$$

(2) Since $mn \in \mathbb{N}$ and $x > 0$, there exists $a > 0$ such that $x = a^{mn}$. Then $x^{1/mn} = a$ and we have

$$(x^p)^q = ((a^{mn})^{h/m})^{k/n} = ((((a^n)^m)^{1/m})^h)^{k/n} = ((a^n)^h)^{k/n}$$
$$= (a^{nh})^{k/n} = (((a^h)^n)^{1/n})^k = (a^h)^k$$
$$= a^{hk} \quad \text{(using Lemma 1.5.4 (2))}$$
$$= (x^{1/mn})^{hk} = x^{hk/mn} = x^{pq}.$$

(3) We have

$$x^p y^p = (x^{1/m})^h (y^{1/m})^h$$
$$= (x^{1/m} y^{1/m})^h \qquad \text{(using Lemma 1.5.4 (3))}$$
$$= (((x^{1/m} y^{1/m})^m)^{1/m})^h$$
$$= ((x^{1/m})^m (y^{1/m})^m)^{h/m} \qquad \text{(using Lemma 1.5.4 (3) again)}$$
$$= (xy)^p. \qquad \qquad \square$$

 Hazard Note that index law (2) in Theorem 1.5.9 requires that both x^p *and* x^q are defined. Otherwise the index law may not hold. For example, since $(-5)^{3/2}$ is not defined, one *cannot* deduce that $((-5)^2)^{3/2} = (-5)^3 = -125$. In fact, $((-5)^2)^{3/2} = 25^{3/2} = 125$.

Corollary 1.5.10 *Let $p, q \in \mathbb{Q}$ and $x, y \in \mathbb{R}$. Then, whenever both sides of the equation are defined,*

$$(1) \quad \frac{1}{x^q} = x^{-q}, \qquad (2) \quad \frac{x^p}{x^q} = x^{p-q}, \qquad (3) \quad \frac{x^p}{y^p} = \left(\frac{x}{y}\right)^p.$$

Proof We have

(1) $x^q x^{-q} = x^{q+(-q)} = x^0 = 1$,

(2) $x^p/x^q = x^p x^{-q} = x^{p+(-q)} = x^{p-q}$,

(3) $x^p/y^p = x^p y^{-p} = x^p(y^{-1})^p = (xy^{-1})^p = (x/y)^p$. $\square$

Remark We shall assume the 'obvious' extensions to the index laws. For example,

$$x^p x^q x^r = x^{p+q+r}, \qquad ((x^p)^q)^r = x^{pqr}, \qquad x^p y^p / z^p = (xy/z)^p,$$

for suitable $p, q, r, \ldots \in \mathbb{Q}$ and $x, y, z, \ldots \in \mathbb{R}$.

Example 1.5.11 Simplify

$$\text{(1) } (50)^{3/4}(5\sqrt{2})^{-1/2}, \qquad \text{(2) } \sqrt{2^{2^{100}}}, \qquad \text{(3) } (8x^2/y^3)^{1/3},$$

where $2^{2^{100}}$ means $2^{(2^{100})}$.

 Hazard Whenever a^{b^c} is written without any brackets it always means $a^{(b^c)}$ not $(a^b)^c$. If the latter meaning was intended it would be written as a^{bc}.

Solution We have

(1) $(50)^{3/4}(5\sqrt{2})^{-1/2} = (2 \times 5^2)^{3/4}(5 \times 2^{1/2})^{-1/2} = 2^{3/4}5^{3/2}5^{-1/2}2^{-1/4} = 2^{1/2}5^1 = 5\sqrt{2}$,

(2) $\sqrt{2^{2^{100}}} = \left(2^{2^{100}}\right)^{1/2} = 2^{(2^{100}/2)} = 2^{(2^{100-1})} = 2^{2^{99}}$,

(3) $(8x^2/y^3)^{1/3} = 8^{1/3}(x^2)^{1/3}/(y^3)^{1/3} = 2x^{2/3}/y$. $\square$

Example 1.5.12 Show that, for $x \in \mathbb{R}$, $\sqrt{x^2} = |x|$. Deduce that, for $n \in \mathbb{N}$ and $x_1, x_2, \ldots, x_n \in \mathbb{R}$,

$$|x_1 x_2 \ldots x_n| = |x_1||x_2|\ldots|x_n|.$$

Solution Since $|x| \geq 0$ and $|x|^2 = x^2$ it follows that $\sqrt{x^2} = |x|$. Then we have

$$|x_1 x_2 \ldots x_n| = \sqrt{(x_1 x_2 \ldots x_n)^2} = \sqrt{x_1^2 x_2^2 \ldots x_n^2}$$

$$= \sqrt{x_1^2}\sqrt{x_2^2}\ldots\sqrt{x_n^2} \qquad \text{(since } x_i \geq 0, \, i = 1, 2, \ldots, n\text{)}$$

$$= |x_1||x_2|\ldots|x_n|.$$ $\square$

1.6   Inequalities

We have already made considerable use of the order relation for real numbers based largely on 'common sense'. Here we put things on a sound footing.

Notation 1.6.1 For $x \in \mathbb{R}$, we write $x > 0$ to mean x is positive and $x < 0$ to mean x is negative.

All inequalities for real numbers stem from four rules.

Rule 1. For each $x \in \mathbb{R}$ *exactly one* of the following holds:

$$x > 0, \quad x = 0, \quad x < 0.$$

Rule 2. Let $x \in \mathbb{R}$. Then $x > 0 \Leftrightarrow -x < 0$.

Rule 3. Let $x, y \in \mathbb{R}$. Then $x, y > 0 \Rightarrow x + y > 0$.

Rule 4. Let $x, y \in \mathbb{R}$. Then $x, y > 0 \Rightarrow xy > 0$.

Theorem 1.6.2 *Let $x \in \mathbb{R}$. Then*

$$x^2 \geq 0 \qquad and \qquad x \neq 0 \Rightarrow x^2 > 0.$$

Proof First note that $0^2 = 0$. Next, by Rule 1, if $x \neq 0$ then $x > 0$ or $x < 0$. When $x > 0$, Rule 4 gives $x^2 > 0$. When $x < 0$, Rule 2 gives $-x > 0$ and then Rule 4 gives $x^2 = (-x)(-x) > 0$. □

Notation 1.6.3 For $x, y \in \mathbb{R}$, we write $x > y$ or $y < x$ to mean $x - y > 0$, and we write $x \geq y$ or $y \leq x$ to mean $x - y > 0$ or $x = y$.

Example 1.6.4 Prove that, for all $a, b \in \mathbb{R}$, $a^2 + b^2 \geq 2ab$.

Solution Let $a, b \in \mathbb{R}$. Using Theorem 1.6.2 we have

$$(a^2 + b^2) - 2ab = (a - b)^2 \geq 0$$

and the result follows immediately. □

From the four rules we can deduce further principles for manipulating inequalities.

Theorem 1.6.5 *Let $a, b, x, y \in \mathbb{R}$ and $q \in \mathbb{Q}$. Then*

(1) $x > y \Rightarrow a + x > a + y$,
(2) $a > b$ *and* $x > y \Rightarrow a + x > b + y$,

(3) $a > 0$ *and* $x > y \Rightarrow ax > ay$,
(4) $a < 0$ *and* $x > y \Rightarrow ax < ay$,
(5) $a > b > 0$ *and* $x > y > 0 \Rightarrow ax > by$,

(6) $x > 0 \Rightarrow 1/x > 0$,
(7) $x > y > 0 \Rightarrow 1/x < 1/y$,

(8) $q \neq 0$ *and* $x > 0 \Rightarrow x^q > 0$,
(9) $q > 0$ *and* $x > y > 0 \Rightarrow x^q > y^q$,
(10) $q < 0$ *and* $x > y > 0 \Rightarrow x^q < y^q$.

Proof The reader is invited to decide which rule is being invoked at each stage.

(1) Let $x > y$. Then $x - y > 0$ so that $(a + x) - (a + y) > 0$, *i.e.* $a + x > a + y$.

(2) Let $a > b$ and $x > y$. Then $a - b > 0$ and $x - y > 0$ so that $(a + x) - (b + y) = (a - b) + (x - y) > 0$, *i.e.* $a + x > b + y$.

(3) Let $a > 0$ and $x > y$. Then $a > 0$ and $x - y > 0$ so that $ax - ay = a(x - y) > 0$, *i.e.* $ax > ay$.

(4) Let $a < 0$ and $x > y$. Then $-a > 0$ and $x - y > 0$ so that $ay - ax = (-a)(x - y) > 0$, *i.e.* $ay > ax$.

(5) Let $a > b > 0$ and $x > y > 0$. Then $a - b > 0$, $a > 0$, $x - y > 0$ and $y > 0$ so that $ax - by = a(x - y) + (a - b)y > 0$, *i.e.* $ax > by$.

(6) Let $x > 0$. Then $1/x \neq 0$ and if $1/x < 0$ then $-1/x > 0$ so that $-1 = x(-1/x) > 0$ which is false. Hence, we must have $1/x > 0$.

(7) Let $x > y > 0$. Then $x - y > 0$ and also $x, y > 0$ so that $xy > 0$, $1/xy > 0$ and $(1/y) - (1/x) = (x - y)(1/xy) > 0$, *i.e.* $1/y > 1/x$.

(8) Let $q = m/n$ where $m, n \in \mathbb{N}$ and let $x > 0$. Then, for any $u > 0$, we have $u^2 = uu > 0$, $u^3 = u^2 u > 0$, $u^4 = u^3 u > 0$ and so on giving $u^m > 0$. This process will be formalised in Section 4.1. Also, by definition, $x^{1/n} > 0$. Hence, putting $u = x^{1/n}$, we get $x^q = (x^{1/n})^m > 0$.

(9) Let $x > y > 0$. The argument is similar to that of (8) except that to get $x^{1/n} > y^{1/n}$ we observe that if $y^{1/n} \geq x^{1/n}$ then $y = (y^{1/n})^n \geq (x^{1/n})^n = x$ which is false.

(10) Let $q = -p < 0$ and $x > y > 0$. Then $p > 0$ so that $x^p > y^p > 0$. Hence $1/x^p < 1/y^p$, *i.e.* $x^q < y^q$. $\qquad\square$

Example 1.6.6 Prove that $x > 1 \Rightarrow x^2 > 1$. Does the converse hold?

Solution Let $x > 1$. Then, using Theorem 1.6.5 (5), $x \times x > 1 \times 1$, *i.e.* $x^2 > 1$ as required.

The converse, $x^2 > 1 \Rightarrow x > 1$, is false. To prove this we produce a *counter-example*, a value of x for which $x^2 > 1$ but $x \not> 1$. It is only necessary to find *one* such example to prove that the implication is false. In this case $x = -2$ will do since $(-2)^2 = 4 > 1$ but $-2 \not> 1$. $\qquad\square$

Remarks (1) We shall assume the 'obvious' extensions to the Rules and the principles of Theorem 1.6.5. For example,

$$w, x, y, z > 0 \quad \Rightarrow \quad wxyz > 0,$$

$$a > b > 0, \ u > v > 0 \ \text{ and } \ x > y > 0 \quad \Rightarrow \quad aux > bvy.$$

(2) Let E be an expression of the form

$$\frac{a_1 a_2 \ldots a_m}{b_1 b_2 \ldots b_n},$$

where all the factors, $a_1, a_2, \ldots, a_m, b_1, b_2, \ldots, b_n$, are real. If all the factors are non-zero, then $E > 0$ if the total number of negative factors is even, and $E < 0$ if the total number of negative factors is odd. One way to use this to determine if or when a given E is positive, negative, zero or undefined, is to draw up a *table of signs* for E showing the sign $(+, -$ or $0)$ of each factor.

Example 1.6.7 Find the values of x for which $E = \dfrac{(1-x)(x+2)}{(x+1)}$ is positive.

Solution As x increases, each factor of E changes from positive to negative or *vice versa* only where it is zero. Hence, E can change sign only at these values of x, *i.e.* at $x = -2, -1$ and 1.

Table of signs for E

x	$\rightarrow$	-2	$\rightarrow$	-1	$\rightarrow$	1	$\rightarrow$
$1-x$	$+$	$+$	$+$	$+$	$+$	0	$-$
$x+2$	$-$	0	$+$	$+$	$+$	$+$	$+$
$x+1$	$-$	$-$	$-$	0	$+$	$+$	$+$
E	$+$	0	$-$	$?$	$+$	0	$-$

Observe that E is undefined when $x = -1$. From the table of signs we see that $E > 0$ for $x < -2$ and for $-1 < x < 1$. $\qquad\square$

When we have to compare two expressions, A and B, in order to decide if or when $A > B$, one strategy is to make use of the equivalence $A > B \Leftrightarrow A - B > 0$.

Step 1.   Consider $A - B$ and factorise it as far as possible.

Step 2.   Determine the sign of each factor of $A - B$.

Step 3.   Determine the sign of $A - B$.

Step 4.   Decide if or when $A - B > 0$, *i.e.* $A > B$.

Steps 2 and 3 might be combined in a table of signs for $A - B$. The strategy is easily adapted to decide when $A \geq B$, $A < B$ or $A \leq B$.

Example 1.6.8 Given that $x > y$ and $a > b > 0$, prove that

$$x > \frac{ax + by}{a + b} > y.$$

Solution For the left-hand inequality we consider

$$x - \frac{ax + by}{a + b} = \frac{x(a + b) - (ax + by)}{a + b} = \frac{b(x - y)}{a + b}.$$

Since $x > y$ and $a > b > 0$, it follows that $x - y > 0$, $b > 0$ and $a + b > 0$. Hence

$$\frac{b(x - y)}{a + b} > 0 \qquad \text{so that} \qquad x > \frac{ax + by}{a + b}$$

as required.

We leave establishing the right-hand inequality in a similar way as an exercise to the reader. $\qquad\square$

Example 1.6.9 Express, as unions of intervals, the sets

$$S = \left\{ x \in \mathbb{R} : x^2 + x + 1 \geq \frac{x-1}{2x-1} \right\},$$

$$T = \left\{ x \in \mathbb{R} : x^2 + x + 1 < \frac{x-1}{2x-1} \right\}.$$

Solution We have

$$x \in S \Leftrightarrow x^2 + x + 1 \geq \frac{x-1}{2x-1}$$

$$\Leftrightarrow x^2 + x + 1 - \frac{x-1}{2x-1} \geq 0$$

$$\Leftrightarrow \frac{(2x-1)(x^2+x+1) - (x-1)}{2x-1} \geq 0$$

$$\Leftrightarrow \frac{2x^3 + x^2}{2x-1} \geq 0 \Leftrightarrow \underbrace{\frac{x^2(2x+1)}{2x-1}}_{E} \geq 0.$$

The factors of E are zero at $x = -1/2$, 0 and $1/2$.

Table of signs for E

x	$\rightarrow$	$-\frac{1}{2}$	$\rightarrow$	0	$\rightarrow$	$\frac{1}{2}$	$\rightarrow$
x^2	+	+	+	0	+	+	+
$2x+1$	−	0	+	+	+	+	+
$2x-1$	−	−	−	−	−	0	+
E	+	0	−	0	−	?	+

Hence, from the table of signs, $S = (-\infty, -1/2] \cup \{0\} \cup (1/2, \infty)$.

The same table of signs can be used for T only this time we want $E < 0$. Hence, $T = (-1/2, 0) \cup (0, 1/2)$. □

Hazard In the solution to Example 1.6.9 we began with $x \in S$ if and only if

$$x^2 + x + 1 \geq \frac{x-1}{2x-1}$$

and it is tempting to continue by assuming that this is equivalent to

$$(2x-1)(x^2+x+1) \geq x - 1.$$

However, this is only true provided $2x - 1 > 0$ so that two different cases would have to be considered. While this approach will work, it is not recommended. It can involve many cases if the cross-multiplication involves more than one factor.

Notes 1.6.10 (1) In the solution to Example 1.6.9 the expression E did *not* change sign at $x = 0$. A factor of E being zero at $x = c$ is a necessary but not a sufficient condition for E to change sign at $x = c$.

(2) In the table of signs in the solution to Example 1.6.9 we should, according to our four-step strategy, have dealt with x^2 as two factors. In fact, there is no objection to combining two or more factors in a table of signs provided it is clear how the sign of the combination behaves. If in doubt, keep them separate.

(3) Let E be a real valued expression involving a real variable x. We say, formally, that E *changes sign* at $x = c$ if there are punctured left and right neighbourhoods, L and R, of c such that E is positive for $x \in L$ and negative for $x \in R$ or *vice versa*.

Example 1.6.11 Let $n \in \mathbb{N}$. Show that as n increases x^n decreases for $x \in (0, 1)$ and increases for $x \in (1, \infty)$.

Solution Let $n \in \mathbb{N}$. In both cases $x > 0$ so that $x^n > 0$. When $x \in (0, 1)$, $x < 1$ and hence $x^n \times x < x^n \times 1$, *i.e.* $x^{n+1} < x^n$. Similarly, when $x \in (1, \infty)$, $x > 1$ and hence $x^{n+1} > x^n$. $\qquad \square$

Remark The relationship between the curves $y = x^m$ $(x > 0)$ and $y = x^n$ $(x > 0)$, where $m, n \in \mathbb{N}$ with $m > n$, follows from Example 1.6.11 and is shown in Figure 1.6.1.

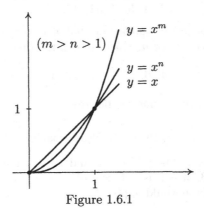

Figure 1.6.1

1.7 Divisibility and Primes

Definitions 1.7.1 Let $a, d \in \mathbb{Z}$ with $d \neq 0$. We say that d *divides* a if

$$a = dk \qquad \text{for some } k \in \mathbb{Z}.$$

When d divides a we write $d \mid a$ and call d a *divisor* or *factor* of a. Otherwise we write $d \nmid a$. For example,

$$20 = 4 \times 5 \quad \text{so} \quad 4 \mid 20,$$

$$20 = (-10) \times (-2) \quad \text{so} \quad -10 \mid 20,$$

$$20 \neq 3k \text{ for any } \textit{integer } k \quad \text{so} \quad 3 \nmid 20.$$

Hazard Do not confuse the *statement* $d \mid a$ with the *number* d/a. For example, $4 \mid 20$ means that 4 is a divisor of 20 while $4/20$ is a number $(= 1/5)$.

Example 1.7.2 Let $a, b, d \in \mathbb{Z}$ with $d \mid a$ and $d \mid b$. Show that $d \mid (2a + 3b)$.

Solution Since $d \mid a$ and $d \mid b$ we have $a = dk$ and $b = dl$ where $k, l \in \mathbb{Z}$. Then

$$2a + 3b = 2dk + 3dl = d(2k + 3l).$$

Since $2k + 3l \in \mathbb{Z}$ it follows that $d \mid (2a + 3b)$ as required.   □

Example 1.7.3 Let $n \in \mathbb{N}$ with $n \geq 2$. Prove that $n - 1$ divides $n^3 - 1$.

Solution Since, by Lemma 1.5.2,

$$n^3 - 1 = (n - 1)(n^2 + n + 1)$$

and $n^2 + n + 1$ is an integer, the result follows immediately.   □

Definitions 1.7.4 We say that p is a *prime number* or that p is *prime* if $p \in \mathbb{N}$, $p \geq 2$ and p has no positive divisors other than 1 and p. For example, the first few primes are

$$2, \ 3, \ 5, \ 7, \ 11, \ 13, \ 17, \ \ldots \ .$$

We say that n is a *composite number* or that n is *composite* if $n \in \mathbb{N}$ and n has a positive divisor other than 1 or n. For example, the first few composite numbers are

$$4, \ 6, \ 8, \ 9, \ 10, \ 12, \ 14, \ \ldots \ .$$

Notice that 1 is neither prime nor composite.

Example 1.7.5 Let $n \in \mathbb{N}$ with $n \geq 3$. Prove that $n^3 - 1$ is composite.

Solution Since $n \in \mathbb{N}$ and $n \geq 3$, $n - 1$ and $n^3 - 1$ are positive integers. From the proof of Example 1.7.3, $n - 1$ is a divisor of $n^3 - 1$. We cannot have $n - 1 = 1$ nor $n - 1 = n^3 - 1$ since these would imply $n = 2$ or $n = -1, 0$ or 1, respectively. Hence, $n^3 - 1$ is composite.   □

We have already referred to integers being even or odd. We can now give the formal definition.

Definitions 1.7.6 Let $k \in \mathbb{Z}$. We say that k is *even* if $2 \mid k$. Otherwise k is *odd*. For example,

$$-12, \ -4, \ 0, \ 2, \ 24 \ \text{ are even}, \qquad -11, \ -9, \ 1, \ 7, \ 15 \ \text{ are odd}.$$

The definitions are equivalent to:

(1) k is even $\Leftrightarrow$ $k = 2h$ for some integer h,

(2) k is odd $\Leftrightarrow$ $k = 2h + 1$ (or $k = 2h - 1$) for some integer h.

Example 1.7.7 Prove that the product of two odd integers is odd.

Solution Let m and n be odd integers. Then there are integers h and k such that $m = 2h + 1$ and $n = 2k + 1$. Hence,

$$mn = (2h + 1)(2k + 1) = 4hk + 2h + 2k + 1 = 2(2hk + h + k) + 1 = 2l + 1,$$

where $l = 2hk + h + k$ is an integer, and the result follows.   □

Example 1.7.8 Prove that the sum and product of two even integers are even.

Solution Let m and n be even integers. Then there are integers h and k such that $m = 2h$ and $n = 2k$. Hence,

$$m + n = 2h + 2k = 2(h + k) \qquad \text{and} \qquad mn = (2h)(2k) = 2(2hk).$$

Since $h + k$ and $2hk$ are integers it follows that $m + n$ and mn are even, as required.   □

Lemma 1.7.9 *Let $k \in \mathbb{Z}$ with k^2 even. Then k is even.*

Proof Suppose that k is odd. Then, by Example 1.7.7, k^2 is odd. Since k^2 cannot be odd *and* even, our assumption that k is odd must be false. Hence k is even.   □

Remark In the proof of Lemma 1.7.9 we began by assuming that the required conclusion was false. From this we deduced something that was impossible. Therefore our initial assumption had to be wrong and the required conclusion true. This is an example of *proof by contradiction.*

1.8   Rationals and Irrationals

Recall that $\mathbb{Q}$, the set of rational numbers, is defined by

$$\mathbb{Q} = \{k/n : k \in \mathbb{Z}, n \in \mathbb{N}\}.$$

Theorem 1.8.1 *The square root of 2 is not a rational number, i.e. $\sqrt{2} \notin \mathbb{Q}$.*

Proof Suppose that $\sqrt{2} \in \mathbb{Q}$. Then there exist $k \in \mathbb{Z}$ and $n \in \mathbb{N}$ such that $\sqrt{2} = k/n$ and k and n have no common factor. [If necessary, we can cancel any common factors to achieve this.] Then

$$2 = \frac{k^2}{n^2}, \qquad \text{i.e. } 2n^2 = k^2.$$

Since $2n^2$ is even, k^2 is even and so, by Lemma 1.7.9, k is even. Then $k = 2h$ for some integer h so that

$$2n^2 = k^2 = (2h)^2 = 4h^2, \qquad \text{i.e. } n^2 = 2h^2.$$

Since $2h^2$ is even, n^2 is even and so, by Lemma 1.7.9, n is even. Thus we have shown that both k and n are even. So they have 2 as a common factor which contradicts the fact that k and n have no common factor. This contradiction shows that our initial assumption is false and the result follows.   □

Remark Since $\sqrt{2}$ is a real number, we may deduce from Theorem 1.8.1 that $\mathbb{Q} \subset \mathbb{R}$. This was mentioned in Section 1.1.

Definition 1.8.2 A real number which is not in $\mathbb{Q}$ is called *irrational*. Some examples are:
$$\sqrt{2}, \ \pi, \ e, \ \log 2.$$
Theorem 1.8.1 tells us that $\sqrt{2}$ is irrational. We shall meet the other examples in later chapters but we shall not prove their irrationality.

Before the next example recall that $\mathbb{Q}$ is a number system, *i.e.* $\mathbb{Q}$ is closed under addition and multiplication. Also, if $q \in \mathbb{Q}$ with $q \neq 0$ then $1/q \in \mathbb{Q}$. See Example 1.1.1.

Example 1.8.3 Let q be rational and c irrational. Prove that $q + c$ is irrational. Further, show that, when $q \neq 0$, qc is irrational.

Solution Suppose that $q + c \in \mathbb{Q}$. Then $c = (q + c) + (-1)q \in \mathbb{Q}$ which contradicts the irrationality of c. Hence $q + c$ is irrational.

Let $q \neq 0$ so that $1/q \in \mathbb{Q}$ and suppose that $qc \in \mathbb{Q}$. Then $c = (qc)(1/q) \in \mathbb{Q}$ which contradicts the irrationality of c. Hence qc is irrational. $\square$

Example 1.8.4 Prove that the set $\mathbb{R} - \mathbb{Q}$ of irrational numbers is *not* closed under either addition *or* multiplication.

Solution We establish each part by means of a counter-example. By Theorem 1.8.1 and Example 1.8.3, $\sqrt{2}$ and $-\sqrt{2}$ are in $\mathbb{R} - \mathbb{Q}$. But $\sqrt{2} + (-\sqrt{2}) = 0 \in \mathbb{Q}$ so that $\mathbb{R} - \mathbb{Q}$ is not closed under addition. Also, $\sqrt{2} \times \sqrt{2} = 2 \in \mathbb{Q}$ so that $\mathbb{R} - \mathbb{Q}$ is not closed under multiplication. $\square$

Remark Either part of Example 1.8.4 is enough to show that $\mathbb{R} - \mathbb{Q}$ is not a number system.

We end this section with a result which shows that $\mathbb{Q}$ and $\mathbb{R} - \mathbb{Q}$ are both distributed 'densely' throughout $\mathbb{R}$. We prove that between *any* two distinct real numbers we can find a rational number and an irrational number.

Theorem 1.8.5 *Let* $a, b \in \mathbb{R}$ *with* $a < b$. *Then there exist* $q \in \mathbb{Q}$ *and* $c \in \mathbb{R} - \mathbb{Q}$ *such that* $a < q < b$ *and* $a < c < b$.

Proof (See Figure 1.8.1.) Choose $n \in \mathbb{N}$ such that $n > 1/(b - a)$. Then $1/n < b - a$.

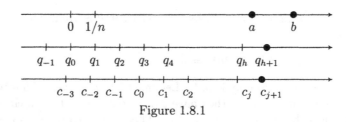

Figure 1.8.1

Consider the rational numbers $q_k = k/n$ $(k \in \mathbb{Z})$. Since $\mathbb{R}$ is the union of the intervals $[q_k, q_{k+1})$ $(k \in \mathbb{Z})$, we must have $a \in [q_h, q_{h+1})$ for some $h \in \mathbb{Z}$. Then $q_h \le a < q_{h+1} < b$. Otherwise, $q_h \le a < b \le q_{h+1}$ which contradicts $1/n < b - a$. Thus we may take $q = q_{h+1}$.

Now consider the numbers $c_k = q_k + \sqrt{2}$ $(k \in \mathbb{Z})$. By Example 1.8.3, these are all irrational. Arguing as above, we must have $c_j \le a < c_{j+1} < b$ for some $j \in \mathbb{Z}$. Thus we may take $c = c_{j+1}$.   □

1.X   Exercises

1. Express each of the following sets as an interval.

$$S_1 = \{x \in \mathbb{R} : |x - 2| < 4\}, \qquad S_2 = \{x \in \mathbb{R} : |x + 5| \le 4\},$$

$$S_3 = \{x \in \mathbb{R} : 3|x - 1| \le 1\}, \qquad S_4 = \{x \in \mathbb{R} : |x - 5| < |x + 5|\},$$

$$S_5 = \{x \in \mathbb{R} : |x| \le |x - 6|\}, \qquad S_6 = \{x \in \mathbb{R} : |x + 1| < |x - 7|\}.$$

2. Sketch the curve $y = x + |x|$.

3. Sketch the set $\{(x, y) \in \mathbb{R}^2 : |x - y| = 2\}$. Hence solve the simultaneous equations

$$\begin{cases} |x - y| = 2, \\ x + y = 4. \end{cases}$$

4. Let $n \in \mathbb{N}$, $n \ge 2$. Sketch the curve $y = (x - 2)^n$ when n is (a) odd, (b) even.

5. Show that

$$(x - y)^3 + (y - z)^3 + (z - x)^3 = 3(x - y)(y - z)(z - x).$$

[Hint. This may be verified by expanding the brackets but there is a more elegant solution.]

6. Solve each of the following equations.

   (a) $\sqrt{x} - \sqrt{x - 5} = 1$,

   (b) $\sqrt{x} - \sqrt{2 - 2x} = 1$,

   (c) $\sqrt{8x + 9} - \sqrt{x + 4} = \sqrt{2x + 6}$,

   (d) $\sqrt{8x + 9} + \sqrt{x + 4} = \sqrt{2x + 6}$,

   (e) $2\sqrt{1 - x} - \sqrt{8x + 3} + \sqrt{7 - 4x} = 0$.

7. Evaluate each of the following.

$$3^1, \quad \pi^0, \quad 3^{-1}, \quad 16^{3/4}, \quad 25^{-1/2}, \quad (4\sqrt{2})^{1/5}.$$

8. Simplify each of the following.

$$\left(x^2\right)^3, \quad \left(\sqrt{x}\right)^4, \quad \frac{x^{3/4}x^{4/3}}{x^{7/12}}, \quad \left(\frac{1}{8x^6}\right)^{-1/3}, \quad \frac{xy}{\sqrt{x/y}}.$$

9. Simplify $\left(\dfrac{x-1}{x+1}\right)^{1/2} - \left(\dfrac{x+1}{x-1}\right)^{1/2}.$

10. Let $n \in \mathbb{N}$. Show that, if $x = \left(1+\dfrac{1}{n}\right)^n$ and $y = \left(1+\dfrac{1}{n}\right)^{n+1}$, then $x^y = y^x$.

11. Express each of the following sets as a union of intervals.

$$S_1 = \left\{x \in \mathbb{R} : x \le \frac{x+3}{x-1}\right\}, \qquad S_2 = \left\{x \in \mathbb{R} : \frac{1}{x+2} \ge \frac{x-3}{x^2+1}\right\},$$

$$S_3 = \left\{x \in \mathbb{R} : \frac{x^3+3}{x^2+1} > \frac{x^3-3}{x^2-1}\right\}, \qquad S_4 = \left\{x \in \mathbb{R} : \frac{x^2-3x+10}{(x-3)(x+1)} < 4\right\}.$$

12. Prove that, if $0 < u < 1$ and $0 < v < 1$ then

$$\frac{u+v}{1+uv} < 1.$$

13. Prove that $\sqrt{2} + \sqrt{6} < \sqrt{15}$.

14. Show that, if $x \in (0,1)$, then

$$\frac{x^3}{x^3+1} > \frac{x^4}{x^4+1}.$$

State and prove a similar inequality valid for $x > 1$.

15. Show that, for all $x \in [0,1]$,

$$\frac{1}{1+2x^2} \le \frac{1}{1+x^2+x^4} \le \frac{1}{1+2x^3}.$$

16. Sketch the region of the plane consisting of those points (x,y) for which

$$x^2 - y^2 + x + y \ge 0.$$

[Hint. First factorise $x^2 - y^2 + x + y$.]

17. Show that, if d divides both a and b then d^2 divides $a^2 + db$.

18. Show that, if d divides x, but d does not divide y, then d does not divide $x + y$.

19. Show that, for $n \in \mathbb{N}$, $n^2 + n$ is even.

20. [Hard] By considering the factorisation of $x^q - 1$, show that, if n is composite then $2^n - 1$ is composite.

   Similarly, by writing n in the form $2^k m$, with m odd, show that if $2^n + 1$ is prime, then n is a power of 2.

21. Prove that, if α is irrational then $\dfrac{\alpha + 1}{\alpha - 1}$ is irrational.

22. Prove that $\sqrt{3}$ is irrational. [Hint. Begin by proving that, if $3 \mid n^2$ then $3 \mid n$.]

Chapter 2

Functions and Inverse Functions

After the ideas of *number* and *set*, the next most important concept in mathematics is that of a *function*. Informally, a function is a rule which associated with each suitable *input* x, an *output* (or *value*) $f(x)$. For example, we have the rule which associates with positive r, the value $f(r) = \pi r^2$ which gives the area of a circle of radius r. Functions occur throughout mathematics and its applications. The typical scientific calculator has many of the most common functions built in (via the 'function keys'). Such a calculator illustrates the need for using a 'suitable input'—if we enter -1 and then press the $\sqrt{\ }$ key, we get an error message. The range of suitable inputs will, of course, vary with the choice of function.

The chapter begins by discussing the general case of a function from one general set to another, introducing the language associated with functions and the ideas of *composition*. In Section 2.5, we consider the 'inverse problem': given the values $f(x)$, can we deduce the input x? We discover that, to achieve this, the function f must satisfy two stringent conditions—one to ensure that *some* suitable x exists, and another to ensure that this x is *unique*.

The other sections apply these ideas to *real functions*—functions where the inputs and outputs are real numbers. Such functions have *graphs*, and these can be used to investigate their properties. We also introduce concepts peculiar to real functions, such as *boundedness* and *monotonicity*.

2.1   Functions and Composition

Definitions 2.1.1 Let A and B be non-empty sets. A *function* or *mapping f from A to B*, written $f : A \to B$, consists of a *domain A* also written dom f, a *codomain B* and a *rule* which assigns to each $x \in A$ a unique element of B denoted by $f(x)$ and called the *value of f at x*. See Figure 2.1.1. For two functions to be equal, they must have the same domain, codomain and rule.

If we say that *f is defined on* a non-empty set C, we mean only that $C \subseteq A$ and so f has a value at every $x \in C$. The set of all values of f is denoted by $f(A)$ and is called the *image* or *image set* of f. Thus

$$f(A) = \{f(x) : x \in A\}.$$

More generally, for $C \subseteq A$, we write

$$f(C) = \{f(x) : x \in C\}.$$

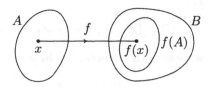

Figure 2.1.1

The *graph* of f is the set

$$\{(x, y) \in A \times B : y = f(x)\},$$

a subset of the cartesian product $A \times B$.

For example, let $f : \{1, 2, 3\} \to \mathbb{Z}$ be defined by $f(x) = x^2 - 4$. Then

- the domain of f is $\{1, 2, 3\}$ and, for example, f is defined on $\{1, 2\}$,

- the codomain of f is $\mathbb{Z}$,

- the image of f is $\{f(1), f(2), f(3)\}$, *i.e.* $\{-3, 0, 5\}$,
  and $f(\{1, 2\}) = \{-3, 0\}$,

- the graph of f is $\{(1, f(1)), (2, f(2)), (3, f(3))\}$,
  i.e. $\{(1, -3), (2, 0), (3, 5)\}$.

Example 2.1.2 Find the image of the function $f : \mathbb{N} \to \mathbb{Z}$ defined by

$$f(n) = \begin{cases} n & \text{if } n \le 4, \\ n + 1 & \text{otherwise.} \end{cases}$$

Solution We have

$$
\begin{aligned}
f(\mathbb{N}) &= \{f(n) : n \in \mathbb{N}\} \\
&= \{f(n) : n \in \mathbb{N}, n \le 4\} \cup \{f(n) : n \in \mathbb{N}, n \ge 5\} \\
&= \{n : n \in \mathbb{N}, n \le 4\} \cup \{n + 1 : n \in \mathbb{N}, n \ge 5\} \\
&= \{1, 2, 3, 4\} \cup \{6, 7, 8, \dots\} = \mathbb{N} - \{5\}.
\end{aligned}
$$

Thus, the image of f is $\mathbb{N} - \{5\}$. $\qquad\qquad\qquad\qquad\qquad\qquad\qquad\square$

It is sometimes useful to break up the rule for a function into a sequence of simpler rules. For example the rule $f(x) = x^2 - 4$ can be thought of as first 'square x' then 'subtract 4'. This idea leads us to the reverse process, the *composition of functions*, which allows us to build up complicated functions from sequences of simple rules.

Definition 2.1.3 Let $f : A \to U$ and $g : B \to V$. The *composition* of f and g (in that order) is defined provided

$$D = \{x \in A : f(x) \in B\} \neq \emptyset.$$

When this is the case, we denote the composition by $g \circ f$ (read 'g circle f') and define $g \circ f : D \to V$ by

$$(g \circ f)(x) = g(f(x)).$$

Note that to evaluate $(g \circ f)(x)$ we apply f first then g.

Roughly speaking, the domain of $g \circ f$ is the largest subset of A whose image is contained in the domain of g. The general situation is illustrated in Figure 2.1.2.

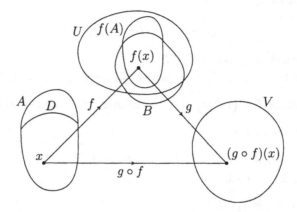

Figure 2.1.2

Example 2.1.4 Let $f : \mathbb{N} \to \mathbb{Z}$ and $g : [0, \infty) \to \mathbb{R}$ be defined by $f(x) = 5 - x$ and $g(x) = x^2 - 3$. Find (a) $g \circ f$ and (b) $f \circ g$.

Solution (a) The function $g \circ f$ is defined since

$$\{x \in \operatorname{dom} f : f(x) \in \operatorname{dom} g\} = \{x \in \mathbb{N} : f(x) \in [0, \infty)\}$$
$$= \{x \in \mathbb{N} : 5 - x \geq 0\}$$
$$= \{x \in \mathbb{N} : x \leq 5\}$$
$$= \{1, 2, 3, 4, 5\} = D \quad \text{(say)}$$

and $D \neq \emptyset$. Then $g \circ f : D \to \mathbb{R}$ is defined by

$$(g \circ f)(x) = g(f(x)) = g(5 - x) = (5 - x)^2 - 3 = x^2 - 10x + 22.$$

(b) The function $f \circ g$ is defined since

$$\{x \in \mathrm{dom}\, g : g(x) \in \mathrm{dom}\, f\} = \{x \geq 0 : g(x) \in \mathbb{N}\}$$
$$= \{x \geq 0 : x^2 - 3 \in \mathbb{N}\}$$
$$= \{x \geq 0 : x^2 - 3 = n \ \text{where} \ n \in \mathbb{N}\}$$
$$= \{x \geq 0 : x^2 = n + 3 \ \text{where} \ n \in \mathbb{N}\}$$
$$= \{\sqrt{n+3} : n \in \mathbb{N}\} = E \ \ (\text{say})$$

and $E \neq \emptyset$. Then $f \circ g : E \to \mathbb{Z}$ is defined by

$$(f \circ g)(x) = f(g(x)) = f(x^2 - 3) = 5 - (x^2 - 3) = 8 - x^2. \qquad \square$$

Example 2.1.5 Let $f : [0,1] \to \mathbb{R}$ and $g : [0,1] \to \mathbb{R}$ be defined by $f(x) = x + 2$ and $g(x) = 1 - x$. Show that $g \circ f$ is not defined.

Solution We have

$$\{x \in \mathrm{dom}\, f : f(x) \in \mathrm{dom}\, g\} = \{x \in [0,1] : x + 2 \in [0,1]\} = \emptyset$$

since

$$x \in [0,1] \ \Rightarrow \ 0 \leq x \leq 1 \ \Rightarrow \ 2 \leq x + 2 \leq 3 \ \Rightarrow \ x + 2 \notin [0,1].$$

Hence, $g \circ f$ is not defined. $\qquad \square$

Problem 2.1.6 Let $a, b, p, q \in \mathbb{R}$. Define $f : \mathbb{R} \to \mathbb{R}$ by $f(x) = ax + b$ and $g : \mathbb{R} \to \mathbb{R}$ by $g(x) = px + q$. When does $g \circ f = f \circ g$?

Definition 2.1.7 For any non-empty set A the *identity* function $i_A : A \to A$ is defined by $i_A(x) = x$.

Example 2.1.8 Let $f : A \to B$. Show that

$$\text{(a)} \quad f \circ i_A = f, \qquad \text{(b)} \quad i_B \circ f = f.$$

Solution We leave (a) to the reader and establish (b).

The function $i_B \circ f$ is defined since

$$\{x \in \mathrm{dom}\, f : f(x) \in \mathrm{dom}\, i_B\} = \{x \in A : f(x) \in B\} = A \neq \emptyset.$$

Then, since $f : A \to B$ and $i_B : B \to B$, it follows that $i_B \circ f : A \to B$. So $i_B \circ f$ and f have the same domain and codomain. Further, for $x \in A$,

$$(i_B \circ f)(x) = i_B(f(x)) = f(x).$$

Hence, $i_B \circ f = f$ as required. $\qquad \square$

Theorem 2.1.9 (Composition is Associative) *Let* $f : A \to U$, $g : B \to V$ *and* $h : C \to W$ *be such that either* $h \circ (g \circ f)$ *or* $(h \circ g) \circ f$ *is defined. Then both are defined and*

$$h \circ (g \circ f) = (h \circ g) \circ f.$$

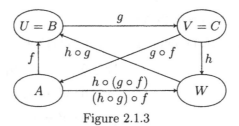

Figure 2.1.3

Proof Figure 2.1.3 illustrates the special case where $U = B$ and $V = C$. In general, we have

$$h \circ (g \circ f) \text{ is defined} \Leftrightarrow \{x \in \text{dom } g \circ f : (g \circ f)(x) \in C\} \neq \emptyset$$
$$\Leftrightarrow \{x \in A : f(x) \in B \text{ and } g(f(x)) \in C\} \neq \emptyset$$

and

$$(h \circ g) \circ f \text{ is defined} \Leftrightarrow \{x \in A : f(x) \in \text{dom } h \circ g\} \neq \emptyset$$
$$\Leftrightarrow \{x \in A : f(x) \in B \text{ and } g(f(x)) \in C\} \neq \emptyset.$$

Thus, $h \circ (g \circ f)$ is defined if and only if $(h \circ g) \circ f$ is defined and when they are defined they both have domain

$$\{x \in A : f(x) \in B \text{ and } g(f(x)) \in C\} = S \text{ (say)}.$$

Also, assuming both are defined, $h \circ (g \circ f)$ and $(h \circ g) \circ f$ have the same codomain W, and, for all $x \in S$,

$$(h \circ (g \circ f))(x) = h((g \circ f)(x)) = h(g(f(x)))$$

and

$$((h \circ g) \circ f)(x) = (h \circ g)(f(x)) = h(g(f(x))).$$

Thus, $h \circ (g \circ f) = (h \circ g) \circ f$ and the proof is complete. $\qquad\square$

Remark The associativity of composition, proved in Theorem 2.1.9, means that we can omit brackets and simply write $h \circ g \circ f$. This extends to $k \circ h \circ g \circ f$ for suitable f, g, h and k, *etc.*

2.2 Real Functions

Definitions 2.2.1 Let $f : A \to B$ where A and B are subsets of $\mathbb{R}$. Then we call f a *real* function. The graph of f is a subset of $\mathbb{R}^2$, the x, y-plane. When A, the domain of f, is a union of non-trivial intervals, the graph of f is often referred to as the *curve* $y = f(x)$ $(x \in A)$. For example, the graph of the function $f : (1, 4) \to \mathbb{R}$ defined by $f(x) = x^2 - 6x + 7$; is shown in Figure 2.2.1. The domain $(1, 4)$ and the image $[-2, 2)$ are also indicated in the figure.

The rule for a real function f is sometimes given without the domain and codomain being specified. When this happens it is assumed that f has the *maximal domain* (*of*

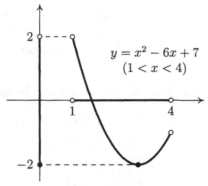

$$y = x^2 - 6x + 7$$
$$(1 < x < 4)$$

Figure 2.2.1

definition), *i.e.* the largest subset of $\mathbb{R}$ for which the rule makes sense. The codomain is taken to be $\mathbb{R}$ and the graph of f is referred to as the *curve* $y = f(x)$. For example, suppose we define a real function g simply by $g(x) = \sqrt{x}$. Then it is assumed that $g : [0, \infty) \to \mathbb{R}$ since the formula for $g(x)$ can be calculated for all $x \in [0, \infty)$ but for no other $x \in \mathbb{R}$. The curve $y = \sqrt{x}$ means the curve $y = \sqrt{x}$ $(x \in [0, \infty))$.

It is sometimes convenient to refer to a real function f as the real function $x \mapsto f(x)$. For example, the real function $x \mapsto \sqrt{x}$ is the function g, above.

Example 2.2.2 Find the maximal domains of the real functions f and g defined by

$$f(x) = \frac{x}{x^2 - 4} \qquad \text{and} \qquad g(x) = \sqrt{4 - x^2}.$$

Solution The formula for f can be calculated provided $x^2 - 4 \neq 0$, *i.e.* $x^2 \neq 4$, *i.e.* $x \neq -2$ and $x \neq 2$. Hence the maximal domain of f is $\mathbb{R} - \{-2, 2\}$ or, expressing it as a union of intervals, $(-\infty, -2) \cup (-2, 2) \cup (2, \infty)$.

The formula for g can be calculated provided $4 - x^2 \geq 0$, *i.e.* $x^2 \leq 4$, *i.e.* $-2 \leq x \leq 2$. Hence the maximal domain of g is $[-2, 2]$. $\square$

Note 2.2.3 Let f and g be functions defined on sets A and B, respectively. Then *both* f and g are defined on $A \cap B$.

More particularly, let f and g be real functions defined on neighbourhoods M and N of c, respectively. Then *both* f and g are defined on $M \cap N$ which is also a neighbourhood of c. See Note 1.2.4 (1).

Definitions 2.2.4 (Function Arithmetic) Let $f : A \to \mathbb{R}$ and $g : B \to \mathbb{R}$ be real functions and let $\lambda \in \mathbb{R}$.

(1) The function $\lambda f : A \to \mathbb{R}$ is defined by

$$(\lambda f)(x) = \lambda \times f(x) \qquad \text{(scalar multiplication)}.$$

(2) Provided $A \cap B \neq \emptyset$, the functions $f + g : A \cap B \to \mathbb{R}$ and $fg : A \cap B \to \mathbb{R}$ are defined by

$$(f + g)(x) = f(x) + g(x) \qquad \text{(sum)},$$
$$(fg)(x) = f(x)g(x) \qquad \text{(product)}.$$

(3) Provided $C = \{x \in A \cap B : g(x) \neq 0\} \neq \emptyset$, the function $f/g : C \to \mathbb{R}$ is defined by

$$\left(\frac{f}{g}\right)(x) = \frac{f(x)}{g(x)} \qquad (quotient).$$

Remark Variations on the above definitions will also be used. For example, for suitable f, g and x,

$$(2f - 3g)(x) = 2f(x) - 3g(x),$$
$$f^2(x) = f(x)^2,$$
$$\left(\frac{3}{g}\right)(x) = \frac{3}{g(x)}.$$

Hazard Given a real function g, do *not* write g^{-1} for the *reciprocal* $1/g$. The notation g^{-1} is reserved for the *inverse* of g. See Section 2.5.

Example 2.2.5 Let $f : [0,3] \to \mathbb{R}$, $g : [1,6] \to \mathbb{R}$ and $h : [-1,2] \to \mathbb{R}$ be defined by

$$f(x) = \sqrt{x^3 + x + 9}, \qquad g(x) = x + 3, \qquad h(x) = x^2 - 2.$$

Express the domain of $F = \dfrac{f^2 - 3g}{h}$ as a union of intervals and simplify the expression for $F(x)$.

Solution Let $G = f^2 - 3g$. So $G = f^2 + (-3)g$. We have

$$\text{dom } f^2 = \text{dom } f \cap \text{dom } f = [0,3]$$

and

$$\text{dom } (-3)g = \text{dom } g = [1,6]$$

so that

$$\text{dom } G = \text{dom } f^2 \cap \text{dom } (-3)g = [0,3] \cap [1,6] = [1,3]$$

and hence,

$$\text{dom } F = (\text{dom } G \cap \text{dom } h) - \{x \in \text{dom } h : h(x) = 0\}$$
$$= ([1,3] \cap [-1,2]) - \{\sqrt{2}\} = [1,2] - \{\sqrt{2}\}$$
$$= [1, \sqrt{2}) \cup (\sqrt{2}, 2] = D \quad (\text{say}).$$

Then, for $x \in D$,

$$F(x) = \frac{f(x)^2 - 3g(x)}{h(x)} = \frac{(x^3 + x + 9) - 3(x + 3)}{x^2 - 2}$$
$$= \frac{x^3 - 2x}{x^2 - 2} = \frac{x(x^2 - 2)}{x^2 - 2} = x. \qquad \square$$

Remark This solution can also be expressed as $F = i_D$ where $D = [1, \sqrt{2}) \cup (\sqrt{2}, 2]$.

Definitions 2.2.6 Let $f : A \to \mathbb{R}$ and $g : B \to \mathbb{R}$ be real functions.

(1) The function $|f| : A \to \mathbb{R}$ is defined by $|f|(x) = |f(x)|$.

(2) Provided $A \cap B \neq \emptyset$, the functions

$$\min\{f, g\} : A \cap B \to \mathbb{R} \text{ and } \max\{f, g\} : A \cap B \to \mathbb{R}$$

are defined by

$$\min\{f, g\}(x) = \min\{f(x), g(x)\},$$
$$\max\{f, g\}(x) = \max\{f(x), g(x)\},$$

For example, Figure 2.2.2 illustrates the situation when $f : [0, 2] \to \mathbb{R}$ and $g : [0, 2] \to \mathbb{R}$ are defined by $f(x) = x^2 - 1$ and $g(x) = 2 - x$.

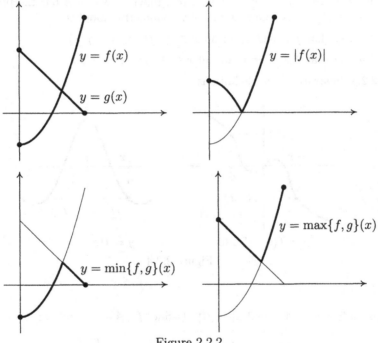

Figure 2.2.2

Example 2.2.7 Let $f : A \to \mathbb{R}$ be a real function. Then $|f| = \max\{f, -f\}$.

Solution This follows directly from Example 1.4.10 and the above definitions. The situation is illustrated in Figure 2.2.3. □

Remark Although we have included the definitions of $|f|$, $\min\{f, g\}$ and $\max\{f, g\}$ in this section on real functions, the definitions are valid when the domains of f and g are not subsets of $\mathbb{R}$.

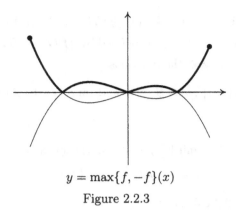

$$y = \max\{f, -f\}(x)$$

Figure 2.2.3

Definitions 2.2.8 Let $f : A \to \mathbb{R}$ be a real function where A has the property that $x \in A \Leftrightarrow -x \in A$, *i.e.* on the real line, A is symmetric about 0.

(1) We say that f is *odd* if, for all $x \in A$, $f(-x) = -f(x)$.

(2) We say that f is *even* if, for all $x \in A$, $f(-x) = f(x)$.

Figure 2.2.4 illustrates these definitions.

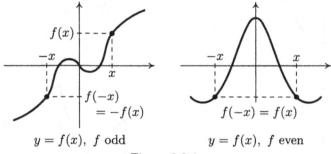

$$y = f(x), \ f \text{ odd} \qquad\qquad y = f(x), \ f \text{ even}$$

Figure 2.2.4

Example 2.2.9 Let $A = [-2, 2] - \{0\}$. Define $f : A \to \mathbb{R}$ and $g : A \to \mathbb{R}$ by

$$f(x) = \frac{1}{x^3 + 2x|x|} \quad \text{and} \quad g(x) = \sqrt{x^4 + \frac{3}{x^2}}.$$

Show that f is odd and g is even.

Solution First note that

$$x \in A \ \Leftrightarrow \ 0 < |x| \le 2 \ \Leftrightarrow \ 0 < |-x| \le 2 \ \Leftrightarrow \ -x \in A.$$

Then, for $x \in A$,

$$f(-x) = \frac{1}{(-x)^3 + 2(-x)|-x|} = \frac{1}{-x^3 - 2x|x|} = -\left(\frac{1}{x^3 + 2x|x|}\right) = -f(x)$$

and

$$g(-x) = \sqrt{(-x)^4 + \frac{3}{(-x)^2}} = \sqrt{x^4 + \frac{3}{x^2}} = g(x).$$

Thus f is odd and g is even as required. $\qquad\square$

Definition 2.2.10 Let $p > 0$ and let $f : A \to \mathbb{R}$ be a real function with the property that p is the smallest positive real number such that, for all $k \in \mathbb{Z}$,

$$x \in A \quad \Rightarrow \quad x + kp \in A \quad \text{and} \quad f(x + kp) = f(x). \qquad (2.2.1)$$

Then we say that f is *periodic* with *period p*.

Roughly speaking, the curve $y = f(x)$ consists of the part between $x = 0$ and $x = p$ repeated infinitely often to left and right.

An alternative to (2.2.1) is given in Problem 2.2.12.

Example 2.2.11 Show that $f : \mathbb{R} - \mathbb{Z} \to \mathbb{R}$, defined by

$$f(x) = (-1)^h (x - h) \quad \text{where } h \in \mathbb{Z} \text{ and } h < x < h + 1,$$

is periodic with period 2.

Solution The graph of f is sketched in Figure 2.2.5.

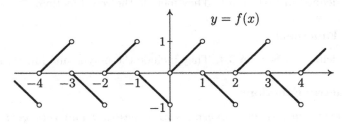

Figure 2.2.5

Let $x \in \mathbb{R} - \mathbb{Z}$ and $k \in \mathbb{Z}$. Then $x + 2k \in \mathbb{R} - \mathbb{Z}$, for if $x + 2k \in \mathbb{Z}$ then $x = (x + 2k) - 2k \in \mathbb{Z}$, since $\mathbb{Z}$ is closed under addition, which is a contradiction. Further, if $h \in \mathbb{Z}$ and $h < x + 2k < h + 1$ then, writing $h' = h - 2k$ so that $h' \in \mathbb{Z}$ and $h' < x < h' + 1$, we have

$$f(x + 2k) = (-1)^h (x + 2k - h) = (-1)^{h'+2k}(x + 2k - h' - 2k)$$
$$= (-1)^{h'}(-1)^{2k}(x - h') = (-1)^{h'}(x - h') = f(x).$$

Thus we have shown that, when $p = 2$, (2.2.1) holds for all $k \in \mathbb{Z}$.

Now let $0 < p < 2$. If $p \neq 1$ then $-p \in \mathbb{R} - \mathbb{Z}$ and $-p + p = 0 \notin \mathbb{R} - \mathbb{Z}$. If $p = 1$ then $f(1/2 + p) = f(3/2) = -1/2$ and $f(1/2) = 1/2$. So, in both cases, (2.2.1) does not hold for $k = 1$

Thus 2 is the smallest positive value of p for which (2.2.1) holds for all $k \in \mathbb{Z}$, *i.e.* f is periodic with period 2. $\qquad\square$

Problem 2.2.12 With the notation of Definition 2.2.10, prove that (2.2.1) holds for all $k \in \mathbb{Z}$ if and only if

$$x \in A \quad \Rightarrow \quad x \pm p \in A \quad \text{and} \quad f(x + p) = f(x). \qquad (2.2.2)$$

2.3   Standard Functions

Real functions are the raw material of differential and integral calculus. Here we list some of the more frequently used functions together with references to their definitions.

The Modulus Function

This is the real function $x \mapsto |x|$. See Section 1.4. Its maximal domain is $\mathbb{R}$.

Power Functions

For $q \in \mathbb{Q}$, the real function $x \mapsto x^q$ is called a *rational power function*. See Section 1.5. Its maximal domain depends on q. See Note 1.5.8. We define more general (*real*) *power functions* in Section 15.3.

The Zero Function

This is the real function $\theta : \mathbb{R} \to \mathbb{R}$ defined by $\theta(x) = 0$.

Polynomial Functions

These are defined in Section 3.1. They include the zero function.

Rational Functions

These are defined in Section 3.4. They include the polynomial functions.

Trigonometric Functions

The functions, *sine, cosine, tangent, secant, cosecant* and *cotangent* are defined in Section 5.1.

Inverse Trigonometric Functions

The *inverse sine, cosine* and *tangent* functions are defined in Section 5.5.

Logarithmic Functions

The *logarithmic* function and also *logarithms to a base b* $(b > 0)$ are defined in Section 15.1.

Exponential Functions

The *exponential* function and also *general exponential* functions are defined in Section 15.2

Hyperbolic Functions

The *hyperbolic cosine, sine, tangent, secant, cosecant* and *cotangent* functions are defined in Section 15.4.

Inverse Hyperbolic Functions

The *inverse hyperbolic cosine*, *sine* and *tangent* functions are defined in Section 15.5.

2.4   Boundedness

In this section, as well as defining the terms 'bounded set' and 'bounded function' we extend the definitions of min and max given in Sections 1.4 and 2.2.

Definitions 2.4.1 Let S be a non-empty subset of $\mathbb{R}$. Then S is *bounded below* by $m \in \mathbb{R}$ and m is a *lower bound* for S if, for all $t \in S$, $m \leq t$. Similarly, S is *bounded above* by $M \in \mathbb{R}$ and M is an *upper bound* for S if, for all $t \in S$, $t \leq M$. For example, $\mathbb{N}$ is bounded below by 1, and by any real number less than 1, but $\mathbb{N}$ has no upper bound.

We say that S is *bounded* if it is bounded below and above. Thus S is bounded if and only if there exist $m, M \in \mathbb{R}$ such that, for all $t \in S$,

$$m \leq t \leq M.$$

For example, since $t \in (0,1)$ implies that $0 < t < 1$, it follows that $(0,1)$ is bounded.

If m is a lower bound for S and $m \in S$ then m is the *minimum* of S, denoted by $\min S$. Similarly, if M is an upper bound for S and $M \in S$ then M is the *maximum* of S, denoted by $\max S$. Thus, when they exist, $\min S$ and $\max S$ belong to S and, for all $t \in S$,

$$\min S \leq t \leq \max S.$$

Every non-empty *finite* subset of $\mathbb{R}$ has a minimum and a maximum. An infinite subset of $\mathbb{R}$ may have neither. For example, $\mathbb{N}$ has a minimum but no maximum while $(0,1)$ has neither a minimum nor a maximum.

Lemma 2.4.2 *Let S be a non-empty subset of $\mathbb{R}$. Then S is bounded if and only if there exists $K \geq 0$ such that, for all $t \in S$, $|t| \leq K$.*

Proof ($\Rightarrow$) Let S be bounded. Then there exist $m, M \in S$ such that, for all $t \in S$, $m \leq t \leq M$. Let $K = \max\{|m|, |M|\}$. Then, for $t \in S$,

$$-t \leq -m \leq |m| \leq K \quad \text{and} \quad t \leq M \leq |M| \leq K.$$

Hence, $|t| \leq K$.

($\Leftarrow$) Let there exist $K \geq 0$ such that, for all $t \in S$, $|t| \leq K$. Then, by Lemma 1.4.4, for all $t \in S$, $-K \leq t \leq K$. Hence S is bounded. $\square$

Definitions 2.4.3 Let $f : A \to B$, where $B \subseteq \mathbb{R}$, and let $C \subseteq A$, $C \neq \emptyset$. Then f is *bounded below on* C if the set $f(C) = \{f(x) : x \in C\}$ is bounded below. Any lower bound for $f(C)$ becomes a *lower bound for f on C*. Similarly, f is *bounded above on* C if $f(C)$ is bounded above. An upper bound for $f(C)$ is an *upper bound for f on C*. We say that f is *bounded on* C if it is bounded below and above on C.

We say that f is *bounded (below/above)* if f is bounded (below/above) on its domain A. Thus, f is bounded if and only if there exist $m, M \in \mathbb{R}$ such that, for all $x \in A$,

$m \le f(x) \le M$. Alternatively, in view of Lemma 2.4.2, f is bounded if and only if there exists $K \ge 0$ such that, for all $x \in A$, $|f(x)| \le K$.

Example 2.4.4 Define $f : (0, \infty) \to \mathbb{R}$ by $f(x) = \dfrac{1}{x}$. Show that

(a) for all $a \in (0, \infty)$, f is bounded on $[a, \infty)$,

(b) f is not bounded.

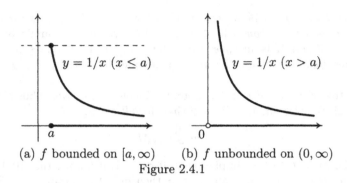

(a) f bounded on $[a, \infty)$ (b) f unbounded on $(0, \infty)$

Figure 2.4.1

Solution (a) See Figure 2.4.1 (a). Let $a \in (0, \infty)$. Then, since $a > 0$,

$$x \in [a, \infty) \;\Rightarrow\; x \ge a \;\Rightarrow\; \frac{1}{x} \le \frac{1}{a} \;\Rightarrow\; f(x) \le \frac{1}{a}.$$

It follows that, for all $x \in [a, \infty)$, $0 < f(x) \le 1/a$. Hence f is bounded on $[a, \infty)$.

(b) See Figure 2.4.1 (b). Suppose that f is bounded. Then f is bounded above by M (say) and $M \ge f(1) = 1 > 0$. Let $x = 1/(M + 1)$. Then $x \in (0, \infty)$ but $f(x) = M + 1 > M$. Thus M is not an upper bound for f. This contradiction establishes that f is not bounded.   □

2.5 Inverse Functions

Consider the function $f : \{1, 2, 3\} \to \{1, 4, 9\}$ defined by $f(x) = x^2$. We can 'reverse' this function to get a new function called the *inverse* of f, denoted by f^{-1}. Specifically, $f^{-1} : \{1, 4, 9\} \to \{1, 2, 3\}$ is defined by $f^{-1}(x) = \sqrt{x}$. In Figure 2.5.1 we see that the representation of f^{-1} simply reverses the arrows in the representation of f.

Not every function can be 'inverted' in this way. If we try to reverse the function $g : \{-2, 2, 3\} \to \{1, 4, 9\}$ defined by $g(x) = x^2$, we run into difficulties. First, since both $g(-2) = 4$ and $g(2) = 4$, what do we define $g^{-1}(4)$ to be? Should it be -2 or 2? Secondly, since $g(x) \ne 1$ for $x = -2$, $x = 2$ or $x = 3$, we do know what to define $g^{-1}(1)$ to be. See Figure 2.5.1 again.

The problems experienced with g are typical of the general situation. Something *can* be done with g, but we begin with functions like f for which the problems do not arise.

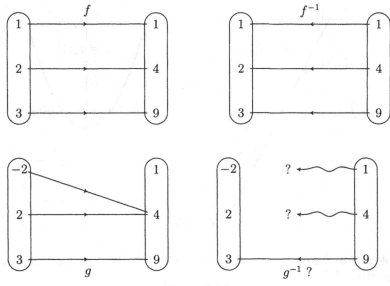

Figure 2.5.1

Definitions 2.5.1 Let $f : A \to B$.

(1) We call f *injective* or an *injection* if, for $s_1, s_2 \in A$,

$$s_1 \neq s_2 \quad \Rightarrow \quad f(s_1) \neq f(s_2)$$

or equivalently, if, for $s_1, s_2 \in A$,

$$f(s_1) = f(s_2) \quad \Rightarrow \quad s_1 = s_2.$$

(2) We call f *surjective* or a *surjection* if the image of f is B, *i.e.* $f(A) = B$, *i.e.* for each $t \in B$, there exists $s \in A$ such that $f(s) = t$.

(3) We call f *bijective* or a *bijection* if it is both injective and surjective.

Example 2.5.2 Define $f : \mathbb{R} \to \mathbb{R}$ by $f(x) = 8 - 2x$. Show that f is bijective.

Solution See Figure 2.5.2 (a)

(1) Let $s_1, s_2 \in \mathbb{R}$. Then

$$f(s_1) = f(s_2) \quad \Rightarrow \quad 8 - 2s_1 = 8 - 2s_2 \quad \Rightarrow \quad s_1 = s_2.$$

Hence, f is injective.

(2) Let $t \in \mathbb{R}$. We want $s \in \mathbb{R}$ with $f(s) = t$. For $f(s) = t$ we must have $8 - 2s = t$, *i.e.* $s = \frac{1}{2}(8 - t)$. Therefore, define $s = \frac{1}{2}(8 - t)$. Then $s \in \mathbb{R}$ and

$$f(s) = 8 - 2 \times \frac{1}{2}(8 - t) = t.$$

Hence, f is surjective.

(3) Since f is injective and surjective, it is bijective.　　　　　　　　□

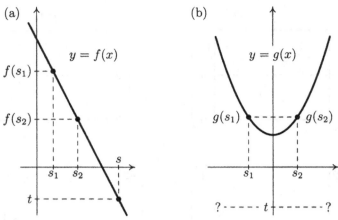

Figure 2.5.2

Example 2.5.3 Define $g : \mathbb{R} \to \mathbb{R}$ by $g(x) = x^2 + 1$. Show that g is neither injective nor surjective.

Solution See Figure 2.5.2 (b)

(1) Since $-1, 1 \in \mathbb{R}$, $-1 \neq 1$ and $g(-1) = g(1)$, it follows that g is not injective.

(2) Since $g(x) = x^2 + 1 \geq 1$ for all $x \in \mathbb{R}$, $0 \notin g(\mathbb{R})$. But $0 \in \mathbb{R}$. Thus $g(\mathbb{R}) \neq \mathbb{R}$ and hence that g is not surjective.   □

Example 2.5.4 Define $f : [0, 2] \to [1, 5]$ by $f(x) = x^2 + 1$. Show that f is bijective.

Solution (1) Let $s_1, s_2 \in [0, 2]$. Then

$$f(s_1) = f(s_2) \Rightarrow s_1^2 + 1 = s_2^2 + 1 \quad \Rightarrow \quad s_1^2 = s_2^2$$
$$\Rightarrow (s_1 - s_2)(s_1 + s_2) = 0$$
$$\Rightarrow s_1 = s_2 \text{ or } s_1 + s_2 = 0.$$

But $0 \leq s_1 \leq 2$ and $0 \leq s_2 \leq 2$. So $s_1 + s_2 = 0$ means that $s_1 = s_2 = 0$. Therefore

$$f(s_1) = f(s_2) \quad \Rightarrow \quad s_1 = s_2.$$

Hence, f is injective.

(2) Let $t \in [1, 5]$. We want $s \in [0, 2]$ with $f(s) = t$. For $f(s) = t$ we must have $s^2 + 1 = t$, i.e. $s = \pm\sqrt{t - 1}$. For $s \in [0, 2]$ we must take the non-negative square root. Therefore, define $s = \sqrt{t - 1}$. Since

$$1 \leq t \leq 5 \Rightarrow 0 \leq t - 1 \leq 4 \Rightarrow 0 \leq s \leq 2,$$

we get $s \in [0, 2]$. Further,

$$f(s) = s^2 + 1 = (t - 1) + 1 = t.$$

Hence, f is surjective.

(3) Since f is injective and surjective, it is bijective.   □

Remark For real functions, injectiveness, surjectiveness and bijectiveness can be interpreted geometrically as follows. Let $f : A \to B$. Then f is injective/surjective/bijective if and only if, for each $k \in B$, the line $y = k$ cuts the curve $y = f(x)$ $(x \in A)$ at most once/at least once/exactly once, respectively. See Figure 2.5.3.

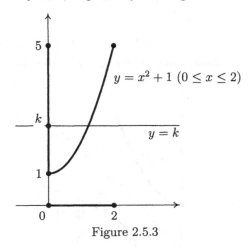

Figure 2.5.3

Definition 2.5.5 Let $f : A \to B$ be a bijection. Since f is surjective, for each $y \in B$, there exists $x \in A$ such that $f(x) = y$. Since f is injective, this x is unique. The *inverse* of f is denoted by f^{-1} and we define $f^{-1} : B \to A$ by

$$f^{-1}(x) = y \text{ where } y \in A \text{ and } f(y) = x.$$

This rule is equivalent to

$$f^{-1}(x) = y \iff f(y) = x$$

but some care should be exercised when using the second formulation as the rule for f may make sense for $y \notin A$.

Properties 2.5.6 The following properties of inverse functions follow directly from the definition. Let $f : A \to B$ be a bijection.

(1) $f^{-1} \circ f = i_A$,   *i.e.* $f^{-1}(f(x)) = x$ $(x \in A)$.

(2) $f \circ f^{-1} = i_B$,   *i.e.* $f(f^{-1}(y)) = y$ $(y \in B)$.

(3) f^{-1} is a bijection.

(4) $(f^{-1})^{-1} = f$.

Example 2.5.7 Find the inverse of the bijection $f : [0, 2] \to [1, 5]$ defined by $f(x) = x^2 + 1$.

Solution From Example 2.5.4, f is a bijection. So f^{-1} exists. We have

$$f^{-1}(x) = y \iff f(y) = x \iff y^2 + 1 = x \iff y^2 = x - 1$$

$$\iff y = \sqrt{x - 1} \quad (\text{since } 0 \le y \le 2).$$

Thus $f^{-1} : [1, 5] \to [0, 2]$ is defined by $f^{-1}(x) = \sqrt{x - 1}$.   □

Note 2.5.8 Let f be a real bijection. Then the graph of f^{-1} is the 'mirror image' of the graph of f, in the line $y = x$. To see this observe that

$$(x, y) \in \text{graph of } f^{-1} \Leftrightarrow f^{-1}(x) = y$$

$$\Leftrightarrow f(y) = x \;\; \Leftrightarrow \;\; (y, x) \in \text{graph of } f.$$

This *mirror property* is illustrated in Figure 2.5.4 for the function f of Example 2.5.7.

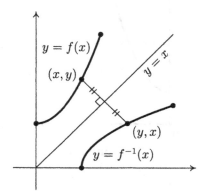

Figure 2.5.4

Finally in this section we show how a 'partial inverse' can be found for a function which is not a bijection.

The idea is to reduce the domain and codomain in such a way as to produce a bijection which can then be inverted. The functions g and f of Examples 2.5.3 and 2.5.4 illustrate the process although in that case we reduced the domain and codomain more than was necessary.

The best we might hope for, in general, would be a restriction which did not reduce the image, thereby keeping as much of the original function as possible. We might still have a choice of domain so there would be no unique 'maximal' bijective restriction and therefore no unique 'maximal partial inverse'. For particular functions, however, there are often 'obvious' domains to choose. We shall see this in Section 5.5 when we define inverse trigonometric functions.

Definition 2.5.9 Let $f : A \to U$ and $g : B \to V$ be such that $A \subseteq B$, $U \subseteq V$ and $f(x) = g(x)$ $(x \in A)$. Then we call f a *restriction* of g.

Example 2.5.10 Let $g : \mathbb{R} \to \mathbb{R}$ be defined by $g(x) = x^2 + 1$. Find two restrictions of g, both bijections and both having the same image as g. Find the inverse of each restriction.

Solution The image I, of g is $[1, \infty)$. To see this, note first that, for all $x \in \mathbb{R}$, $x^2 + 1 \geq 1$ so that $I \subseteq [1, \infty)$. Also, for each $y \in [1, \infty)$, $\sqrt{y - 1} \in \mathbb{R}$ and $g(\sqrt{y - 1}) = y \in [1, \infty)$ so that $[1, \infty) \subseteq I$.

Two suitable restrictions of g are

$$f_1 : [0, \infty) \to [1, \infty) \text{ defined by } f_1(x) = x^2 + 1$$

and

$$f_2 : (-\infty, 0] \to [1, \infty) \text{ defined by } f_2(x) = x^2 + 1.$$

See Figure 2.5.5. Showing that f_1 and f_2 are bijections is similar to the solution of Example 2.5.4 and is left to the reader.

Then

$$f_1^{-1} : [1, \infty) \to [0, \infty) \text{ is defined by } f_1^{-1}(x) = \sqrt{x - 1}$$

and

$$f_2^{-1} : [1, \infty) \to (-\infty, 0] \text{ is defined by } f_2^{-1}(x) = -\sqrt{x - 1}.$$

This verification also is left to the reader. See the solution to Example 2.5.7.   □

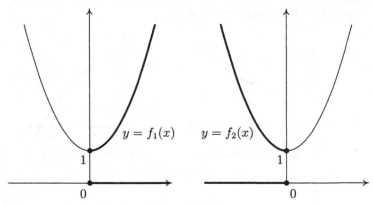

Figure 2.5.5

 Hazard The process of finding a bijective restriction f for a non-bijective function g and then inverting the restriction, does *not* produce an inverse for the *original* function g. However, as we shall see in Section 5.5, the notation is sometimes abused to the extent that we write g^{-1} instead of f^{-1}. We shall *not* do this without warning.

Problem 2.5.11 In Example 2.5.10 we found two bijective restrictions of the function g each having the same image as g. Can you find a third?

2.6   Monotonic Functions

Definitions 2.6.1 Let f be a real function whose domain contains the set A. Then f is *increasing on* A if, for $s_1, s_2 \in A$,

$$s_1 < s_2 \implies f(s_1) \le f(s_2)$$

and f is *strictly increasing* on A if, for $s_1, s_2 \in A$,

$$s_1 < s_2 \implies f(s_1) < f(s_2).$$

Similarly, f is *decreasing* on A if, for $s_1, s_2 \in A$,

$$s_1 < s_2 \implies f(s_1) \geq f(s_2)$$

and f is *strictly decreasing* on A if, for $s_1, s_2 \in A$,

$$s_1 < s_2 \implies f(s_1) > f(s_2).$$

We call f (*strictly*) *monotonic* on A if it is either (strictly) increasing or (strictly) decreasing on A.

For each of these properties, when no set A is mentioned the property is assumed to hold on the domain of f.

Figure 2.6.1 illustrates the definitions of strictly increasing/decreasing. Geometrically, the difference between monotonic and strictly monotonic is that the former allows the curve to include horizontal line segments.

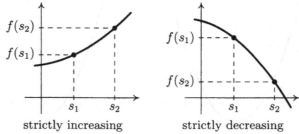

strictly increasing   strictly decreasing

Figure 2.6.1

Example 2.6.2 Show that the real function defined by $f(x) = x^3 - 3x$ is strictly increasing on $[1, \infty)$.

Solution The curve $y = f(x)$ is sketched in Figure 2.6.2.

Let $s_1, s_2 \in [1, \infty)$ with $s_1 < s_2$. Then

$$\begin{aligned}
f(s_1) - f(s_2) &= (s_1^3 - 3s_1) - (s_2^3 - 3s_2) \\
&= s_1^3 - s_2^3 - 3s_1 + 3s_2 \\
&= (s_1 - s_2)(s_1^2 + s_1 s_2 + s_2^2) - 3(s_1 - s_2) \\
&= (s_1 - s_2)(s_1^2 + s_1 s_2 + s_2^2 - 3).
\end{aligned}$$

Since $1 \leq s_1 < s_2$ we have

$$s_1 - s_2 < 0 \quad \text{and} \quad s_1^2 + s_1 s_2 + s_2^2 - 3 > 1 + 1 + 1 - 3 = 0.$$

Hence

$$f(s_1) - f(s_2) < 0, \qquad \textit{i.e.} \ f(s_1) < f(s_2).$$

Thus f is strictly increasing on $[1, \infty)$.   □

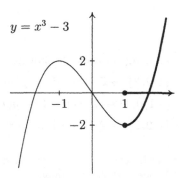

$y = x^3 - 3$

Figure 2.6.2

Hazard In Example 2.6.2, f is strictly increasing on $(-\infty, -1]$ and on $[1, \infty)$. The verification for $(-\infty, -1]$ is left to the reader. However, f is *not* strictly increasing (or even increasing) on $(-\infty, -1] \cup [1, \infty)$ since $-1 < 1$ but $f(-1) = 2 > -2 = f(1)$.

Theorem 2.6.3 *Let* $f : A \to B$ *be a strictly monotonic real function. Then* f *is injective.*

Proof Let f be strictly increasing [strictly decreasing] and let $s_1, s_2 \in A$ with $s_1 \neq s_2$. If $s_1 < s_2$ then $f(s_1) < f(s_2)$ $[f(s_1) > f(s_2)]$ and if $s_2 < s_1$ then $f(s_2) < f(s_1)$ $[f(s_2) > f(s_1)]$. So $f(s_1) \neq f(s_2)$ and it follows that f is injective. $\qquad\square$

Problem 2.6.4 Find an example of a real function $f : [0, 1] \to \mathbb{R}$ which is injective but not monotonic.

Theorem 2.6.5 *Let* $f : A \to B$ *be a strictly increasing [strictly decreasing] real bijection. Then* $f^{-1} : B \to A$ *is strictly increasing [strictly decreasing].*

Proof We prove the case when f is strictly increasing. Let $t_1, t_2 \in B$ with $t_1 < t_2$. Let $s_1 = f^{-1}(t_1)$ and $s_2 = f^{-1}(t_2)$. If $s_1 = s_2$ then $t_1 = f(s_1) = f(s_2) = t_2$ contradicting $t_1 < t_2$. If $s_2 < s_1$ then, since f is strictly increasing, $t_2 = f(s_2) < f(s_1) = t_1$, again contradicting $t_1 < t_2$. Hence $s_1 < s_2$, *i.e.* $f^{-1}(t_1) < f^{-1}(t_2)$, and it follows that f^{-1} is strictly increasing. $\qquad\square$

Example 2.6.6 Show that the real function $x \mapsto x^2$ is strictly increasing on $[0, \infty)$. Deduce that the real function $x \mapsto \sqrt{x}$ is strictly increasing on $[0, \infty)$.

Solution (Outline) Show, directly from the definition (as in Example 2.6.2), that $x \mapsto x^2$ is strictly increasing on $[0, \infty)$. Exhibit $x \mapsto \sqrt{x}$ as the inverse of a restriction of $x \mapsto x^2$. Deduce from Theorem 2.6.5 that $x \mapsto \sqrt{x}$ is strictly increasing on $[0, \infty)$. $\quad\square$

2.X   Exercises

1.   Find the graph and the image of each of following functions.

      (a) $f : \{1, 2, 3\} \to \mathbb{Z}$ defined by $f(x) = 3x - 7$,

      (b) $g : \{-1, 0, 1\} \to \mathbb{Z}$ defined by $g(x) = x^2 - 3x + 7$,

      (c) $h : \{1, 2, 3\} \to \mathbb{Z}$ defined by $h(x) = x^2 - 3x + 7$.

2.   For each of the following functions, draw a rough sketch of the graph as a subset of the x, y-plane and hence deduce the image.

      (a) $p : [1, 2] \to \mathbb{R}$ defined by $p(x) = 1/x$,

      (b) $q : (0, 1] \to \mathbb{R}$ defined by $q(x) = 1/x$,

      (c) $u : \mathbb{R} \to \mathbb{R}$ defined by $u(x) = 2 - x^2$,

      (d) $v : \mathbb{R} \to \mathbb{R}$ defined by $v(x) = x^3 + 1$.

3.   Find $f \circ f$ when $f : [-1, 1] \to \mathbb{R}$ is defined by

$$f(x) = \sqrt{1 - x^2}.$$

   Simplify the expression for $(f \circ f)(x)$.

4.   Find the maximal domain of the real function f when $f(x)$ is

      (a) $\dfrac{1}{(2x + 1)}$,   (b) $\sqrt{4x - 1}$,

      (c) $\dfrac{1}{(2x^2 - 2x - 5)}$,   (d) $(x^2 + x - 2)^{1/2}$,

      (e) $(x^2 + x - 2)^{-1/2}$,   (f) $(x^2 + x - 2)^{-1/3}$.

5.   Find $g \circ f$ and $f \circ g$ when f and g are the real functions defined by

      (a) $f(x) = 2x - 1$, $g(x) = 3x^2 + 8x + 14$,

      (b) $f(x) = \dfrac{x}{x + 2}$, $g(x) = \dfrac{x + 1}{x - 1}$,

      (c) $f(x) = 4x + 3$, $g(x) = f(2x + 1)$,

      (d) $f(x) = \begin{cases} x - 1 & \text{if } x < 0, \\ x + 1 & \text{if } x \geq 0, \end{cases}$ $g(x) = x - 2$.

   In each case determine the set $\{x : (g \circ f)(x) = (f \circ g)(x)\}$.

6.   The real function f is defined by

$$f(x) = \begin{cases} x + 1 & \text{if } x \neq 0, \\ 0 & \text{if } x = 0. \end{cases}$$

   Prove that if $g : \mathbb{R} \to \mathbb{R}$ is such that $g \circ f = f \circ g$ then $g(0) = 0$.

7. Which of the functions in Exercise 2.X.2 are (a) bounded below, (b) bounded above?

8. Let functions f, g and h be as defined in Exercise 2.X.1.

    (a) Find a restriction of f which is a bijection and has the same image as f. Determine the graph of the inverse of this restriction.

    (b) Find a restriction of g which is a bijection and has the same image as g. Determine the graph of the inverse of this restriction.

    (c) Find two restriction of h, both bijections and both having the same image as h. Determine the graphs of the inverses of these restriction.

9. Show that each of the following real functions is a bijection and determine its inverse

    (a) $f : \mathbb{R} \to \mathbb{R}$ defined by $f(x) = 3x + 5$,

    (b) $g : [0, 3] \to [1, 2]$ defined by $g(x) = \sqrt{4 - x}$,

    (c) $h : [2, 3] \to [1, 2]$ defined by $h(x) = \dfrac{2}{x - 1}$,

    (d) $u : (0, 1) \to (1, \infty)$ defined by $u(x) = \dfrac{1 + x}{1 - x}$,

    (e) $v : [-1, 1] \to [-1, 3]$ defined by $v(x) = x^2 - 2x$.

10. Let $f : A \to B$ and $g : B \to A$. Prove that, if $g \circ f = i_A$ then f is injective and g is surjective. Deduce that, if $g \circ f = i_A$ and $f \circ g = i_B$ then f is a bijection and $f^{-1} = g$. [Note. See Definition 2.1.7 for the definitions of i_A and i_B.]

11. Show, directly from the definition, that the real function $x \mapsto \sqrt{x}$ is strictly increasing on $[0, \infty)$. [Hint. Use the fact that $(\sqrt{s_1} - \sqrt{s_2})(\sqrt{s_1} + \sqrt{s_2}) = s_1 - s_2$.]

12. Show that the real function $x \mapsto x^3$ is strictly increasing on $\mathbb{R}$. *Deduce* that the real function $x \mapsto x^{1/3}$ is strictly increasing on $\mathbb{R}$. [Hint. See Example 2.6.6.]

13. Let f and g be the real functions defined by

$$f(x) = \frac{x^2 + 2}{2x + 1} \quad \text{and} \quad g(x) = x^2(8 - x^2).$$

Show that f is strictly increasing on $[1, \infty)$ and g is strictly decreasing on $[2, \infty)$.

Chapter 3

Polynomials and Rational Functions

The reader will be familiar with expressions such as $2x^2 + 3x + 4$ which are combinations of powers of x. These are known as *polynomials*. In the first two sections, we prove the basis results about polynomials. In many respects, these are reminiscent of results about $\mathbb{Z}$. Trivially, polynomials may be added and multiplied to produce new polynomials (*i.e.* the set of polynomials is closed under addition and multiplication). More significantly, we have ideas of *division with remainder* and *factorisation*. The former generalises the process of synthetic division with which the reader may be familiar. Just as we define rational numbers as quotients of integers, we define *rational functions* as quotients of polynomials. We show how a rational function can be written as a sum of simpler functions. This has an important application to the integration of rational functions in Chapter 16.

If $p(x)$ is a polynomial, and c is a real number, then $p(c)$ is a real number. Thus a polynomial gives rise to a real function. We show how the general theory of polynomials can be applied to investigate these *polynomial functions*. The key result is the *remainder theorem* which shows that the zeros of a polynomial function are associated with the linear factors of the polynomial. We illustrate the theory by a thorough discussion of quadratic polynomials and the corresponding functions.

3.1 Polynomials

Definitions 3.1.1 Let $\mathbb{F}$ be a number system. A *polynomial in x over $\mathbb{F}$* is an expression of the form

$$p(x) = a_n x^n + a_{n-1} x^{n-1} + \cdots + a_1 x + a_0$$

where $a_n, a_{n-1}, \ldots, a_1, a_0 \in \mathbb{F}$ are the *coefficients* of $p(x)$.

The set of all polynomials in x over $\mathbb{F}$ is denoted by $\mathbb{F}[x]$. The definition requires only that the coefficients belong to $\mathbb{F}$. Thus a polynomial may be described as being in $\mathbb{F}[x]$ for any $\mathbb{F}$ which contains all the coefficients. For example, $(-4)x^2 + 5x + 16$ is in $\mathbb{Z}[x]$, in $\mathbb{Q}[x]$ and in $\mathbb{R}[x]$. It is not in $\mathbb{N}[x]$, as $(-4) \notin \mathbb{N}$.

If $p(x) \in \mathbb{F}[x]$ has all its coefficients equal to 0 then $p(x)$ is the *zero polynomial*. Otherwise it is *non-zero*.

For a non-zero polynomial $p(x)$, the greatest i for which the coefficient $a_i \neq 0$ is the *degree* of $p(x)$, denoted by $\deg p(x)$, and the coefficient a_i is the *leading* coefficient. For example, $(-4)x^2 + 5x + 16$ has degree 2 and leading coefficient -4. No degree is defined for the zero polynomial. A non-zero polynomial is *monic* if its leading coefficient is 1.

The zero polynomial and those of degree 0 are *constant* polynomials. Polynomials of degree 1, 2 and 3 are, respectively, *linear, quadratic* and *cubic*.

Remark Let $p(x) \in \mathbb{F}[x]$. The symbol x is assumed to behave arithmetically as if it were an element of $\mathbb{F}$. Thus, for example, a term $0x^i$ simplifies to 0 and so may be omitted when writing $p(x)$. Similarly, we use the normal arithmetic abbreviations. Thus, for example, the polynomial $0x^4 + 1x^3 + 0x^2 + (-3)x + (-5)$ would normally be written as $x^3 - 3x - 5$. Finally, the assumption about x allows us to perform arithmetic with polynomials. For example,

$$(-4x^2 + 5x + 16) + (x^3 - 3x - 5) = x^3 - 4x^2 + 2x + 11,$$

$$(-4x^2 + 5x + 16)(x^3 - 3x - 5) = -4x^5 + 5x^4 + 28x^3 + 5x^2 - 73x - 80.$$

Definition 3.1.2 Suppose that $p(x) = a_n x^n + a_{n-1} x^{n-1} + \cdots + a_1 x + a_0 \in \mathbb{F}[x]$. Then, for $c \in \mathbb{F}$, as $\mathbb{F}$ is a number system, $a_n c^n + a_{n-1} c^{n-1} + \cdots + a_1 c + a_0 \in \mathbb{F}$. We denote this number by $p(c)$. Thus we have a function $p : \mathbb{F} \to \mathbb{F}$ whose value at c is $p(c)$. This is the *polynomial function* p associated with $p(x)$. When $\mathbb{F} = \mathbb{R}$, p is a *real* polynomial function and has maximal domain $\mathbb{R}$.

Hazard The dual use of the notation $p(c)$, introduced in the above definition, is ambiguous. When c is a symbol, $p(c)$ denotes a polynomial in c. When c is an element of $\mathbb{F}$, then $p(c)$ denotes a number in $\mathbb{F}$. With this warning in mind, we see that the polynomial function associated with $p(x) \in \mathbb{F}[x]$ is defined by $x \mapsto p(x)$.

For the remainder of the chapter we take $\mathbb{F} = \mathbb{R}$.

Theorem 3.1.3 *Let* $p(x), q(x) \in \mathbb{R}[x]$ *be non-zero. Then*

$$\deg(p(x)q(x)) = \deg p(x) + \deg q(x).$$

Proof Let $n = \deg p(x)$ and $m = \deg q(x)$. Suppose that $m, n > 0$. Then there exist $a, b \in \mathbb{R}$ with $a, b \neq 0$ such that

$$p(x) = ax^m + \text{terms of lower degree},$$

$$q(x) = bx^n + \text{terms of lower degree}.$$

Hence
$$p(x)q(x) = abx^{m+n} + \text{terms of lower degree}.$$

Since $a, b \neq 0$, $ab \neq 0$ and hence $\deg(p(x)q(x)) = m + n$, as required.

The reader is invited to complete the proof by considering cases where at least one of m and n is zero. $\qquad\square$

Definitions 3.1.4 Let $p(x) \in \mathbb{R}[x]$. If $c \in \mathbb{R}$ is such that $p(c) = 0$, then c is called a *zero* of the polynomial $p(x)$, a *zero* of the polynomial function p and a *root* of the equation $p(x) = 0$.

Theorem 3.1.5 Let $p(x) = a_n x^n + a_{n-1} x^{n-1} + \cdots + a_1 x + a_0 \in \mathbb{R}[x]$ have degree $n > 0$. Let

$$M = \max\{1, |a_n|^{-1}(|a_{n-1}| + \cdots + |a_1| + |a_0|)\}.$$

Then all zeros of p lie in the set $\{x \in \mathbb{R} : |x| \le M\}$, i.e. in the interval $[-M, M]$.

Proof Suppose that p has a zero $x \in \mathbb{R}$ such that $|x| > M$. Then, since $p(x) = 0$ and $|x| > M \ge 1$,

$$
\begin{aligned}
|a_n||x|^n = |a_n x^n| &= |-a_{n-1} x^{n-1} - \cdots - a_1 x - a_0| \\
&\le |a_{n-1}||x|^{n-1} + \cdots + |a_1||x| + |a_0| \\
&\le |a_{n-1}||x|^{n-1} + \cdots + |a_1||x|^{n-1} + |a_0||x|^{n-1},
\end{aligned}
$$

since $|x| > 1$,

$$= (|a_{n-1}| + \cdots + |a_1| + |a_0|)|x|^{n-1}.$$

This contradicts the assumption that $|x| > M \ge |a_n|^{-1}(|a_{n-1}| + \cdots + |a_1| + |a_0|)$ and hence establishes the result. □

Corollary 3.1.6 Let $p(x) = a_n x^n + a_{n-1} x^{n-1} + \cdots + a_1 x + a_0 \in \mathbb{R}[x]$ and let there exists $K \ge 0$ such that, for all $|x| > K$, $p(x) = 0$. Then, for $i = 0, 1, \ldots, n$, $a_i = 0$, i.e. $p(x)$ is the zero polynomial.

Proof Suppose that $\deg p(x) > 0$. By Theorem 3.1.5, there exists $M > 0$ such that all the zeros of p lie in $[-M, M]$. Choose $x > \max\{K, M\}$. Then $|x| > K$ so that $p(x) = 0$ and $|x| > M$ so that $x \notin [-M, M]$. This contradicts the choice of M. It follows that $\deg p(x) = 0$ so that $a_n = a_{n-1} = \cdots = a_1 = 0$. Thus $p(x) = a_0$. Since p has at least one zero, for example $K + 1$, we must have $a_0 = 0$ and the proof is complete. □

Theorem 3.1.7 (Equating Coefficients) Let $p(x), q(x) \in \mathbb{R}[x]$ with

$$
\begin{aligned}
p(x) &= a_n x^n + a_{n-1} x^{n-1} + \cdots + a_1 x + a_0, \\
q(x) &= b_n x^n + b_{n-1} x^{n-1} + \cdots + b_1 x + b_0
\end{aligned}
$$

and let there exist $K \ge 0$ such that, for all $|x| > K$, $p(x) = q(x)$. Then, for $i = 0, 1, \ldots, n$, $a_i = b_i$.

Proof Let $r(x) = p(x) - q(x)$ so that $r(x) \in \mathbb{R}[x]$ and

$$r(x) = (a_n - b_n)x^n + (a_{n-1} - b_{n-1})x^{n-1} + \cdots + (a_1 - b_1)x + (a_0 - b_0).$$

For all $|x| > K$, $r(x) = 0$. So it follows from Corollary 3.1.6 that, for $i = 0, 1, \ldots, n$, $a_i - b_i = 0$, i.e. $a_i = b_i$. □

Corollary 3.1.8 Let $c \in \mathbb{R}$ and let $p(x), q(x) \in \mathbb{R}[x]$ be such that, for all $x \in \mathbb{R}$,

$$(x - c)p(x) = (x - c)q(x).$$

Then, for all $x \in \mathbb{R}$, $p(x) = q(x)$.

Proof Let $K = |c|$. Then, for all $|x| > K$, $x - c \neq 0$, so we may divide by $x - c$ to get $p(x) = q(x)$. The result now follows from Theorem 3.1.7.   □

Remark By Theorem 3.1.7, two different polynomials (in the sense that not all the coefficients of one are equal to the corresponding coefficients of the other) give rise to different polynomial functions. Hence we may describe polynomial functions using polynomial terminology. For example, the real function $x \mapsto -4x^3 + 5x + 16$ has *degree* 3 and so is a *cubic function*.

3.2   Division and Factors

When we divide one positive integer by another we get a *quotient* and a *remainder*. For example,

$$\frac{4321}{999} = 4 + \frac{325}{999}, \qquad i.e. \; 4321 = 4 \times 999 + 325,$$

where 4 is the quotient and 325 the remainder. In general, starting with $p, d \in \mathbb{N}$, we can find $q, r \in \mathbb{N} \cup \{0\}$ with $0 \leq r < d$ such that $p = qd + r$.

The same process can be carried out for polynomials.

Theorem 3.2.1 *Let* $p(x), d(x) \in \mathbb{R}[x]$. *Then there exist* $q(x), r(x) \in \mathbb{R}[x]$ *with* $r(x)$ *the zero polynomial or* $\deg r(x) < \deg d(x)$ *such that*

$$p(x) = q(x)d(x) + r(x).$$

Proof The proof is omitted. A method for finding $q(x)$ and $r(x)$ is demonstrated in Example 3.2.3.   □

Definitions 3.2.2 With the notation of Theorem 3.2.1, where

$$p(x) = q(x)d(x) + r(x),$$

we say that the *dividend* $p(x)$ is *divided* by the *divisor* $d(x)$ giving the *quotient* $q(x)$ and the *remainder* $r(x)$.

When $r(x)$ is the zero polynomial, so that $p(x) = q(x)d(x)$, and $\deg d(x) \geq 1$ we say that $d(x)$ is a *factor* of $p(x)$ or that $d(x)$ *divides* $p(x)$.

Example 3.2.3 (Long Division of Polynomials) Find the quotient and the remainder when

$$6x^4 + 7x^3 - 2x^2 + 9x - 21 \quad \text{is divided by} \quad 2x^2 + 3x - 1.$$

Solution We shall go through the process step by step but we begin with the completed calculation so that the reader can see where the steps are leading.

$$3x^2 - x + 2$$
$$2x^2 + 3x - 1 \,\big)\, 6x^4 + 7x^3 - 2x^2 + 9x - 21$$
$$\underline{6x^4 + 9x^3 - 3x^2}$$
$$-2x^3 + x^2 + 9x$$
$$\underline{-2x^3 - 3x^2 + x}$$
$$4x^2 + 8x - 21$$
$$\underline{4x^2 + 6x - 2}$$
$$2x - 19$$

From this, the quotient is $q(x) = 3x^2 - x + 2$ and the remainder is $r(x) = 2x - 19$ so that

$$6x^4 + 7x^3 - 2x^2 + 9x - 21 = (3x^2 - x + 2)(2x^2 + 3x - 1) + (2x - 19).$$

This can be verified directly.

The above calculation is carried out as follows. In Step 1 we *set up* the *long division*. After that there are four basic steps—*divide, multiply, subtract* and *carry down*—which are repeated in sequence until there is nothing left to 'carry down'.

The dots represent those terms which, at each stage, are known but not used.

Step 1. Set up the (long) division of $6x^4 + 7x^3 - 2x^2 + 9x - 21$ by $2x^2 + 3x - 1$.

$$2x^2 + 3x - 1 \,\big)\, 6x^4 + 7x^3 - 2x^2 + 9x - 21$$

Step 2. Divide $6x^4$ by $2x^2$ to get $3x^2$.

$$3x^2$$
$$2x^2 \quad \cdot \quad \cdot \,\big)\, 6x^4 \quad \cdot \quad \cdot \quad \cdot \quad \cdot$$

Step 3. Multiply $2x^2 + 3x - 1$ by $3x^2$ to get $6x^4 + 9x^3 - 3x^2$.

$$3x^2$$
$$2x^2 + 3x - 1 \,\big)\, \cdot \quad \cdot \quad \cdot \quad \cdot \quad \cdot$$
$$6x^4 + 9x^3 - 3x^2$$

Step 4. Subtract $6x^4 + 9x^3 - 3x^2$ from $6x^4 + 7x^3 - 2x^2$ to get $-2x^3 + x^2$.

$$\cdot$$
$$\cdot \quad \cdot \quad \cdot \,\big)\, 6x^4 + 7x^3 - 2x^2 \quad \cdot \quad \cdot$$
$$\underline{6x^4 + 9x^3 - 3x^2}$$
$$-2x^3 + x^2$$

Step 5. Carry down $+9x$.

$$
\begin{array}{r}
\cdot \quad \cdot \quad \cdot + 9x \\
\hline
- 2x^3 + x^2 + 9x
\end{array}
$$

Step 6. Divide $-2x^3$ by $2x^2$ to get $-x$.

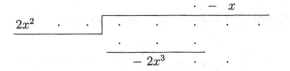

Step 7. Multiply $2x^2 + 3x - 1$ by $-x$ to get $-2x^3 - 3x^2 + x$.

$$
2x^2 + 3x - 1 \overline{\big)\ \begin{array}{r}
\cdot \quad - x \\
\cdot \quad \cdot \quad \cdot \quad \cdot \quad \cdot \\
\hline
\cdot \quad \cdot \quad \cdot \\
\hline
- 2x^3 - 3x^2 + x
\end{array}}
$$

Step 8. Subtract $-2x^3 - 3x^2 + x$ from $-2x^3 + x^2 + 9x$ to get $4x^2 + 8x$.

$$
\begin{array}{r}
- 2x^3 + x^2 + 9x \\
- 2x^3 - 3x^2 + x \\
\hline
4x^2 + 8x
\end{array}
$$

Step 9. Carry down -21.

$$
\begin{array}{r}
\cdot \quad \cdot \quad \cdot \quad - 21 \\
\hline
\\
\hline
4x^2 + 8x - 21
\end{array}
$$

Step 10. Divide $4x^2$ by $2x^2$ to get 2.

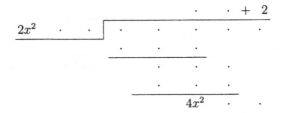

Step 11. Multiply $2x^2 + 3x - 1$ by 2 to get $4x^2 + 6x - 2$.

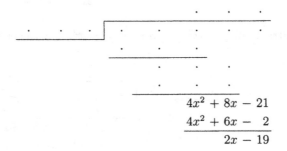

Step 12. Subtract $4x^2 + 6x - 2$ from $4x^2 + 8x - 21$ to get $2x - 19$.

$$
\begin{array}{r}
4x^2 + 8x - 21 \\
4x^2 + 6x - 2 \\
\hline
2x - 19
\end{array}
$$

This completes the division process since the remainder $2x - 19$ has degree less than that of the divisor $2x^2 + 3x - 1$. $\quad\square$

Example 3.2.4 Find polynomials $q(x)$ and $r(x)$ with $\deg r(x) < 3$ such that

$$2x^5 - 2x^4 + 3x^3 - 4x + 5 = q(x)\,(2x^3 - 2x^2 + 3) + r(x).$$

Solution We have

$$
\begin{array}{r}
x^2 + 0x + \frac{3}{2} \\
2x^3 - 2x^2 + 0x + 3 \enclose{longdiv}{2x^5 - 2x^4 + 3x^3 + 0x^2 - 4x + 5} \\
2x^5 - 2x^4 + 0x^3 + 3x^2 \\
\hline
0x^4 + 3x^3 - 3x^2 - 4x \qquad * \\
0x^4 + 0x^3 + 0x^2 + 0x \qquad * \\
\hline
3x^3 - 3x^2 - 4x + 5 \\
3x^3 - 3x^2 + 0x + \frac{9}{2} \\
\hline
0x^2 - 4x + \frac{1}{2}
\end{array}
$$

Hence,

$$2x^5 - 2x^4 + 3x^3 - 4x + 5 = (x^2 + \tfrac{3}{2})(2x^3 - 2x^2 + 3) + (-4x + \tfrac{1}{2})$$

which is in the required form.   □

Hazard In the solution of Example 3.2.4 we 'padded out' the dividend and the divisor with terms having zero coefficients so that all powers of x were represented up to the degree of each polynomial. While this is not essential it is strongly recommended as a precaution against error in the calculation.

On the other hand, the two lines marked $*$ would normally be omitted by someone familiar with the long division process.

Example 3.2.5 Show that $(x-1)^2$ is a factor of $p(x) = x^5 - 3x^3 - x^2 + 6x - 3$.

Solution We have $(x-1)^2 = x^2 - 2x + 1$ and as a consequence consider

$$
\begin{array}{r}
x^3 + 2x^2 + 0x - 3 \\
x^2 - 2x + 1 \enclose{longdiv}{x^5 + 0x^4 - 3x^3 - x^2 + 6x - 3} \\
x^5 - 2x^4 + x^3 \\
\hline
2x^4 - 4x^3 - x^2 \\
2x^4 - 4x^3 + 2x^2 \\
\hline
-3x^2 + 6x - 3 \\
-3x^2 + 6x - 3
\end{array}
$$

So the remainder is the zero polynomial and hence $(x-1)^2$ is a factor of $p(x)$ as required. We have also shown that

$$x^5 - 3x^3 - x^2 + 6x - 3 = (x-1)^2(x^3 + 2x^2 - 3).$$

So $x^3 + 2x^2 - 3$ is another factor of $p(x)$.   □

Theorem 3.2.6 (The Remainder Theorem) *Let $p(x), d(x) \in \mathbb{R}[x]$ with $d(x) = ax + b$, $a \neq 0$ (linear). Then, when $p(x)$ is divided by $d(x)$, the remainder is $p(-b/a)$ (constant).*

Proof Let $q(x)$ and $r(x)$ be the quotient and remainder, respectively. Since $\deg d(x) = 1$, $r(x)$ must be constant. Suppose $r(x) = k$. Then

$$p(x) = q(x)d(x) + r(x) = q(x)d(x) + k.$$

Now replace x with $-b/a$. Since $d(-b/a) = 0$ we get

$$p(-b/a) = k$$

and the result follows immediately. $\qquad\square$

Corollary 3.2.7 *Let $p(x) \in \mathbb{R}[x]$ and let $c \in \mathbb{F}$. Then*

    (1) *when $p(x)$ is divided by $x - c$ the remainder is $p(c)$,*

    (2) *$x - c$ is a factor of $p(c)$ if and only if $p(c) = 0$.*

Proof (1) This is the special case of Theorem 3.2.6 where $a = 1$ and $b = -c$.

    (2) This follows from (1) and Definitions 3.2.2. $\qquad\square$

An example of the use of Corollary 3.2.7 (2) is the following widely used result.

Lemma 3.2.8 *Let $a \in \mathbb{R}$. Then,*

    (1) *for $n \in \mathbb{N}$,*

$$x^n - a^n = (x - a)(x^{n-1} + ax^{n-2} + a^2 x^{n-3} + \cdots + a^{n-2}x + a^{n-1}), \qquad (3.2.1)$$

    (2) *for odd $n \in \mathbb{N}$,*

$$x^n + a^n = (x + a)(x^{n-1} - ax^{n-2} + a^2 x^{n-3} - \cdots - a^{n-2}x + a^{n-1}). \qquad (3.2.2)$$

Proof (1) It is clear that a is a zero of the polynomial $x^n - a^n$ and hence, by Corollary 3.2.7 (2), $x - a$ is a factor. It is left as an exercise to the reader to multiply the factors of the right hand side of (3.2.1) to verify the factorisation.

    (2) Let $b = -a$. When n is odd, $a^n = (-b)^n = -b^n$ and so (3.2.2) follows from (3.2.1) with $a = -b$. $\qquad\square$

Example 3.2.9 Given that $x + 1$ is a factor of $p(x) = 2x^3 - 5x^2 - x + a$, find a and complete the factorisation of $p(x)$.

Solution From Corollary 3.2.7, $p(-1) = 0$, *i.e.* $2(-1)^3 - 5(-1)^2 - (-1) + a = 0$. So $a = 6$. Hence

$$
\begin{aligned}
p(x) &= 2x^3 - 5x^2 - x + 6 \\
&= (x + 1)(2x^2 - 7x + 6) \quad \text{(by long division)} \\
&= (x + 1)(2x - 3)(x - 2) \quad \text{(see Section 3.3).} \qquad\square
\end{aligned}
$$

3.3 Quadratics

Definitions 3.3.1 Let $f(x)$ be a real quadratic polynomial, *i.e.*

$$f(x) = ax^2 + bx + c, \qquad (3.3.1)$$

where $a, b, c \in \mathbb{R}$ and $a \neq 0$. The *discriminant* of $f(x)$ is $\Delta = b^2 - 4ac$. For example, $2x^2 + 3x - 4$ has $\Delta = 3^2 - (4)(2)(-4) = 41$. As we proceed, we shall see how significant the discriminant is.

Since $a \neq 0$ in (3.3.1), we can rewrite $f(x)$ as follows.

$$\begin{aligned}
f(x) &= a\left(x^2 + \frac{b}{a}x\right) + c \\
&= a\left(\left(x + \frac{b}{2a}\right)^2 - \left(\frac{b}{2a}\right)^2\right) + c \\
&= a\left(x + \frac{b}{2a}\right)^2 - \frac{b^2 - 4ac}{4a} \\
&= a\left(x + \frac{b}{2a}\right)^2 - \frac{\Delta}{4a} \qquad (3.3.2) \\
&= a\left(\left(x + \frac{b}{2a}\right)^2 - \frac{\Delta}{4a^2}\right). \qquad (3.3.3)
\end{aligned}$$

In the forms (3.3.2) and (3.3.3), x occurs only in the square $(x + b/(2a))^2$. We refer to the above process as *completing the square in x*.

Example 3.3.2 Find the maximum value of $g(x) = 1 + x - x^2 \ (x \in \mathbb{R})$, and the value of x for which the maximum is attained.

Solution We use the above technique of completing the square as far as (3.3.2). We have

$$g(x) = -x^2 + x + 1 = -(x^2 - x) + 1$$

$$= -\left(\left(x - \frac{1}{2}\right)^2 - \frac{1}{4}\right) + 1 = -\left(x - \frac{1}{2}\right)^2 + \frac{5}{4}.$$

Since $(x - 1/2)^2 \geq 0$ and $(x - 1/2)^2 = 0$ only when $x = 1/2$, it follows that $g(x) \leq 5/4$, with equality only when $x = 1/2$. Thus $g(x)$ has maximum value $5/4$, attained when $x = 1/2$. $\quad\square$

Theorem 3.3.3 *The real quadratic equation* $f(x) = ax^2 + bx + c = 0$ *has*

    (1) *no real root if* $\Delta < 0$,

    (2) *one real root if* $\Delta = 0$,

    (3) *two real roots if* $\Delta > 0$.

When $\Delta \geq 0$, *the real roots of* $f(x) = 0$ *are* $\dfrac{-b \pm \sqrt{\Delta}}{2a}$.

Proof Using form (3.3.3) we see that $f(x) = 0$ if and only if

$$\left(x + \frac{b}{2a}\right)^2 = \frac{\Delta}{4a^2}. \tag{3.3.4}$$

Since $a \neq 0$, $a^2 > 0$, so that the right-hand side of (3.3.4) has the same sign as Δ, or is zero if $\Delta = 0$.

Case 1. Let $\Delta < 0$. Then the right-hand side of (3.3.4) is negative. So there can be no solution of (3.3.4) and hence no root of $f(x) = 0$. This establishes (1).

Case 2. Let $\Delta \geq 0$. Then the right-hand side of (3.3.4) is non-negative and we can take its square root. The solutions of (3.3.4), and hence the roots of $f(x) = 0$, are given by

$$x + \frac{b}{2a} = \pm\frac{\sqrt{\Delta}}{2a}, \qquad i.e.\ x = \frac{-b \pm \sqrt{\Delta}}{2a}.$$

This establishes (2) and (3) and the last part is immediate.　　　　　□

Definition 3.3.4 A real quadratic polynomial with $\Delta < 0$ is said to be *irreducible*.

From Theorem 3.3.3, being irreducible is equivalent to having no real zero. Then, from Corollary 3.2.7, we see that a real quadratic polynomial is irreducible if and only if it cannot be written as a product of linear factors—hence the term 'irreducible'.

Theorem 3.3.5 *Let $f(x) = ax^2 + bx + c$ be a real quadratic polynomial with $\Delta = b^2 - 4ac$. Let $x \in \mathbb{R}$. Then*

(1) If $\Delta < 0$ then $f(x)$ has the same sign as a.

(2) If $\Delta = 0$ then $f(x)$ has the same sign as a except when $x = -b/(2a)$, the zero of $f(x)$.

(3) If $\Delta > 0$, so that $f(x)$ has distinct zeros α and β with $\alpha < \beta$ (say), then $f(x)$ has the same sign as a when $x \notin [\alpha, \beta]$ and the opposite sign when $x \in (a, b)$.

Proof If we write $f(x)$ in the form (3.3.3) then (1) and (2) follow at once from the fact that $(x - b/(2a))^2 \geq 0$.

To prove (3), observe that we can write

$$f(x) = a(x - \alpha)(x - \beta).$$

This follows from Corollary 3.2.7 since $x - \alpha$ and $x - \beta$ are factors.

Table of signs for $f(x)$, $a < 0$.

x	$\rightarrow$	α	$\rightarrow$	β	$\rightarrow$
a	$-$	$-$	$-$	$-$	$-$
$x - \alpha$	$-$	0	$+$	$+$	$+$
$x - \beta$	$-$	$-$	$-$	0	$+$
$f(x)$	$-$	0	$+$	0	$-$

Table of signs for $f(x)$, $a > 0$.

x	$\rightarrow$	α	$\rightarrow$	β	$\rightarrow$
a	$+$	$+$	$+$	$+$	$+$
$x - \alpha$	$-$	0	$+$	$+$	$+$
$x - \beta$	$-$	$-$	$-$	0	$+$
$f(x)$	$+$	0	$-$	0	$+$

Now (3) follows from the above tables of signs for $f(x)$.　　　　　□

Theorem 3.3.5 allows us to determine the sign of a real quadratic without resorting to a table of signs. It is particularly useful (and easy) when $\Delta < 0$. It also allows us to sketch the graphs of quadratic functions. See Figure 3.3.1.

$a > 0, \ \Delta > 0$ $\quad a > 0, \ \Delta = 0$ $\quad a > 0, \ \Delta < 0$

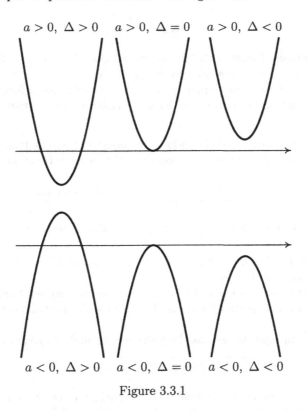

$a < 0, \ \Delta > 0$ $\quad a < 0, \ \Delta = 0$ $\quad a < 0, \ \Delta < 0$

Figure 3.3.1

Example 3.3.6 For $k \in \mathbb{R}$, let

$$F_k(x) = x^2 + (3k + 2)x + (2k^2 + 8).$$

Find the range of values of k for which $F_k(x)$ has no real zero.

Solution For fixed $k \in \mathbb{R}$, $F_k(x)$ is a real quadratic polynomial in x. As such it has discriminant

$$\Delta = (3k + 2)^2 - 4 \times 1 \times (2k^2 + 8) = k^2 + 12k - 28 = (k + 14)(k - 2).$$

By Theorem 3.3.3, $F_k(x)$ has no real zero if and only if $\Delta < 0$. But $\Delta = k^2 + 12k - 28$ is itself a real quadratic polynomial in k with positive leading coefficient and real zeros -14 and 2. Hence, it follows from Theorem 3.3.5 that $\Delta < 0$ if and only if $k \in (-14, 2)$. Thus the required range is $-14 < k < 2$. $\qquad\square$

Problem 3.3.7 Let $F_k(x)$ be defined as in Example 3.3.6. Find the values of k for which $F_k(x)$ has real zeros which differ by 6.

3.4   Rational Functions

Definitions 3.4.1 Let $f(x), g(x) \in \mathbb{R}[x]$ with $g(x)$ non-zero. We can define a real function h by

$$h(x) = \frac{f(x)}{g(x)}. \tag{3.4.1}$$

We call h a *rational* function. It has (maximal) domain $\mathbb{R} - \{x \in \mathbb{R} : g(x) = 0\}$. For example, $x \mapsto x/(x^2 - 1)$ has domain $\mathbb{R} - \{-1, 1\}$.

By analogy with rational numbers, we say that the rational function h is *proper* if h is the zero function or $\deg f(x) < \deg g(x)$. Otherwise h is *improper*.

Hazard Some care must be taken when the numerator and denominator of a rational function have a common factor. For example, the rational functions

$$x \mapsto \frac{x}{x^2 - x} = \frac{x}{x(x - 1)} \quad \text{and} \quad x \mapsto \frac{1}{x - 1}$$

are not equal. The first has domain $\mathbb{R} - \{0, 1\}$ while the second has domain $\mathbb{R} - \{1\}$.

Definitions 3.4.2 Let h be the rational function defined by (3.4.1) and let $f(x)$ and $g(x)$ have no common factor.

When h is proper and not the zero function, we can express $h(x)$ as a sum of (one or more) terms called *partial fractions*. This sum is the *partial fraction decomposition* of $h(x)$.

When h is improper we can use long division to find $q(x), r(x) \in \mathbb{R}[x]$ such that

$$f(x) = q(x)g(x) + r(x)$$

where $r(x)$ is not the zero polynomial (since $f(x)$ and $g(x)$ have no common factor) and $\deg r(x) < \deg g(x)$. For $x \in \text{dom } h$, so that $g(x) \neq 0$, we can divide by $g(x)$ to get

$$h(x) = \frac{f(x)}{g(x)} = q(x) + \frac{r(x)}{g(x)},$$

where $r(x)/g(x)$ is a proper rational function. In this case, the *partial fraction decomposition* of $h(x)$ is the sum of $q(x)$ and the partial fractions for $r(x)/g(x)$.

The first step in the decomposition process is to factorise the denominator $g(x)$ in $\mathbb{R}[x]$. As we shall see in Section 6.4, the real polynomial $g(x)$ can be written as a product of some or all of the following:

(1) a constant a, the leading coefficient of $g(x)$,

(2) real linear factors, *i.e.* factors of the form $x - \alpha$ where $\alpha \in \mathbb{R}$,

(3) irreducible real quadratic factors, *i.e.* factors of the form $x^2 + \beta x + \gamma$ where $\beta, \gamma \in \mathbb{R}$ with $\Delta = \beta^2 - 4\gamma < 0$.

Example 3.4.3 Factorise $g(x) = x^4 + 3x^3 + 4x^2 + 3x + 1$ in $\mathbb{R}[x]$.

Solution Since $g(-1) = 0$, $x + 1$ is a factor of $g(x)$ (see Corollary 3.2.7). Hence, by long division,

$$g(x) = (x + 1)(x^3 + 2x^2 + 2x + 1).$$

Similarly, since -1 is a zero of the second factor,

$$g(x) = (x + 1)(x + 1)(x^2 + x + 1) = (x + 1)^2(x^2 + x + 1). \qquad (3.4.2)$$

Since $x^2 + x + 1$ has $\Delta = 1^2 - 4 \times 1 \times 1 = -3 < 0$, the quadratic factor is irreducible and the factorisation is complete. $\qquad \square$

The process of combining repeated factors into powers, which we used in equations (3.4.2), is adopted more generally.

Let h be as in Definition 3.4.2. In general, we group together any repeated factors of $g(x)$ (linear or quadratic) into powers of the factors. When this is done, $g(x)$ can be expressed as a product or some of all of the following:

(1) a constant a,

(2) terms of the form $(x - \alpha)^m$,

(3) terms of the form $(x^2 + \beta x + \gamma)^n$,

where the underlying real linear or real irreducible quadratic factors are distinct. With this notation, we construct the partial fractions for $h(x)$ as follows. For each factor $(x - \alpha)^m$ of $g(x)$, include

$$\frac{A_1}{x - \alpha} + \frac{A_2}{(x - \alpha)^2} + \cdots + \frac{A_m}{(x - \alpha)^m} \qquad (3.4.3)$$

and for each factor $(x^2 + \beta x + \gamma)^n$, include

$$\frac{B_1 x + C_1}{x^2 + \beta x + \gamma} + \frac{B_2 x + C_2}{(x^2 + \beta x + \gamma)^2} + \cdots + \frac{B_n x + C_n}{(x^2 + \beta x + \gamma)^n}, \qquad (3.4.4)$$

where the constants in the numerators are real numbers still to be determined.

The denominator of each partial fraction is a factor of $g(x)$. So, if we multiply the sum of the partial fractions by $g(x)$, then in each fraction the denominator will cancel with part of $g(x)$. The result will therefore be a polynomial in x. This is the key to determining the real constants.

Example 3.4.4 Find the partial fraction decomposition of $f(x)/g(x)$ where

$$f(x) = 4x^2 + 3x + 2, \quad g(x) = x^4 + 3x^3 + 4x^2 + 3x + 1.$$

Solution First we factorise $g(x)$. This was done in Example 3.4.3 where we found that

$$g(x) = (x + 1)^2(x^2 + x + 1).$$

Following the outline preceding this example, we write

$$\frac{4x^2 + 3x + 2}{(x + 1)^2(x^2 + x + 1)} = \frac{A_1}{x + 1} + \frac{A_2}{(x + 1)^2} + \frac{Bx + C}{x^2 + x + 1}. \qquad (3.4.5)$$

Multiplying both sides by $g(x)$ gives

$$4x^2 + 3x + 2 = (x+1)(x^2+x+1)A_1 + (x^2+x+1)A_2$$
$$+ (x+1)^2(Bx+C). \tag{3.4.6}$$

We consider two ways of determining A_1, A_2, B and C.

Method 1. (Substituting Values)

As $x+1$ occurs in two terms on the right-hand side of (3.4.6), we put $x = -1$ (so these terms will vanish). This gives

$$4(-1)^2 + 3(-1) + 2 = 0 \times A_1 + ((-1)^2 + (-1) + 1)A_2 + 0 \times (Bx+C),$$

giving $A_2 = 3$. No other real value of x will isolate a constant (since $g(x)$ has no other real zero). However, we can substitute 3 for A_2 in (3.4.6) to get

$$4x^2 + 3x + 2 = (x+1)(x^2+x+1)A_1 + 3(x^2+x+1)$$
$$+ (x+1)^2(Bx+C).$$

Writing the terms with known coefficients on the left, this becomes

$$x^2 - 1 = (x+1)(x^2+x+1)A_1 + (x+1)^2(Bx+C). \tag{3.4.7}$$

Each term on the right-hand side of (3.4.7) has a factor of $x+1$, so the left-hand side must also have $x+1$ as a factor. In fact, the left-hand side is $(x+1)(x-1)$. [Note. This is a useful check on our arithmetic. If our value for A_2 is wrong then it is unlikely that $x+1$ would be a factor.] Dividing (3.4.7) by $x+1$ we get

$$x - 1 = (x^2+x+1)A_1 + (x+1)(Bx+C). \tag{3.4.8}$$

Now we can put $x = -1$ again (to eliminate the second term on the right-hand side). This gives

$$(-1) - 1 = ((-1)^2 + (-1) + 1)A_1,$$

giving $A_1 = -2$. Once more no other real value of x will isolate a constant, so we substitute -2 for A_1 in (3.4.8) to get

$$x - 1 = -2(x^2+x+1) + (x+1)(Bx+C).$$

Then, writing the terms with known coefficients on the left,

$$2x^2 + 3x + 1 = (x+1)(Bx+C). \tag{3.4.9}$$

The left-hand side of (3.4.9) must have $x+1$ as a factor. In fact, the left-hand side is $(x+1)(2x+1)$. Dividing by $x+1$ and equating coefficients we get

$$2x + 1 = Bx + C, \quad i.e. \ B = 2 \text{ and } C = 1.$$

Thus, (3.4.5) becomes

$$\frac{4x^2 + 3x + 2}{(x+1)^2(x^2+x+1)} = \frac{-2}{x+1} + \frac{3}{(x+1)^2} + \frac{2x+1}{x^2+x+1}. \tag{3.4.10}$$

This is the required partial fraction decomposition.

Method 2. (Equating Coefficients)

Here, we multiply out the terms on the right-hand side of (3.4.5) and regroup them to obtain coefficients for the powers of x. Thus

$$4x^2 + 3x + 2 = (x^3 + 2x^2 + 2x + 1)A_1 + (x^2 + x + 1)A_2$$
$$+ (x^2 + 2x + 1)(Bx + C)$$
$$= (A_1 + B)x^3 + (2A_1 + A_2 + 2B + C)x^2$$
$$+ (2A_1 + A_2 + B + 2C)x + (A_1 + A_2 + C).$$

Then, equating coefficients, we get

$$\left.\begin{array}{rcl} 0 &=& A_1 \qquad\quad + B \\ 4 &=& 2A_1 + A_2 + 2B + C \\ 3 &=& 2A_1 + A_2 + B + 2C \\ 2 &=& A_1 + A_2 \qquad + C \end{array}\right\}.$$

which must be solved simultaneously for A_1, A_2, B and C. A systematic way for solving such a system of equations is given in Chapter 11. For the present, we ask the reader to verify that $A_1 = -2$, $A_2 = 3$, $B = 2$ and $C = 1$ satisfy all four equations. Thus (as expected) we arrive at the same partial fraction decomposition (3.4.10) that we obtained by Method 1. $\qquad\square$

Note 3.4.5 In the solution to Example 3.4.4, the constant A_2 can be found using the *cover-up rule*, as follows. Consider equation (3.4.5):

$$\frac{4x^2 + 3x + 2}{(x + 1)^2(x^2 + x + 1)} = \frac{A_1}{x + 1} + \frac{A_2}{(x + 1)^2} + \frac{Bx + C}{x^2 + x + 1}.$$

Take the left hand side, 'cover up' the factor $(x + 1)^2$ then substitute $x = -1$. This gives 3 which is the value of A_2. Thus:

$$\frac{4x^2 + 3x + 2}{\boxed{}(x^2 + x + 1)} \qquad \xrightarrow{\ x = -1\ } \qquad \frac{4(-1)^2 + 3(-1) + 2}{(-1)^2 + (-1) + 1} = 3.$$

This is essentially the same as Method 1 in the solution to Example 3.4.4. However, the rule cannot be used to find A_1, B or C.

For the general situation, consider the expressions (3.4.3) and (3.4.4) which give partial fractions for the rational function $f(x)/g(x)$. The cover-up rule can be used to find A_m but none of the other constants $A_1, \ldots, A_{m-1}$, $B_1, \ldots, B_n$ or $C_1, \ldots C_n$. To find A_m, 'cover-up' the factor $(x - a)^m$ in the denominator of $f(x)/g(x)$ then substitute $x = a$.

When $g(x) = (x - \alpha_1)(x - \alpha_2) \ldots (x - \alpha_k)$, where $\alpha_1, \alpha_2, \ldots, \alpha_k$ are distinct, the method will give all the constants. See Example 3.4.8.

Remarks (1) At the stage of writing down the form of the partial fractions, the number of constants introduced is equal to the degree of the original denominator. This is a useful check. Thus, in Example 3.4.4, there were four constants, A_1, A_2, B and C, and the original denominator, $x^4 + 3x^3 + 4x^2 + 3x + 1$, had degree 4.

(2) In the discussion preceding Example 3.4.4, we stipulated that the polynomials $f(x)$ and $g(x)$ should have no common factor. While this is desirable in order to keep the calculations to a minimum, it is not essential. The process for finding the partial fractions is still valid and any 'extra' constants will be found to be 0.

(3) The reader may have noticed some potential problems in the solution to Example 3.4.4.

First, we assumed that (3.4.6) held for $x = -1$, a value of x for which (3.4.5) is undefined. In fact, the assumption is justified by Theorems 3.1.5 and 3.1.7. These theorems also justify the equating of coefficients in Method 2.

Secondly, at two points in Method 1, we divided by $x + 1$ and then put $x = -1$ in the resulting equation. This process is justified by Corollary 3.1.8.

Example 3.4.6 Find the partial fraction decomposition of

$$h(x) = \frac{x^4 + 3x^3 - 1}{x^4 + 3x^3 + 4x^2 + 3x + 1}.$$

Solution Here h is improper so we begin by using long division.

$$
\begin{array}{r}
1 \\
x^4 + 3x^3 + 4x^2 + 3x + 1 \overline{\smash{\big)}\ x^4 + 3x^3 + 0x^2 + 0x - 1} \\
\underline{x^4 + 3x^3 + 4x^2 + 3x + 1} \\
-4x^2 - 3x - 2
\end{array}
$$

Thus,

$$h(x) = 1 - \frac{4x^2 + 3x + 2}{x^4 + 3x^3 + 4x^2 + 3x + 1}.$$

Hence, using the partial fractions from Example 3.4.4, we have

$$h(x) = 1 + \frac{2}{x+1} - \frac{3}{(x+1)^2} - \frac{2x+1}{x^2+x+1}.$$

This is the required partial fraction decomposition. $\qquad\qquad \square$

Example 3.4.7 Find the partial fraction decomposition of

$$h(x) = \frac{x}{x^4 + x^2 + 1}.$$

Solution To factorise $x^4 + x^2 + 1$ we use the 'trick' of rewriting it as the difference of squares. Thus,

$$x^4 + x^2 + 1 = (x^4 + 2x^2 + 1) - x^2 = (x^2 + 1)^2 - x^2$$
$$= ((x^2 + 1) - x)((x^2 + 1) + x) = (x^2 - x + 1)(x^2 + x + 1).$$

Each of the quadratic factors has $\Delta = -3$ and so is irreducible. Then

$$\frac{x}{x^4 + x^2 + 1} = \frac{Ax + B}{x^2 - x + 1} + \frac{Cx + D}{x^2 + x + 1}.$$

Multiplying both sides by $x^4 + x^2 + 1$, we get

$$x = (x^2 + x + 1)(Ax + B) + (x^2 - x + 1)(Cx + D). \qquad (3.4.11)$$

As $x^4 + x^2 + 1$ has no real zero, it has no linear factor. So Method 1 of Example 3.4.4 cannot be used. We use Method 2. Multiplying out the right-hand side of 3.4.11 and regrouping, we obtain

$$x = (A + C)x^3 + (A + B - C + D)x^2 + (A + B + C - D)x + (B + D).$$

Then, equating coefficients, we get

$$\left.\begin{array}{l} 0 = A \quad\;\;\; + C \\ 0 = A + B - C + D \\ 1 = A + B + C - D \\ 0 = \quad\;\; B \quad\;\; + D \end{array}\right\}.$$

These are relatively easy to solve. Observe that the first equation gives $C = -A$, and the fourth gives $D = -B$. We substitute these into the second and third equations to get

$$\left.\begin{array}{l} 0 = 2A \\ 1 = 2B \end{array}\right\}.$$

Hence, $A = 0$, $B = 1/2$, $C = -A = 0$ and $D = -B = -1/2$ so that

$$h(x) = \frac{1/2}{x^2 - x + 1} - \frac{1/2}{x^2 + x + 1} = \frac{1}{2(x^2 - x + 1)} - \frac{1}{2(x^2 + x + 1)}$$

is the required partial fraction decomposition. $\qquad\qquad\qquad\qquad\qquad\qquad$ □

Example 3.4.8 Find the partial fraction decomposition of

$$h(x) = \frac{1}{x^2 + 3x + 2}.$$

Deduce that, for $n \in \mathbb{N}$,

$$\frac{1}{1 \times 2} + \frac{1}{2 \times 3} + \frac{1}{3 \times 4} + \cdots + \frac{1}{n(n+1)} = \frac{n}{n+1}. \qquad (3.4.12)$$

Solution Factorising $x^2 + 3x + 2$ gives

$$x^2 + 3x + 2 = (x + 1)(x + 2).$$

Then

$$\frac{1}{x^2 + 3x + 2} = \frac{A}{x + 1} + \frac{B}{x + 2}$$

and multiplying both sides by $x^2 + 3x + 2$ gives

$$1 = (x + 2)A + (x + 1)B.$$

Putting $x = -1$ and $x = -2$ in turn, we get $A = 1$ and $B = -1$. Alternatively, use the cover-up rule (see Note 3.4.5). Hence

$$h(x) = \frac{1}{(x+1)(x+2)} = \frac{1}{x+1} - \frac{1}{x+2} \qquad (3.4.13)$$

is the required partial fraction decomposition.

The terms on the left-hand side of equation (3.4.12) are

$$\frac{1}{(x+1)(x+2)} \qquad (x = 0, 1, 2, \ldots, n-1).$$

So each may be replaced, using equation (3.4.13). Thus the left-hand side of equation (3.4.12) becomes

$$\left(\frac{1}{1} - \frac{1}{2}\right) + \left(\frac{1}{2} - \frac{1}{3}\right) + \left(\frac{1}{3} - \frac{1}{4}\right) + \cdots + \left(\frac{1}{n} - \frac{1}{n+1}\right). \qquad (3.4.14)$$

Now observe that the second term in the first bracket cancels with the first term in the second bracket, the second term in the second bracket cancels with the first term in the third bracket, and so on. Thus (3.4.14) 'telescopes', leaving only the first term in the first bracket and the second term in the last bracket. Hence

$$\text{LHS (3.4.12)} = \frac{1}{1} - \frac{1}{n+1} = \frac{(n+1) - 1}{n+1} = \frac{n}{n+1} = \text{RHS (3.4.12)},$$

as required. □

Note 3.4.9 The partial fraction decomposition of Example 3.4.8 is a special case of the decomposition

$$\frac{1}{(x-\alpha)(x-\beta)} = \frac{1}{\alpha - \beta}\left(\frac{1}{x-\alpha} - \frac{1}{x-\beta}\right)$$

where $\alpha \neq \beta$, the verification of which is left to the reader. The reader is also invited to *deduce* from this that if $ad - bc \neq 0$ then

$$\frac{1}{(ax+b)(cx+d)} = \frac{1}{ad - bc}\left(\frac{a}{ax+b} - \frac{c}{cx+d}\right).$$

3.X   Exercises

1. By using long division, find the quotient $q(x)$ and remainder $r(x)$ such that

    (a) $6x^4 + 11x^3 + 15x^2 + 14x - 4 = (3x^2 + 4x + 1)q(x) + r(x)$,

    (b) $3x^4 - 2x^3 - 17x^2 - 5x + 3 = (3x^2 + x - 1)q(x) + r(x)$,

    (c) $2x^5 - 5x^4 - 6x^3 + 15x^2 + 2x - 1 = (x^2 - 3x + 1)q(x) + r(x)$,

    (d) $x^4 + 3x^2 + 1 = (x^2 + x + 1)q(x) + r(x)$,

    (e) $x^4 + 7 = (2x^3 - 3)q(x) + r(x)$.

2.   Express $\dfrac{4x^3 + 1}{(x-1)^2}$ in the form $ax + b + \dfrac{cx+d}{(x-1)^2}$, where $a, b, c, d \in \mathbb{R}$.

3.   What is the remainder when $x^5 + 4x^3 + 2x^2 - 3x + 2$ is divided by $2x + 1$? [Hint. Long division need not be used.]

4.   Find all the real zeros of each of the following polynomials.

   (a) $x^3 - 6x^2 + 7x + 2$,

   (b) $x^3 + 2x^2 + 6x + 5$,

   (c) $12x^4 - 31x^3 + 8x^2 + 31x - 20$,

   (d) $x^4 - 3x^2 - 4$,

   (e) $x^6 + 3x^4 + 4x^2 + 2$.

5.   Using the 'trick' described in Example 3.4.7, factorise

   (a) $x^4 + 3x^2 + 4$,   (b) $x^4 + 4$.

6.   Verify that $x^2 + x + 2$ is a factor of the polynomial

$$p(x) = 2x^4 - x^3 + 5x^2 - 2x + 8.$$

   Deduce that $p(x)$ has no real zero.

7.   Find the minimum value of

$$f(x, y) = x^2 + y^2 + 4x - 6y + 20 \quad (x, y \in \mathbb{R}).$$

   For which x and y is this minimum obtained?

8.   Let $a \in \mathbb{R}$. Verify that $(x - 1)^2$ is a factor of

$$p(x) = x^4 - ax^2 + (2a - 4)x + (3 - a).$$

   Prove that all the zeros of $p(x)$ are real if and only if $a \geq 2$.

9.   The sum of the zeros of $f(x) = x^2 - 3kx - 14$ is 3. Find k.

10.   Find the real values of x for which

   (a) $x^2 - 5x + 6 < 0$,

   (b) $x^2 + 2x \leq 3$,

   (c) $2x^2 - 3x + 5 > 0$,

   (d) $x(9x - 10) \leq 5(4x - 5)$,

   (e) $(x^3 - 27) - 2(x^2 - 9) - 2(x - 3) > 0$.

11. Let $f(x) = x^2 - (k+1)x + k^2$, where k is a real constant. Show that, if $f(x) > 0$ for all real x, then $k \notin [-1/3, 1]$. If $k \in (-1/3, 1)$, find the *length* of the range of values of x for which $f(x) \leq 0$.

12. Find the partial fraction decomposition of each of the following.

   (a) $\dfrac{x+3}{x^2 - 2x - 3}$,

   (b) $\dfrac{x^2 + 2x + 4}{x(x+1)^2}$,

   (c) $\dfrac{x^2}{(x+3)^3}$,

   (d) $\dfrac{2x^3 + 3x^2 - 16x - 4}{x^3 - 4x}$,

   (e) $\dfrac{1}{x^3 + 1}$,

   (f) $\dfrac{22x^4}{(x^2 - 1)(2x^2 + 9)}$,

   (g) $\dfrac{5x^2 - 7x + 18}{x^3 - 2x^2 + 4x - 8}$,

   (h) $\dfrac{50x^3 - 50x^2 + 100x + 50}{(x^2 + 1)^2(x - 2)}$.

Chapter 4

Induction and the Binomial Theorem

The reader will recall that $(a+b)^2 = a^2 + 2ab + b^2$, and may well know that $(a+b)^3 = a^3 + 3a^2b + 3ab^2 + b^3$. In this chapter, we will obtain a formula for $(a+b)^n$ which holds for *any* $n \in \mathbb{N}$. The formula consists of a sum of integer multiples of terms of the form $a^{n-r}b^r$, in which the coefficients are called *binomial coefficients*. The main objective of this chapter is to obtain the values of these coefficients. To achieve this, we need the technique of *proof by induction*. This is a method which allows us to prove a result for *all* integers $n \in \mathbb{N}$ in just two steps. It has many important applications in mathematics.

We are also interested in the number of ways of performing certain tasks, such as arranging n objects in order or choosing r out of n objects. Binomial coefficients also appear in connection with the second of these problems. For example, to win the UK National Lottery, we must choose 6 numbers from the set $\{1, 2, \ldots, 49\}$. As we shall see in Section 4.2, this can be done in $13,983,816$ ways. Thus the odds of winning are just one in this very large number!

4.1 The Principle of Induction

Consider the following statements, each of which concerns a positive integer n.

$$A(n) \; : \; \text{there exists } k \in \mathbb{N} \text{ such that } n = 2k;$$
$$B(n) \; : \; 1^2 + 2^2 + 3^2 + \cdots + n^2 = \tfrac{1}{6}n(n+1)(2n+1);$$
$$C(n) \; : \; 2^n < n(n-1) \cdots \times 2 \times 1.$$

For a *fixed* value of n, it is easy to check whether a given statement is true or false. If we consider values of n up to 5, we find that $A(1)$, $A(3)$ and $A(5)$ are false while $A(2)$ and $A(4)$ are true; $B(1), \ldots, B(5)$ are all true; $C(1)$, $C(2)$ and $C(3)$ are false while $C(4)$ and $C(5)$ are true. It is not difficult to see that $A(n)$ is false if n is odd, and true if n is even. Further experimentation might convince the reader that $B(n)$ is true for all $n \in \mathbb{N}$, and $C(n)$ is true for $n \geq 4$. We seek a method for proving statements which are true for n greater than or equal to some starting value.

The key is the following feature of the positive integers which distinguishes $\mathbb{N}$ from $\mathbb{Z}$, $\mathbb{Q}$ and $\mathbb{R}$.

Assumption (The Well Ordering Principle) Each non-empty subset S of $\mathbb{N}$ has a least element, *i.e.* if $S \subseteq \mathbb{N}$ and $S \neq \emptyset$ then $\min S$ exists.

Intuition would suggest that this is 'obvious' even when $\mathbb{N}$ is replaced with any subset of $\mathbb{Z}$, bounded below. On the other hand there is no corresponding property for $\mathbb{Z}$ or larger number systems. For example, let T be the set of negative integers. Then T is a non-empty subset of $\mathbb{Z}$ but it has no least element, for if $k \in T$ then $k - 1 \in T$ and $k - 1 < k$.

Assuming the well ordering principle enables us to find a method of the type sought above.

Theorem 4.1.1 (The Principle of Induction) *Let $h \in \mathbb{Z}$ and let $P(n)$ be a statement such that*

(1)   *$P(h)$ is true, and*

(2)   *for each integer $k \geq h$, if $P(k)$ is true then $P(k + 1)$ is true also.*

Then $P(n)$ is true for all integers $n \geq h$.

Proof Let $S = \{n \in \mathbb{Z} : n \geq h \text{ and } P(n) \text{ is false}\}$. Suppose that S is non-empty. We will show that this leads to a contradiction.

By the well ordering principle, S has a least element m. Since $m \in S$, $m \geq h$ and $P(m)$ is false. By (1), $P(h)$ is true and so $m > h$. Thus $m - 1 \geq h$ and, since $m = \min S$, $m - 1 \notin S$. Hence $P(m - 1)$ is true and therefore, by (2) (with $k = m - 1$), $P(m)$ is true also.

We have shown that $P(m)$ is both false and true, which is a contradiction. It follows that S is empty, *i.e.* there is no $n \in \mathbb{N}$ such that $n \geq h$ and $P(n)$ is false. Hence $P(n)$ is true for all integers $n \geq h$.   □

The Principle of Induction can be used in a *proof by induction* as follows.

Step 1. Verify that $P(h)$ is true.

Step 2. Assume that $P(k)$ is true (where $k \geq h$ is an integer).

Step 3. Deduce (from Step 2) that $P(k + 1)$ is true.

Step 4. Conclude that $P(n)$ is true for all integers $n \geq h$.

Remarks (1) The assumption in Step 2 is called the *induction hypothesis*.

(2) Steps 2 and 3 together establish condition (2) of Theorem 4.1.1.

Definitions 4.1.2 Let $f : \mathbb{Z} \to \mathbb{R}$ and let $h, k \in \mathbb{Z}$ with $h \leq k$. Then

$$\sum_{r=h}^{k} f(r) \quad \text{denotes the sum} \quad f(h) + f(h + 1) + \cdots + f(k),$$

i.e. the sum of the values of f at the integers r in the range $h \le r \le k$. Similarly,

$$\prod_{r=h}^{k} f(r) \quad \text{denotes the product} \quad f(h)f(h+1)\cdots f(k).$$

For example,

$$\sum_{r=3}^{5} r^3 = 3^3 + 4^3 + 5^3 \ (= 216), \qquad \prod_{r=3}^{5} r^3 = 3^3 \times 4^3 \times 5^3 \ (= 216,000).$$

This use of $\sum$ and $\prod$ is referred to as *sigma notation* and *pi notation*, respectively. The forms $\sum_{r=h}^{k} f(r)$ and $\prod_{r=h}^{k} f(r)$ are also used.

Remarks (1) Both $\sum_{r=h}^{h} f(r)$ and $\prod_{r=h}^{h} f(r)$ equal $f(h)$.

(2) In both $\sum_{r=h}^{k} f(r)$ and $\prod_{r=h}^{k} f(r)$, r is a dummy variable. So, for example,

$$\sum_{r=3}^{5} r^3 = \sum_{s=3}^{5} s^3 = \sum_{t=3}^{5} t^3. \tag{4.1.1}$$

Note also that, for example,

$$\sum_{r=3}^{5} r^3 = \sum_{r=4}^{6} (r-1)^3 = \sum_{r=0}^{2} (r+3)^3.$$

This is equivalent to using the substitutions $s = r - 1$ and $t = r + 3$ in (4.1.1).

(3) If the formula for f involves more than one 'unknown' then it is important to be clear which is the function variable, also called the *summation* or *product variable*. For example,

$$\sum_{r=3}^{5} r^m = 3^m + 4^m + 5^m \quad \text{while} \quad \sum_{m=3}^{5} r^m = r^3 + r^4 + r^5.$$

(4) Elementary properties will be assumed. For example,

$$\sum_{r=h}^{k} \lambda f(r) = \lambda \sum_{r=h}^{k} f(r), \qquad \sum_{r=h}^{k} [f(r) + g(r)] = \sum_{r=h}^{k} f(r) + \sum_{r=h}^{k} g(r),$$

$$\prod_{r=h}^{k} [f(r)g(r)] = \left[\prod_{r=h}^{k} f(r) \right] \left[\prod_{r=h}^{k} g(r) \right].$$

Note 4.1.3 Statements which involve $\sum_{r=1}^{n} f(r)$ or $\prod_{r=1}^{n} f(r)$ often yield to induction techniques, since

$$\sum_{r=1}^{k+1} f(r)) = \underbrace{f(1) + f(2) + \cdots + f(k)} + f(k+1) = \left[\sum_{r=1}^{k} f(r) \right] + f(k+1)$$

and

$$\prod_{r=1}^{k+1} f(r)) = \underbrace{f(1)f(2)\ldots f(k)} f(k+1) = \left[\prod_{r=1}^{k} f(r) \right] f(k+1).$$

Example 4.1.4 Prove that, for all $n \in \mathbb{N}$,

$$\sum_{r=1}^{n} r^2 = \tfrac{1}{6}n(n+1)(2n+1).$$

Solution Let $P(n)$ be the statement

$$\sum_{r=1}^{n} r^2 = \tfrac{1}{6}n(n+1)(2n+1).$$

First,

$$\text{LHS } P(1) = \sum_{r=1}^{1} r^2 = 1^2 = 1$$

and

$$\text{RHS } P(1) = \tfrac{1}{6}(1)(1+1)(2.1+1) = 1.$$

Thus $P(1)$ is true.

Next, let $k \in \mathbb{N}$ and assume that $P(k)$ is true, *i.e.* that

$$\sum_{r=1}^{k} r^2 = \tfrac{1}{6}k(k+1)(2k+1).$$

Now consider $P(k+1)$.

$$\begin{aligned}
\text{LHS } P(k+1) &= \sum_{r=1}^{k+1} r^2 = \sum_{r=1}^{k} r^2 + (k+1)^2 \quad \text{(see Note 4.1.3)} \\
&= \tfrac{1}{6}k(k+1)(2k+1) + (k+1)^2 \quad \text{(by induction hypothesis)} \\
&= \tfrac{1}{6}(k+1)[k(2k+1) + 6(k+1)] \\
&= \tfrac{1}{6}(k+1)(2k^2 + 7k + 6) = \tfrac{1}{6}(k+1)(k+2)(2k+3) \\
&= \tfrac{1}{6}(k+1)[(k+1)+1][2(k+1)+1] = \text{RHS } P(k+1).
\end{aligned}$$

Thus, if $P(k)$ is true then $P(k+1)$ is true also.

Hence, by induction, $P(n)$ is true for all $n \in \mathbb{N}$. $\square$

Theorem 4.1.5 *For all $n \in \mathbb{N}$,*

$$(1) \quad \sum_{r=1}^{n} r = \tfrac{1}{2}n(n+1),$$

$$(2) \quad \sum_{r=1}^{n} r^2 = \tfrac{1}{6}n(n+1)(2n+1),$$

$$(3) \quad \sum_{r=1}^{n} r^3 = \tfrac{1}{4}n^2(n+1)^2.$$

Proof We proved (2) in Example 4.1.4. The proofs of (1) and (3) follow the same pattern and are left to the reader. $\square$

Using Theorem 4.1.5, we can evaluate sums of the form $\sum_{r=1}^{n} f(r)$ where $f(r)$ is a polynomial in r of degree at most 3.

Example 4.1.6 Obtain a formula for

$$2 \times 3 + 3 \times 5 + 4 \times 7 + \cdots + (n+1)(2n+1),$$

where $n \in \mathbb{N}$.

Solution Using sigma notation we have

$$\sum_{r=1}^{n}(r+1)(2r+1) = \sum_{r=1}^{n}(2r^2 + 3r + 1)$$

$$= 2\sum_{r=1}^{n} r^2 + 3\sum_{r=1}^{n} r + \sum_{r=1}^{n} 1$$

$$= 2\left[\tfrac{1}{6}n(n+1)(2n+1)\right] + 3\left[\tfrac{1}{2}n(n+1)\right] + n$$

$$= \tfrac{1}{6}n[2(n+1)(2n+1) + 9(n+1) + 6]$$

$$= \tfrac{1}{6}n(4n^2 + 15n + 17),$$

as required.   □

Remark In Example 4.1.6, we did not know in advance the formula for the sum. So we could *not* use induction.

 Hazard In Example 4.1.6, we met the sum $\sum_{r=1}^{n} 1$. This sum consists of n terms, each of which is 1. Hence the sum equals n (as stated in the example).

Induction is often used to deal with objects which have *recursive* definitions. For example, a function $f : \mathbb{N} \to \mathbb{R}$ where $f(n+1)$ is calculated from $f(n)$.

Definition 4.1.7 For $n \in \mathbb{N} \cup \{0\}$, $n!$ (read 'n factorial') is defined by

$$0! = 1 \quad \text{and, for } n \in \mathbb{N}, \quad n! = n \times (n-1)!.$$

For example,

$$
\begin{aligned}
3! &= 3 \times 2! \\
&= 3 \times (2 \times 1!) \\
&= 3 \times 2 \times (1 \times 0!) \\
&= 3 \times 2 \times 1 \times 1 \qquad \text{(by the definition of } 0!) \\
&= 6.
\end{aligned}
$$

Following the above method, it is not difficult to see that for $n \in \mathbb{N}$,

$$n! = n(n-1)\cdots \times 2 \times 1,$$

i.e. $n!$ is the product of the positive integers $1, 2, \ldots, n$.

Notes 4.1.8 (1) An important example of a recursive definition is that of the nth derivative of y with respect to x. This is dealt with in Note 8.4.2 but for the reader already familiar with derivative notation we state a version here.

$$\frac{d^1 y}{dx^1} = \frac{dy}{dx} \quad \text{and, for } n \in \mathbb{N}, \quad \frac{d^{n+1} y}{dx^{n+1}} = \frac{d}{dx}\left(\frac{d^n y}{dx^n}\right).$$

(2) The definitions of $\sum_{r=1}^{n} f(r)$ and $\prod_{r=1}^{n} f(r)$ can be made recursive. Define $\sum_{r=1}^{1} f(r) = \prod_{r=1}^{1} f(r) = f(1)$ and use the equations in Note 4.1.3.

Problem 4.1.9 Write down a recursive definition of the nth power of a real number, where $n \in \mathbb{N}$. Hint: consider the solution to Example 4.1.10.

Example 4.1.10 Prove that, for all $n \in \mathbb{N}$ with $n \geq 4$, $2^n < n!$.

Solution Let $P(n)$ be the statement

$$2^n < n!.$$

First,
$$\text{LHS } P(4) = 2^4 = 16 \quad \text{and} \quad \text{RHS } P(4) = 4! = 24.$$

Thus $P(4)$ is true.

Next, let $k \in \mathbb{N}$, $k \geq 4$ and assume that $P(k)$ is true, *i.e.* $2^k < k!$, and consider $P(k+1)$.

$$\begin{aligned}
\text{LHS } P(k+1) = 2^{k+1} &= 2 \times 2^k \\
&< 2 \times k! && \text{(by inductive hypothesis)} \\
&< (k+1)k! && \text{(as } k \geq 4 \text{ so that } k+1 > 2) \\
&= (k+1)! = \text{RHS } P(k+1).
\end{aligned}$$

Thus, if $P(k)$ is true then $P(k+1)$ is true also.

Hence, by induction, $P(n)$ is true for all $n \in \mathbb{N}$, $n \geq 4$. □

Remark In Example 4.1.10, $P(1), \ldots, P(3)$ are all false.

Example 4.1.11 Prove that, for all $n \in \mathbb{N}$, 7 divides $4 \times 6^{2n} + 3 \times 2^{3n}$.

Solution Let $P(n)$ be the statement

$$7 \mid (4 \times 6^{2n} + 3 \times 2^{3n}).$$

First, with $n = 1$,

$$4 \times 6^{2n} + 3 \times 2^{3n} = 4 \times 36 + 3 \times 8 = 168 = 7 \times 24.$$

Thus $P(1)$ is true.

Next, let $k \in \mathbb{N}$ and assume that $P(k)$ is true, *i.e.* $7 \mid (4 \times 6^{2k} + 3 \times 2^{3k})$. We can rewrite this as $4 \times 6^{2k} + 3 \times 2^{3k} = 7h$ where $h \in \mathbb{Z}$ so that

$$4 \times 6^{2k} = 7h - 3 \times 2^{3k}. \tag{4.1.2}$$

Now consider $P(k+1)$. We have

$$\begin{aligned}
4 \times 6^{2k+2} + 3 \times 2^{3k+3} &= 4 \times 6^2 \times 6^{2k} + 3 \times 2^3 \times 2^{3k} \\
&= 6^2(7h - 3 \times 2^{3k}) + 8 \times 3 \times 2^{3k} \quad \text{(by (4.1.2))} \\
&= 7 \times 36h - 28 \times 3 \times 2^{3k} \\
&= 7(36h - 12 \times 2^{3k}),
\end{aligned}$$

and so since $36h - 12 \times 2^{3k}$ is an integer, $7 \mid (4 \times 6^{2k+2} + 3 \times 2^{3k+3})$. Thus, if $P(k)$ is true then $P(k+1)$ is true also.

Hence, by induction, $P(n)$ is true for all $n \in \mathbb{N}$. $\square$

Problem 4.1.12 Let $P(n)$ be a statement such that

(1) $P(1)$ is true, and

(2) if $P(1), \ldots, P(k)$ are all true then $P(k+1)$ is true also.

Prove that $P(n)$ is true for all $n \in \mathbb{N}$.

4.2 Picking and Choosing

Theorem 4.2.1 (The Multiplication Principle) *Let P be a process which can be broken into n stages ($n \in \mathbb{N}$) in such a way that the rth stage can be done in p_r ways and this number is independent of the way in which the first $(r-1)$ stages are done. Then P may be done in $p_1 p_2 \ldots p_n$ ways.*

Proof The proof is omitted. $\square$

Example 4.2.2 A computer system uses passwords consisting of two letters followed by two digits. How many different passwords are possible?

Solution Consider the choice of each character in turn.

Stage 1:	choose the first character	26 possibilities.
Stage 2:	choose the second character	26 possibilities.
Stage 3:	choose the first digit	10 possibilities.
Stage 4:	choose the second digit	10 possibilities.

Using the multiplication principle, we find that the number of passwords is $26 \times 26 \times 10 \times 10 = 67,600$. $\square$

Example 4.2.3 Consider the problem in Example 4.2.2. How many passwords are there in which the letters and the digits are all different?

Solution We may use the same four stage process as before, but now, at Stage 2, there are only 25 possibilities since the second letter must be different from the first. Similarly, at Stage 4 there are only 9 possibilities. Thus the number of suitable passwords is $26 \times 25 \times 10 \times 9 = 58,500$. $\square$

As the next example shows, we must be careful about the 'independence' clause in the multiplication principle.

Example 4.2.4 Using the digits $1, 2, 3, 4, 5, 6$, how many even 2-digit numbers can be formed in which the two digits are different?

Solution A simple 2-stage approach would be to choose the first digit and *then* the second. On this basis, consider the choice of the *second* digit. Since the resulting number is to be even, we must choose 2, 4 or 6. The restriction that the two digits must be different means that there are 3 possibilities if the first digit is odd, and 2 possibilities if the first digit is even. Thus the multiplication principle cannot be invoked.

The way to solve this problem is to make the choice of the second digit at Stage 1. As the result must be even, there are three possibilities (2, 4 or 6). Then at Stage 2 we choose the first digit. Since it must be different from the second digit, there are 5 possibilities *whatever* the choice of the second digit. Thus the multiplication principle applies to give a total of $3 \times 5 = 15$ suitable numbers.   □

Definition 4.2.5 A *permutation* of a finite non-empty set is an arrangement of its elements *in order*. For example, the 3-element set $S = \{a, b, c\}$ has six permutations:

$$abc, \quad acb, \quad bca, \quad bac, \quad cab, \quad cba.$$

Theorem 4.2.6 *For $n \in \mathbb{N}$, an n-element set has $n!$ permutations.*

Proof Choosing a permutation is an n-stage process. At the rth stage (choosing the rth element) there are $n - (r - 1)$ possibilities, $(r - 1)$ elements already having been chosen. Thus the multiplication principle applies to give

$$n(n - 1) \cdots \times 2 \times 1 = n!$$

permutations as required.   □

Definition 4.2.7 Let $n \in \mathbb{N}$ and $r \in \mathbb{N} \cup \{0\}$ with $0 \leq r \leq n$. Then $\binom{n}{r}$ (read 'n choose r') is the number of r-element subsets of an n-element set. Here we are interested only in subsets—the order of the elements is immaterial. For example, The 3-element set $\{a, b, c\}$ has 2-element subsets $\{a, b\}$, $\{b, c\}$ and $\{c, a\}$. Thus $\binom{3}{2} = 3$. More generally, it is easy to see that

$$\binom{n}{n} = 1 \qquad \text{and} \qquad \binom{n}{1} = n.$$

Theorem 4.2.8 *Let $n \in \mathbb{N}$ and $r \in \mathbb{N} \cup \{0\}$ with $0 \leq r \leq n$. Then*

$$\binom{n}{r} = \binom{n}{n - r}.$$

Proof The complement of an r-element subset of an n-element set is an $(n-r)$-element subset. Hence the number of subsets of each size is the same. $\square$

Corollary 4.2.9 *Let $n \in \mathbb{N}$. Then*

$$\binom{n}{0} = 1 \qquad and \qquad \binom{n}{n-1} = n.$$

Proof This follows immediately from Theorem 4.2.8 and the examples at the end of Definition 4.2.7. $\square$

Theorem 4.2.10 *Let $n \in \mathbb{N}$ and $r \in \mathbb{N} \cup \{0\}$ with $0 \le r \le n$. Then*

$$\binom{n}{r} = \frac{n!}{r!(n-r)!}.$$

Proof Since $0! = 1$ and $\binom{n}{0} = \binom{n}{n} = 1$, the result holds for $r = 0$ and $r = n$.

Now let $0 < r < n$ and let S be an n-element set. From Theorem 4.2.6, S has $n!$ permutations. We can also count these permutations using the following 3-stage process.

Stage 1: choose an r-element subset T of S $\binom{n}{r}$ possibilities.

Stage 2: choose a permutation of T $r!$ possibilities.

Stage 3: choose a permutation of $S - T$ $(n-r)!$ possibilities.

Hence, by the multiplication principle, the number of permutations of S is $\binom{n}{r} r!(n-r)!$ and since the two counts must give the same result,

$$n! = \binom{n}{r} r!(n-r)!$$

from which the result follows immediately. $\square$

Example 4.2.11 Use Theorem 4.2.10 to give an alternative proof of Theorem 4.2.8.

Solution Let $n \in \mathbb{N}$ and $r \in \mathbb{N} \cup \{0\}$ with $0 \le r \le n$. Then, using Theorem 4.2.10, we have

$$\binom{n}{r} = \frac{n!}{r!(n-r)!} = \frac{n!}{(n-r)![n-(n-r)]!} = \binom{n}{n-r}$$

as required. $\square$

Example 4.2.12 Let $n \in \mathbb{N}$ with $n \ge 3$. Express $\binom{n}{3}$ as a polynomial in n.

Solution By Theorem 4.2.10,

$$\binom{n}{3} = \frac{n!}{3!(n-3)!} = \frac{n(n-1)(n-2)(n-3)(n-4) \cdots \times 2 \times 1}{3 \times 2 \times 1 \times (n-3)(n-4) \cdots \times 2 \times 1}$$

$$= \frac{n(n-1)(n-2)}{3 \times 2 \times 1} = \tfrac{1}{6}n^3 - \tfrac{1}{2}n^2 + \tfrac{1}{3}n$$

which is in the required form. $\square$

Note 4.2.13 The cancellation of $(n-3)(n-4)\cdots\times 2\times 1$ in the solution to Example 4.2.12 is quite typical. More generally,

$$\binom{n}{r} = \frac{n(n-1)\cdots(n-r+1)(n-r)(n-r-1)\cdots\times 2\times 1}{r(r-1)\cdots 2\times 1\times(n-r)(n-r-1)\cdots\times 2\times 1},$$

and cancelling $(n-r)(n-r-1)\ldots 2\times 1$ gives

$$\binom{n}{r} = \frac{n(n-1)\cdots(n-r+1)}{r(r-1)\cdots\times 2\times 1}.$$

An easy way to remember this result is that there are r factors on the top and bottom, those on top begin with n and those on the bottom begin with r. This is 'hinted at' by the notation $\binom{n}{r}$.

Example 4.2.14 Evaluate $\binom{7}{2}$ and $\binom{18}{15}$.

Solution   Using Note 4.2.13,

$$\binom{7}{2} = \frac{7\times 6}{2\times 1} = 21.$$

Using Note 4.2.13 directly to evaluate $\binom{18}{15}$ would initially produce an expression involving 30 factors. Instead we use Theorem 4.2.8 first, as follows.

$$\binom{18}{15} = \binom{18}{18-15} = \binom{18}{3} = \frac{18\times 17\times 16}{3\times 2\times 1} = 816. \qquad \square$$

Example 4.2.15 Find the value of $n \in \mathbb{N}$ such that an n-element set has 70 4-element subsets.

Solution   We require

$$70 = \binom{n}{4} = \frac{n(n-1)(n-2)(n-3)}{4\times 3\times 2\times 1}.$$

Multiplying by 4! gives

$$24\times 70 = n(n-1)(n-2)(n-3).$$

Since the right-hand side increases with n for $n > 3$ it follows that there is at most one solution. Trial and error gives $n = 8$. $\square$

Theorem 4.2.16 *Let* $n, r \in \mathbb{N}$ *with* $1 \le r \le n$. *Then*

$$\binom{n}{r-1} + \binom{n}{r} = \binom{n+1}{r}.$$

Proof Let $S = \{0, 1, \ldots, n\}$. Then S has $(n + 1)$ elements. By definition, S has $\binom{n+1}{r}$ r-element subsets. We now count separately those subsets of S which contain 0 and those which do not.

If a subset contains 0 then it also contains $(r - 1)$ elements of $T = \{1, 2, \ldots, n\}$. Since T has n elements, there are $\binom{n}{r-1}$ such subsets.

If a subset does not contain 0 then it consists of r elements of T. There are $\binom{n}{r}$ such subsets.

The total of the two types must be $\binom{n+1}{r}$ since all r-element subsets have been counted. Hence

$$\binom{n}{r-1} + \binom{n}{r} = \binom{n+1}{r}$$

as required. $\qquad\square$

Example 4.2.17 Let $n, r \in \mathbb{N}$ with $1 \le r \le n$. Show that

$$\binom{n}{r} = \frac{n}{r}\binom{n-1}{r-1}.$$

Solution We have

$$\begin{aligned}
\text{RHS} &= \frac{n}{r}\frac{(n-1)!}{(r-1)!(n-r)!} &&\text{(by Theorem 4.2.10)} \\
&= \frac{n!}{r!(n-r)!} &&\text{(since } k(k-1)! = k!\text{)} \\
&= \text{LHS} &&\text{(by Theorem 4.2.10)}
\end{aligned}$$

as required. $\qquad\square$

4.3 The Binomial Theorem

Example 4.3.1 Without carrying out the complete expansion of $(x+1)(x+2)(x+3)$, find the coefficient of x^2.

Solution Each term in the expansion is a product of three factors, one factor being taken from each bracket. In particular, the x^2 terms arise by choosing the x term from two brackets and the constant term from the third. The possibilities are $x \times x \times 3$, $x \times 2 \times x$ and $1 \times x \times x$, in total $6x^2$. Hence the required coefficient is 6. $\qquad\square$

The idea of extracting a particular term from a product of factors is at the heart of the proof of the most important theorem in this chapter.

Theorem 4.3.2 (The Binomial Theorem) *Let* $n \in \mathbb{N}$ *and* $a, b \in \mathbb{R}$. *Then*

$$(a+b)^n = \sum_{r=0}^{n}\binom{n}{r}a^{n-r}b^r.$$

The RHS is called the *binomial expansion*.

Proof In the expansion of $(a + b)^n$ each term is a product of n factors, one from each bracket. For the term $a^{n-r}b^r$, we must choose the b from any r brackets and the a from the others. For a fixed r, the number of possibilities is the number of ways of choosing r brackets from n. This is $\binom{n}{r}$. Hence the coefficient of $a^{n-r}b^r$ is $\binom{n}{r}$ and the result follows immediately. $\quad\square$

Corollary 4.3.3 *Let* $n \in \mathbb{N}$ *and* $x \in \mathbb{R}$. *Then*

$$(1 + x)^n = \sum_{r=0}^{n} \binom{n}{r} x^r.$$

Proof Put $a = 1$ and $b = x$ in Theorem 4.3.2. $\quad\square$

Example 4.3.4 Find the coefficient of x^{15} in the expansion of $(1 + x)^{18}$.

Solution From Corollary 4.3.3, the required coefficient is $\binom{18}{15}$. Using Example 4.2.14 this equals 816. $\quad\square$

Example 4.3.5 Find the coefficient of x^4 in the expansion of $\left(2x^3 - \dfrac{1}{x^2}\right)^8$.

Solution We use Theorem 4.3.2 with $a = 2x^3$, $b = -1/x^2$ and $n = 8$. The *typical* term in the expansion is

$$\binom{8}{r}(2x^3)^{8-r}\left(-\frac{1}{x^2}\right)^r = (-1)^r \binom{8}{r} 2^{8-r} x^{24-5r},$$

where $r \in \{0, 1, \ldots, 8\}$. Note that we have separated the constant parts of the factor from the powers of x, and combined the latter into a single power of x.

For the term in x^4 we must have

$$24 - 5r = 4, \qquad \textit{i.e. } r = 4.$$

Thus the required coefficient is

$$(-1)^4 \binom{8}{4} 2^4 = 1 \times \frac{8 \times 7 \times 6 \times 5}{4 \times 3 \times 2 \times 1} \times 16 = 1120. \qquad\square$$

Remark If we required the coefficient of x^5 (instead of x^4) in Example 4.3.5 then the reasoning would be slightly different.

The power of x in the typical term is x^{24-5r}. For a term in x^5 we must have

$$24 - 5r = 5, \qquad \textit{i.e. } r = \frac{19}{5}.$$

Since $r \notin \{0, 1, \ldots, 8\}$, there is no term in x^5. Hence the coefficient of x^5 is 0.

Example 4.3.6 Prove that, for $n \in \mathbb{N}$,

$$\sum_{r=0}^{n} \binom{n}{r} = 2^n \quad \text{and} \quad \sum_{r=0}^{n} (-1)^r \binom{n}{r} = 0.$$

Solution Let $n \in \mathbb{N}$. From Corollary 4.3.3 we have, for $x \in \mathbb{R}$,

$$(1+x)^n = \sum_{r=0}^{n} \binom{n}{r} x^r. \tag{4.3.1}$$

Putting $x = 1$ in (4.3.1) gives

$$(1+1)^n = \sum_{r=0}^{n} \binom{n}{r} 1^r, \quad \text{i.e. } \sum_{r=0}^{n} \binom{n}{r} = 2^n.$$

Putting $x = -1$ in (4.3.1) gives

$$(1-1)^n = \sum_{r=0}^{n} \binom{n}{r} (-1)^r, \quad \text{i.e. } \sum_{r=0}^{n} (-1)^r \binom{n}{r} = 0. \qquad \square$$

Example 4.3.7 By expanding $(1+x)(1+x)^n$ in two ways, give an alternative proof of Theorem 4.2.16.

Solution Let $n, s \in \mathbb{N}$ with $1 \le s \le n$ and let $x \in \mathbb{R}$. Then, by Corollary 4.3.3,

$$(1+x)(1+x)^n = (1+x)^{n+1} = \sum_{r=0}^{n+1} \binom{n+1}{r} x^r.$$

The coefficient of x^s is $\binom{n+1}{s}$. On the other hand, expanding $(1+x)^n$ first, gives

$$(1+x)(1+x)^n = (1+x) \sum_{r=0}^{n} \binom{n}{r} x^r$$

$$= \sum_{r=0}^{n} \binom{n}{r} x^r + x \sum_{r=0}^{n} \binom{n}{r} x^r$$

$$= \sum_{r=0}^{n} \binom{n}{r} x^r + \sum_{r=0}^{n} \binom{n}{r} x^{r+1}.$$

Now the term in x^s is obtained by taking $r = s$ in the first sum and $r = s - 1$ in the second. Hence the coefficient of x^s is $\binom{n}{s} + \binom{n}{s-1}$. Comparing the results of the two methods gives

$$\binom{n}{s} + \binom{n}{s-1} = \binom{n+1}{s}$$

as required. $\qquad \square$

While the techniques of Section 4.3, and in particular Example 4.2.14, are useful when we wish to find a particular term in a binomial expansion, they are poor when we need all the terms. For modest values of n, *Pascal's triangle* provides a better method. For $n \in \mathbb{N}$, the nth row of the triangle consists of the coefficients $\binom{n}{r}$ where $r = 0, 1, \ldots, n$.

$$\text{Row } n: \qquad \binom{n}{0} \qquad \binom{n}{1} \qquad \cdots \qquad \binom{n}{n}$$

The first five rows of Pascal's triangle are shown below with the binomial coefficients evaluated.

$$
\begin{array}{lccccccccc}
\text{Row 1:} &   &   &   &   & 1 &   & 1 &   &   \\
\text{Row 2:} &   &   &   & 1 &   & 2 &   & 1 &   \\
\text{Row 3:} &   &   & 1 &   & 3 &   & 3 &   & 1 \\
\text{Row 4:} &   & 1 &   & 4 &   & 6 &   & 4 &   & 1 \\
\text{Row 5:} & 1 &   & 5 &   & 10 &   & 10 &   & 5 &   & 1
\end{array}
$$

Given the nth row, we can obtain the $(n+1)^{\text{th}}$ row as follows.

(1) The first and last terms are both 1.

(2) Otherwise, each term of the $(n+1)^{\text{th}}$ row is the sum of the two terms in the nth row, nearest to it.

For example, the 6th row may be obtained from the 5th row as shown below.

$$
\begin{array}{lccccccccccc}
\text{Row 5:} &   & 1 &   & 5 &   & 10 &   & 10 &   & 5 &   & 1 \\
\text{Row 6:} & 1 &   & 6 &   & 15 &   & 20 &   & 15 &   & 6 &   & 1
\end{array}
$$

To see that this process works in general consider the nth and $(n+1)^{\text{th}}$ rows.

$$\text{Row } n: \qquad \binom{n}{0} \quad \cdots \quad \binom{n}{r-1} \qquad \binom{n}{r} \quad \cdots \quad \binom{n}{n}$$

$$\text{Row } (n+1): \quad \binom{n+1}{0} \qquad \cdots \qquad \binom{n+1}{r} \qquad \cdots \qquad \binom{n+1}{n+1}$$

Then observe that $\binom{n+1}{0} = \binom{n+1}{n+1} = 1$ and, by Theorem 4.2.16, $\binom{n+1}{r} = \binom{n}{r-1} + \binom{n}{r}$.

Example 4.3.8 Expand $(2x + 3y)^5$.

Solution Using the binomial theorem and the 5th row of Pascal's triangle we have

$$(2x + 3y)^5 = \binom{5}{0}(2x)^5(3y)^0 + \binom{5}{1}(2x)^4(3y)^1 + \binom{5}{2}(2x)^3(3y)^2$$

$$+ \binom{5}{3}(2x)^2(3y)^3 + \binom{5}{4}(2x)^1(3y)^4 + \binom{5}{5}(2x)^0(3y)^5$$

$$= 1 \times (2x)^5(3y)^0 + 5 \times (2x)^4(3y)^1 + 10 \times (2x)^3(3y)^2$$

$$+ 10 \times (2x)^2(3y)^3 + 5 \times (2x)^1(3y)^4 + 1 \times (2x)^0(3y)^5$$

$$= 1 \times 2^5 x^5 + 5 \times 2^4 \times 3x^4 y + 10 \times 2^3 \times 3^2 x^3 y^2 + 10 \times 2^2 \times 3^3 x^2 y^3$$

$$+ 5 \times 2 \times 3^4 xy^4 + 1 \times 3^5 y^5$$

$$= 32x^5 + 240x^4 y + 720x^3 y^2 + 1080x^2 y^3 + 810xy^4 + 243y^5. \qquad \square$$

Problem 4.3.9 Let $k, n \in \mathbb{N}$ with $k < n$. Prove that

$$\binom{n}{k} < \frac{n^n}{k^k(n-k)^{n-k}}.$$

4.X Exercises

1. Prove each of the following statements by induction.

   (a) For all $n \in \mathbb{N}$, $\displaystyle\sum_{r=1}^{n} r(3r + 1) = n(n + 1)^2$,

   (b) For all $n \in \mathbb{N}$, $\displaystyle\sum_{r=1}^{n} \frac{r}{2^r} = 2 - \frac{n+2}{2^n}$,

   (c) For all $n \in \mathbb{N}$, $\displaystyle\sum_{r=1}^{2n-1} (-1)^{r+1} r^3 = n^2(4n - 3)$,

   (d) For all $n \in \mathbb{N} \cup \{0\}$, $7 \mid (3^{2n+1} + 2^{n+2})$,

   (e) For all $n \in \mathbb{N}$, $n \geq 2$, $\displaystyle\prod_{r=2}^{n} \left(1 - \frac{1}{r^2}\right) = \frac{n+1}{2n}$.

2. Using Theorem 4.1.5, deduce that

   $$\sum_{r=1}^{n} (3r^2 - 5) = \frac{n(n+3)(2n-3)}{2}.$$

   Find also a formula for the sum of the cubes of the first n odd integers.

3. The sequence $u_1, u_2, u_3, \ldots$, is defined recursively by

   $$u_1 = 1, \qquad 3u_{n+1} = 2u_n - 1 \ (n \geq 1).$$

   Prove by induction that, for all $n \in \mathbb{N}$, $u_n = 3\left(\dfrac{2}{3}\right)^n - 1$.

4. Two sequences $x_1, x_2, x_3, \ldots$ and $y_1, y_2, y_3, \ldots$ are defined recursively by

$$x_1 = 1, \qquad x_{n+1} = x_n + 2y_n \ (n \geq 1)$$

$$y_1 = 1, \qquad y_{n+1} = x_n + y_n \ (n \geq 1).$$

Prove that, for all $n \in \mathbb{N}$, $x_n^2 - 2y_n^2 = (-1)^n$.

5. Find $n \in \mathbb{N}$ such that
$$(n+2)! = 4(n! + (n+1)!).$$

6. Find all $n \in \mathbb{N}$ and $r \in \mathbb{N} \cup \{0\}$ such that

(a) $\dbinom{n}{4} = 210$,   (b) $\dbinom{10}{r} = 210$.

7. Prove that, if $0 \leq s \leq k \leq r \leq n$ then

$$\binom{n}{r}\binom{r}{k}\binom{k}{s} = \binom{n}{s}\binom{n-s}{k-s}\binom{n-k}{r-k}.$$

8. Prove by induction that, for $n \in \mathbb{N}$,

$$\sum_{r=1}^{n} r(r!) = (n+1)! - 1.$$

9. Using the first ten letters of the alphabet, how many 5-letter words can be made in which no letter is repeated?

10. Using the digits $1, 2, \ldots, 7$, how many 5-digit numbers can be made (a) if repetitions are not allowed, and (b) if repetitions are allowed? In each case, how many of the numbers are even?

11. Write out in full the binomial expansions of

(a) $(a^2 - b^3)^4$,   (b) $\left(x^2 - \dfrac{3}{x}\right)^5$,   (c) $\left(z + \dfrac{1}{z^2}\right)^6$.

12. Find the term independent of x in the expansion of $\left(2x^3 - \dfrac{1}{2x^2}\right)^{10}$.

13. Find the coefficients of x and x^2 in the expansion of $\left(x - \dfrac{2}{x^2}\right)^7$.

14. Find the coefficient of $x^2 y^3 z^2$ in the expansion of $(x + y + z)^7$.

15.   Let $n \in \mathbb{N}$. Find $n \geq 10$ such that the coefficients of x^9 and x^{10} in the expansion of $(5x + 3)^n$ are equal.

16.   Express $\displaystyle\sum_{r=1}^{n} 2^{r-1} \binom{n}{r}$ in the form $\dfrac{(a^n - b)}{c}$, where $a, b, c \in \mathbb{N}$.

17.   Use the fact that, for $1 \leq r \leq n$, $\binom{n+1}{r} = \binom{n}{r} + \binom{n}{r-1}$, to prove that

$$\sum_{k=0}^{r} \binom{n-k}{r-k} = \binom{n+1}{r}.$$

Chapter 5

Trigonometry

The standard trigonometric functions, sin, cos and tan are first encountered in calculations involving right-angled triangles, but they have much more general applications. For example, a process which involves repeated patterns (such as the heartbeat) can be modelled using combinations of such functions. This chapter is a general introduction to these functions, though not to such applications.

We expect that the reader is familiar with these functions and with many of their properties, but may not have met the associated functions cosec, sec and cot. We consider the relationships between these six functions and obtain their graphs. We investigate the relations between the values for related inputs. For example, $\sin 2\theta = 2 \sin \theta \cos \theta$. The reader should be aware that more complicated results, such as that for $\sin 4\theta$ and $\cos 4\theta$, are more easily obtained using complex numbers (see Section 6.3). We also consider the solutions of equations involving trigonometric functions. Unlike polynomial equations, these usually have infinite families of solutions.

In Section 5.4, we introduce the t-formulae. These express the trigonometric functions of θ as rational functions of $t = \tan \frac{1}{2}\theta$. While of little interest in themselves, they allow us to transform integrals involving trigonometric functions into integrals of rational functions. As we shall see in Chapter 16, integrals of the latter type can be evaluated in a systematic way.

The final section introduces the so-called *inverse trigonometric functions*. The alert reader will observe that each trigonometric function is periodic and so is certainly *not* injective. As in Chapter 2, we can restrict the domain and codomain to obtain a bijection. It is the inverses of these restrictions that we call the inverse trigonometric functions.

5.1   Trigonometric Functions

The familiar formula $2\pi r$ for the circumference of circle with radius r exhibits an important fact, namely that the circumference divided by the radius is the same for *all* circles. Establishing this result is not a trivial matter. Instead we shall accept it without proof, or rather a generalisation of it which will enable us to measure angles by means of circular arcs of any radius.

Consider a circle, radius r, with an angle θ subtended at the centre by an arc of length a (see Figure 5.1.1). The ratio of the arc length to the radius measures the

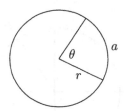

Figure 5.1.1

angle in *radians*, *i.e.*

$$\theta = \frac{a}{r} \text{ radians.}$$

This measure is independent of the circle.

We usually omit the word 'radians' except when we wish to make clear that we are dealing with an angle rather than simply a real number. When $a = r$ we have an angle of 1 radian which is approximately $57°$. The number of radians in half a revolution is denoted by π. So π radians is $180°$. The relationship between radians and degrees for some commonly used angles is shown in the table.

Radians	0	$\pi/6$	$\pi/4$	$\pi/3$	$\pi/2$	π	2π
Degrees	$0°$	$30°$	$45°$	$60°$	$90°$	$180°$	$360°$

Since 2π radians is $360°$ (one revolution), the formula, mentioned above, for the circumference of a circle is easily deduced.

 Hazard In this book, and in mathematics generally, all angles are measured in radians unless otherwise stated.

Definitions 5.1.1 Consider the circle, centre the origin O, radius 1. Let A be the point $(1, 0)$ and let $\theta \in \mathbb{R}$. Measure from A a distance $|\theta|$ round the circle, counterclockwise if $\theta > 0$ or clockwise if $\theta < 0$, to reach the point P. We say that θ *terminates* at P. Figure 5.1.2 shows the case where $0 < \theta < \pi/2$. When $|\theta| > 2\pi$, it may be necessary to 'wind' round the circle several times to reach P.

The real functions *sine*, *cosine* and *tangent* are denoted by sin, cos and tan, respectively. They are defined as follows.

$$\sin \theta = \text{the } y\text{-coordinate of } P,$$
$$\cos \theta = \text{the } x\text{-coordinate of } P,$$
$$\tan \theta = \frac{\sin \theta}{\cos \theta}.$$

So P is the point $(\cos \theta, \sin \theta)$.

The sine and cosine functions both have maximal domain $\mathbb{R}$. From the definitions

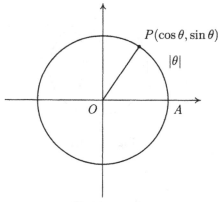

Figure 5.1.2

we get

$$\sin \theta = 0 \iff \theta = \ldots, -3\pi, -2\pi, -\pi, 0, \pi, 2\pi, 3\pi, \ldots$$
$$\iff \theta = k\pi \text{ for some } k \in \mathbb{Z},$$
$$\cos \theta = 0 \iff \theta = \ldots, -\tfrac{5\pi}{2}, -\tfrac{3\pi}{2}, -\tfrac{\pi}{2}, \tfrac{\pi}{2}, \tfrac{3\pi}{2}, \tfrac{5\pi}{2}, \ldots$$
$$\iff \theta = (2k+1)\tfrac{\pi}{2} = \tfrac{\pi}{2} + k\pi \text{ for some } k \in \mathbb{Z}.$$

Since $\tan \theta$ is defined if and only if $\cos \theta \neq 0$, the tangent function has maximal domain $\mathbb{R} - \{\tfrac{\pi}{2} + k\pi : k \in \mathbb{Z}\}$ and

$$\tan \theta = 0 \iff \theta = k\pi \text{ for some } k \in \mathbb{Z}.$$

We define three more trigonometric functions. The real functions *cosecant, secant* and *cotangent* are denoted by cosec, sec and cot, respectively. They are defined as follows.

$$\operatorname{cosec} \theta = \frac{1}{\sin \theta}, \qquad \sec \theta = \frac{1}{\cos \theta}, \qquad \cot \theta = \frac{\cos \theta}{\sin \theta}.$$

We leave the reader to determine their maximal domains.

Problem 5.1.2 When does $\cot \theta = 1/\tan \theta$?

Notes 5.1.3 (1) We have defined $\sin \theta$, $\cos \theta$, *etc.* for θ a *real number*. When f is a real function and we refer to $f(\theta)$ where θ is an *angle*, we mean that the angle is θ *radians*. With this convention, when $0 < \theta < \pi/2$, our definitions agree with the usual 'right–angled triangle' definitions. See Figure 5.1.3.

(2) In the definition of radian measurement and the definitions of the trigonometric functions, we made use of the lengths of circular arcs. We have not defined precisely what such a length is and for the present it must remain intuitive. The formal definition of arc length is dealt with in Chapter 16.

Properties 5.1.4 The following properties of trigonometric functions follow directly from the definitions. Let $\theta \in \mathbb{R}$. Equations are assumed to hold where both sides are defined.

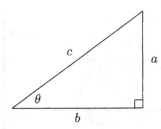

$$\sin\theta = \frac{a}{c},$$

$$\cos\theta = \frac{b}{c},$$

$$\tan\theta = \frac{a}{b}.$$

Figure 5.1.3

(1) $-1 \le \sin\theta \le 1$, $-1 \le \cos\theta \le 1$.

(2) $\sin^2\theta + \cos^2\theta = 1$, $\tan^2\theta + 1 = \sec^2\theta$, $1 + \cot^2\theta = \operatorname{cosec}^2\theta$.

(3) For $k \in \mathbb{Z}$, $\sin(\theta + 2k\pi) = \sin\theta$ (period 2π),
$$\cos(\theta + 2k\pi) = \cos\theta \quad \text{(period } 2\pi\text{)},$$
$$\tan(\theta + k\pi) = \tan\theta \quad \text{(period } \pi\text{)}.$$

(4) $\sin(-\theta) = -\sin\theta$, $\cos(-\theta) = \cos\theta$, $\tan(-\theta) = -\tan\theta$.

Thus sin and tan are odd functions while cos is an even function.

(5) The diagram shows in which quadrants lie the angles for which sin, cos and tan are *positive*.

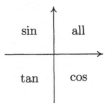

(6) The sines, cosines and tangents of some commonly used angles are given in the table.

θ	0	$\pi/6$	$\pi/4$	$\pi/3$	$\pi/2$
$\sin\theta$	0	$1/2$	$1/\sqrt{2}$	$\sqrt{3}/2$	1
$\cos\theta$	1	$\sqrt{3}/2$	$1/\sqrt{2}$	$1/2$	0
$\tan\theta$	0	$1/\sqrt{3}$	1	$\sqrt{3}$	?

Theorem 5.1.5 *The functions* sin *and* cos *both have image* $[-1, 1]$. *The function* tan *has image* $\mathbb{R}$.

Proof (Non-rigorous) From Property 5.1.4 (1), the images of both sin and cos are subsets of $[-1, 1]$. By definition, the image of tan is a subset of $\mathbb{R}$. Now consider the diagrams in Figure 5.1.4. To see that $\tan\theta$ is the y-coordinate of R, use Notes 5.1.3 (1) and Property 5.1.4 (4).

(1) As θ covers the interval $[-\frac{\pi}{2}, \frac{\pi}{2}]$, P ranges over all points on the semicircular arc and $\sin\theta$ takes all values in the interval $[-1, 1]$.

(2) As θ covers the interval $[0, \pi]$, Q ranges over all points on the semicircular arc and $\cos\theta$ takes all values in the interval $[-1, 1]$.

(3) As θ covers the interval $(-\frac{\pi}{2}, \frac{\pi}{2})$, R ranges over all points on the line $x = 1$ and $\tan\theta$ takes all values in the interval $(-\infty, \infty)$, *i.e.* $\mathbb{R}$. $\square$

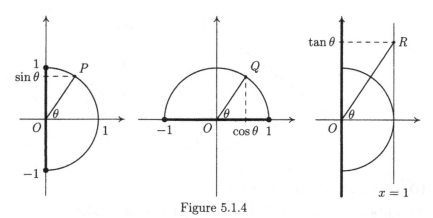

Figure 5.1.4

Theorem 5.1.6 *Let $\theta \in (0, \pi/2)$. Then $\sin\theta < \theta < \tan\theta$.*

Proof (Non-rigorous) In Figure 5.1.5 the circle has centre O and radius 1, AB and QP are perpendicular to OA, and BC is a tangent.

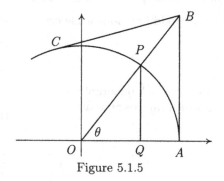

Figure 5.1.5

Considering the right-angled triangle OQP we see that $\sin\theta = |QP|$. Hence

$$\sin\theta = |QP| < |AP| < |\operatorname{arc} AP| = \theta.$$

Considering the right-angled triangle OAB, and using Notes 5.1.3 (1), we see that $\tan\theta = |AB|$. If we imagine an elastic string stretched round the circle, then taking the string at point P we must stretch it further to reach point B. While this does not *prove* anything, it intuitively justifies writing $|\operatorname{arc} APC| < |AB| + |BC|$ from which we deduce that $|\operatorname{arc} AP| < |AB|$. Hence

$$\theta = |\operatorname{arc} AP| < |AB| = \tan\theta.$$

The proof is complete. $\qquad\qquad\square$

Note 5.1.7 Putting $\theta = \pi/6$ in Theorem 5.1.6 gives $3 < \pi < 2\sqrt{3} = 3{\cdot}4641\ldots$. In fact π lies between $3{\cdot}14159$ and $3{\cdot}14160$. Observe that $22/7 = 3{\cdot}1428\ldots$, which is often used as an approximation to π, lies outside the latter range.

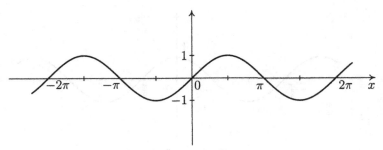

Figure 5.1.6 : Sine

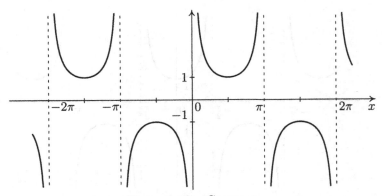

Figure 5.1.7 : Cosecant

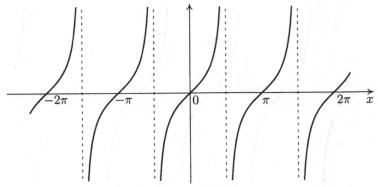

Figure 5.1.8 : Tangent

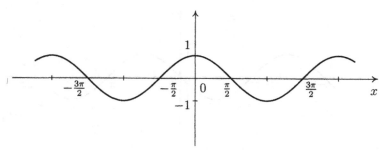

Figure 5.1.9 : Cosine

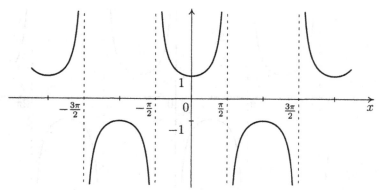

Figure 5.1.10 : Secant

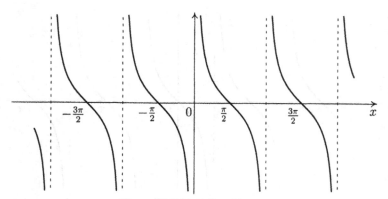

Figure 5.1.11 : Cotangent

Corollary 5.1.8 (1) *Let* $\theta \geq 0$. *Then* $\sin \theta \leq \theta$.

(2) *Let* $\theta \in \mathbb{R}$. *Then* $|\sin \theta| \leq |\theta|$.

Proof (1) We have $\sin 0 = 0$ and, by Theorem 5.1.6, for $0 < \theta < \pi/2$, $\sin \theta < \theta$. Now let $\theta \geq \pi/2$. Then, by Note 5.1.7, $1 < \pi/2$ so that $\sin \theta \leq 1 < \pi/2 \leq \theta$.

(2) We consider three cases. For $0 \leq \theta < \pi/2$, $|\sin \theta| = \sin \theta \leq \theta = |\theta|$. For, $-\pi/2 < \theta < 0$, $|\sin \theta| = |\sin(-\theta)| = \sin \theta \leq -\theta = |\theta|$. For $|\theta| \geq \pi/2$, $|\sin \theta| \leq 1 < \pi/2 \leq |\theta|$. $\qquad\square$

We now have enough information from the definitions and properties to sketch the graphs of the six trigonometric functions. They are shown in Figures 5.1.6 to 5.1.11. As is customary we use x as the function variable. Curve sketching techniques are discussed in Chapter 10.

Example 5.1.9 Sketch the curve $y = \dfrac{1}{1 + \sin 2x}$ $(0 \leq x \leq \pi)$.

Solution We begin with $y = \sin x$ in the range $0 \leq x \leq 2\pi$ and then deduce successively $y = \sin 2x$, $y = 1 + \sin 2x$ and $y = 1/(1 + \sin 2x)$ in the range $0 \leq x \leq \pi$. The sketches are shown in Figure 5.1.12. $\qquad\square$

5.2 Identities

Consider the equation

$$\tan (\theta + \phi) = \frac{\tan \theta + \tan \phi}{1 - \tan \theta \tan \phi}.$$

This does not hold for all θ and ϕ. For example, the right-hand side is not defined when $\theta = \pi/4$ and $\phi = \pi/2$. However, as we shall see later, it does hold for all real numbers θ and ϕ for which both sides of the equation are defined. Such an equation is called a trigonometric *identity*.

Convention When we are given or asked to establish a trigonometric equation involving variables $\theta, \phi, \dots$ and no restriction on the variables is given or implied, we shall *assume* that the equation is an identity, *i.e.* it holds or is to be established for all real numbers $\theta, \phi, \dots$ for which both sides of the equation are defined.

Many trigonometric identities are based on the next theorem and its corollaries. First, we prove two lemmas.

Lemma 5.2.1 *Let* $\theta, \phi \in \mathbb{R}$. *Then*

$$\cos (\theta - \phi) = \cos \theta \cos \phi + \sin \theta \sin \phi.$$

Proof

In Figure 5.2.1 the circles both have centre O and radius 1. Let θ terminate at P, ϕ terminate at Q and $\theta - \phi$ terminate at R. Then, P, Q and R are the points $(\cos \theta, \sin \theta)$,

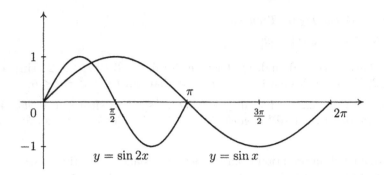

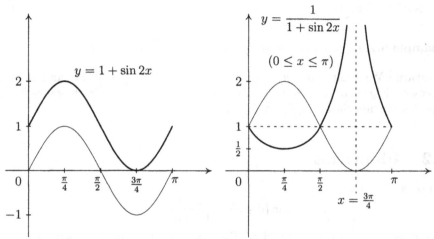

Figure 5.1.12

$(\cos\phi, \sin\phi)$ and $(\cos(\theta - \phi), \sin(\theta - \phi))$, respectively. Since the triangles OQP and OAR are congruent, $|QP|^2 = |AR|^2$. Hence,

$$(\cos\theta - \cos\phi)^2 + (\sin\theta - \sin\phi)^2 = (\cos(\theta - \phi) - 1)^2 + (\sin(\theta - \phi) - 0)^2,$$

i.e.

$$\cos^2\theta - 2\cos\theta\cos\phi + \cos^2\phi + \sin^2\theta - 2\sin\theta\sin\phi + \sin^2\phi$$
$$= 1 - 2\cos(\theta - \phi) + \cos^2(\theta - \phi) + \sin^2(\theta - \phi),$$

i.e.

$$2 - 2\cos\theta\cos\phi - 2\sin\theta\sin\phi = 2 - 2\cos(\theta - \phi),$$

i.e.

$$\cos(\theta - \phi) = \cos\theta\cos\phi + \sin\theta\sin\phi.$$

$\square$

Lemma 5.2.2 *Let* $\theta \in \mathbb{R}$. *Then*

$$\cos\left(\tfrac{\pi}{2} - \phi\right) = \sin\phi \quad and \quad \sin\left(\tfrac{\pi}{2} - \phi\right) = \cos\phi.$$

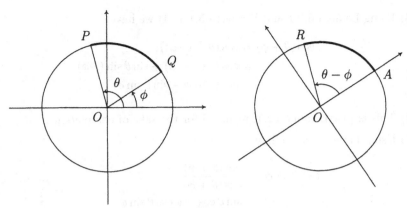

Figure 5.2.1

Proof Putting $\theta = \frac{\pi}{2}$ in Lemma 5.2.1 gives

$$\cos\left(\tfrac{\pi}{2} - \phi\right) = \cos\tfrac{\pi}{2}\cos\phi + \sin\tfrac{\pi}{2}\sin\phi = 0 \times \cos\phi + 1 \times \sin\phi = \sin\phi.$$

Replacing ϕ with $\frac{\pi}{2} - \phi$ in $\cos\left(\frac{\pi}{2} - \phi\right) = \sin\phi$ completes the proof. $\qquad\qquad$ $\square$

Theorem 5.2.3 *Let* $\theta, \phi \in \mathbb{R}$. *Then*

(1) $\sin(\theta + \phi) = \sin\theta\cos\phi + \cos\theta\sin\phi,$
(2) $\sin(\theta - \phi) = \sin\theta\cos\phi - \cos\theta\sin\phi,$
(3) $\cos(\theta + \phi) = \cos\theta\cos\phi - \sin\theta\sin\phi,$
(4) $\cos(\theta - \phi) = \cos\theta\cos\phi + \sin\theta\sin\phi$

and, whenever both sides of the equation are defined,

(5) $\tan(\theta + \phi) = \dfrac{\tan\theta + \tan\phi}{1 - \tan\theta\tan\phi},$

(6) $\tan(\theta - \phi) = \dfrac{\tan\theta - \tan\phi}{1 + \tan\theta\tan\phi}.$

Proof (1) Using Lemmas 5.2.2 and 5.2.1 and Property 5.1.4 (4) we have

$$\begin{aligned}
\sin(\theta + \phi) &= \cos\left(\tfrac{\pi}{2} - \theta - \phi\right) \\
&= \cos\left(\tfrac{\pi}{2} - \theta\right)\cos\phi + \sin\left(\tfrac{\pi}{2} - \theta\right)\sin\phi \\
&= \sin\theta\cos\phi + \cos\theta\sin\phi.
\end{aligned}$$

(2) Using (1) and Property 5.1.4 (4) we have

$$\begin{aligned}
\sin(\theta - \phi) &= \sin(\theta + (-\phi)) \\
&= \sin\theta\cos(-\phi) + \cos\theta\sin(-\phi) \\
&= \sin\theta\cos\phi - \cos\theta\sin\phi.
\end{aligned}$$

(3) Using Lemma 5.2.1 and Property 5.1.4 (4) we have

$$\cos(\theta + \phi) = \cos(\theta - (-\phi))$$
$$= \cos\theta\cos(-\phi) + \sin\theta\sin(-\phi)$$
$$= \cos\theta\cos\phi - \sin\theta\sin\phi.$$

(4) This is just Lemma 5.2.1, included for the sake of completeness.

(5) Using (1) and (3) we have

$$\tan(\theta + \phi) = \frac{\sin(\theta + \phi)}{\cos(\theta + \phi)}$$
$$= \frac{\sin\theta\cos\phi + \cos\theta\sin\phi}{\cos\theta\cos\phi - \sin\theta\sin\phi}$$
$$= \frac{\left(\dfrac{\sin\theta\cos\phi}{\cos\theta\cos\phi}\right) + \left(\dfrac{\cos\theta\sin\phi}{\cos\theta\cos\phi}\right)}{\left(\dfrac{\cos\theta\cos\phi}{\cos\theta\cos\phi}\right) - \left(\dfrac{\sin\theta\sin\phi}{\cos\theta\cos\phi}\right)}$$
$$= \frac{\tan\theta + \tan\phi}{1 - \tan\theta\tan\phi}.$$

(6) Using (5) and Property 5.1.4 (4) we have

$$\tan(\theta - \phi) = \tan(\theta + (-\phi)) = \frac{\tan\theta + \tan(-\phi)}{1 - \tan\theta\tan(-\phi)} = \frac{\tan\theta - \tan\phi}{1 + \tan\theta\tan\phi}. \qquad \square$$

Corollary 5.2.4 *Let* $\theta \in \mathbb{R}$. *Then*

(1) $\sin 2\theta = 2\sin\theta\cos\theta$,

(2) $\cos 2\theta = \cos^2\theta - \sin^2\theta = 1 - 2\sin^2\theta = 2\cos^2\theta - 1$,

(3) $\sin^2\theta = \frac{1}{2}(1 - \cos 2\theta)$ *and* $\cos^2\theta = \frac{1}{2}(1 + \cos 2\theta)$,

and whenever both sides of the equation are defined,

(4) $\tan 2\theta = \dfrac{2\tan\theta}{1 - \tan^2\theta}.$

Proof For (1), (2) and (4) put $\phi = \theta$ in Theorem 5.2.3 (1), (3) and (5), respectively. The identities in (3) follow directly from (2). $\qquad \square$

Corollary 5.2.5 *Let* $\theta, \phi, \alpha, \beta \in \mathbb{R}$. *Then*

(1) $\sin\theta\cos\phi = \frac{1}{2}[\sin(\theta + \phi) + \sin(\theta - \phi)]$,

(2) $\cos\theta\sin\phi = \frac{1}{2}[\sin(\theta + \phi) - \sin(\theta - \phi)]$,

(3) $\cos\theta\cos\phi = \frac{1}{2}[\cos(\theta + \phi) + \cos(\theta - \phi)]$,

(4) $\sin\theta\sin\phi = \frac{1}{2}[\cos(\theta - \phi) - \cos(\theta + \phi)]$

and

(5) $\quad \sin \alpha + \sin \beta \quad = 2 \sin \left(\dfrac{\alpha + \beta}{2} \right) \cos \left(\dfrac{\alpha - \beta}{2} \right),$

(6) $\quad \sin \alpha - \sin \beta \quad = 2 \cos \left(\dfrac{\alpha + \beta}{2} \right) \sin \left(\dfrac{\alpha - \beta}{2} \right),$

(7) $\quad \cos \alpha + \cos \beta \quad = 2 \cos \left(\dfrac{\alpha + \beta}{2} \right) \cos \left(\dfrac{\alpha - \beta}{2} \right),$

(8) $\quad \cos \alpha - \cos \beta \quad = -2 \sin \left(\dfrac{\alpha + \beta}{2} \right) \sin \left(\dfrac{\alpha - \beta}{2} \right).$

Proof From Theorem 5.2.3 (1) and (2) we have

$$\sin (\theta + \phi) = \sin \theta \cos \phi + \cos \theta \sin \phi,$$

$$\sin (\theta - \phi) = \sin \theta \cos \phi - \cos \theta \sin \phi.$$

Adding these gives

$$\sin (\theta + \phi) + \sin (\theta - \phi) = 2 \sin \theta \cos \phi$$

from which (1) follows. The proofs of (2), (3) and (4) are similar and are left to the reader.

For (5), (6), (7) and (8) put $\theta = \frac{1}{2}(\alpha + \beta)$ and $\phi = \frac{1}{2}(\alpha - \beta)$ in (1), (2), (3) and (4), respectively. Note that $\theta + \phi = \alpha$ and $\theta - \phi = \beta$. □

Example 5.2.6 Simplify $\dfrac{\sin 10\theta + \sin 2\theta}{\cos 7\theta + \cos \theta}$.

Solution Using Corollary 5.2.5 (5) and (7) and Corollary 5.2.4 (1) we have

$$\frac{\sin 10\theta + \sin 2\theta}{\cos 7\theta + \cos \theta} = \frac{2 \sin 6\theta \cos 4\theta}{2 \cos 4\theta \cos 3\theta} = \frac{\sin 6\theta}{\cos 3\theta} = 2 \sin 3\theta. \qquad □$$

Example 5.2.7 Find a formula for $\sin 3\theta$ in terms of $\sin \theta$.

Solution Using Properties 5.1.4, Theorem 5.2.3 (1) and Corollary 5.2.4 (1) and (2) we have

$$\begin{aligned}
\sin 3\theta &= \sin (2\theta + \theta) \\
&= \sin 2\theta \cos \theta + \cos 2\theta \sin \theta \\
&= 2 \sin \theta \cos \theta \cos \theta + (1 - 2 \sin^2 \theta) \sin \theta \\
&= 2 \sin \theta (1 - \sin^2 \theta) + (1 - 2 \sin^2 \theta) \sin \theta \\
&= 2 \sin \theta - 2 \sin^3 \theta + \sin \theta - 2 \sin^3 \theta.
\end{aligned}$$

Hence,

$$\sin 3\theta = 3 \sin \theta - 4 \sin^3 \theta. \qquad □$$

Remark An alternative approach to the solution of Example 5.2.7 and similar problems can be found in Section 6.3.

Example 5.2.8 Use $\cos \dfrac{\pi}{6} = \dfrac{\sqrt{3}}{2}$ to evaluate $\sin \dfrac{\pi}{12}$ and $\cos \dfrac{\pi}{12}$.

Solution   From Corollary 5.2.4 (3) we have

$$\sin^2 \theta = \tfrac{1}{2}(1 - \cos 2\theta) \qquad \text{and} \qquad \cos^2 \theta = \tfrac{1}{2}(1 + \cos 2\theta).$$

Putting $\theta = \pi/12$ we have

$$\sin^2 \frac{\pi}{12} = \frac{1}{2}\left(1 - \cos\frac{\pi}{6}\right) = \frac{1}{2}\left(1 - \frac{\sqrt{3}}{2}\right) = \frac{2 - \sqrt{3}}{4},$$

$$\cos^2 \frac{\pi}{12} = \frac{1}{2}\left(1 + \cos\frac{\pi}{6}\right) = \frac{1}{2}\left(1 + \frac{\sqrt{3}}{2}\right) = \frac{2 + \sqrt{3}}{4}.$$

Since $\pi/12 \in (0, \pi/2)$, $\sin(\pi/12) > 0$ and $\cos(\pi/12) > 0$. Hence,

$$\sin\frac{\pi}{12} = \frac{\sqrt{2 - \sqrt{3}}}{2} \qquad \text{and} \qquad \cos\frac{\pi}{12} = \frac{\sqrt{2 + \sqrt{3}}}{2}. \qquad \square$$

Problem 5.2.9 Find real numbers a and b such that $2 - \sqrt{3} = (a + b\sqrt{3})^2$. Hence, using the solution to Example 5.2.8, show that

$$\sin\frac{\pi}{12} = \frac{\sqrt{3} - 1}{2\sqrt{2}} \qquad \text{and} \qquad \cos\frac{\pi}{12} = \frac{\sqrt{3} + 1}{2\sqrt{2}}.$$

Problem 5.2.10 Evaluate $\tan(\pi/12)$ and, using a calculator to approximate $\sqrt{2}$ and $\sqrt{3}$, show that $3{\cdot}1058 < \pi < 3{\cdot}2154$.

Theorem 5.2.11 (Equating Coefficients) *Let $a, b, p, q \in \mathbb{R}$ and let*

$$a\cos x + b\sin x = p\cos x + q\sin x,$$

for all $x \in \mathbb{R}$. Then $a = p$ and $b = q$.

Proof   Putting $x = 0$ gives $a = p$ and putting $x = \pi/2$ gives $b = q$. $\qquad \square$

Problem 5.2.12 Let $a, b, p, q, \theta, \phi \in \mathbb{R}$ where $\theta - \phi \neq k\pi$ for any $k \in \mathbb{Z}$ and let

$$a\cos\theta + b\sin\theta = p\cos\theta + q\sin\theta,$$
$$a\cos\phi + b\sin\phi = p\cos\phi + q\sin\phi.$$

Prove that $a = p$ and $b = q$. [Note that here the above equations hold for fixed values of θ and ϕ whereas the corresponding equation in Theorem 5.2.11 hold for *all* x.

5.3   General Solutions of Equations

Consider the equation

$$\sin x = \tfrac{1}{2}.$$

It is clear from what we already know that $x = \pi/6$ is a solution, but it is not the only solution. In fact, for any $k \in \mathbb{Z}$, $x = \pi/6 + 2k\pi$ and $x = 5\pi/6 + 2k\pi$ satisfy the equation. As we shall see later, these are all possible solutions and for that reason we say this is the *general solution* of the equation.

Lemma 5.3.1 *We have*

    (1) $\sin x = 0 \ \Leftrightarrow x = k\pi \ \ for\ some\ k \in \mathbb{Z}$,

    (2) $\cos x = 0 \ \Leftrightarrow x = \dfrac{\pi}{2} + k\pi \ \ for\ some\ k \in \mathbb{Z}$,

    (3) $\tan x = 0 \ \Leftrightarrow x = k\pi \ \ for\ some\ k \in \mathbb{Z}$.

Proof   We have seen these results earlier. They follow directly from the definitions of sin, cos and tan. See Section 5.1.   □

Theorem 5.3.2 *Let $\alpha \in \mathbb{R}$. Then*

    (1) $\sin x = \sin \alpha \ \ \Leftrightarrow \ \ x = \alpha + 2k\pi \ \ or\ x = \pi - \alpha + 2k\pi \ \ for\ some\ k \in \mathbb{Z}$,

    (2) $\cos x = \cos \alpha \ \ \Leftrightarrow \ \ x = \alpha + 2k\pi \ \ or\ x = -\alpha + 2k\pi \ \ for\ some\ k \in \mathbb{Z}$,

    (3) $\tan x = \tan \alpha \ \ \Leftrightarrow \ \ x = \alpha + k\pi \ \ for\ some\ k \in \mathbb{Z}$.

Proof   (1) We have

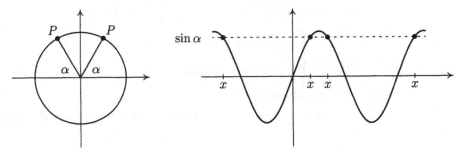

Figure 5.3.1 : $\sin x = \sin \alpha$

$$\sin x = \sin \alpha \Leftrightarrow \sin x - \sin \alpha = 0$$

$$\Leftrightarrow 2\cos\left(\frac{x+\alpha}{2}\right)\sin\left(\frac{x-\alpha}{2}\right) = 0$$

$$\Leftrightarrow \cos\left(\frac{x+\alpha}{2}\right) = 0 \ \ \text{or} \ \ \sin\left(\frac{x-\alpha}{2}\right) = 0$$

$$\Leftrightarrow \begin{cases} \dfrac{x+\alpha}{2} = \dfrac{\pi}{2} + k\pi \ \text{for some } k \in \mathbb{Z} \ \ \text{or} \\ \dfrac{x-\alpha}{2} = k\pi \ \text{for some } k \in \mathbb{Z} \end{cases}$$

(by Lemma 5.3.1 (2) and (1))

$$\Leftrightarrow x = \pi - \alpha + 2k\pi \ \ \text{or} \ \ x = \alpha + 2k\pi \ \ \text{for some } k \in \mathbb{Z}.$$

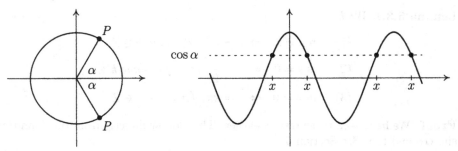

Figure 5.3.2 : $\cos x = \cos \alpha$

(2) We have

$$\cos x = \cos \alpha \Leftrightarrow \cos x - \cos \alpha = 0$$

$$\Leftrightarrow -2\sin\left(\frac{x+\alpha}{2}\right)\sin\left(\frac{x-\alpha}{2}\right) = 0$$

$$\Leftrightarrow \sin\left(\frac{x+\alpha}{2}\right) = 0 \quad \text{or} \quad \sin\left(\frac{x-\alpha}{2}\right) = 0$$

$$\Leftrightarrow \begin{cases} \dfrac{x+\alpha}{2} = k\pi \text{ for some } k \in \mathbb{Z} \text{ or} \\ \dfrac{x-\alpha}{2} = k\pi \text{ for some } k \in \mathbb{Z} \end{cases}$$

(by Lemma 5.3.1 (1))

$$\Leftrightarrow x = -\alpha + 2k\pi \quad \text{or} \quad x = \alpha + 2k\pi \quad \text{for some } k \in \mathbb{Z}.$$

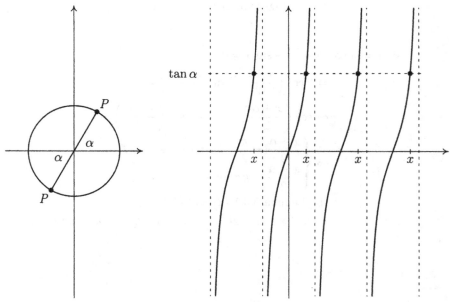

Figure 5.3.3 : $\tan x = \tan \alpha$

(3) Let $\tan \alpha$ be defined. Then $\cos \alpha \neq 0$ and we have

$$\tan x = \tan \alpha \Leftrightarrow \frac{\sin x}{\cos x} = \frac{\sin \alpha}{\cos \alpha}$$

$$\Leftrightarrow \sin x \cos \alpha = \cos x \sin \alpha \quad (\text{since } \cos \alpha \neq 0)$$

$$\Leftrightarrow \sin x \cos \alpha - \cos x \sin \alpha = 0$$

$$\Leftrightarrow \sin (x - \alpha) = 0$$

$$\Leftrightarrow x - \alpha = k\pi \quad \text{for some } k \in \mathbb{Z} \quad (\text{by Lemma 5.3.1 (1)})$$

$$\Leftrightarrow x = \alpha + k\pi \quad \text{for some } k \in \mathbb{Z}. \qquad \square$$

Remark The diagrams in Figures 5.3.1, 5.3.2 and 5.3.3 illustrate the general solutions given in Theorem 5.3.2. For example, consider the cosine diagrams in Figure 5.3.2. The general solution of the equation $\cos x = \cos \alpha$ consists of those $x \in \mathbb{R}$ which terminate at one of the points P on the left-hand diagram. These are also the numbers marked x on the right-hand diagram.

We can now solve more complicated trigonometric equations by first reducing them to a form in which Theorem 5.3.2 can be applied.

Example 5.3.3 Find those solutions of the equation

$$\sin 5x - \sin x = \cos 3x.$$

which lie in the interval $(0, \pi)$.

Solution We have

$$\sin 5x - \sin x = \cos 3x$$

$$\Leftrightarrow \sin 5x - \sin x - \cos 3x = 0$$

$$\Leftrightarrow 2 \sin 2x \cos 3x - \cos 3x = 0$$

$$\Leftrightarrow \cos 3x(2 \sin 2x - 1) = 0$$

$$\Leftrightarrow \cos 3x = 0 \quad \text{or} \quad \sin 2x = \frac{1}{2}$$

$$\Leftrightarrow 3x = \frac{\pi}{2} + 2k\pi \quad \text{or} \quad 3x = -\frac{\pi}{2} + 2k\pi \quad \text{for some } k \in \mathbb{Z} \quad \text{or}$$

$$2x = \frac{\pi}{6} + 2k\pi \quad \text{or} \quad 2x = -\frac{\pi}{6} + (2k+1)\pi \quad \text{for some } k \in \mathbb{Z}$$

$$\Leftrightarrow x = \frac{\pi}{6} + \frac{2k\pi}{3} \quad \text{or} \quad x = -\frac{\pi}{6} + \frac{2k\pi}{3} \quad \text{or}$$

$$x = \frac{\pi}{12} + k\pi \quad \text{or} \quad x = \frac{5\pi}{12} + k\pi \quad \text{for some } k \in \mathbb{Z}.$$

Thus we have found the general solution which is

$$x = \frac{\pi}{6} + \frac{2k\pi}{3}, \quad -\frac{\pi}{6} + \frac{2k\pi}{3}, \quad \frac{\pi}{12} + k\pi \quad \text{or} \quad \frac{5\pi}{12} + k\pi \quad (k \in \mathbb{Z}).$$

For the solutions in $(0, \pi)$ we check each case in turn, choosing those k in $\mathbb{Z}$ which give $0 < x < \pi$. Hence we obtain

$$x = \frac{\pi}{6}, \frac{5\pi}{6}, \frac{\pi}{2}, \frac{\pi}{12}, \frac{5\pi}{12}. \qquad \square$$

Example 5.3.4 Express $\sqrt{3}\sin x + \cos x$ in the form $r\sin(x+\theta)$ where $r,\theta \in \mathbb{R}$. Hence find the general solution of the equation

$$\sqrt{3}\sin x + \cos x = -\sqrt{2}.$$

Solution   We have

$$r\sin(x+\theta) = r(\sin x\cos\theta + \cos x\sin\theta)$$
$$= (r\cos\theta)\sin x + (r\sin\theta)\cos x.$$

So we require

$$r\cos\theta = \sqrt{3} \qquad \text{and} \qquad r\sin\theta = 1. \tag{5.3.1}$$

Squaring and adding gives

$$r^2(\cos^2\theta + \sin^2\theta) = (\sqrt{3})^2 + (1)^2, \qquad \textit{i.e. } r^2 = 4.$$

We take $r = 2$. Now, to satisfy (5.3.1) we require

$$\cos\theta = \frac{\sqrt{3}}{2} \qquad \text{and} \qquad \sin\theta = \frac{1}{2}.$$

We take $\theta = \pi/6$. Hence the given equation becomes

$$2\sin\left(x + \frac{\pi}{6}\right) = -\sqrt{2},$$

i.e.

$$\sin\left(x + \frac{\pi}{6}\right) = -\frac{1}{\sqrt{2}} = \sin\left(-\frac{\pi}{4}\right).$$

Applying Theorem 5.3.2 (1) gives

$$x + \frac{\pi}{6} = -\frac{\pi}{4} + 2k\pi \quad \text{or} \quad x + \frac{\pi}{6} = \frac{\pi}{4} + (2k+1)\pi \quad \text{for some } k \in \mathbb{Z}.$$

Hence the required general solution is

$$x = -\frac{5\pi}{12} + 2k\pi \quad \text{or} \quad x = \frac{13\pi}{12} + 2k\pi \quad (k \in \mathbb{Z}). \qquad \square$$

Remark   The general solution of an equation may be expressed in different ways. For example, the general solution of the equation solved in Example 5.3.4 could be written

$$x = \frac{19\pi}{12} + 2k\pi \quad \text{or} \quad x = -\frac{11\pi}{12} + 2k\pi \quad (k \in \mathbb{Z})$$

or

$$x = \frac{\pi}{3} \pm \frac{3\pi}{4} + 2k\pi \quad (k \in \mathbb{Z}).$$

5.4   The t-formulae

The *t-formulae* express $\sin\theta$, $\cos\theta$ and $\tan\theta$ in terms of $\tan\frac{1}{2}\theta$ which is often abbreviated to t. The formulae have various applications. In particular, as we shall see in Chapter 16, they can be used to evaluate certain trigonometric integrals.

Theorem 5.4.1 (*t*-formulae) *Let $\theta \in \mathbb{R}$ such that $t = \tan\frac{1}{2}\theta$ is defined. Then*

$$(1)\ \ \sin\theta = \frac{2t}{1+t^2} \qquad and \qquad (2)\ \ \cos\theta = \frac{1-t^2}{1+t^2},$$

and whenever $\tan\theta$ is defined (so that $t^2 \neq 1$),

$$(3)\ \ \tan\theta = \frac{2t}{1-t^2}.$$

Proof   First note that

$$\cos^2\tfrac{1}{2}\theta = \frac{1}{\sec^2\frac{1}{2}\theta} = \frac{1}{1+\tan^2\frac{1}{2}\theta} = \frac{1}{1+t^2}.$$

Then we have

$$(1)\ \ \sin\theta = \sin 2(\tfrac{1}{2}\theta) = 2\sin\tfrac{1}{2}\theta\cos\tfrac{1}{2}\theta = 2\tan\tfrac{1}{2}\theta\cos^2\tfrac{1}{2}\theta = \frac{2t}{1+t^2},$$

$$(2)\ \ \cos\theta = \cos 2(\tfrac{1}{2}\theta) = 2\cos^2\tfrac{1}{2}\theta - 1 = \frac{2}{1+t^2} - 1 = \frac{1-t^2}{1+t^2},$$

$$(3)\ \ \tan\theta = \frac{\sin\theta}{\cos\theta} = \frac{2t}{1+t^2}\frac{1+t^2}{1-t^2} = \frac{2t}{1-t^2},$$

as required. $\square$

Note 5.4.2   It is easily verified that $\tan\frac{1}{2}\theta$ is undefined if and only if $\theta = (2k+1)\pi$, ($k \in \mathbb{Z}$). At such values of θ, $\cos\theta = -1$ and $\sin\theta = 0$.

Example 5.4.3   Evaluate $\tan x$ when $4\cos x - \sin x + 3 = 0$.

Solution   First note that $\tan\frac{1}{2}x$ is defined when x satisfies this equation. If not then, by Note 5.4.2, $\cos x = -1$ and $\sin x = 0$ so that $4\cos x - \sin x + 3 \neq 0$.

Now let $t = \tan\frac{1}{2}x$. Using the t-formulae to substitute for $\sin x$ and $\cos x$, the equation becomes

$$4\frac{1-t^2}{1+t^2} - \frac{2t}{1+t^2} + 3 = 0.$$

Since $1+t^2 \neq 0$, this is equivalent to

$$4(1-t^2) - 2t + 3(1+t^2) = 0,$$

i.e.

$$t^2 + 2t - 7 = 0.$$

So $t = -1 \pm 2\sqrt{2}$ and hence, using the t-formula for $\tan x$,

$$\tan x = \frac{2t}{1 - t^2} = \frac{2(-1 \pm 2\sqrt{2})}{1 - (-1 \pm 2\sqrt{2})^2} = \frac{-1 \pm 2\sqrt{2}}{2(-2 \pm \sqrt{2})} = \frac{-2 \mp 3\sqrt{2}}{4},$$

where the final result is obtained by multiplying both the denominator and numerator by $-2 \mp \sqrt{2}$.   □

Remark There is a further t-formula, for the derivative of θ with respect to t where $t = \tan \frac{1}{2}\theta$. We shall meet this in Section 16.3.

5.5   Inverse Trigonometric Functions

None of the functions sin, cos and tan is a bijection and so none has an inverse. For each, we shall restrict the domain and, where necessary, the codomain to produce a bijection with the same image as the original function. The inverses of these bijections are denoted by $\sin^{-1}$, $\cos^{-1}$ and $\tan^{-1}$ and are called the *inverse sine*, *inverse cosine* and *inverse tangent* functions, respectively. Note that, despite the notation, these are *not* true inverses of the sine, cosine and tangent functions.

The graphs of the bijective restrictions and their inverses are sketched in Figures 5.5.1, 5.5.2 and 5.5.3.

Theorem 5.5.1 *Define*

$$f : [-\tfrac{\pi}{2}, \tfrac{\pi}{2}] \to [-1, 1] \ \ by \ \ f(x) = \sin x,$$

$$g : [0, \pi] \to [-1, 1] \ \ by \ \ g(x) = \cos x,$$

$$h : (-\tfrac{\pi}{2}, \tfrac{\pi}{2}) \to \mathbb{R} \ \ by \ \ h(x) = \tan x.$$

Then f, g and h are bijective.

Proof We must show that f, g and h are both injective and surjective.

The diagrams in Figure 5.1.4 provide the basis for an intuitive view of the injectiveness of the three functions. A rigorous approach is available *via* Theorem 5.3.2. We shall leave f and g to the reader and illustrate the method by considering h.

Let $\theta_1, \theta_2 \in (-\tfrac{\pi}{2}, \tfrac{\pi}{2})$. Without loss of generality suppose that $\theta_1 \geq \theta_2$. Then $0 \leq \theta_1 - \theta_2 < \pi$. If $\theta_1 - \theta_2 = k\pi$ for some $k \in \mathbb{Z}$ then $k = 0$ and $\theta_1 = \theta_2$. Hence, using Theorem 5.3.2 (3),

$$h(\theta_1) = h(\theta_2) \Rightarrow \tan \theta_1 = \tan \theta_2$$
$$\Rightarrow \theta_1 = \theta_2 + k\pi \ \ \text{for some } k \in \mathbb{Z}$$
$$\Rightarrow \theta_1 - \theta_2 = k\pi \ \ \text{for some } k \in \mathbb{Z}$$
$$\Rightarrow \theta_1 = \theta_2.$$

Thus h is injective.

The surjectivity of f, g and h follows directly from Theorem 5.1.5.   □

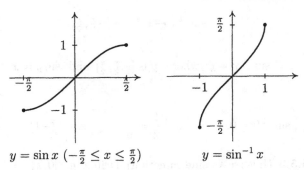

$y = \sin x \ (-\frac{\pi}{2} \le x \le \frac{\pi}{2})$ $y = \sin^{-1} x$

Figure 5.5.1 : Inverse Sine

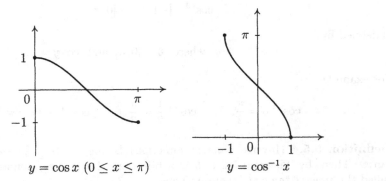

$y = \cos x \ (0 \le x \le \pi)$ $y = \cos^{-1} x$

Figure 5.5.2 : Inverse Cosine

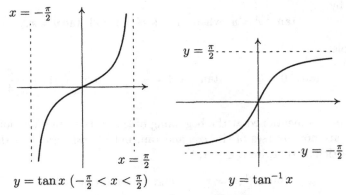

$y = \tan x \ (-\frac{\pi}{2} < x < \frac{\pi}{2})$ $y = \tan^{-1} x$

Figure 5.5.3 : Inverse Tangent

Definition 5.5.2 (Inverse Sine Function) Define $f : [-\frac{\pi}{2}, \frac{\pi}{2}] \to [-1, 1]$ by $f(x) = \sin x$. Then, by Theorem 5.5.1, f is a bijection. Its inverse, denoted by $\sin^{-1}$, is called the *inverse sine* function. Thus,

$$\sin^{-1} : [-1, 1] \to [-\tfrac{\pi}{2}, \tfrac{\pi}{2}]$$

is defined by

$$\sin^{-1} x = y \quad \text{where} \quad y \in [-\tfrac{\pi}{2}, \tfrac{\pi}{2}] \text{ and } \sin y = x.$$

For example,

$$\sin^{-1} 0 = 0, \qquad \sin^{-1} \frac{1}{2} = \frac{\pi}{6}, \qquad \sin^{-1}(-1) = -\frac{\pi}{2}.$$

Definition 5.5.3 (Inverse Cosine Function) Define $g : [0, \pi] \to [-1, 1]$ by $f(x) = \cos x$. Then, by Theorem 5.5.1, g is a bijection. Its inverse, denoted by $\cos^{-1}$, is called the *inverse cosine* function. Thus,

$$\cos^{-1} : [-1, 1] \to [0, \pi]$$

is defined by

$$\cos^{-1} x = y \quad \text{where} \quad y \in [0, \pi] \text{ and } \cos y = x.$$

For example,

$$\cos^{-1} 0 = \frac{\pi}{2}, \qquad \cos^{-1} \frac{1}{2} = \frac{\pi}{3}, \qquad \cos^{-1}(-1) = \pi.$$

Definition 5.5.4 (Inverse Tangent Function) Define $h : (-\frac{\pi}{2}, \frac{\pi}{2}) \to \mathbb{R}$ by $f(x) = \tan x$. Then, by Theorem 5.5.1, h is a bijection. Its inverse, denoted by $\tan^{-1}$, is called the *inverse tangent* function. Thus,

$$\tan^{-1} : \mathbb{R} \to (-\tfrac{\pi}{2}, \tfrac{\pi}{2})$$

is defined by

$$\tan^{-1} x = y \quad \text{where} \quad y \in (-\tfrac{\pi}{2}, \tfrac{\pi}{2}) \text{ and } \tan y = x.$$

For example,

$$\tan^{-1} 0 = 0, \qquad \tan^{-1} \sqrt{3} = \frac{\pi}{3}, \qquad \tan^{-1}(-1) = -\frac{\pi}{4}.$$

 Hazard As was mentioned at the beginning of the section, the functions $\sin^{-1}$, $\cos^{-1}$ and $\tan^{-1}$ are not inverses of sin, cos and tan, but of *restrictions* of these functions. Note in particular that, for $x \in [-1, 1]$,

$$\sin(\sin^{-1} x) = x \qquad \text{and} \qquad \cos(\cos^{-1} x) = x$$

but, for example,

$$\sin^{-1}(\sin(2\pi)) = 0 \qquad \text{and} \qquad \cos^{-1}(\cos(-\pi)) = \pi.$$

Similarly, for $x \in \mathbb{R}$, $\tan(\tan^{-1} x) = x$ but, for example, $\tan^{-1}(\tan \pi) = 0$.

Example 5.5.5 Evaluate

$$(a)\ \cos\left(\sin^{-1}(-4/5)\right),\qquad (b)\ \tan^{-1}2+\tan^{-1}3.$$

Solution (a) Let $\sin^{-1}(-4/5)=\theta$. Then

$$\sin\theta = -\frac{4}{5}\qquad\text{and}\qquad -\frac{\pi}{2}<\theta<\frac{\pi}{2}.$$

Hence,

$$\cos^2\theta = 1 - \sin^2\theta = \frac{9}{25}.$$

Since cosine is positive in $(-\frac{\pi}{2},\frac{\pi}{2})$, taking square roots we get

$$\cos\left(\sin^{-1}\left(-\frac{4}{5}\right)\right) = \cos\theta = \frac{3}{5}$$

(b) Let $\tan^{-1}2=\alpha$ and $\tan^{-1}3=\beta$. Then $\tan\alpha = 2$ and $\tan\beta = 3$ so that

$$\tan(\alpha+\beta) = \frac{\tan\alpha+\tan\beta}{1-\tan\alpha\tan\beta} = \frac{2+3}{1-6} = -1.$$

Hence,

$$\alpha+\beta = -\frac{\pi}{4}+k\pi\quad\text{for some } k\in\mathbb{Z}.$$

But, since $\alpha = \tan^{-1}2$ and $\beta = \tan^{-1}3$, we must have

$$0<\alpha<\frac{\pi}{2}\qquad\text{and}\qquad 0<\beta<\frac{\pi}{2}.$$

Adding gives

$$0<\alpha+\beta<\pi$$

so that k must be 1. Hence,

$$\tan^{-1}2+\tan^{-1}3 = \alpha+\beta = \frac{3\pi}{4}.\qquad\square$$

Problem 5.5.6 By considering $\sin\theta$, simplify $\theta = \sin^{-1}x + \cos^{-1}x$ where $x\in[-1,1]$.

5.X   Exercises

1. Let $\tan\theta = 5/12$ and let $\cos\theta$ be negative. Find the values of $\sin\theta$, $\cos\theta$, $\cot\theta$, $\sec\theta$ and $\operatorname{cosec}\theta$.

2. Let $n\in\mathbb{N}$ and let $\theta_1,\theta_2,\ldots,\theta_n\in\mathbb{R}$. Show that

$$|\sin\theta_1 + \sin\theta_2 + \cdots + \sin\theta_n| \le n.$$

3. Find the maximal domain of the real function f when $f(x)$ is

    (a) $\tan(\sin x)$,   (b) $\sqrt{\cos x}$.

4. Find $g \circ f$ when $f : \mathbb{R} \to \mathbb{R}$ and $g : [-1, 1] \to \mathbb{R}$ are defined by

$$f(x) = \cos x \qquad \text{and} \qquad g(x) = \sqrt{1 - x^2}.$$

    Simplify the expression for $(g \circ f)(x)$.

5. Sketch the curve $y = f(x)$ in the range $0 \le x \le \pi$ when $f(x)$ is

    (a) $4 \sin x$,   (b) $3 - \sin x$,   (c) $\sin 2x$,

    (d) $\tan(x/2)$,   (e) $\tan(x - \frac{\pi}{4})$,   (f) $\tan(\frac{\pi}{4} - x)$,

    (g) $|\cos x|$,   (h) $|\frac{1}{2} + \cos x|$,   (i) $(\frac{1}{2} + \cos x)^{-1}$.

6. Define $P : [0, 2\pi) \to \mathbb{R}^2$ by $P(\theta) = (\cos \theta, \sin \theta)$. Use the formula for $\cos(\alpha - \beta)$ to prove that P is injective. Describe the image of P as a subset of the x, y-plane.

7. Establish each of the following identities.

    (a) $\cos^2 \theta - \sin^2 \phi = \cos^2 \phi - \sin^2 \theta$,

    (b) $\cot \theta - \tan \theta = 2 \cot 2\theta$,

    (c) $\cot \theta + \tan \theta = 2 \operatorname{cosec} 2\theta$,

    (d) $\tan 4\theta = \dfrac{4 \tan \theta - 4 \tan^3 \theta}{1 - 6 \tan^2 \theta + \tan^4 \theta}$,

    (e) $\dfrac{1 - \sin \theta}{1 + \sin \theta} = (\sec \theta - \tan \theta)^2$,

    (f) $\dfrac{\sin y + \sin 3y}{\cos y + \cos 3y} = \tan 2y$,

    (g) $\sin \alpha \sin 2\alpha + \sin 2\alpha \sin 5\alpha = \sin 3\alpha \sin 4\alpha$,

    (h) $(\operatorname{cosec} \theta - 1)^{-1} - (\operatorname{cosec} \theta + 1)^{-1} = 2 \tan^2 \theta$.

8. Prove that $\cos \theta = \dfrac{\cos 3\theta}{2 \cos 2\theta - 1}$. Hence find the value of $\cos(\pi/12)$ and deduce the value of $\sin(\pi/12)$.

9. Show that whenever the point

$$(x, y) = (\tan \theta + \sin \theta, \tan \theta - \sin \theta),$$

    is defined, it lies on the curve $(x^2 - y^2)^2 = 16xy$.

10. Find the general solution of each of the following equations.

$$\text{(a)} \quad \sin 2x = \cos(3\pi/4),$$

$$\text{(b)} \quad \cos 2x + \cos 4x = \cos 3x,$$

$$\text{(c)} \quad \sin x - \sqrt{3}\cos x = 1.$$

11. Evaluate

   (a) $\sin^{-1} 1$,   (b) $\cos^{-1}(-1/2)$,

   (c) $\tan^{-1}(1/\sqrt{3})$,   (d) $\sin^{-1}(\sin(3\pi/4))$,

   (e) $\cos^{-1}(\sin(-5\pi/6))$,   (f) $\cot(\sin^{-1}(5/13))$,

   (g) $\tan(\tan^{-1} 2 + \tan^{-1}(-3))$,

   (h) $\tan(\tan^{-1} a - \tan^{-1}(1/a))$ $(a \neq 0)$,

   (i) $\sin(\cos^{-1}(1/3) + \tan^{-1}\sqrt{2})$,

   (j) $\tan^{-1} 3 - \tan^{-1}(1/2)$.

   For (j), observe that $0 < \tan^{-1}(1/2) < \pi/4 < \tan^{-1} 3 < \pi/2$.

12. Find the maximal domain of the real function f when $f(x)$ is

$$\text{(a)} \quad \tan(\sin^{-1} x), \qquad \text{(b)} \quad \cos^{-1}(2/x).$$

13. Sketch the graph of the real function f when $f(x)$ is defined by

   (a) $\sin^{-1}(x/2)$,   (b) $\cos^{-1}(1-x)$,   (c) $|\tan^{-1} x|$.

14. Show that, for $-1 \leq t \leq 1$, $\sin(\cos^{-1} t) = \sqrt{1-t^2}$. Hence solve the equation

$$\sin^{-1}\frac{x}{3} = \cos^{-1}\frac{x}{4}.$$

Chapter 6

Complex Numbers

As we have seen in Section 3.3, there are some polynomials which have no real roots. For example, there is no real number x which satisfies $x^2 + 1 = 0$. The aim of this chapter is to introduce and study a number system $\mathbb{C}$—the complex numbers—in which this, and all other quadratic equations can be solved. The reader should see this as a natural step in the evolution of number systems outlined in the introduction the Chapter 1 and amplified in the paragraph following Problem 1.1.2.

In the introduction to Chapter 1, we indicated that everything would be based on real numbers. In keeping with this philosophy, we define a complex number as a *ordered pair* (a, b) of real numbers. The reader will be familiar with the use of such a pair to describe a point of the plane. The ability to regard the pair (a, b) as either a complex number or a point is extremely fruitful. In particular, the geometrical interpretation of a complex number indicates a connection with trigonometric functions. Here the key result is de Moivre's theorem (Theorem 6.3.1). This allows us to extend work on the trigonometric identities in Chapter 5.

Although $\mathbb{C}$ was introduced to allow us to solve one polynomial equation *viz.* $x^2 + 1 = 0$), it is a remarkable fact that it contains roots of *all* polynomial equations. Although the proof of this is beyond the scope of this book, we apply the result to justify the claim in Chapter 3 that all *real* polynomials may be factorised into real linear and irreducible real quadratic factors.

Section 6.5 is concerned with the *complex roots of unity*, *i.e.* roots in $\mathbb{C}$ of an equation of the form $z^n - 1 = 0$, and of general n^{th} roots of elements of $\mathbb{C}$. In accordance with fundamental fheorem of algebra, we will find that there are always exactly n roots.

In the final section, the ability to regard points on the plane as complex numbers is exploited in the study of rotations and translations.

6.1 The Complex Plane

Definitions 6.1.1 Let $z = (x, y)$ and $w = (u, v)$ be elements of $\mathbb{R}^2$. We define addition and multiplication by

$$z + w = (x, y) + (u, v) = (x + u, y + v), \tag{6.1.1}$$

$$zw = (x, y)(u, v) = (xu - yv, xv + yu). \tag{6.1.2}$$

With these operations $\mathbb{R}^2$ becomes a number system called the *complex numbers* which we now denote by $\mathbb{C}$.

When points on the plane $\mathbb{R}^2$ are to be regarded as complex numbers we refer to the plane as the *complex plane*.

The complex numbers of the form $(x, 0)$ form a number system in their own right, since

$$(x, 0) + (u, 0) = (x + u, 0) \qquad \text{and} \qquad (x, 0)(u, 0) = (xu, 0).$$

Observe the parallel with the addition and multiplication of real numbers. For this reason, we regard $\mathbb{R}$ as a subset of $\mathbb{C}$ and write x for the complex number $(x, 0)$. Thus the real line is embedded in the complex plane in the form of the x-axis. This is illustrated in Figure 6.1.1.

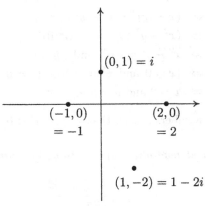

Figure 6.1.1

Theorem 6.1.2 $(0, 1)^2 = -1$.

Proof Put $x = u = 0$ and $y = v = 1$ in equation (6.1.2). $\square$

Notation 6.1.3 The complex number $(0, 1)$ is denoted by i.

Remarks (1) Theorem 6.1.2 tells us that $i^2 = -1$. Hence i is *not* a real number. For every $x \in \mathbb{R}$, $x^2 \geq 0$, whereas $i^2 < 0$.

(2) Observe that, for $a, b \in \mathbb{R}$, we can write

$$(a, b) = (a, 0) + (b, 0)(0, 1) = a + ib.$$

Then equations (6.1.1) and (6.1.2) can be written as follows. Let $x + iy$, $u + iv \in \mathbb{C}$, where $x, y, u, v \in \mathbb{R}$. Then

$$(x + iy) + (u + iv) = (x + u) + i(y + v),$$

$$(x + iy)(u + iv) = (xu - yv) + i(xv + yu).$$

(3) It follows from (1) and Theorem 6.1.2 that complex addition and multiplication can be carried out as if only real numbers were involved, if we remember to make the substitution $i^2 = -1$ when it is appropriate. The next example demonstrates this.

Example 6.1.4 Evaluate the product of complex numbers $6 + 3i$ and $5 - i$.

Solution We have

$$(6 + 3i)(5 - i) = 30 - 6i + 15i - 3i^2 = 30 + 9i + 3 = 33 + 9i. \qquad \square$$

Lemma 6.1.5 *For $z \in \mathbb{C}$ and $\alpha \neq 0$,*

$$z^2 = -\alpha^2 \; \Leftrightarrow \; z = \pm i\alpha.$$

Proof Let $z = x + iy$, where $x, y \in \mathbb{R}$.

$$
\begin{aligned}
z^2 = -\alpha^2 \; &\Leftrightarrow \; (x + iy)^2 = x^2 - y^2 + 2ixy = -\alpha^2 \\
&\Leftrightarrow \; (x^2 - y^2, 2xy) = (-\alpha^2, 0) \\
&\Leftrightarrow \; x^2 - y^2 = -\alpha^2 \text{ and } xy = 0 \\
&\Leftrightarrow \; (x = 0 \text{ and } y^2 = \alpha^2) \text{ or } (y = 0 \text{ and } x^2 = -\alpha^2) \\
&\Leftrightarrow \; x = 0 \text{ and } y = \pm\alpha,
\end{aligned}
$$

since there are no real numbers x such that $x^2 = -\alpha^2 < 0$. $\qquad \square$

Theorem 6.1.6 *The real quadratic $q(z) = az^2 + bz + c$, with discriminant $b^2 - 4ac < 0$, has (complex) zeros*

$$\frac{-b \pm i\sqrt{4ac - b^2}}{2a}. \tag{6.1.3}$$

Proof We complete the square of the quadratic to get

$$q(z) = a\big(z^2 + (b/a)z\big) + c = a(z + b/2a)^2 + (4ac - b^2)/4a.$$

Note that since $b^2 - 4ac < 0$, $4ac - b^2 > 0$. Hence,

$$
\begin{aligned}
q(z) = 0 \; &\Leftrightarrow \; (z + b/2a)^2 = -(4ac - b^2)/4a^2 \\
&\Leftrightarrow \; (z + b/2a)^2 = -\big(\sqrt{4ac - b^2}/2a\big)^2 \\
&\Leftrightarrow \; z + b/2a = \pm i\sqrt{4ac - b^2}/2a \quad \text{(by Lemma 6.1.5)} \\
&\Leftrightarrow \; z = \frac{-b \pm i\sqrt{4ac - b^2}}{2a},
\end{aligned}
$$

as required. $\qquad \square$

Example 6.1.7 Find the (complex) roots of

$$z^2 + z + 1 = 0.$$

Solution The discriminant is $1^2 - 4(1)(1) = -3 < 0$. We use (6.1.3) to obtain

$$z = \frac{-1 \pm i\sqrt{3}}{2}. \qquad \square$$

Definitions 6.1.8 Let $z = (x, y) = x + iy \in \mathbb{C}$. Then x is the *real part* of z and y is the *imaginary part* of z. We write $\operatorname{Re} z = x$ and $\operatorname{Im} z = y$.

If $\operatorname{Im} z = 0$ we say that z is *real*, and if $z \neq 0$ and $\operatorname{Re} z = 0$ we say that z is *purely imaginary*.

In the complex plane, the x-axis consists of all of the real complex numbers and is called the *real axis*. The y-axis consists of all purely imaginary complex numbers together with 0 and is called the *imaginary axis*. See Figure 6.1.1

Hazard The imaginary part of $(x, y) = x + iy \in \mathbb{C}$ is y. It is *not* iy.

Properties 6.1.9 The following properties of real and imaginary parts are easily verified.

Let $z, w \in \mathbb{C}$ and $\lambda \in \mathbb{R}$. Then

(1) $z = w \iff \operatorname{Re} z = \operatorname{Re} w$ and $\operatorname{Im} z = \operatorname{Im} w$.

(2) $\operatorname{Re}(\lambda z) = \lambda \operatorname{Re} z$ and $\operatorname{Im}(\lambda z) = \lambda \operatorname{Im} z$.

(3) $\operatorname{Re}(z + w) = \operatorname{Re} z + \operatorname{Re} w$ and $\operatorname{Im}(z + w) = \operatorname{Im} z + \operatorname{Im} w$.

Remark If we know that two complex numbers z and w are equal then we can use Property 6.1.9 (1) to deduce that $\operatorname{Re} z = \operatorname{Re} w$ and $\operatorname{Im} z = \operatorname{Im} w$. This is called *equating real and imaginary parts* and is used frequently.

Example 6.1.10 Determine the real and imaginary parts of the following complex numbers:

(a) $z_1 = \sqrt{2} - i$,   (b) $z_2 = (1 + i)^2$,

(c) $z_3 = (a + ib)(a - ib)$ where $a, b \in \mathbb{R}$.

Solution (a) We have $\operatorname{Re} z_1 = \sqrt{2}$ and $\operatorname{Im} z_1 = -1$.

(b) Here $z_2 = 1 + 2i + i^2 = 2i$. So $\operatorname{Re} z_2 = 0$ and $\operatorname{Im} z_2 = 2$. Thus z_2 is purely imaginary.

(c) Now $z_3 = a^2 - i^2 b^2 = a^2 + b^2 \in \mathbb{R}$ since $a, b \in \mathbb{R}$. So $\operatorname{Re} z_3 = a^2 + b^2$, $\operatorname{Im} z_3 = 0$ and hence z_3 is real. $\square$

Definition 6.1.11 Let $z = (x, y) = x + iy \in \mathbb{C}$. Then the *modulus* of z is the real number

$$|z| = \sqrt{x^2 + y^2}.$$

Geometrically, $|z|$ is the distance of z from the origin in the complex plane.

Note that, for $z = (x, 0) = x \in \mathbb{R}$, this definition of $|z|$ agrees with our earlier definition of $|x|$ (see Example 1.5.12).

Extending the notion of a step on the real number line (Section 1.4) to the complex plane, we may think of a complex number $z = (x, y)$ as a step from $(0, 0)$ to (x, y). The length of this step is $|z| = \sqrt{x^2 + y^2}$. In this way, step $z + w$ is step z (from 0) followed by step w (from z) and, as shown in Figure 6.1.2, these steps form the sides of a triangle.

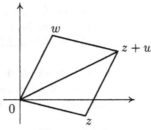

Figure 6.1.2

Using this geometrical interpretation of addition we get the following Theorems.

Theorem 6.1.12 *For $z, w \in \mathbb{C}$, $|w - z| = |z - w|$ is the distance between z and w.*

Proof We have
$$w = z + (w - z),$$
and therefore $w - z$ is the step from z to w. The length of this step is $|w - z|$ and so the distance from z to w is $|w - z|$. □

Theorem 6.1.13 (The Triangle Inequality for Complex Numbers) *For $z, w \in \mathbb{C}$,*
$$|z + w| \leq |z| + |w|.$$

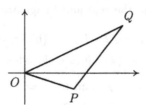

Figure 6.1.3

Proof Let O, P and Q be points in the complex plane corresponding to the complex numbers 0, z and $z + w$, respectively. In any triangle OPQ, $|OQ| \leq |QP| + |PO|$, as no side is longer than the sum of the other two. Using the definition of modulus and Theorem 6.1.12 we have $|OQ| = |z + w|$, $|QP| = |w|$ and $|PO| = |z|$ and the result follows. □

Definition 6.1.14 Let $z = (x, y) = x + iy \in \mathbb{C}$. Then the (*complex*) *conjugate* of z is $\overline{z} = (x, -y) = x - iy$.

Remarks The geometrical interpretation of the complex conjugate is that $\overline{z}$ is the reflection of z in the real axis and it follows from this that

(1) $\overline{(\overline{z})}$ is the reflection of $\overline{z}$ and hence $\overline{(\overline{z})} = z$.

(2) The only complex numbers that are unchanged by conjugation are those which are their own reflection in the real axis. Thus $z = \overline{z}$ if and only if z is real.

Theorem 6.1.15 *Let $z, w \in \mathbb{C}$. Then*

   (1) $\overline{(\overline{z})} = z$,

   (2) $z = \overline{z} \iff z$ *is real,*

   (3) $z + \overline{z} = 2 \operatorname{Re} z$,

   (4) $z\overline{z} = |z|^2$,

   (5) $|\overline{z}| = |z|$,

   (6) $\overline{z + w} = \overline{z} + \overline{w}$

   (7) $\overline{zw} = \overline{z}\,\overline{w}$.

Proof   (1) and (2) follow from the geometric description above.

In the following let $z = x + iy$ and $w = u + iv$ where $x, y, u, v \in \mathbb{R}$.

   (3) $z + \overline{z} = (x + iy) + (x - iy) = 2x = 2 \operatorname{Re} z$.

   (4) $z\overline{z} = (x + iy)(x - iy) = x^2 + y^2 = (\sqrt{x^2 + y^2})^2 = |z|^2$.

   (5) $|\overline{z}| = \sqrt{x^2 + (-y)^2} = \sqrt{x^2 + y^2} = |z|$.

   (6) $\overline{z + w} = \overline{x + u + i(y + v)} = x + u - i(y + v) = (x - iy) + (u - iv) = \overline{z} + \overline{w}$.

   (7) $\overline{zw} = \overline{xu - yv + i(xv + yu)} = xu - yv - i(xv + yu) = (x - iy)(u - iv)$
           $= \overline{z}\,\overline{w}$ $\square$

Remark   The results of Theorem 6.1.13 and Theorem 6.1.15 (6), (7) may be extended to sums and products of any finite number of terms. Thus, for $n \in \mathbb{N}$ and $z_1, \ldots, z_n \in \mathbb{C}$,

$$|z_1 + \cdots + z_n| \le |z_1| + \cdots + |z_n|,$$

$$\overline{z_1 + \cdots + z_n} = \overline{z_1} + \cdots + \overline{z_n},$$

$$\overline{z_1 \ldots z_n} = \overline{z_1} \ldots \overline{z_n}.$$

Example 6.1.16   Prove that, for non-zero $z \in \mathbb{C}$,

$$z \text{ is purely imaginary} \iff z = -\overline{z}.$$

Solution   Let $z = x + iy$, where $x, y \in \mathbb{R}$. Then

$$\begin{aligned}
z = -\overline{z} &\iff x + iy = -(x - iy) \\
&\iff x + iy = -x + iy \\
&\iff x = 0 \\
&\iff \operatorname{Re} z = 0, \\
&\iff z \text{ is purely imaginary,}
\end{aligned}$$

since $z \ne 0$. $\square$

Theorem 6.1.17 *For non-zero $z \in \mathbb{C}$, the reciprocal $1/z$ of z exists and*

$$\frac{1}{z} = \frac{1}{|z|^2}\overline{z}.$$

Proof Since $z \neq 0$, $|z| \neq 0$. Also, by Theorem 6.1.15 (4), $z\bar{z} = |z|^2$, and hence

$$z\left(\frac{1}{|z|^2}\bar{z}\right) = \frac{z\bar{z}}{|z|^2} = 1.$$

Thus z has a reciprocal and this is as stated. $\square$

Definitions 6.1.18 As for real numbers in Section 1.5, we define *integer powers* of complex numbers as follows. Let $z \in \mathbb{C}$ and $n \in \mathbb{N}$. Then

$$z^n = \underbrace{zz \ldots z}_{n \text{ factors}}$$

and when z is non-zero,

$$z^0 = 1 \qquad \text{and} \qquad z^{-n} = \frac{1}{z^n}.$$

Now let $z, w \in \mathbb{C}$ with $w \neq 0$. We define the *quotient* z/w by

$$\frac{z}{w} = z\frac{1}{w} = \frac{1}{|w|^2}z\bar{w},$$

using Theorem 6.1.17. A practical method for determining such a quotient is demonstrated in Example 6.1.19.

Example 6.1.19 Find the real and imaginary parts of

$$u = \frac{2+i}{3-i}.$$

Solution We multiply the numerator and denominator of u by the complex conjugate of the denominator:

$$u = \frac{(2+i)}{(3-i)} \times \frac{(3+i)}{(3+i)} = \frac{6+5i+i^2}{|3+i|^2} = \frac{5+5i}{10} = \frac{1}{2} + \frac{1}{2}i.$$

Thus $\operatorname{Re} u = 1/2$ and $\operatorname{Im} u = 1/2$. $\square$

Example 6.1.20 Determine the set S of points that are equidistant from i and 2.

Solution The set S may be expressed as

$$S = \{z \in \mathbb{C} : |z - i| = |z - 2|\}.$$

Let $z = x + iy$ where $x, y \in \mathbb{R}$ and then

$$\begin{aligned}
z \in S \quad &\Leftrightarrow \quad |z - i| = |z - 2|, \\
&\Leftrightarrow \quad x^2 + (y-1)^2 = (x-2)^2 + y^2, \\
&\Leftrightarrow \quad x^2 + y^2 - 2y + 1 = x^2 - 4x + 4 + y^2, \\
&\Leftrightarrow \quad y = 2x - \frac{3}{2}.
\end{aligned}$$

Thus $S = \{z \in \mathbb{C} : z = a + (2a - 3/2)i, \text{ for some } a \in \mathbb{R}\}$. Geometrically, it is the perpendicular bisector of the line joining i and 2. $\square$

Example 6.1.21 For $z \neq i$, define

$$w = \frac{z + i}{z - i}.$$

Prove that

(a) w is real $\Leftrightarrow$ z is zero or purely imaginary.

(b) $|w| < 1$ $\Leftrightarrow$ $\operatorname{Im} z < 0$.

Solution In each case we write a set of equivalent statements.

(a) We have

$$
\begin{aligned}
w \text{ is real} \quad &\Leftrightarrow \quad w = \overline{w}, \\
&\Leftrightarrow \quad \frac{z + i}{z - i} = \frac{\overline{z} - i}{\overline{z} + i}, \\
&\Leftrightarrow \quad (z + i)(\overline{z} + i) = (z - i)(\overline{z} - i), \\
&\Leftrightarrow \quad z\overline{z} - 1 + (z + \overline{z})i = z\overline{z} - 1 - (z + \overline{z})i, \\
&\Leftrightarrow \quad z + \overline{z} = 0, \\
&\Leftrightarrow \quad \operatorname{Re} z = 0, \\
&\Leftrightarrow \quad z \text{ is zero or purely imaginary.}
\end{aligned}
$$

(b) We have

$$
\begin{aligned}
|w| < 1 \quad &\Leftrightarrow \quad \frac{|z + i|}{|z - i|} < 1, \\
&\Leftrightarrow \quad |z + i| < |z - i|, \\
&\Leftrightarrow \quad z \text{ is closer to } -i \text{ than } i, \\
&\Leftrightarrow \quad \operatorname{Im} z < 0. \qquad \square
\end{aligned}
$$

6.2 Polar Form and Complex Exponentials

The complex number $z = (x, y) = x + iy$ is the point in the complex plane with *Cartesian coordinates* (x, y). When $z \neq 0$ the point can also be specified by means of its *polar coordinates* (r, θ), defined below. Polar coordinates are also discussed later, in Section 10.6.

Definitions 6.2.1 Let $z \in \mathbb{C}$ be non-zero. Then the angle $\theta \in (-\pi, \pi]$, measured from the positive real axis Ox to the line Oz joining the origin to z, is called the *principal argument of z* and is denoted $\arg z$. See Figure 6.2.1 for an example. As described in Definition 5.1.1, if $\theta > 0$ the angle is measured anticlockwise and if $\theta < 0$ it is measured clockwise.

Any angle which differs from $\arg z$ by an integer multiple of 2π is called a *determination of the argument of z*, or simply the *argument of z*, and is denoted $\operatorname{Arg} z$.

The pair $(|z|, \operatorname{Arg} z)$, for any determination of the argument, forms *polar coordinates* for z.

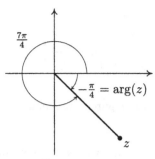

Figure 6.2.1

Example 6.2.2 Find the principal argument and all determinations of the argument of $z = 1 - i$.

Solution From Figure 6.2.1 we see that the angle in $(-\pi, \pi]$ measured anticlockwise from Ox to Oz is $-\pi/4$. Hence, the principal argument is $\arg z = -\pi/4$ and the determinations of the argument are $\text{Arg } z = -\pi/4 + 2k\pi$ $(k \in \mathbb{Z})$. In particular the choice $\text{Arg } z = 7\pi/4$ $(k = 1)$ is illustrated.   □

Note 6.2.3 It follows from its definition that, for non-zero $z \in \mathbb{C}$,

$$\arg z = \begin{cases} 0 & \text{if } z \text{ lies on the positive real axis,} \\ \pi & \text{if } z \text{ lies on the negative axis,} \\ \pi/2 & \text{if } z \text{ lies on the positive imaginary axis,} \\ -\pi/2 & \text{if } z \text{ lies on the negative imaginary axis.} \end{cases}$$

If z does not lie on the real or imaginary axis then, writing $\alpha = \tan^{-1} |y/x|$, we have

$$\arg z = \begin{cases} \alpha & \text{if } x > 0 \text{ and } y > 0, \\ -\alpha & \text{if } x > 0 \text{ and } y < 0, \\ \pi - \alpha & \text{if } x < 0 \text{ and } y > 0, \\ -\pi + \alpha & \text{if } x < 0 \text{ and } y < 0. \end{cases}$$

Alternative formulations are given in the corollary to the next theorem.

Theorem 6.2.4 *Let $z = (x, y) \in \mathbb{C}$ be non-zero with polar coordinates (r, θ). Then*

$$\begin{cases} x = r \cos \theta, \\ y = r \sin \theta \end{cases} \quad \text{and} \quad \begin{cases} r = \sqrt{x^2 + y^2}, \\ \cos \theta = x/r, \\ \sin \theta = y/r. \end{cases}$$

Proof This result follows from elementary geometry (see Figure 6.2.2).   □

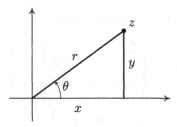

Figure 6.2.2

Corollary 6.2.5 *Let $z = (x, y) \in \mathbb{C}$ be non-zero with z not on the negative real axis, i.e. $\arg z \neq \pi$. Then*

$$
\arg z = \begin{cases}
\tan^{-1}(y/x) & \text{if } x > 0, \\
\dfrac{\pi}{2} - \tan^{-1}(x/y) & \text{if } y > 0, \\
-\dfrac{\pi}{2} - \tan^{-1}(x/y) & \text{if } y < 0.
\end{cases}
$$

Proof Let z have polar coordinates (r, θ), where $\theta = \arg z$, for $\theta \in (-\pi, \pi)$. We consider three cases.

- If $x > 0$ then $\theta \in (-\pi/2, \pi/2)$. By Theorem 6.2.4, $\tan \theta = y/x$. Hence $\theta = \tan^{-1}(y/x)$.

- If $y > 0$ then $\theta \in (0, \pi)$ so that $\pi/2 - \theta \in (-\pi/2, \pi/2)$. Using Lemma 5.2.2 and Theorem 6.2.4,
$$
\tan\left(\frac{\pi}{2} - \theta\right) = \cot \theta = \frac{x}{y}.
$$
Hence $\pi/2 - \theta = \tan^{-1}(x/y)$, *i.e.* $\theta = \pi/2 - \tan^{-1}(x/y)$.

- If $y < 0$ then $\theta \in (-\pi, 0)$ so that $-\pi/2 - \theta \in (-\pi/2, \pi/2)$. Since $\tan$ has period π,
$$
\tan\left(-\frac{\pi}{2} - \theta\right) = \tan\left(\frac{\pi}{2} - \theta\right) = \cot \theta = \frac{x}{y}.
$$
Hence $-\pi/2 - \theta = \tan^{-1}(x/y)$, *i.e.* $\theta = -\pi/2 - \tan^{-1}(x/y)$. $\square$

Example 6.2.6 Let $z = (x, y) \in \mathbb{C}$ be non-zero with z not on the negative real axis. Use Corollary 6.2.5 to prove that

$$
\arg \overline{z} = -\arg z
$$

Solution By definition, $\overline{z} = (x, -y)$.

If z lies on the positive real axis then so does $\overline{z}$. So $\arg z = 0$, $\arg \overline{z} = 0$ and the result follows.

Otherwise, $y \neq 0$ and, from Corollary 6.2.5,

$$\arg \overline{z} = \begin{cases} \dfrac{\pi}{2} - \tan^{-1}\left(\dfrac{x}{-y}\right) & \text{if } -y > 0, \\[2ex] -\dfrac{\pi}{2} - \tan^{-1}\left(\dfrac{x}{-y}\right) & \text{if } -y < 0 \end{cases}$$

$$= \begin{cases} \dfrac{\pi}{2} + \tan^{-1}\left(\dfrac{x}{y}\right) & \text{if } y < 0, \\[2ex] -\dfrac{\pi}{2} + \tan^{-1}\left(\dfrac{x}{y}\right) & \text{if } y > 0 \end{cases}$$

$$= -\arg z,$$

completing the proof. □

Definition 6.2.7 For $\theta \in \mathbb{R}$, we define

$$e^{i\theta} = \cos\theta + i\sin\theta.$$

This is called a *complex exponential*.

Notes 6.2.8 (1) The complex number $e^{i\theta}$ has coordinates $(\cos\theta, \sin\theta)$, as shown in Figure 6.2.3. Since $|e^{i\theta}|^2 = \cos^2\theta + \sin^2\theta = 1$, $e^{i\theta}$ lies on the unit circle. Further, by the definition of cos and sin in Section 5.1, θ is a determination of $\operatorname{Arg} e^{i\theta}$.

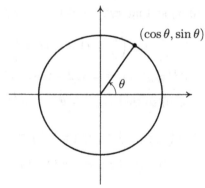

Figure 6.2.3

(2) Observe that $e^{i\theta} = (\cos\theta, \sin\theta)$ and $e^{i\phi} = (\cos\phi, \sin\phi)$ are the same point if and only if $\phi = \theta + 2k\pi$ for some $k \in \mathbb{Z}$. In other words, $e^{i\theta} = e^{i\phi}$ if and only if θ and ϕ are determinations of the argument of the same complex number.

Theorem 6.2.9 *Let $\theta, \phi \in \mathbb{R}$. Then*

(1) $e^{i\theta}e^{i\phi} = e^{i(\theta+\phi)}$,

(2) $\overline{e^{i\theta}} = 1/e^{i\theta} = e^{-i\theta}$.

Proof (1) We have

$$e^{i\theta}e^{i\phi} = (\cos\theta + i\sin\theta)(\cos\phi + i\sin\phi),$$
$$= (\cos\theta\cos\phi - \sin\theta\sin\phi) + i(\sin\theta\cos\phi + \cos\theta\sin\phi),$$
$$= \cos(\theta + \phi) + i\sin(\theta + \phi),$$
$$= e^{i(\theta + \phi)}.$$

(2) Since $e^{i\theta}\overline{e^{i\theta}} = |e^{i\theta}|^2 = 1$, $\overline{e^{i\theta}} = 1/e^{i\theta}$. Also,

$$\overline{e^{i\theta}} = \cos\theta - i\sin\theta = \cos(-\theta) + i\sin(-\theta) = e^{-i\theta}, \qquad \square$$

Remark Theorem 6.2.9 shows that complex exponentials obeys two of the index laws for real exponentiation (see Section 15.2) and hence their definition is a 'natural' one.

Example 6.2.10 Express each of 1, -1, i and $-i$ as a complex exponential.

Solution From Figure 6.2.4 we see that $\text{Arg}(1) = 0$, $\text{Arg}(-1) = \pi$, $\text{Arg}(i) = \pi/2$ and $\text{Arg}(-i) = 3\pi/2$ or $-\pi/2$. Therefore

$$1 = e^{i0}, \quad -1 = e^{i\pi}, \quad i = e^{i\pi/2}, \quad -i = e^{3i\pi/2} = e^{-i\pi/2}.$$

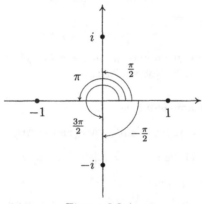

Figure 6.2.4

Alternatively, we may use Theorem 6.2.9 (1) to show that $-i = (-1)i = e^{i\pi}e^{i\pi/2} = e^{3i\pi/2}$ or, by Theorem 6.2.9 (2), $-i = \overline{i} = \overline{e^{i\pi/2}} = e^{-i\pi/2}$. $\qquad \square$

Lemma 6.2.11 *Let $z \in \mathbb{C}$ be non-zero. Then*

$$z = |z|e^{i\,\text{Arg}\,z},$$

for any determination of the argument.

Proof Let $|z| = r$ and let $\text{Arg}\,z = \theta$. Then, from Theorem 6.2.4,

$$z = (r\cos\theta) + i(r\sin\theta) = r(\cos\theta + i\sin\theta) = re^{i\theta},$$

as required. $\qquad \square$

Definition 6.2.12 For non-zero $z \in \mathbb{C}$,

$$|z|e^{i\,\mathrm{Arg}\,z},$$

for any determination of the argument, is called the *polar form* of z.

Example 6.2.13 Find the polar form of $z = -\sqrt{3} + i$.

Solution We have
$$|z| = \sqrt{(-\sqrt{3})^2 + 1^2} = 2$$
and, using Note 6.2.3 for the case $\mathrm{Re}\,z < 0$ and $\mathrm{Im}\,z > 0$,
$$\mathrm{Arg}\,z = \arg z = \pi - \tan^{-1}(1/\sqrt{3}) = 5\pi/6.$$

Thus $2e^{i5\pi/6}$ is the required polar form.   □

Theorem 6.2.14 *Let $z = re^{i\theta} \in \mathbb{C}$ where $r > 0$ and $\theta \in \mathbb{R}$. Then*
$$|z| = r \quad and \quad \mathrm{Arg}\,z = \theta.$$

Proof Since $r > 0$,
$$|z| = \sqrt{(r\cos\theta)^2 + (r\sin\theta)^2} = \sqrt{r^2(\cos^2\theta + \sin^2\theta)} = \sqrt{r^2} = r.$$

Also, by Lemma 6.2.11,
$$z = |z|e^{i\,\mathrm{Arg}\,z} = re^{i\,\mathrm{Arg}\,z}.$$

Comparing this with the given expression for z we get $e^{i\,\mathrm{Arg}\,z} = e^{i\theta}$ and hence θ is a determination of $\mathrm{Arg}\,z$. (See Notes 6.2.8 (2)).   □

Corollary 6.2.15 *For non-zero $z, w \in \mathbb{C}$,*
$$|zw| = |z||w| \quad and \quad \mathrm{Arg}(zw) = \mathrm{Arg}\,z + \mathrm{Arg}\,w.$$

Proof By Lemma 6.2.11 we have $z = |z|e^{i\,\mathrm{Arg}\,z}$ and $w = |w|e^{i\,\mathrm{Arg}\,w}$. So
$$zw = |z||w|e^{i\,\mathrm{Arg}\,z}e^{i\,\mathrm{Arg}\,w}$$
$$= |z||w|e^{i(\mathrm{Arg}\,z + \mathrm{Arg}\,w)} \quad \text{(by Theorem 6.2.9)},$$

in which $|z||w| > 0$. From Theorem 6.2.14, we may now deduce the required expressions for the modulus and argument of zw.   □

Remark The above results generalise to any finite number of terms as follows. For $n \in \mathbb{N}$ and non-zero $z_1, z_2, \ldots, z_n \in \mathbb{C}$,
$$|z_1 z_2 \ldots z_n| = |z_1||z_2| \ldots |z_n|,$$
$$\mathrm{Arg}(z_1 z_2 \ldots z_n) = \mathrm{Arg}\,z_1 + \mathrm{Arg}\,z_2 + \cdots + \mathrm{Arg}\,z_n,$$
and, in particular, for $n \in \mathbb{N}$ and non-zero $z \in \mathbb{C}$,
$$|z^n| = |z|^n \quad and \quad \mathrm{Arg}\,z^n = n\,\mathrm{Arg}\,z.$$

The result for the argument of z^n will be proved in Corollary 6.3.2.

Example 6.2.16 Let $z \in \mathbb{C}$ be non-zero and let $x \in \mathbb{R}$ be non-zero. Express the modulus and argument of xz in terms of x and the modulus and argument of z.

Solution Let $|z| = r$ and $\text{Arg } z = \theta$. Then

$$xz = (xr)e^{i\theta}.$$

There are two cases.

    Case 1. ($x > 0$)   Here $xr > 0$. So, by Theorem 6.2.14, $|xz| = xr$ and $\text{Arg}(xz) = \theta$.

    Case 2. ($x < 0$)   Here $-xr > 0$. Also, from Theorem 6.2.9 (1), we have

$$-e^{i\theta} = e^{i\pi}e^{i\theta} = e^{i(\theta+\pi)}.$$

So

$$xz = (-xr)(-e^{i\theta}) = (-xr)e^{i(\theta+\pi)}.$$

Theorem 6.2.14 now gives $|xz| = -xr$ and $\text{Arg}(xz) = \theta + \pi$. $\square$

6.3   De Moivre's Theorem and Trigonometry

Theorem 6.3.1 (De Moivre's Theorem) *For $\theta \in \mathbb{R}$ and $n \in \mathbb{Z}$,*

$$(e^{i\theta})^n = e^{in\theta}, \tag{6.3.1}$$

i.e.

$$(\cos\theta + i\sin\theta)^n = \cos n\theta + i\sin n\theta. \tag{6.3.2}$$

Proof First we note that the result is true when $n = 0$, since the left hand side of (6.3.1) is 1 by convention and the right hand side is $e^{i0} = \cos 0 + i\sin 0 = 1$.

    For integers $n \geq 1$ we establish (6.3.1) by induction. For $n = 1$,

$$\text{LHS} = (e^{i\theta})^1 = e^{i\theta} = \text{RHS}.$$

We now suppose that (6.3.1) holds for some $n = k \geq 1$ and then

$$\begin{aligned}
(e^{i\theta})^{k+1} &= (e^{i\theta})^k(e^{i\theta})^1 \\
&= e^{ik\theta}e^{i\theta} \quad \text{(by induction hypothesis)} \\
&= e^{i(k+1)\theta} \quad \text{(by Theorem 6.2.9)},
\end{aligned}$$

which shows that (6.3.1) then holds for $n = k + 1$. Hence, by induction, (6.3.1) holds for all $n \in \mathbb{N}$.

    For the remaining integers $n < 0$, let $m = -n$ so that $m \in \mathbb{N}$. Then

$$(e^{i\theta})^n = (e^{i\theta})^{-m} = 1/(e^{i\theta})^m.$$

As $m \in \mathbb{N}$, we may write this as

$$1/e^{im\theta} = e^{-im\theta} = e^{in\theta},$$

as required. $\square$

Remarks (1) Equation (6.3.2) leads to a number of applications in trigonometry. For $n \in \mathbb{N}$, the left hand side can be expanded using the (complex form of the) binomial theorem which is stated as follows. Let $n \in \mathbb{N}$ and $z, w \in \mathbb{C}$. Then

$$(z + w)^n = \sum_{r=0}^{n} \binom{n}{r} z^{n-r} w^r.$$

(2) The following simple consequence of de Moivre's theorem can be useful. For $\theta \in \mathbb{R}$ and $n \in \mathbb{Z}$,

$$(\cos \theta + i \sin \theta)^{-n} = \cos(-n\theta) + i \sin(-n\theta) = \cos n\theta - i \sin n\theta. \qquad (6.3.3)$$

Corollary 6.3.2 *Let $z \in \mathbb{C}$, z non-zero with polar form $re^{i\theta}$ and let $n \in \mathbb{Z}$. Then*

$$z^n = r^n e^{in\theta}.$$

Proof We have $z^n = (re^{i\theta})^n = r^n(e^{i\theta})^n$, then using de Moivre's theorem $(e^{i\theta})^n = e^{in\theta}$ and the result follows. $\qquad \square$

Example 6.3.3 Find the real and imaginary parts of $(i - \sqrt{3})^7$.

Solution Let $z = -\sqrt{3} + i$. The polar form of z is $2e^{i5\pi/6}$ (see Example 6.2.13). We have

$$z^7 = 2^7 e^{i7(5\pi/6)} \quad \text{(by Corollary 6.3.2)}$$
$$= 2^7 e^{i35\pi/6} = 2^7 e^{i(6\pi - \pi/6)} = 2^7 e^{-i\pi/6}$$
$$= 2^7(\cos(-\pi/6) + i\sin(-\pi/6)) = 2^6(\sqrt{3} - i).$$

Therefore $\operatorname{Re} z^7 = 64\sqrt{3}$ and $\operatorname{Im} z^7 = -64$. $\qquad \square$

Example 6.3.4 Express $\cos 5\theta$ as a polynomial in $\cos \theta$.

Solution By de Moivre's theorem, $\cos 5\theta + i \sin 5\theta = (\cos \theta + i \sin \theta)^5$. Using the shorthand notation $c = \cos \theta$ and $s = \sin \theta$, from the binomial theorem we have

$$\cos 5\theta + i \sin 5\theta = (c + is)^5,$$
$$= c^5 + 5c^4(is) + 10c^3(is)^2 + 10c^2(is)^3 + 5c(is)^4 + (is)^5,$$
$$= c^5 - 10c^3 s^2 + 5cs^4 + i(5c^4 s - 10c^2 s^3 + s^5).$$

Equating real parts, we get

$$\cos 5\theta = c^5 - 10c^3 s^2 + 5cs^4$$
$$= c^5 - 10c^3(1 - c^2) + 5c(1 - c^2)^2$$
$$= 16\cos^5 \theta - 20\cos^3 \theta + 5\cos \theta, \qquad (6.3.4)$$

as required.

As a check we observe that for $\theta = 0$, the left hand side of (6.3.4) equals 1 and the right hand side of (6.3.4) equals $16 - 20 + 5 = 1$ also. $\qquad \square$

Example 6.3.5 Express $\tan 3\theta$ in terms of $\tan\theta$.

Solution We first determine $\cos 3\theta$ and $\sin 3\theta$. Using the notation $c = \cos\theta$ and $s = \sin\theta$, we get, from de Moivre's theorem and the binomial theorem,

$$\cos 3\theta + i\sin 3\theta = (c + is)^3 = c^3 + 3ic^2s - 3cs^2 - is^3.$$

Then, equating real and imaginary parts,

$$\cos 3\theta = c^3 - 3cs^2 \quad \text{and} \quad \sin 3\theta = 3c^2s - s^3.$$

Hence

$$\tan 3\theta = \frac{\sin 3\theta}{\cos 3\theta} = \frac{3c^2s - s^3}{c^3 - 3cs^2} = \frac{3(s/c) - (s/c)^3}{1 - 3(s/c)^2}$$
$$= \frac{3\tan\theta - \tan^3\theta}{1 - 3\tan^2\theta}$$

as required. □

Problem 6.3.6 Let $n \in \mathbb{N}$. By considering $(1 + e^{2i\theta})^n$, establish the identity

$$\tan n\theta = \frac{\displaystyle\sum_{r=0}^{n} \binom{n}{r} \sin 2r\theta}{\displaystyle\sum_{r=0}^{n} \binom{n}{r} \cos 2r\theta}.$$

For which $\theta \in \mathbb{R}$ is this identity valid?

Theorem 6.3.7 *Let $z = e^{i\theta}$, where $\theta \in \mathbb{R}$ and let $n \in \mathbb{Z}$. Then*

$$2\cos n\theta = z^n + z^{-n} \quad \text{and} \quad 2i\sin n\theta = z^n - z^{-n}.$$

Proof By de Moivre's theorem,

$$z^n = \cos n\theta + i\sin n\theta, \tag{6.3.5}$$

and, from (6.3.3),

$$z^{-n} = \cos n\theta - i\sin n\theta. \tag{6.3.6}$$

Adding and subtracting (6.3.5) and (6.3.6) gives

$$z^n + z^{-n} = 2\cos n\theta \quad \text{and} \quad z^n - z^{-n} = 2i\sin n\theta,$$

respectively. □

Example 6.3.8 Express $\sin^6\theta$ in the form

$$a\cos 6\theta + b\cos 4\theta + c\cos 2\theta + d,$$

where $a, b, c, d \in \mathbb{R}$.

Solution Let $z = e^{i\theta}$, $\theta \in \mathbb{R}$. Then, using Theorem 6.3.7, for $n = 1$, we have

$$(2i\sin\theta)^6 = (z - z^{-1})^6$$
$$= z^6 - 6z^5z^{-1} + 15z^4z^{-2} - 20z^3z^{-3} + 15z^2z^{-4} - 6zz^{-5} + z^{-6}$$
$$= (z^6 + z^{-6}) - 6(z^4 + z^{-4}) + 15(z^2 + z^{-2}) - 20.$$

Using Theorem 6.3.7, for $n = 6$, 4 and 2 we get

$$-2^6\sin^6\theta = (2\cos 6\theta) - 6(2\cos 4\theta) + 15(2\cos 2\theta) - 20,$$

and so

$$\sin^6\theta = -\frac{1}{32}\cos 6\theta + \frac{3}{16}\cos 4\theta - \frac{15}{32}\cos 2\theta + \frac{5}{16}. \qquad \square$$

Example 6.3.9 Express $\sin^3\theta\cos^5\theta$ in terms of sines of multiples of θ.

Solution Let $z = e^{i\theta}$, $\theta \in \mathbb{R}$. Then

$$(2i\sin\theta)^3(2\cos\theta)^5 = (z - z^{-1})^3(z + z^{-1})^5,$$
$$= (z - z^{-1})^3(z + z^{-1})^3(z + z^{-1})^2,$$
$$= (z^2 - z^{-2})^3(z + z^{-1})^2,$$
$$= (z^6 - 3z^2 + 3z^{-2} - z^{-6})(z^2 + 2 + z^{-2}),$$
$$= (z^8 - z^{-8}) + 2(z^6 - z^{-6}) - 2(z^4 - z^{-4}) - 6(z^2 - z^{-2}).$$

Therefore

$$\sin^3\theta\cos^5\theta = -\frac{1}{128}\sin 8\theta - \frac{1}{64}\sin 6\theta + \frac{1}{64}\sin 4\theta + \frac{3}{64}\sin 2\theta. \qquad \square$$

Problem 6.3.10 Let $S = z + z^2 + z^3 + \cdots + z^n$, where $z \in \mathbb{C}$ and $n \in \mathbb{N}$. By considering $S - zS$, show that, when $z \neq 1$,

$$S = \frac{z(1 - z^n)}{1 - z}.$$

Deduce that

$$|\sin 1 + \sin 2 + \sin 3 + \cdots + \sin n| \leq M,$$

where M is independent of n. This result will help with Problem 18.3.25.

6.4   Complex Polynomials

In this section we will consider $\mathbb{C}[z]$, the set of polynomials in z with complex coefficients. The ultimate goal will be a justification of the claim, made in Section 3.4, that all real polynomials may be factorised into linear and irreducible quadratic polynomials.

Theorem 6.4.1 (The Fundamental Theorem of Algebra) *Let $f(z) \in \mathbb{C}[z]$ be non-zero with $\deg f(z) > 0$. Then $f(z) = 0$ has a root $\alpha \in \mathbb{C}$. Equivalently, $f(z)$ has a linear factor $z - \alpha$.*

Proof The proof of this is beyond the scope of this book.   □

Theorem 6.4.2 *Let $f(z) \in \mathbb{C}[z]$ be non-zero with $\deg f(z) = n > 0$. Then there exist $\lambda, \alpha_1, \alpha_2, \ldots, \alpha_n \in \mathbb{C}$ such that*

$$f(z) = \lambda(z - \alpha_1)(z - \alpha_2)\ldots(z - \alpha_n). \tag{6.4.1}$$

and the polynomial $f(z)$ has at most n distinct zeros.

Proof The proof is by induction. For $n = 1$, $f(z) = az + b$ for some $a, b \in \mathbb{C}$ where $a \neq 0$. Thus we may write

$$f(z) = \lambda(z - \alpha), \quad \text{where } \lambda = a \text{ and } \alpha = -\frac{b}{a}.$$

Hence (6.4.1) holds for $n = 1$.

The induction hypothesis is that (6.4.1) holds for all polynomials of degree $k \geq 1$. Now consider a polynomial $f(z)$ with degree $k + 1$. By the fundamental theorem of algebra there exist $\alpha \in \mathbb{C}$ such that $z - \alpha$ is a factor of $f(z)$, *i.e.*

$$f(z) = (z - \alpha)g(z), \tag{6.4.2}$$

where $g(z)$ is a polynomial of degree k. By the inductive hypothesis we have

$$g(z) = \lambda(z - \alpha_1)(z - \alpha_2)\ldots(z - \alpha_k), \tag{6.4.3}$$

for some constants $\lambda, \alpha_1, \alpha_2, \ldots, \alpha_k \in \mathbb{C}$. Combining (6.4.2) and (6.4.3), we obtain

$$f(z) = \lambda(z - \alpha_1)(z - \alpha_2)\ldots(z - \alpha_k)(z - \alpha),$$

so that (6.4.1) holds for the degree $k + 1$ polynomial $f(z)$.

Hence, by induction, (6.4.1) holds for all polynomials of degree n where $n \in \mathbb{N}$.

Further, from (6.4.1),

$$\begin{aligned}
f(\alpha) = 0 &\Leftrightarrow \lambda(\alpha - \alpha_1)(\alpha - \alpha_2)\ldots(\alpha - \alpha_n) = 0, \\
&\Leftrightarrow (\alpha - \alpha_1)(\alpha - \alpha_2)\ldots(\alpha - \alpha_n) = 0, \text{ as } \ell(f) \neq 0, \\
&\Leftrightarrow \alpha = \alpha_i, \text{ for some } i = 1, 2, \ldots, n.
\end{aligned}$$

Since the α_i are not necessarily distinct, there are at most n distinct zeros.   □

Notes 6.4.3 (1) Since $\mathbb{R} \subset \mathbb{C}$, $\mathbb{R}[z] \subset \mathbb{C}[z]$ and so the above results apply to real polynomials in particular.

(2) The long division process and results on factorisation of polynomials in $\mathbb{R}[z]$, developed in Section 3.2, also apply to polynomials in $\mathbb{C}[z]$.

Theorem 6.4.4 *Let $f(z) \in \mathbb{R}[z]$ be non-constant and let $\alpha \in \mathbb{C}$ be a zero of $f(z)$. Then $\overline{\alpha}$ is a zero of $f(z)$.*

Proof Suppose that $\deg f(z) = n > 0$ (since $f(x)$ is non-constant) then

$$f(z) = c_n z^n + \cdots + c_1 z + c_0,$$

for some coefficients $c_i \in \mathbb{R}$. If α is a zero of $f(z)$ then $f(\alpha) = 0$ and consequently, $\overline{f(\alpha)} = 0$. Now

$$\begin{aligned}
0 = \overline{f(\alpha)} &= \overline{c_n \alpha^n + \cdots + c_1 \alpha + c_0}, \\
&= \overline{c_n}\, \overline{\alpha}^n + \cdots + \overline{c_1}\, \overline{\alpha} + \overline{c_0}, \\
&= c_n \overline{\alpha}^n + \cdots + c_1 \overline{\alpha} + c_0, \text{ as } c_i \in \mathbb{R}, \\
&= f(\overline{\alpha}).
\end{aligned}$$

Thus $\overline{\alpha}$ is a zero of $f(z)$. $\square$

Corollary 6.4.5 *Let $f(z) \in \mathbb{R}[z]$ and let $\alpha \in \mathbb{C}$ be a non-real zero of $f(z)$. Then $f(z)$ has a real quadratic factor*

$$(z - \alpha)(z - \overline{\alpha}) = z^2 - 2(\operatorname{Re}\alpha)z + |\alpha|^2.$$

Proof By Theorem 6.4.4, $\overline{\alpha}$ is a zero and moreover, as α is not real, $\overline{\alpha} \neq \alpha$. Thus, $f(z)$ has factors $(z - \alpha)$ and $(z - \overline{\alpha})$ and hence the quadratic factor

$$(z - \alpha)(z - \overline{\alpha}) = z^2 - (\alpha + \overline{\alpha})z + \alpha\overline{\alpha} = z^2 - 2(\operatorname{Re}\alpha)z + |\alpha|^2,$$

which has real coefficients. $\square$

Example 6.4.6 Given that $-2 + 3i$ is a zero of

$$f(z) = z^4 + 7z^2 - 12z + 130,$$

find all zeros of $f(z)$ in $\mathbb{C}$.(

Solution Since $-2 + 3i$ is a zero, $\overline{-2 + 3i} = -2 - 3i$ is also a zero and so

$$\begin{aligned}
(z - (-2 + 3i))(z - (-2 - 3i)) &= ((z + 2) - 3i)((z + 2) + 3i) \\
&= (z + 2)^2 + 3^2 = z^2 + 4z + 13,
\end{aligned}$$

is a factor of $f(z)$.

By long division (or by inspection) we find that

$$f(z) = (z^2 + 4z + 13)(z^2 - 4z + 10),$$

and the remaining factor, $z^2 - 4z + 10$, has discriminant $\Delta = -24 < 0$ and hence zeros

$$\frac{4 \pm i\sqrt{24}}{2} = 2 \pm i\sqrt{6}.$$

Thus the zeros are $-2 \pm 3i$ and $2 \pm i\sqrt{6}$. $\square$

Theorem 6.4.7 *Every real non-constant polynomial can be factorised into real linear factors and real quadratic factors.*

Proof Let $f(z) \in \mathbb{R}[z]$ have degree $n > 0$. By Theorem 6.4.2 f has n (complex) linear factors $z - \alpha_1, \ldots, z - \alpha_n$ where $\alpha_1, \ldots, \alpha_n \in \mathbb{C}$ are the zeros of f.

These factors are of two types. For each α_i which is real, $z - \alpha_i$ is a real linear factor of $f(z)$. For each α_i which non-real, $\overline{\alpha}_i$ is also a zero (Theorem 6.4.4) and hence $\overline{\alpha}_i = \alpha_j$, for some $j = 1, \ldots, n$ and $j \neq i$. The product of the pair of factors $z - \alpha_i$ and $z - \overline{\alpha}_i$ gives a real quadratic factor (see Corollary 6.4.5). $\qquad\square$

6.5   Roots of Unity

We now consider the zeros of one particular type of complex polynomial

$$f(z) = z^n - a,$$

where $a \in \mathbb{C}$. From Theorem 6.4.2, $f(z)$ has at most n distinct zeros in $\mathbb{C}$ and we will see that if $a \neq 0$ then $f(z)$ has precisely n distinct zeros. Since each is a root of the equation

$$z^n = a,$$

they are called the *complex n^{th} roots of a*.

Example 6.5.1 Let $\omega = e^{i2\pi/7}$. Mark the complex numbers ω^0, ω^1, ω^2, ..., ω^7 on a sketch of the complex plane. Express ω^{12} in terms of one of these powers of ω.

Solution By de Moivre's theorem, $\omega^k = e^{i2k\pi/7}$ for $k \in \mathbb{Z}$ and hence

$$|\omega^k| = 1 \quad \text{and} \quad \operatorname{Arg} \omega^k = \frac{2k\pi}{7}.$$

Thus $\omega^0 = e^{i0} = 1$, $\omega^7 = e^{i2\pi} = 1$ and $\omega^1, \omega_2, \ldots, \omega^6$ are evenly placed around the unit circle in the complex plane as shown in Figure 6.5.1.

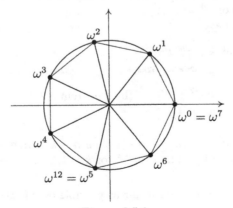

Figure 6.5.1

Also $\omega^{12} = \omega^7 \omega^5 = 1 \times \omega^5 = \omega^5$. $\qquad\square$

Remarks (1) Example 6.5.1 demonstrates that raising $e^{i\alpha}$ to the kth power is equivalent to rotating 1 (on the unit circle) anticlockwise though an angle $k\alpha$.

(2) The points $\omega^0, \omega^1, \ldots, \omega^6$ are the vertices of a regular septagon.

Theorem 6.5.2 *The complex n^{th} roots of 1 are the n distinct numbers*

$$e^{i2k\pi/n} \quad (k = 0, 1, 2, \ldots, n-1).$$

[These are also called the nth *roots of unity*.]

Proof The points $\omega^0, \omega^1, \ldots, \omega^{n-1}$ lie on the unit circle and are each separated from the next by the angle $2\pi/n$. Hence they are distinct.

Also, for each $k \in \mathbb{Z}$, using de Moivre's theorem $(e^{i2k\pi/n})^n = e^{i2k\pi} = 1$ and so ω^0, $\omega^1, \ldots, \omega^{n-1}$ are n distinct zeros of $z^n - 1$. Since, by Theorem 6.4.2, $z^n - 1$ has at most n distinct zeros, we have all the roots. $\qquad\square$

Definition 6.5.3 An nth root of unity α such that $\alpha^k \neq 1$ for $k = 1, \ldots, n-1$ is called a *primitive n^{th} root of unity*.

With the notation $\omega = e^{i2\pi/n}$ the nth roots of unity are $\omega^0, \omega^1, \ldots, \omega^{n-1}$. Hence ω is a primitive root.

Example 6.5.4 Find the primitive 6th roots of unity.

Solution Let $\omega = e^{i2\pi/6} = e^{i\pi/3}$. The 6th roots of unity are $1, \omega, \omega^2, \omega^3, \omega^4, \omega^5$. Clearly 1 is not primitive and ω is primitive. Now $(\omega^2)^3 = \omega^6 = 1$, $(\omega^3)^2 = \omega^6 = 1$ and $(\omega^4)^3 = (\omega^6)^2 = 1$ and so ω^2, ω^3 and ω^4 are not primitive. Finally, $(\omega^5)^2 = \omega^{10} = \omega^4$, $(\omega^5)^3 = \omega^{15} = \omega^3$, $(\omega^5)^4 = \omega^{20} = \omega^2$ and $(\omega^5)^5 = \omega^{25} = \omega$ and so ω^5 is primitive.

Hence the primitive 6th roots of unity are $e^{i\pi/3}$ and $e^{i5\pi/3}$. $\qquad\square$

Note 6.5.5 Each ω^k is a zero of the real polynomial $z^n - 1$. Consequently, $\overline{\omega^k}$ is also a root, *i.e.* it is one of $\omega^0, \omega^1, \ldots, \omega^{n-1}$.

For all $k \in \mathbb{Z}$ we have

$$
\begin{aligned}
\overline{\omega^k} &= e^{-i2k\pi/n} \\
&= e^{i2n\pi/n} e^{-i2k\pi/n} \quad (\text{since } e^{i2n\pi/n} = e^{i2\pi} = 1) \\
&= e^{i2(n-k)\pi/n} \quad (\text{by Theorem 6.2.9}) \\
&= \omega^{n-k}.
\end{aligned}
$$

Observe that if $k = 0, 1, \ldots, n-2, n-1$ then $n - k = n, n-1, \ldots, 2, 1$, respectively. Hence, $\overline{\omega^0} = \omega^n = \omega^0 = 1$ and $\overline{\omega^k} = \omega^{n-k}$ for $k = 1, 2, \ldots, n-1$.

Example 6.5.6 Factorise $z^5 - 1$ as far as possible over $\mathbb{R}[z]$.

Solution The zeros of $z^5 - 1$ are ω^k for $k = 0, 1, 2, 3, 4$, where $\omega = e^{i2\pi/5}$. We have $\omega^0 = 1$, giving the real linear factor $z - 1$. Also ω is non-real and has complex conjugate $\overline{\omega} = \omega^4$. Hence,

$$(z - \omega)(z - \omega^4) = z^2 - 2(\text{Re}\,\omega)z + |\omega|^2 = z^2 - 2\cos(2\pi/5)z + 1$$

is an irreducible real quadratic factor. Similarly, the root ω^2 is not real and is such that $\overline{\omega^2} = \omega^3$ and gives the irreducible factor

$$
\begin{aligned}
(z - \omega^2)(z - \omega^3) &= z^2 - 2(\operatorname{Re}\omega^2)z + |\omega^2|^2 \\
&= z^2 - 2(\operatorname{Re} e^{i4\pi/5})z + 1 \\
&= z^2 - 2\cos(4\pi/5)z + 1.
\end{aligned}
$$

Hence the full factorisation of $z^5 - 1$ over $\mathbb{R}[z]$ is

$$
z^5 - 1 = (z - 1)(z^2 - 2\cos(2\pi/5)z + 1)(z^2 - 2\cos(4\pi/5)z + 1). \qquad \square
$$

Theorem 6.5.7 *Let $\xi \neq 1$ be an nth root of unity then*

$$
1 + \xi + \xi^2 + \cdots + \xi^{n-1} = 0.
$$

Proof We have the factorisation (see Lemma 3.2.8 and Note 6.4.3 (2))

$$
z^n - 1 = (z - 1)(z^{n-1} + z^{n-2} + \cdots + z + 1).
$$

Since $\xi \neq 1$ it follows that

$$
\xi^n - 1 = 0 \iff 1 + \xi + \xi^2 + \cdots + \xi^{n-1} = 0
$$

as required. $\qquad \square$

Example 6.5.8 Let $\omega = e^{i2\pi/5}$ and define $\alpha = \omega + \omega^4$ and $\beta = \omega^2 + \omega^3$. Find a polynomial whose roots are α and β and deduce the value of $\cos(2\pi/5)$.

Solution The quadratic

$$
(z - \alpha)(z - \beta) = z^2 - (\alpha + \beta)z + \alpha\beta
$$

has roots α and β. Now

$$
\begin{aligned}
\alpha + \beta &= \omega + \omega^4 + \omega^2 + \omega^3 \\
&= -1 \quad \text{(by Theorem 6.5.7)}
\end{aligned}
$$

and

$$
\begin{aligned}
\alpha\beta &= (\omega + \omega^4)(\omega^2 + \omega^3) \\
&= \omega^3 + \omega^4 + \omega^6 + \omega^7 \\
&= \omega^3 + \omega^4 + \omega + \omega^2 \quad \text{(since } \omega^5 = 1) \\
&= -1 \quad \text{(by Theorem 6.5.7).}
\end{aligned}
$$

Therefore the required polynomial is $z^2 + z - 1$.

From Example 6.5.6, we have $\alpha = 2\cos(2\pi/5)$ and, since $2\pi/5 \in (0, \pi/2)$, $\alpha > 0$. Thus α is a positive root of $z^2 + z - 1$. The roots of this polynomial may also be determined as

$$
\frac{-1 + \sqrt{5}}{2} > 0 \quad \text{and} \quad \frac{-1 - \sqrt{5}}{2} < 0.
$$

Hence $\alpha = (\sqrt{5} - 1)/2$. Thus

$$
\cos\frac{2\pi}{5} = \frac{\sqrt{5} - 1}{4}. \qquad \square
$$

Theorem 6.5.9 *Let $a \in \mathbb{C}$ be non-zero and polar form $re^{i\theta}$ and let $n \in \mathbb{N}$. Then the complex n^{th} roots of a are*

$$\alpha_k = r^{1/n} e^{i(\theta + 2k\pi)/n}, \quad (k = 0, 1, \ldots, n-1).$$

Proof For all $k \in \mathbb{Z}$, we have

$$\alpha_k^n = (r^{1/n})^n e^{i(\theta + 2k\pi)} \quad \text{(by de Moivre's theorem)}$$
$$= re^{i\theta} = a.$$

So α_k is an nth root of a.

Also,

$$\alpha_j = \alpha_k \iff r^{1/n} e^{i(\theta + 2j\pi)/n} = r^{1/n} e^{i(\theta + 2k\pi)/n}$$
$$\iff e^{i2(j-k)\pi/n} = 1 \quad \text{(by Theorem 6.2.9)}$$
$$\iff (j-k)/n \in \mathbb{Z}.$$

If j and k are integers in the interval $[0, n-1]$ then

$$(j-k)/n \in \mathbb{Z} \iff (j-k)/n = 0$$

and so $\alpha_j = \alpha_k \iff j = k$. Hence the roots α_k for $k = 0, 1, \ldots, n-1$ are distinct and so by Theorem 6.4.2 these are all of the roots. $\square$

Example 6.5.10 Find the complex 4th roots of $2(\sqrt{3} - i)$.

Solution Let $2(\sqrt{3} - i) = re^{i\theta}$. So

$$r = \sqrt{(2\sqrt{3})^2 + (-2)^2} = 4 \quad \text{and} \quad \theta = \tan^{-1}(-1/\sqrt{3}) = -\pi/6.$$

Hence the 4th roots of $2(\sqrt{3} - i)$ are

$$\alpha_k = 4^{1/4} e^{i(-\pi/6 + 2k\pi)/4} \quad (k = 0, 1, 2, 3)$$
$$= \sqrt{2} e^{i(12k-1)\pi/24} \quad (k = 0, 1, 2, 3),$$

i.e. $\sqrt{2}e^{-i\pi/24}$, $\sqrt{2}e^{i11\pi/24}$, $\sqrt{2}e^{i23\pi/24}$ and $\sqrt{2}e^{i35\pi/24}$. $\square$

6.6   Rigid Transformations of the Plane

Since the time of Euclid, geometers have been interested in *rigid transformations* of the plane. These are mappings of the plane which preserve distance. Familiar examples are *rotations* and *translations*.

We shall see that, if t is a rigid transformation and P is a point, then the coordinates of $t(P)$ are related to those of P. Further, if $\mathcal{C}$ is a plane curve, then we can derive an equation for the transformed curve, $t(\mathcal{C})$, from an equation for $\mathcal{C}$.

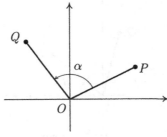

Figure 6.6.1

Rotations about the origin

Definition 6.6.1 For $\alpha \in \mathbb{R}$, *rotation through α about O* is the mapping $\rho_\alpha : \mathbb{R}^2 \to \mathbb{R}^2$ which maps O to O and $P \neq O$ to the point Q obtained by rotating OP (anticlockwise) through angle α about O (see Figure 6.6.1).

Note that since we are rotating rigidly about O, $|OQ| = |OP|$.

Although we could derive the results that follow using purely geometrical techniques, it is easier to exploit the fact that points may be regarded as complex numbers. We derive relationships between complex numbers and then deduce the relationships between the corresponding points.

Theorem 6.6.2 *Let $\alpha \in \mathbb{R}$ and $P(x, y)$, $Q(X, Y)$ be points such that $Q = \rho_\alpha(P)$.*

(1) *If P corresponds to $z \in \mathbb{C}$, then Q corresponds to $Z = e^{i\alpha}z$.*

(2) *The relationships between the coordinates of P and Q are*

$$X = x \cos \alpha - y \sin \alpha$$
$$Y = x \sin \alpha + y \cos \alpha \tag{6.6.1}$$

and

$$x = X \cos \alpha + Y \sin \alpha$$
$$y = -X \sin \alpha + Y \cos \alpha \tag{6.6.2}$$

Proof (1) If $P = O$, then $Q = O$ by the definition of ρ_α. Then $z = Z = 0$ and so $Z = e^{i\alpha}z$ as required.

If $P \neq O$, then z has polar form $re^{i\theta}$, where $r = |OP|$ and $\theta = \arg(z)$. Since we are rotating about O, $|OQ| = |OP| = r$. From Figure 6.6.2, $\arg(Z) = \alpha + \theta$, so

$$\begin{aligned} Z &= re^{i(\alpha+\theta)} \\ &= re^{i\alpha}e^{i\theta} \qquad \text{(by Theorem 6.3.1)} \\ &= e^{i\alpha}(re^{i\theta}) = e^{i\alpha}z, \end{aligned}$$

as required.

(2) The points P, Q correspond to complex numbers $z = x + iy$ and $Z = X + iY$ respectively. Also, $e^{i\theta} = \cos \alpha + i \sin \alpha$. Taking real and imaginary in $Z = e^{i\alpha}z$ gives (6.6.1).

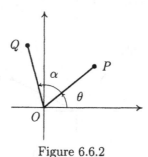

Figure 6.6.2

While we *could* obtain (6.6.2) from (6.6.1) by solving for x, y in terms of X, Y, it is more elegant to observe that we may rewrite $Z = e^{i\alpha}z$ as

$$z = e^{-i\alpha}Z.$$

Since $\cos(-\alpha) = \cos\alpha$ and $\sin(-\alpha) = -\sin\alpha$, the real imaginary parts of this give (6.6.2). $\qquad\square$

Equations (6.6.1) are used in the obvious way—to obtain the coordinates of a point obtained by rotation about O.

Example 6.6.3 Let P be the point $(1, 1)$. Find a point Q such that $\triangle OPQ$ is equilateral.

Solution By elementary geometry,

$$\triangle OPQ \text{ is equilateral} \iff |OP| = |OQ|, \text{ and } \angle POQ \text{ has magnitude } \pi/3.$$

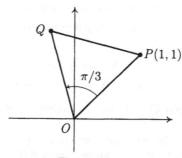

Figure 6.6.3

Figure 6.6.3 shows the case $\angle POQ = +\pi/3$. Then $Q = \rho_{\pi/3}(P)$. [Note that there is a second solution Q', where $Q' = \rho_{-\pi/3}(P)$.]

By (6.6.1), Q has coordinates (X, Y) where

$$X = 1 \times \cos(\pi/3) - 1 \times \sin(\pi/3) = \tfrac{1}{2}(1 - \sqrt{3})$$
$$Y = 1 \times \sin(\pi/3) + 1 \times \cos(\pi/3) = \tfrac{1}{2}(1 + \sqrt{3})$$

i.e. Q has coordinates $(\tfrac{1}{2}(1 - \sqrt{3}), \tfrac{1}{2}(1 + \sqrt{3}))$. $\qquad\square$

Equations (6.6.2) are used to find an equation for a curve obtained by rotation about O.

Example 6.6.4 Let $\mathcal{L}$ be the line $y + x = 1$. Find an equation for $\rho_{\pi/2}(\mathcal{L})$.

Solution Consider a point $Q(X,Y) \in \rho_{\pi/2}(\mathcal{L})$, i.e. $Q = \rho_{\pi/2}(P)$ for some $P(x,y) \in \mathcal{L}$. We have

$$
\begin{aligned}
Q \in \rho_{\pi/2}(\mathcal{L}) \quad &\Leftrightarrow \quad P \in \mathcal{L} \\
&\Leftrightarrow \quad y + x = 1 &&\text{(since } \mathcal{L} \text{ is } y + x = 1) \\
&\Leftrightarrow \quad (-X \times 1 + Y \times 0) + (X \times 0 + Y \times 1) = 1 &&\text{(using (6.6.2))} \\
&\Leftrightarrow \quad Y - X = 1.
\end{aligned}
$$

Reverting to the coordinates x, y, $\rho_{\pi/2}(\mathcal{L})$ is the line $y - x = 1$. $\square$

Remark The angle between the lines $\mathcal{L} : y + x = 1$ and $\rho_{\pi/2}(\mathcal{L}) : y - x = 1$ is $\pi/2$, i.e. these lines are perpendicular. We observe that the product of their gradients is $-1 \times 1 = -1$ as expected.

Example 6.6.5 Find an equation for the curve obtained by rotating $xy = 1$ through $-\pi/4$.

Solution From (6.6.2), the rotation $\rho_{-\pi/4}$ maps $P(x,y)$ to $Q(X,Y)$ where

$$
\begin{aligned}
x &= (1/\sqrt{2})X + (-1/\sqrt{2})Y \\
y &= -(-1/\sqrt{2})X + (1/\sqrt{2})Y.
\end{aligned}
$$

Let $\mathcal{C} : xy = 1$. Then $\rho_{-\pi/4}(\mathcal{C})$ has equation $((X - Y)/\sqrt{2}\,(X + Y)/\sqrt{2}) = 1$, i.e.

$$
\frac{X^2}{2} - \frac{Y^2}{2} = 1.
$$
$\square$

Note 6.6.6 The rotated curve found in the above example is a *hyperbola* (see Section 10.5). Example 6.6.5 shows that the original curve $xy = 1$ is also a hyperbola.

Translations

Definition 6.6.7 For $(u, v) \in \mathbb{R}^2$, *translation by* (u, v) is the mapping $\tau_{(u,v)} : \mathbb{R}^2 \to \mathbb{R}^2$ which moves each point by u in the x-direction and by v in the y-direction. In these moves the signs of u and v must be taken into account. If $u > 0$, the point moves to the *right*, while if $u < 0$ then it moves to the *left*. Similarly, if $v > 0$, the point moves *up*, while if $v < 0$ then it moves *down*. Figure 6.6.4 shows points P and $Q = \tau_{(u,v)}(P)$ with $u > 0$ and $v < 0$.

Theorem 6.6.8 *Let $(u, v) \in \mathbb{R}^2$ and $\beta = u + iv$ and let $P(x,y), Q(X,Y)$ be points such that $Q = \tau_{(u,v)}(P)$.*

(1) *If P corresponds to z, then Q corresponds to $Z = z + \beta$.*

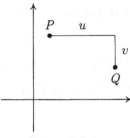

Figure 6.6.4

(2) *The relationships between the coordinates of P and Q are*

$$X = x + u$$
$$Y = y + v \tag{6.6.3}$$

and

$$x = X - u$$
$$y = Y - v \tag{6.6.4}$$

Proof We first establish (6.6.3). Since P has coordinates (x, y), from Figure 6.6.5, R has coordinates $(x + u, y)$ and Q has coordinates $(x + u, y + v)$, giving (6.6.3). Solving (6.6.3) for x and y gives (6.6.4).

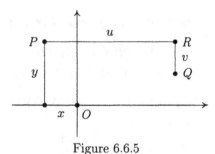

Figure 6.6.5

Finally, we have $z = x + iy$ and $Z = X + iY$. Thus, from (6.6.3)

$$Z = (x + u) + i(y + v) = (x + iy) + (u + iv) = z + \beta,$$

as required.   □

Example 6.6.9 Let $\mathcal{L}$ be the line $x + 2y = 1$. For $(u, v) \in \mathbb{R}^2$, find an equation for $\tau_{(u,v)}(\mathcal{L})$. Let $\mathcal{L}'$ be the line $x + 2y = 0$. Show that

$$\tau_{(u,v)}(\mathcal{L}) = \mathcal{L} \iff (u, v) \in \mathcal{L}'.$$

Solution Let $P(x, y)$, $Q(X, Y)$ be points with $Q = \tau_{(u,v)}(P)$.

$$(X, Y) \in \tau_{(u,v)}(\mathcal{L}) \iff (x, y) \in \mathcal{L}$$
$$\iff x + 2y = 1$$
$$\iff (X - u) + 2(Y - v) = 1 \quad \text{(using (6.6.4))}$$
$$\iff X + 2Y = u + 2v + 1.$$

Thus, in general, $\tau_{(u,v)}(\mathcal{L})$ has equation $x + 2y = u + 2v + 1$ and so

$$\begin{aligned}\tau_{(u,v)}(\mathcal{L}) = \mathcal{L} &\Leftrightarrow u + 2v + 1 = 1 \\ &\Leftrightarrow u + 2v = 0 \\ &\Leftrightarrow (u, v) \in \mathcal{L}',\end{aligned}$$

as required.   □

6.X   Exercises

1.   For which $n \in \mathbb{N}$ does

   (a) $i^n = 1$,   (b) $i^n = -1$,   (c) $i^n = i$,   (d) $i^n = -i$.

2.   Let $z = 3 + 4i$ and $w = -2 + i$. Express each of the following in the form $x + iy$ where $x, y \in \mathbb{R}$.

$$z + w, \quad z - w, \quad zw, \quad z/w, \quad iz + \overline{w}.$$

3.   Express each of the following in the form $x + iy$ where $x, y \in \mathbb{R}$.

   (a) $\dfrac{1}{1 + 2i}$,   (b) $\dfrac{i}{1 + 2i}$,   (c) $\dfrac{3 - 4i}{3 + i}$,

   (d) $\dfrac{5 - i}{4 + 3i}$,   (e) $\dfrac{2 + 3i}{2i - 5}$,   (f) $\dfrac{3 + 2i}{(1 + 2i)(3 - i)}$.

4.   Let $a, b, c \in \mathbb{C}$ with $|a| = |b| = |c| = 1$. Show that

$$\overline{a + b + c} = \frac{1}{a} + \frac{1}{b} + \frac{1}{c}.$$

   Deduce that $|a + b + c| = |ab + bc + ca|$.

5.   Solve each of the following equations for z.

   (a) $\dfrac{iz - 2}{z + 3i} = 1 + 2i$,   (b) $\dfrac{1}{z} + \dfrac{1}{2 + i} = \dfrac{1}{1 + 3i}$.

6.   Find all $z \in \mathbb{C}$ which satisfy the equation

$$z^2 + 2\overline{z} + 1 = 0.$$

7.   Let $z \in \mathbb{C}$, $z \neq \pm i$. Prove that $\dfrac{z}{1 + z^2}$ is real if and only if z is real or $|z| = 1$.

8. Let $z \in \mathbb{C}$, $z \neq -i$, and let $w = \dfrac{z-i}{z+i}$. Prove that

    (a) $\operatorname{Re} w > 0 \iff |z| > 1$,

    (b) w is real $\iff$ z lies on the imaginary axis.

9. Let $z, w \in \mathbb{C}$ be such that $z = \dfrac{i(w+2)}{w+2(1-i)}$. Prove that $|z+1| = 1$ if and only if $|w| = 2$.

10. Express in polar form each of the following complex numbers.
$$1+i, \quad 1-i, \quad -2+2i, \quad -3, \quad -1-i\sqrt{3}, \quad -4i.$$

11. Evaluate $|4+3i|$. Mark each of the following complex numbers on a sketch of the complex plane.
$$4+3i, \quad 4-3i, \quad -4+3i, \quad -4-3i, \quad 3-4i.$$

    Let $\alpha = \tan^{-1}(3/4)$. Express in terms of α the principal value of the argument of each of the above complex numbers.

12. Let $z \in \mathbb{C}$ with $0 \leq \arg z \leq \pi/3$. Find, by geometrical considerations, the minimum value of $|z - 2i|$.

13. Find the real and imaginary parts of $\dfrac{e^{i\theta}}{e^{i2\theta} - 2i}$ where $\theta \in \mathbb{R}$.

14. Express $(1+i)/(\sqrt{3}+i)$ in the form $x+iy$ where $x, y \in \mathbb{R}$. Then express each of $1+i$ and $\sqrt{3}+i$ in polar form and deduce that
$$\cos\frac{\pi}{12} = \frac{\sqrt{3}+1}{2\sqrt{2}} \quad \text{and} \quad \sin\frac{\pi}{12} = \frac{\sqrt{3}-1}{2\sqrt{2}}.$$

15. Use de Moivre's theorem to find the real and imaginary parts of
$$\text{(a)} \ (\sqrt{3}-i)^{10}, \qquad \text{(b)} \ (\sqrt{3}-i)^{-7}.$$

    Determine also, for which values of $n \in \mathbb{Z}$, $(\sqrt{3} - i)^n$ is (c) real, (d) purely imaginary.

16. Express $\sin 5\theta$ and $\cos 6\theta$ as polynomials in $\sin \theta$.

17.   Use complex numbers to express

   (a) $\sin^5 \theta$ in the form $a \sin 5\theta + b \sin 3\theta + c \sin \theta$,

   (b) $\sin^3 \theta \cos^4 \theta$ in the form $a \sin 7\theta + b \sin 5\theta + c \sin 3\theta + d \sin \theta$,

   (c) $\cos^6 \theta$ in the form $a \cos 6\theta + b \cos 4\theta + c \cos 2\theta + d$,

   where $a, b, c, d \in \mathbb{R}$.

18.   Find the complex roots of the quadratic equations

$$\text{(a)} \quad z^2 + 6 = 0, \qquad \text{(b)} \quad z^2 + 6z + 34 = 0.$$

19.   Given that $-1 + 4i$ is a root of the equation

$$2z^4 + 6z^3 + 43z^2 + 44z + 85 = 0,$$

   find all its roots.

20.   Find all the roots of each of the following equations.

   (a) $z^3 = 1$,   (b) $z^3 = -1$,   (c) $z^3 = i$,   (d) $z^3 = -i$,

   (e) $z^3 = 2(-1 + i)$,   (f) $z^4 = -8(1 + \sqrt{3}i)$,   (g) $z^2 = \dfrac{1+i}{1-i}$.

   In each case mark the roots on a sketch of the complex plane.

21.   Factorise $z^5 + 1$ as far as possible over $\mathbb{R}[z]$.

22.   Let $n \in \mathbb{N}$, $n \geq 2$, and let $\xi \neq 1$ be an nth root of unity. Prove that, for $r = 0, 1, \ldots, n$, $\overline{\xi^r} = \xi^{n-r}$. Use the binomial theorem to show that $\overline{(1 - \xi)^n} = (\xi - 1)^n$. Deduce that $(1 - \xi)^{2n}$ is real.

23.   Use the full factorisation of $x^5 - 1$ over $\mathbb{R}[x]$ (see Example 6.5.6) to deduce the values of $\cos(2\pi/5)$ and $\cos(4\pi/5)$.

24.   Prove that the product of the gradients of any perpendicular lines is -1. [Hint. See Example 6.6.4.]

25.   By considering an appropriate translation and an appropriate rotation, show that the curve

$$y = \frac{x + 1}{x - 1},$$

   is a hyperbola. [See Note 6.6.6.]

Chapter 7

Limits and Continuity

Here we begin the study of calculus, one of the most useful branches of mathematics. In this introductory chapter, we ask how the value $f(x)$ of a function varies as we vary the input. This involves the concept of *limit*, on which all of calculus depends. Here, it arises when we discuss the following question: if f is a real function and x is approximately equal to a, then must $f(x)$ be approximately equal to $f(a)$? This is an important question with significant implications. For example if we enter 1.4142 (an approximation to $\sqrt{2}$) on a calculator, and press the 'sin' key, then we *assume* that the displayed result is an approximation to $\sin\sqrt{2}$. We will see that, for sin, this is the correct assumption, but that there are functions for which it is false. The functions for which the assumption is valid are the *continuous* functions. All of the standard function listed in Section 2.3 are continuous. Indeed most of the functions we meet in applications are also continuous, but not all. For example, the cost per minute of a telephone call jumps abruptly as we pass from standard rate time to off-peak, so is not a continuous function of time.

One good reason for studying continuous functions is that there are useful results for them which do not hold in general. We shall state several such theorems (generally without proof). Perhaps the most important is the intermediate value theorem (Theorem 7.3.13) which allows us to *estimate* the zeros of functions, even if we cannot calculate them exactly.

In the final section, we extend the concept of limit to the case where the input or the value becomes arbitrarily large (limits involving ∞). This allows us to discuss long term trends—does a population die out or grow out of control?

7.1 Function Limits

Throughout this chapter our approach to limits is informal. The reader who would like to see a rigorous treatment is referred to Appendix C where the definition of function limit is followed by proofs of a selection of the properties stated in the next section.

Consider the real function g defined by $g(x) = \dfrac{\sin \pi x}{\pi |x - 2|}$. The curve $y = g(x)$ is sketched in Figure 7.1.1

The function g is not defined at $x = 2$, but as x approaches 2 from the left, the corresponding point on the graph of g approaches $(2, -1)$. So $g(x)$ approaches -1. We

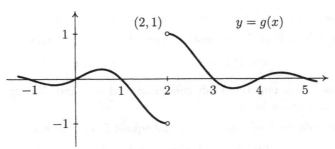

Figure 7.1.1

call -1 the *left limit of g at* 2 and write

$$\lim_{x \to 2^-} g(x) = -1 \quad \text{or} \quad g(x) \to -1 \text{ as } x \to 2^-.$$

Similarly, as x approaches 2 from the right, the corresponding point on the graph of g approaches $(2, 1)$. So $g(x)$ approaches 1. We call 1 the *right limit of g at* 2 and write

$$\lim_{x \to 2^+} g(x) = 1 \quad \text{or} \quad g(x) \to 1 \text{ as } x \to 2^+.$$

Now consider the real function h defined by $h(x) = \dfrac{\sin \pi x}{\pi(x-2)}$. The curve $y = h(x)$ is sketched in Figure 7.1.2

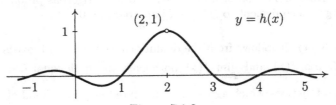

Figure 7.1.2

This time, as x approaches 2 from either side, the corresponding point on the graph of h approaches $(2, 1)$. So $h(x)$ approaches 1 and

$$\lim_{x \to 2^-} h(x) = \lim_{x \to 2^+} h(x) = 1.$$

We call 1 the *limit of h at* 2 and write

$$\lim_{x \to 2} h(x) = 1 \quad \text{or} \quad h(x) \to 1 \text{ as } x \to 2.$$

Observe that, although g has a left limit and a right limit at 2, since they are not equal, g does not have a limit at 2.

Remark When the symbol $\to$ is used in connection with a limiting process it is usually read as 'tends to'. Thus, $h(x) \to 1$ as $x \to 2$ would be read '$h(x)$ tends to 1 as x tends to 2'.

Definitions 7.1.1 Let f be a real function and let $c, L \in \mathbb{R}$.

(1) The *left limit of f at c* exists and equals L, and we write

$$\lim_{x \to c^-} f(x) = L \quad \text{or} \quad f(x) \to L \text{ as } x \to c^-,$$

if f is defined on a punctured left neighbourhood of c and $f(x)$ approaches L as x approaches c from the left.

(2) The *right limit of f at c* exists and equals L, and we write

$$\lim_{x \to c^+} f(x) = L \quad \text{or} \quad f(x) \to L \text{ as } x \to c^+,$$

if f is defined on a punctured right neighbourhood of c and $f(x)$ approaches L as x approaches c from the right.

(3) The *limit of f at c* exists and equals L, and we write

$$\lim_{x \to c} f(x) = L \quad \text{or} \quad f(x) \to L \text{ as } x \to c,$$

if f is defined on a punctured neighbourhood of c and $f(x)$ approaches L as x approaches c from either side.

Lemma 7.1.2 *Let f be a real function and let $c, L \in \mathbb{R}$. Then*

$$\lim_{x \to c} f(x) = L \quad \Leftrightarrow \quad \lim_{x \to c} |f(x) - L| = 0.$$

The result remains true if both limits are replaced with left [right] limits.

Proof A formal definition of limit is required for a rigorous proof. Intuitively, $f(x)$ approaching L is equivalent to $|f(x) - L|$ approaching 0.   □

Notes 7.1.3 (1) It follows from the definitions that $\lim_{x \to c} f(x)$ exists and equals L if and only if $\lim_{x \to c^-} f(x)$ and $\lim_{x \to c^+} f(x)$ both exist and both equal L.

(2) Two functions which agree on a punctured neighbourhood of c must have the same (left/right) limit at c whenever it exists.

(3) Lemma 7.1.2 tells us, in particular, that

$$\lim_{x \to c} f(x) = 0 \quad \Leftrightarrow \quad \lim_{x \to c} |f(x)| = 0.$$

(4) In the expression $\lim_{x \to c} f(x)$, x is a 'dummy' variable. Any symbol will do. Thus

$$\lim_{x \to c} f(x) = \lim_{t \to c} f(t) = \lim_{h \to c} f(h) = \lim_{\theta \to c} f(\theta) = \cdots.$$

(5) A function can have a (left/right) limit at c whether or not it is defined at c. The function value at c, when it exists, may or may not coincide with the (left/right) limit at c when it exists.

(6) A function defined on a punctured neighbourhood of c need not have a limit or even one-sided limits at c. Consider the real function $x \mapsto \sin(\pi/x)$. This function has neither a left nor a right limit at 0. As x approaches 0 from either side, $f(x)$ oscillates more and more rapidly between the values -1 and 1. See Figure 7.1.3.

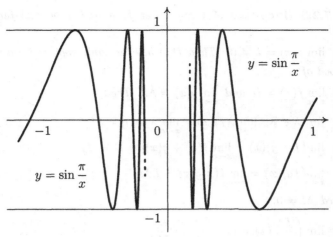

Figure 7.1.3

7.2   Properties of Limits

Proofs of the first three theorems in this section require a formal definition of limit. We therefore omit them and, where possible, use diagrams to reflect an intuitive view of the results.

Theorem 7.2.1 *Let f be a real function and let $c \in \mathbb{R}$. Then if $\lim_{x \to c} f(x)$ exists, it is unique.*

Proof A proof is given in Appendix C, Theorem C.1.5.   □

Theorem 7.2.2 *Let $c, K \in \mathbb{R}$. Then*

$$(1) \ \lim_{x \to c} K = K, \qquad (2) \ \lim_{x \to c} x = c.$$

Proof The proof is omitted. See Figure 7.2.1.   □

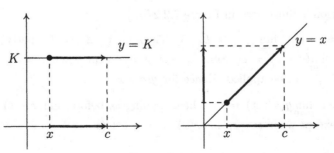

Figure 7.2.1

Theorem 7.2.3 (Properties of Limits) *Let f, g and h be real functions and let $c, L, M \in \mathbb{R}$.*

(1)　*Let $\lim_{x \to c} f(x) = L \neq 0$. Then $f(x)$ has the same sign as L on some punctured neighbourhood of c.*

(2)　*Let $\lim_{x \to c} f(x) = L$ and $\lim_{x \to c} g(x) = M$. Then*

(a)　*for $\lambda \in \mathbb{R}$, $\lim_{x \to c}(\lambda f)(x) = \lim_{x \to c}[\lambda f(x)] = \lambda L$,*

(b)　$\lim_{x \to c}(f + g)(x) = \lim_{x \to c}[f(x) + g(x)] = L + M$,

(c)　$\lim_{x \to c}(fg)(x) = \lim_{x \to c} f(x)g(x) = LM$

and, provided $M \neq 0$,

(d)　$\lim_{x \to c}\left(\dfrac{f}{g}\right)(x) = \lim_{x \to c}\dfrac{f(x)}{g(x)} = \dfrac{L}{M}.$

(3)　*Let $\lim_{x \to c} f(x) = L$ and $\lim_{x \to L} g(x) = M$. If either $M = g(L)$ or $f(x) \neq L$ for all x in a punctured neighbourhood of c then $\lim_{x \to c}(g \circ f)(x) = \lim_{x \to c} g(f(x)) = M$.*

(4)　*Let $f(x) \leq g(x)$ for all x in a punctured neighbourhood of c, let $\lim_{x \to c} f(x) = L$ and let $\lim_{x \to c} g(x) = M$. Then $L \leq M$.*

(5)　**(The Sandwich Principle)**　*Let $f(x) \leq g(x) \leq h(x)$ for all x in a punctured neighbourhood of c and let $\lim_{x \to c} f(x) = \lim_{x \to c} h(x) = L$. Then $\lim_{x \to c} g(x) = L$.*

Proof　The proof is omitted. Properties (1), (4) and (5) are illustrated in Figure 7.2.2 (a), (b) and (c), respectively.

Proofs of (1), for $L > 0$, and (2)(b) are given in Appendix C, Theorems C.1.6 and C.1.7.　　　　　　　　　　　　　　　　　　　　　　　　　　$\square$

Corollary 7.2.4 *Let g be a real function and let $c, L \in \mathbb{R}$. Then*

$$\lim_{x \to c} g(x) = L \iff \lim_{x \to 0} g(c + x) = L.$$

[The situation is illustrated in Figure 7.2.2 (d).]

Proof　$(\Rightarrow)$　Let $\lim_{x \to c} g(x) = L$. By Theorem 7.2.2 and Property (2)(b) of Theorem 7.2.3, $\lim_{x \to 0}(c + x) = c$. Since $c + x \neq c$ except when $x = 0$, Property (3) of Theorem 7.2.3 can be applied. Hence $\lim_{x \to 0} g(c + x) = L$.

$(\Leftarrow)$　Let $\lim_{x \to 0} g(c + x) = L$. Then, arguing as before, $\lim_{x \to c}(x - c) = 0$ and, since $x - c \neq 0$ except when $x = c$, $\lim_{x \to c} g(x) = \lim_{x \to c} g(c + (x - c)) = L$.　　　$\square$

Notes 7.2.5 (1)　Theorem 7.2.1, Properties (1), (2), (4) and (5) of Theorem 7.2.3, and Corollary 7.2.4, remain true if we replace all limits with left [right] limits and all punctured neighbourhoods with punctured left [right] neighbourhoods. We shall assume these properties also.

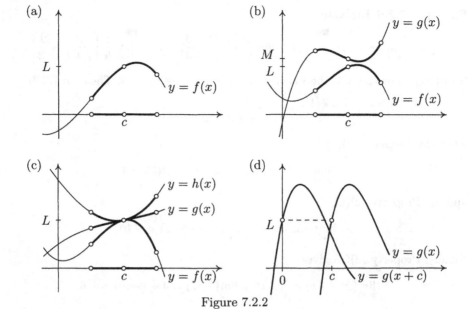

Figure 7.2.2

(2) In Property (3), knowing only that $\lim_{x \to c} f(x) = L$ and $\lim_{x \to L} g(x) = K$ is not sufficient to deduce that $\lim_{x \to c} (g \circ f)(x) = K$. Consider, for example, the real functions f and g defined by

$$f(x) = 2, \qquad g(x) = \begin{cases} x - 1 & \text{if } x \neq 2, \\ 0 & \text{if } x = 2. \end{cases}$$

Then $\lim_{x \to 3} f(x) = 2$, $\lim_{x \to 2} g(x) = 1$ but $\lim_{x \to 3} (g \circ f)(x) = 0$.

(3) It is implied by Property (3) that, under the stated conditions, $g \circ f$ is defined on a deleted neighbourhood of c. If also $f(c) = L$ and $g(L)$ is defined, then $g \circ f$ is defined on a neighbourhood of c.

(4) Putting $g(x) = M$ and $f(x) = L$ in turn, in Property (4), gives the following special cases.

(a) Let $f(x) \leq M$ for all x in a punctured neighbourhood of c and let $\lim_{x \to c} f(x) = L$. Then $L \leq M$.

(b) Let $g(x) \geq L$ for all x in a punctured neighbourhood of c and let $\lim_{x \to c} g(x) = M$. Then $M \geq L$.

(5) The conclusion of Corollary 7.2.4 can be written

$$\lim_{x \to c} g(x) = L \iff \lim_{h \to 0} g(c + h) = L.$$

We are, in effect, making the substitution $x = c + h$ and using the fact that $x \to c$ is equivalent to $h \to 0$.

Example 7.2.6 Evaluate

(a) $\lim_{x\to 2}(x^2 - 5x)$,   (b) $\lim_{x\to -1}\dfrac{x^2 + 4x + 3}{x^2 + 5x + 4}$,   (c) $\lim_{h\to 0}\dfrac{1}{h}\left[\dfrac{1}{2+h} - \dfrac{1}{2}\right]$.

Solution Throughout, the Properties referred to are those of Theorem 7.2.3

(a) By Theorem 7.2.2 (1),

$$\lim_{x\to 2} x = 2.$$

Then, by Property (2)(c),

$$\lim_{x\to 2} x^2 = (\lim_{x\to 2} x)(\lim_{x\to 2} x) = (2)(2) = 4$$

and, by Property (2)(a),

$$\lim_{x\to 2}(-5x) = -5\lim_{x\to 2} x = (-5)(2) = -10.$$

Finally, Property (2)(b) gives

$$\lim_{x\to 2}(x^2 - 5x) = \lim_{x\to 2} x^2 + \lim_{x\to 2}(-5x) = 4 + (-10) = -6.$$

(b) Here, for $x \neq -1$, we have

$$\frac{x^2 + 4x + 3}{x^2 + 5x + 4} = \frac{(x + 1)(x + 3)}{(x + 1)(x + 4)} = \frac{x + 3}{x + 4}. \tag{7.2.1}$$

This, therefore, holds for all x in a punctured neighbourhood of -1.
By Theorem 7.2.2 (1),

$$\lim_{x\to -1} x = -1, \qquad \lim_{x\to -1} 3 = 3 \quad \text{and} \quad \lim_{x\to -1} 4 = 4.$$

Then, by Property (2)(b),

$$\lim_{x\to -1}(x + 3) = \lim_{x\to -1} x + \lim_{x\to -1} 3 = -1 + 3 = 2$$

and

$$\lim_{x\to -1}(x + 4) = \lim_{x\to -1} x + \lim_{x\to -1} 4 = -1 + 4 = 3.$$

Since $\lim_{x\to -1}(x + 4) \neq 0$, Property (2)(d) gives

$$\lim_{x\to -1}\frac{x + 3}{x + 4} = \frac{\lim_{x\to -1}(x + 3)}{\lim_{x\to -1}(x + 4)} = \frac{2}{3}.$$

So by Note 7.1.3 (2),

$$\lim_{x\to -1}\frac{x^2 + 4x + 3}{x^2 + 5x + 4} = \lim_{x\to -1}\frac{x + 3}{x + 4}.$$

Hence, by (7.2.1),

$$\lim_{x\to -1}\frac{x^2 + 4x + 3}{x^2 + 5x + 4} = \frac{2}{3}.$$

Leaving out the detailed reasoning, we could write the above solution as follows.

$$\frac{x^2 + 4x + 3}{x^2 + 5x + 4} = \frac{(x+1)(x+3)}{(x+1)(x+4)} = \frac{x+3}{x+4} \quad (x \neq -1)$$

$$\to \frac{2}{3} \text{ as } x \to -1.$$

(c) The maximal domain of the function

$$h \mapsto \frac{1}{h}\left[\frac{1}{2+h} - \frac{1}{2}\right]$$

is $\mathbb{R} - \{-2, 0\}$. Since we seek the limit at 0 and we can chose a punctured neighbourhood of 0 which does not contain -2, we may assume throughout that $h \neq -2$. We have

$$\frac{1}{h}\left[\frac{1}{2+h} - \frac{1}{2}\right] = \frac{1}{h}\left[\frac{2-(2+h)}{2(2+h)}\right] = \frac{-h}{2h(2+h)} = \frac{-1}{2(2+h)} \quad (h \neq 0)$$

$$\to \frac{-1}{2(2)} = -\frac{1}{4} \text{ as } h \to 0. \qquad \square$$

Theorem 7.2.7 *Let f be a real function, let $K \geq 0$ and let N be a punctured neighbourhood of $c \in \mathbb{R}$ such that, for all $x \in N$,*

$$|f(x) - L| \leq K|x - c|.$$

Then $\lim_{x \to c} f(x) = L$.

Proof By Theorem 7.2.2 (2) and Note 7.1.3 (3) we have $\lim_{x \to c} |x - c| = 0$. Then, by Property (2)(a), $\lim_{x \to c} K|x - c| = 0$. For all $x \in N$, $0 \leq |f(x) - L| \leq K|x - c|$. Hence, the sandwich principle gives $\lim_{x \to c} |f(x) - L| = 0$ and Lemma 7.1.2 gives the result. $\square$

Example 7.2.8 Show that

$$\text{(a)} \quad \lim_{x \to 0} \cos x = 1, \qquad \text{(b)} \quad \lim_{x \to 4} \sqrt{x} = 2.$$

Solution (a) By Corollary 5.1.8, for $\theta \in \mathbb{R}$, $|\sin \theta| \leq |\theta|$. So, for $x \in \mathbb{R}$, using $\cos x = 1 - 2\sin^2(x/2)$, we have

$$|\cos x - 1| = \left|-2\sin^2\left(\frac{x}{2}\right)\right| = 2\left|\sin\frac{x}{2}\right|\left|\sin\frac{x}{2}\right| \leq 2(1)\left|\frac{x}{2}\right| = |x - 0|.$$

In particular, for all x in a punctured neighbourhood of 0, $|\cos x - 1| \leq |x - 0|$. Hence, by Theorem 7.2.7, $\lim_{x \to 0} \cos x = 1$.

(b) Let $N = (1, 7) - \{4\}$. Then N is a punctured neighbourhood of 4. For $x \in N$, $x > 1$ so that $\sqrt{x} > 1$, and

$$|\sqrt{x} - 2| = \left|\frac{(\sqrt{x} - 2)(\sqrt{x} + 2)}{\sqrt{x} + 2}\right| = \frac{|x - 4|}{|\sqrt{x} + 2|} < \frac{|x - 4|}{3} = \frac{1}{3}|x - 4|.$$

Hence, Lemma 7.2.7 applies to give $\lim_{x \to 4} \sqrt{x} = 2$. $\square$

Theorem 7.2.9 $\displaystyle\lim_{x\to 0}\frac{\sin x}{x} = 1.$

Proof We show separately that the right and left limits equal 1.

(1) Let $x \in (0, \pi/2)$, a punctured right neighbourhood of 0. Then, using Theorem 5.1.6,

$$0 < \sin x < x < \frac{\sin x}{\cos x}.$$

Hence

$$\frac{\cos x}{\sin x} < \frac{1}{x} < \frac{1}{\sin x}$$

and, since $\sin x > 0$,

$$\cos x < \frac{\sin x}{x} < 1.$$

By Example 7.2.8 (a), $\displaystyle\lim_{x\to 0^+}\cos x = 1$ and by Theorem 7.2.2, $\displaystyle\lim_{x\to 0^+} 1 = 1$. Hence, by the sandwich principle, $\displaystyle\lim_{x\to 0^+}\frac{\sin x}{x} = 1$.

(2) Let $x < 0$. Then $u = -x > 0$ and $u \to 0$ as $x \to 0^-$. So, using Part (1),

$$\frac{\sin x}{x} = \frac{\sin (-u)}{-u} = \frac{\sin u}{u} \to 1 \text{ as } x \to 0^-. \qquad\qquad \square$$

Example 7.2.10 Evaluate

(a) $\displaystyle\lim_{x\to\pi}\frac{\sin(\pi - x)}{\pi - x}$, (b) $\displaystyle\lim_{x\to 0}\frac{\cos 4x - 1}{x}$, (c) $\displaystyle\lim_{\theta\to 0}\frac{\sin 5\theta}{\sin 3\theta}$.

Solution (a) By Corollary 7.2.4,

$$\lim_{x\to\pi}\frac{\sin(\pi - x)}{\pi - x} = \lim_{x\to 0}\frac{\sin(\pi - (x + \pi))}{\pi - (x + \pi)} = \lim_{x\to 0}\frac{\sin(-x)}{-x} = \lim_{x\to 0}\frac{\sin x}{x} = 1.$$

(b) Here and in (c) we omit the detailed reasoning. For $x \neq 0$, using the fact that $\cos 4x = 1 - 2\sin^2 2x$, we have

$$\frac{\cos 4x - 1}{x} = \frac{-2\sin^2 2x}{x} = -4\sin 2x\frac{\sin 2x}{2x}$$

$$\to -4(0)(1) = 0 \text{ as } x \to 0.$$

(c) For $\theta \neq 0$,

$$\frac{\sin 5\theta}{\sin 3\theta} = \frac{\sin 5\theta}{5\theta}\cdot\frac{5}{3}\cdot\frac{3\theta}{\sin 3\theta} = \frac{5}{3}\cdot\frac{\sin 5\theta}{5\theta}\Big/\frac{\sin 3\theta}{3\theta}$$

$$\to \frac{5}{3}(1)/(1) = \frac{5}{3} \text{ as } x \to 0. \qquad\qquad \square$$

Hazard If f and g are real functions defined on a punctured neighbourhood of $c \in \mathbb{R}$ and $\displaystyle\lim_{x\to c} f(x) = 0$ we cannot conclude that $\displaystyle\lim_{x\to c} f(x)g(x) = 0$. For example, consider $f(x) = x$, $g(x) = 1/x$ and $c = 0$. However, the situation improves if g is bounded as we see in Problem 7.2.11.

Problem 7.2.11 Let f and g be real functions defined on a punctured neighbourhood N of $c \in \mathbb{R}$ such that $\lim_{x \to c} f(x) = 0$ and g is bounded on N. Use Note 7.1.3 (3) and the sandwich principle to prove that $\lim_{x \to c} f(x)g(x) = 0$.

7.3 Continuity

As with function limits, continuity is dealt with formally in Appendix C, Section C.2. However, the definitions in this section *do* provide a rigorous introduction to continuity once the reader has dealt formally with limits.

Roughly speaking, a real function is continuous where its graph forms an unbroken line. Consider the function h whose graph is sketched in Figure 7.1.2. This function is not defined at 2 and its graph has a 'hole' at $(2, 1)$. Intuitively, we could extend the definition of h in a continuous way by defining $h(2) = 1$. This would 'plug the hole'. More fundamentally, this would mean defining

$$h(2) = \lim_{x \to 2} h(x).$$

The following definitions of continuity are based on this idea.

Definitions 7.3.1 Let f be a real function and let $c \in \mathbb{R}$. Then f is *continuous at* c provided $\lim_{x \to c} f(x)$ exists and equals $f(c)$. Note that this requires f to be defined on a neighbourhood of c.

If f is continuous at every $c \in A$, we say that f is *continuous on* A. If the domain of f contains $[a, b]$, we say that f is *continuous on* $[a, b]$ provided f is continuous on (a, b) and

$$\lim_{x \to a^+} f(x) = f(a) \quad \text{and} \quad \lim_{x \to b^-} f(x) = f(b),$$

i.e. f is *right continuous at* a and *left continuous at* b. Similarly definitions apply to $[a, b)$, $(a, b]$, $[a, \infty)$ and $(-\infty, a]$.

We call f *continuous* if it is continuous on its domain.

Example 7.3.2 Show that the real function f defined by

$$f(x) = \frac{3x + 1}{x^2 - 2}$$

is continuous (on its maximal domain).

Solution The maximal domain of f is $A = \mathbb{R} - \{-\sqrt{2}, \sqrt{2}\}$. Then, for $c \in A$, f is defined on a neighbourhood of c (see Note 1.2.4) and

$$\lim_{x \to c} f(x) = \frac{3 \lim_{x \to c} x + \lim_{x \to c} 1}{\left(\lim_{x \to c} x\right)^2 + \lim_{x \to c} (-2)} = \frac{3c + 1}{c^2 - 2} = f(c).$$

Thus f is continuous at c and hence on A. □

Note 7.3.3 The method of Example 7.3.2 will show that any rational function is continuous on its maximal domain.

Example 7.3.4 Let f be the real function defined by

$$f(x) = \begin{cases} 1+x & \text{if } x \le -1, \\ 1-x^2 & \text{if } -1 < x \le 2, \\ \dfrac{1}{1-x} & \text{if } x > 2. \end{cases}$$

Show that f is continuous at -1 but not at 2.

Solution The curve $y = f(x)$ is sketched in Figure 7.3.1.

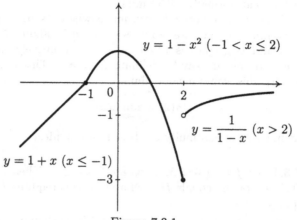

$y = 1 - x^2 \ (-1 < x \le 2)$

$y = \dfrac{1}{1-x} \ (x > 2)$

$y = 1 + x \ (x \le -1)$

Figure 7.3.1

We have

$$\lim_{x \to -1^-} f(x) = \lim_{x \to -1^-} (1+x) = 0, \qquad \lim_{x \to -1^+} f(x) = \lim_{x \to -1^+} (1-x^2) = 0$$

and $f(-1) = 0$. Hence, $\lim_{x \to -1} f(x)$ exists and equals $f(-1)$ so that f is continuous at -1.

However,

$$\lim_{x \to 2^+} f(x) = \lim_{x \to 2^+} \frac{1}{1-x} = -1$$

while $f(2) = 1 - 2^2 = -3$. So f cannot be continuous at 2. In fact, the limit of f at 2 does not exist. $\qquad \square$

Theorem 7.3.5 *Let f be a real function, let $K \ge 0$ and let N be a neighbourhood of $c \in \mathbb{R}$ such that, for all $x \in N$,*

$$|f(x) - f(c)| \le K|x - c|.$$

Then f is continuous at c.

Proof By Theorem 7.2.7, $\lim_{x \to c} f(x) = f(c)$. So f is continuous at c. $\qquad \square$

The next two examples generalise Example 7.2.8.

Example 7.3.6 Prove that the modulus function and the cosine function are continuous on $\mathbb{R}$.

Solution Let $c \in \mathbb{R}$. Then, for $x \in \mathbb{R}$ (and hence for all x in a neighbourhood of c),

$$\big|\, |x| - |c| \,\big| \le |x - c|$$

and

$$|\cos x - \cos c| = \left| -2 \sin\left(\frac{x+c}{2}\right) \sin\left(\frac{x-c}{2}\right) \right| = 2 \left| \sin\frac{x+c}{2} \right| \left| \sin\frac{x-c}{2} \right|$$

$$\le 2(1) \left| \frac{x-c}{2} \right| = |x - c|.$$

Theorem 7.3.5 applies. Hence the modulus function and the cosine function are continuous at c and hence on $\mathbb{R}$. $\qquad\square$

Example 7.3.7 Prove that the real function f defined by $f(x) = \sqrt{x}$ is continuous on $(0, \infty)$. Deduce that

$$\lim_{h \to 0} \frac{\sqrt{1+h} - 1}{h} = \frac{1}{2}.$$

Solution Let $c \in (0, \infty)$. Then $(0, \infty)$ contains a neighbourhood N of c and, for $x \in N$, $(\sqrt{x} - \sqrt{c})(\sqrt{x} + \sqrt{c}) = x - c$ so that

$$|f(x) - f(c)| = |\sqrt{x} - \sqrt{c}| = \frac{|x - c|}{|\sqrt{x} + \sqrt{c}|} \le \frac{|x - c|}{\sqrt{c}} = \frac{1}{\sqrt{c}}|x - c|.$$

Theorem 7.3.5 can be applied. Thus f is continuous at c and hence on $(0, \infty)$.

In particular, f is continuous at 1. So $\lim_{x \to 1} f(x) = f(1)$ and, by Note 7.2.5 (5), $\lim_{h \to 0} \sqrt{1+h} = 1$. Hence, for $h > -1$ and $h \ne 0$,

$$\frac{\sqrt{1+h} - 1}{h} = \frac{\sqrt{1+h} - 1}{h} \cdot \frac{\sqrt{1+h} + 1}{\sqrt{1+h} + 1} = \frac{(1+h) - 1}{h(\sqrt{1+h} + 1)}$$

$$= \frac{1}{\sqrt{1+h} + 1} \to \frac{1}{2} \quad \text{as } h \to 0$$

as required. $\qquad\square$

Note 7.3.8 With f as in Example 7.3.7, it can be shown that $\lim_{x \to 0^+} f(x) = f(0)$ so that, in fact, the real function $x \mapsto \sqrt{x}$ is continuous on its maximal domain $[0, \infty)$. For a formal proof, see Examples C.2.3 and C.2.7.

Theorem 7.3.9 (Properties of Continuous Functions) *Let f and g be real functions and let $c \in \mathbb{R}$.*

(1) *Let f be continuous at c with $f(c) \ne 0$. Then $f(x)$ has the same sign as $f(c)$ on some neighbourhood of c.*

(2) *Let f and g be continuous at c and let $\lambda \in \mathbb{R}$. Then λf, $f + g$ and fg are continuous at c, and, provided $g(c) \neq 0$, f/g is continuous at c.*

(3) *Let f be continuous at c and let g be continuous at $f(c)$. Then $g \circ f$ is continuous at c.*

Properties (1) and (2) remain true if 'continuous' is replaced with 'left [right] continuous' and 'neighbourhood' with 'left [right'$,$] neighbourhood'.

Proof These properties follow directly from Properties (1), (2) and (3) in Theorem 7.2.3, Note 7.2.5 (1) and the definitions of continuity (Definitions 7.3.1). To illustrate this we shall prove (3). Since f is continuous at c, $\lim_{x \to c} f(x) = f(c)$. Since g is continuous at $f(c)$, $\lim_{x \to f(c)} g(x) = g(f(c))$. So Theorem 7.2.3 (3) applies to give $\lim_{x \to c}(g \circ f)(x) = (g \circ f)(c)$ and (3) follows.

A proof of (3) is also given in Appendix C, Theorem C.2.5.   □

Problem 7.3.10 Let g be a real function, defined and bounded on a neighbourhood of $c \in \mathbb{R}$. Show that the real function h defined by $h(x) = (x - c)g(x)$ is continuous at c. Hint: use Problem 7.2.11.

Now let $g : \mathbb{R} \to \mathbb{R}$ be defined by

$$g(x) = \begin{cases} 1 & \text{if } x \in \mathbb{Q}, \\ 0 & \text{if } x \in \mathbb{R} - \mathbb{Q} \end{cases}$$

It is *given* that g is discontinuous at every point of $\mathbb{R}$. Deduce that $x \mapsto xg(x)$ is continuous *only* at 0. Can you find a real function, defined on $\mathbb{R}$, which is continuous only at -2 and 2?

The next two theorems are of fundamental importance in the study of continuous real functions. Their proofs, however, are outwith the scope of this book.

Theorem 7.3.11 (Extreme Value Theorem) *Let f be a real function, continuous on $[a, b]$. Then there exist $c_1, c_2 \in [a, b]$ such that, for all $x \in [a, b]$,*

$$f(c_1) \leq f(x) \leq f(c_2).$$

Proof The proof is omitted. See Figure 7.3.2.   □

Problem 7.3.12 It follows from Theorem 7.3.11 that if $f : [a, b] \to \mathbb{R}$ is continuous then it is bounded. Find a real function $g : (a, b] \to \mathbb{R}$ which is continuous but not bounded.

Theorem 7.3.13 (Intermediate Value Theorem) *Let f be a real function, continuous on $[a, b]$, with*

$$f(a) < d < f(b) \quad or \quad f(a) > d > f(b).$$

Then there exists $c \in (a, b)$ such that $f(c) = d$.

Proof The proof is omitted. See Figure 7.3.3   □

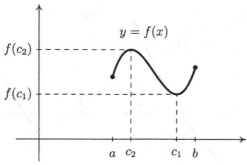

Figure 7.3.2

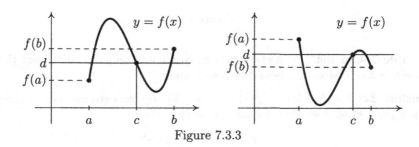

Figure 7.3.3

Example 7.3.14 Given that the equation

$$x^3 + 2x - 7 = 0$$

has exactly one real root, find the best integer approximation to the root.

Solution Let f be the real function defined by

$$f(x) = x^3 + 2x - 7.$$

Then f is continuous on $\mathbb{R}$ (see Note 7.3.3) and hence on any bounded closed interval (see Definitions 7.3.1).

Trying integer values of x we find that $f(1) = -4$ and $f(2) = 5$. Since f is continuous on $[1, 2]$ and $f(1) < 0 < f(2)$, the intermediate value theorem gives $f(c) = 0$ for some $c \in (1, 2)$, *i.e.* the real root lies between 1 and 2. See Figure 7.3.4 (a). To decide whether 1 or 2 is nearer to the root we evaluate $f(3/2)$. This gives $-5/8$. Since f is continuous on $[3/2, 2]$ and $f(3/2) < 0 < f(2)$, the intermediate value theorem gives the real root in $(3/2, 2)$. See Figure 7.3.4 (b). Hence 2 is the best integer approximation to the root. □

Problem 7.3.15 Let f be a real function which is continuous with $|f|$ constant on $\mathbb{R}$. Prove that f is constant on $\mathbb{R}$.

Remark In the solution of Example 7.3.14 it is the signs of f at 1, 2 and 3/2 that are relevant rather than the values.

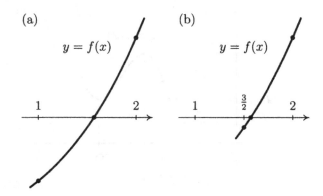

Figure 7.3.4

Example 7.3.16 Let f be a real function, continuous on $[a, b]$. Show that the image, $f([a, b])$, is a singleton or a bounded closed interval.

Solution Let $I = f([a, b]) = \{f(x) : a \le x \le b\}$. By the extreme value theorem there exist $c_1, c_2 \in [a, b]$ such that, for all $x \in [a, b]$,

$$f(c_1) \le f(x) \le f(c_2).$$

If $f(c_1) = f(c_2) = k$ then f is constant and $I = \{k\}$. Otherwise $I \subseteq [f(c_1), f(c_2)]$. In this case, by the intermediate value theorem, for each $d \in (f(c_1), f(c_2))$, there exists $c \in (a, b)$ such that $f(c) = d$. Hence $(f(c_1), f(c_2)) \subseteq I$. Since $f(c_1) \in I$ and $f(c_2) \in I$ it follows that $I = [f(c_1), f(c_2)]$.   □

Problem 7.3.17 Find continuous real functions f, g and h, each with domain $(0, 1)$, such that their images are $(0, 1)$, $(0, 1]$ and $(0, \infty)$, respectively.

Theorem 7.3.18 *Let $f : A \to B$ be a continuous real bijection where A is an interval. Then*

(1) *f is strictly monotonic on A,*

(2) *$f^{-1} : B \to A$ is continuous.*

Proof (1) We prove only the case where $A = [a, b]$ and $f(a) < f(b)$.

First let $x, y, z \in A$ with $x < y < z$ and $f(x) < f(z)$. If $f(y) < f(x)$ then $f(y) < f(x) < f(z)$ and, by the intermediate value theorem, there exists $u \in (y, z)$ such that $f(u) = f(x)$. But this contradicts the injectiveness of f. Similarly, if $f(z) < f(y)$ then $f(x) < f(z) < f(y)$ and there exists $v \in (x, y)$ such that $f(v) = f(z)$, contradicting the injectiveness of f. Hence, $f(x) \le f(y) \le f(z)$ and the injectiveness of f gives $f(x) < f(y) < f(z)$.

Now let $s_1, s_2 \in A$ with $s_1 < s_2$. Applying the above argument to a, s_1, b gives $f(a) < f(s_1) < f(b)$. Then applying it again to s_1, s_2, b gives $f(s_1) < f(s_2) < f(b)$ and it follows that f is strictly increasing on A.

(2) The proof is omitted. The mirror property for inverse functions (see Note 2.5.8) suggests intuitively that the results holds.   □

Corollary 7.3.19 *Let f be a real bijection, continuous on a neighbourhood N of c. Then*

(1) $f(N)$ *contains a neighbourhood of* $f(c)$,

(2) f^{-1} *is continuous at* $f(c)$.

Proof (1) This can be deduced from Theorem 7.3.18. Alternatively, we can find $a, b \in N$ such that $a < c < b$. By Example 7.3.16, $f([a,b]) = [u,v]$ (say). Since f is injective, $u < f(c) < v$. So $f(c) \in (u,v) \subseteq f(N)$ and it follows (see Note 1.2.4 (2)) that $f(N)$ contains a neighbourhood of $f(c)$.

(2) (Outline) Let $g : N \to f(N)$ be a restriction of f. Then, by Theorem 7.3.18, g^{-1} is continuous. Since g^{-1} is a restriction of f^{-1} it follows that f^{-1} is continuous at $f(c)$. $\qquad\qquad\qquad\qquad\qquad\qquad\qquad\qquad\qquad\qquad\qquad\qquad\qquad\qquad$ $\square$

Assumption Henceforth we shall assume that polynomial functions, rational functions, rational power functions the modulus function, the six (basic) trigonometric functions and the three (basic) inverse trigonometric functions are all continuous on their maximal domains.

7.4   Approaching Infinity

In this section we investigate function variables and function values which become arbitrarily large.

If $u \in \mathbb{R}$ becomes arbitrarily large we say that u *approaches* ∞ (or $+\infty$). If $-u$ becomes arbitrarily large we say that u *approaches* $-\infty$.

Consider the real function f defined by $f(x) = \dfrac{2x^2 - x - 9}{|x+1|(x-3)}$. The curve $y = f(x)$ is sketched in Figure 7.4.1.

As x approaches ∞, $f(x)$ approaches 2. We call 2 the *limit of f at ∞* and write

$$\lim_{x \to \infty} f(x) = 2 \quad \text{or} \quad f(x) \to 2 \text{ as } x \to \infty.$$

As x approaches $-\infty$, $f(x)$ approaches -2. We call -2 the *limit of f at $-\infty$* and write

$$\lim_{x \to -\infty} f(x) = -2 \quad \text{or} \quad f(x) \to -2 \text{ as } x \to -\infty.$$

The lines $y = 2$ and $y = -2$ are called *horizontal asymptotes* for the curve $y = f(x)$. They are examples of *non-vertical asymptotes* which are dealt with in Chapter 10.

As x approaches 3 from the right, $f(x)$ approaches ∞, and we write

$$f(x) \to \infty \text{ as } x \to 3^+.$$

As x approaches 3 from the left, $f(x)$ approaches $-\infty$, and we write

$$f(x) \to -\infty \text{ as } x \to 3^-.$$

As x approaches -1 from either side, $f(x)$ approaches ∞, and we write

$$f(x) \to \infty \text{ as } x \to -1.$$

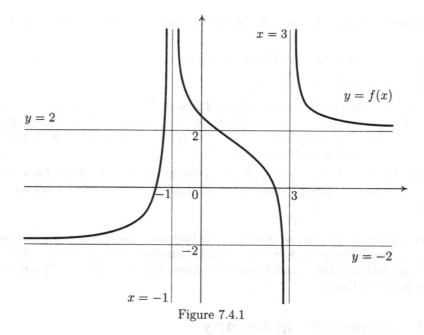

Figure 7.4.1

The lines $x = -1$, and $x = 3$ are called *vertical asymptotes* for the curve $y = f(x)$. Vertical asymptotes are dealt with in Chapter 10.

 Hazard The symbolism $u \to \pm\infty$ may be read 'u tends to $\pm\infty$'. But, when $f(x) \to \infty$ or $f(x) \to -\infty$, we shall *not* refer to ∞ or $-\infty$ as the limit of f. If we say that f has a limit L, it is assumed that $L \in \mathbb{R}$.

We now treat separately the situations where $x \to \pm\infty$ and $f(x) \to \pm\infty$. In both cases, relating the definitions to the intuitive view in the example above relies on the following assumption. If $u, v \in \mathbb{R}$ and $uv = 1$ then u approaching ∞ $[-\infty]$ is equivalent to v approaching 0 from the right [left]. See Figure 7.4.2.

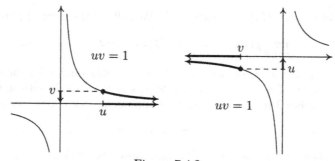

Figure 7.4.2

Definitions 7.4.1 Let f be a real function and let $L \in \mathbb{R}$.

(1) The *limit of f at* ∞ exists and equals L, and we write

$$\lim_{x \to \infty} f(x) = L \quad \text{or} \quad f(x) \to L \text{ as } x \to \infty,$$

if $\lim\limits_{t \to 0^+} f(1/t)$ exists and equals L.

(2) The *limit of f at* $-\infty$ exists and equals L, and we write

$$\lim_{x \to -\infty} f(x) = L \quad \text{or} \quad f(x) \to L \text{ as } x \to -\infty,$$

if $\lim\limits_{t \to 0^-} f(1/t)$ exists and equals L.

Notes 7.4.2 (1) For $\lim\limits_{t \to 0^+} f(1/t)$ to exist, the function $t \mapsto f(1/t)$ must be defined on a punctured right neighbourhood of 0. For $\delta > 0$, $t \in (0, \delta)$ if and only if $1/t \in (1/\delta, \infty)$. It follows that for $\lim\limits_{x \to \infty} f(x)$ to exist, f must be defined on an interval of the form (K, ∞). Similarly, for $\lim\limits_{x \to -\infty} f(x)$ to exist, f must be defined on an interval of the form $(-\infty, K)$.

(2) Corresponding to Theorem 7.2.2 we have the following result which follows directly from the definitions. Let $K \in \mathbb{R}$. Then

$$\lim_{x \to \pm\infty} K = K \quad \text{and} \quad \lim_{x \to \pm\infty} \frac{1}{x} = 0.$$

(3) The properties stated in Theorem 7.2.3 remain true if we replace all occurrences of $x \to c$ with $x \to \infty$ $[x \to -\infty]$ and all punctured neighbourhoods of c with intervals of the form (K, ∞) [intervals of the form $(-\infty, K)$]. Property (3)(b) (for $x \to \infty$) is proved in Example 7.4.3. Proofs of the other properties are similar and are left to the reader.

Example 7.4.3 Let $\lim\limits_{x \to \infty} f(x) = L$ and $\lim\limits_{x \to \infty} g(x) = M$. Show that

$$\lim_{x \to \infty} (f + g)(x) = L + M.$$

Solution From the definition, we have $\lim\limits_{t \to 0^+} f(1/t) = L$ and $\lim\limits_{t \to 0^+} g(1/t) = M$. Hence, using the definition and Theorem 7.2.3 (2)(b),

$$\lim_{x \to \infty} (f + g)(x) = \lim_{t \to 0^+} (f + g)(1/t) = L + M. \qquad \square$$

Example 7.4.4 Evaluate

(a) $\lim\limits_{x \to \infty} x \sin \dfrac{\pi}{x},$ (b) $\lim\limits_{x \to -\infty} \dfrac{\sqrt{x^2 + 5}}{7x - 3},$ (c) $\lim\limits_{x \to \infty} \dfrac{3x^2 - 4x + 5}{9x^2 - 1}.$

Solution (a) Let f be the real function defined by $f(x) = x \sin(\pi/x)$. Then, for $t > 0$,

$$f(1/t) = (1/t) \sin \pi t = \pi \frac{\sin \pi t}{\pi t} \to \pi(1) = \pi \quad \text{as} \quad t \to 0^+.$$

Hence, $\lim_{x \to \infty} f(x) = \pi$.

(b) Let g be the real function defined by $g(x) = \sqrt{x^2 + 5}/(7x - 3)$. Then, for $t < 0$,

$$g(1/t) = \frac{\sqrt{1/t^2 + 5}}{7/t - 3} = \frac{-\sqrt{1 + 5t^2}/t}{(7 - 3t)/t} = -\frac{\sqrt{1 + 5t^2}}{7 - 3t} \to -\frac{1}{7} \quad \text{as} \quad t \to 0^-.$$

Hence, $\lim_{x \to -\infty} g(x) = -1/7$.

(c) Here we shall use the properties stated in Notes 7.4.2 rather than the definition. We have, for $x > 1/3$,

$$\frac{3x^2 - 4x + 5}{9x^2 - 1} = \frac{3 - 4/x + 5/x^2}{9 - 1/x^2} \to \frac{3 - 0 + 0}{9 - 0} = \frac{1}{3} \quad \text{as} \quad x \to \infty.$$

Note that we began by dividing the numerator and denominator by x^2, the highest power of x appearing in either polynomial.   □

Definitions 7.4.5 Let f be a real function and let $c \in \mathbb{R}$.

(1) We say that $f(x)$ *tends to* ∞ $[-\infty]$ *as* x *tends to* c *from the right*, and write

$$f(x) \to \infty \; [-\infty] \text{ as } x \to c^+,$$

if $f(x) > 0$ $[f(x) < 0]$ for all x in a punctured right neighbourhood of c and $\lim_{x \to c^+} \frac{1}{f(x)} = 0$.

(2) We say that $f(x)$ *tends to* ∞ $[-\infty]$ *as* x *tends to* c *from the left*, and write

$$f(x) \to \infty \; [-\infty] \text{ as } x \to c^-,$$

if $f(x) > 0$ $[f(x) < 0]$ for all x in a punctured left neighbourhood of c and $\lim_{x \to c^-} \frac{1}{f(x)} = 0$.

(3) We say that $f(x)$ *tends to* ∞ $[-\infty]$ *as* x *tends to* c, and write

$$f(x) \to \infty \; [-\infty] \text{ as } x \to c,$$

if $f(x) \to \infty$ $[-\infty]$ as $x \to c^+$ and as $x \to c^-$.

Example 7.4.6 Determine the behaviour of

$$f(x) = \frac{(x + 1)(x - 3)}{x - 2}$$

as $x \to 2^+$.

Solution Since $f(x)$ is undefined for $x = 2$ we 'suspect' that $f(x) \to \pm\infty$ as $x \to 2^+$.

We first consider a table of signs for $f(x)$ in the range $2 \le x \le 3$.

x	2	$\rightarrow$	3
$x+1$	+	+	+
$x-3$	$-$	$-$	0
$x-2$	0	+	+
$f(x)$	?	$-$	0

This shows that $f(x) < 0$ for all x in $(2,3)$, a punctured right neighbourhood of 2. Further, the real function g defined by

$$g(x) = \frac{x-2}{(x+1)(x-3)},$$

is continuous at 2 so that, for $x \in (2,3)$,

$$\frac{1}{f(x)} = g(x) \rightarrow g(2) = 0 \text{ as } x \rightarrow 2^{+}.$$

Hence, $f(x) \rightarrow -\infty$ as $x \rightarrow 2^{+}$. $\square$

Note 7.4.7 We list a selection of properties which can be deduced from the above definitions together with Theorem 7.2.3.

Let f and g be real functions and let $c, L \in \mathbb{R}$. Throughout, we suppose that x tends to c (from the right / from the left).

(1) Let $f(x) \le g(x)$ for all x in a punctured (left/right) neighbourhood of c. Then

$$f(x) \rightarrow \infty \implies g(x) \rightarrow \infty \qquad \text{and} \qquad g(x) \rightarrow -\infty \implies f(x) \rightarrow -\infty.$$

For example, suppose that $f(x) \rightarrow \infty$. Then on some punctured neighbourhood of c we have $0 < f(x) \le g(x)$ and hence $0 \le 1/g(x) \le 1/f(x)$. Since $1/f(x) \rightarrow 0$, the result follows by the sandwich principle (see Theorem 7.2.3 (5)).

(2) Let $f(x) \rightarrow \infty$ and $g(x) \rightarrow L$. Then

$$f(x) + g(x) \rightarrow \infty \quad \text{and} \quad f(x)g(x) \rightarrow \begin{cases} \infty & \text{if } L > 0, \\ -\infty & \text{if } L < 0. \end{cases}$$

These may be proved by considering

$$\frac{1}{f(x) + g(x)} = \frac{1}{f(x)}\left(\frac{1}{1 + (1/f(x))g(x)}\right) \quad \text{and} \quad \frac{1}{f(x)g(x)} = \frac{1}{f(x)} \cdot \frac{1}{g(x)}.$$

Note that nothing is claimed for the product when $L = 0$.

(3) Let $f(x) \rightarrow \infty$ and $g(x) \rightarrow \infty$. Then

$$f(x) + g(x) \rightarrow \infty \quad \text{and} \quad f(x)g(x) \rightarrow \infty.$$

For the sum, note that, for all x in some punctured (right/left) neighbourhood of c, $f(x) \le f(x) + g(x)$ so that (1) can be applied.

We shall also assume 'obvious' variations on these properties. For example, if $f(x) \rightarrow \infty$ and $L > 0$ then, since $L \rightarrow L$, it follows from (2) that $f(x)L \rightarrow \infty$. See also Problem 7.4.8.

Problem 7.4.8 Let f and g be a real function and let $c, L \in \mathbb{R}$. Throughout, suppose that $x \to c$. Prove that $f(x) \to -\infty$ if and only if $-f(x) \to \infty$. Deduce that if $f(x) \to -\infty$ and $g(x) \to L$ then $f(x) + g(x) \to -\infty$.

The final definitions in this section essentially combine the earlier definitions.

Definitions 7.4.9 Let f be a real function.

(1) We say that $f(x)$ *tends to* ∞ $[-\infty]$ *as* x *tends to* ∞, and write

$$f(x) \to \infty \ [-\infty] \text{ as } x \to \infty,$$

if $f(x) > 0$ $[f(x) < 0]$ for all x in an interval of the form (K, ∞) and $1/f(x) \to 0$ as $x \to \infty$. This is equivalent to $\lim_{t \to 0^+} 1/f(1/t) = 0$.

(2) We say that $f(x)$ *tends to* ∞ $[-\infty]$ *as* x *tends to* $-\infty$, and write

$$f(x) \to \infty \ [-\infty] \text{ as } x \to -\infty,$$

if $f(x) > 0$ $[f(x) < 0]$ for all x in an interval of the form $(-\infty, K)$ and $1/f(x) \to 0$ as $x \to -\infty$. This is equivalent to $\lim_{t \to 0^-} 1/f(1/t) = 0$.

Example 7.4.10 Determine the behaviour of

$$\text{(a)} \ f(x) = \frac{(x+1)(x-3)}{x-2}, \qquad \text{(b)} \ g(x) = x \cos(1/x),$$

as $x \to \infty$.

Solution (a) Trying 'large' values of x suggests that $f(x) \to \infty$ as $x \to \infty$.

We first consider a table of signs for $f(x)$ in the range $x \geq 3$.

x	3	$\to$
$x+1$	+	+
$x-3$	0	+
$x-2$	+	+
$f(x)$	0	+

This shows that $f(x) > 0$ for all x in $(3, \infty)$. Further, for $x \in (3, \infty)$,

$$\frac{1}{f(x)} = \frac{x-2}{x^2 - 2x - 3} = \frac{1/x - 2/x^2}{1 - 2/x - 3/x^2} \to \frac{0-0}{1-0-0} = 0 \text{ as } x \to \infty.$$

Hence, $f(x) \to \infty$ as $x \to \infty$ as required.

(b) For $t > 0$,

$$\frac{1}{g(1/t)} = \frac{1}{(1/t)\cos t} = \frac{t}{\cos t} \to \frac{0}{1} = 0 \text{ as } t \to 0^+.$$

Hence $g(x) \to \infty$ as $x \to \infty$. $\qquad\square$

Note 7.4.11 The reader is invited to list a selection of properties which can be deduced from the above definitions together with Theorem 7.2.3. Where necessary such properties will be assumed. As an example, we give a version of the sandwich principle which will be used later.

Let f and g be real functions such that $f(x) \le g(x)$ $(x > K)$ and $f(x) \to \infty$ as $x \to \infty$. Then $g(x) \to \infty$ as $x \to \infty$.

To prove this, suppose, without loss, that $K > 0$ and $0 < f(x) \le g(x)$ $(x > K)$. Then $0 < 1/g(x) \le 1/f(x)$ $(x > K)$. Since $1/f(x) \to 0$ as $x \to \infty$, the result follows from Theorem 7.2.3 (5) and Note 7.4.2 (3).

7.X   Exercises

1. Evaluate

    (a) $\displaystyle\lim_{x \to 1} \frac{x-1}{x^2-1}$,

    (b) $\displaystyle\lim_{x \to -1} \frac{x^2-1}{x^2+x}$,

    (c) $\displaystyle\lim_{x \to -3} \frac{x^2-x-12}{x^2+x-6}$,

    (d) $\displaystyle\lim_{x \to a} \frac{x^2-a^2}{x^3-a^3}$ $(a \ne 0)$,

    (e) $\displaystyle\lim_{h \to 0} \frac{\sqrt{2-h}-\sqrt{2}}{h}$,

    (f) $\displaystyle\lim_{h \to 0} \frac{1}{h}\left(\frac{1}{x+h}-\frac{1}{x}\right)$ $(x \ne 0)$,

    (g) $\displaystyle\lim_{x \to 0} \frac{\sin 5x}{5x}$,

    (h) $\displaystyle\lim_{x \to 0} \frac{\sin 5x}{3x}$,

    (i) $\displaystyle\lim_{x \to 0} \frac{\sin 5x}{\sin 3x}$,

    (j) $\displaystyle\lim_{x \to 0} \frac{\tan 5x}{\tan 3x}$,

    (k) $\displaystyle\lim_{x \to 1} \frac{\sin(x-1)}{x^2-1}$,

    (l) $\displaystyle\lim_{\theta \to \pi/2} (\pi - 2\theta)\sec\theta$.

2. Define $f : \mathbb{R} \to \mathbb{R}$ by

    $$f(x) = \begin{cases} 1-x & \text{if } x \le -2, \\ 2x^2+3x+1 & \text{if } -2 < x < -1, \\ 1/(1-x) & \text{if } -1 \le x < 0, \\ \cos 2x & \text{if } 0 \le x \le \pi/2, \\ \cos x/(x-\pi/2) & \text{if } x > \pi/2. \end{cases}$$

    Determine whether or not f is continuous at $-2, -1, 0$ or $\pi/2$.

3. A real function f is continuous on $\mathbb{R} - \{1\}$ and

    $$f(x) = \frac{x-5+6/x}{x-3+2/x} \quad (x \in \mathbb{R} - \{0,1,2\}).$$

    What are the values of $f(0)$ and $f(2)$?

4. Let f and g be real functions, continuous at $c \in \mathbb{R}$. Prove that $\max\{f, g\}$ is continuous at c.

5. Find the best integer approximations to the real roots of the following equations.

   (a) $x^5 + 5x + 4 = 0$ (1 real root),

   (b) $x^3 + x^2 + x = 25$ (1 real root),

   (c) $4x^3 - 8x^2 - x + 3 = 0$ (3 real roots),

   (d) $\cos(\pi x/2) = 4x - 7$ (1 real root).

6. Let $f : [a, b] \to [a, b]$ be continuous. By applying the intermediate value theorem to $g : [a, b] \to \mathbb{R}$ defined by $g(x) = f(x) - x$, show that there exists $c \in [a, b]$ such that $f(c) = c$. We call c a *fixed point for* f.

7. Let $T : [0, 2\pi] \to \mathbb{R}$ be continuous with $T(0) = T(2\pi)$. By applying the intermediate value theorem to $f : [0, \pi] \to \mathbb{R}$, defined by $f(x) = T(x) - T(x + \pi)$, show that there exists $c \in [0, \pi]$ such that $T(c) = T(c + \pi)$. What conclusion might be drawn about temperatures on the equator?

8. Evaluate

   (a) $\lim\limits_{x \to \infty} \dfrac{2 - 3x}{4x - 5}$,

   (b) $\lim\limits_{x \to \infty} \dfrac{3x^2 + 2x + 1}{x^2 + 3x + 5}$,

   (c) $\lim\limits_{x \to -\infty} \dfrac{x^3 + 2x^{-1}}{5x^3 + 4x}$,

   (d) $\lim\limits_{x \to \infty} \dfrac{7x - 3}{x^2 + 3x + 5}$,

   (e) $\lim\limits_{x \to \infty} \dfrac{2x - 1}{\sqrt{x^2 + 4}}$,

   (f) $\lim\limits_{x \to -\infty} x \tan \dfrac{2\pi}{x}$.

9. Determine the behaviour of $f(x)$ as $x \to 3^-$ and as $x \to 3^+$ when $f(x)$ is

   (a) $\dfrac{x - 2}{(x - 1)(x - 3)}$,

   (b) $\dfrac{x^2 + x + 1}{(x - 3)^2}$.

10. By evaluating

   (a) $\lim\limits_{t \to 0^-} \tan^{-1}(1/t)$ and (b) $\lim\limits_{t \to 0^+} \tan^{-1}(1/t)$,

   show that $\lim\limits_{t \to 0} \tan^{-1}(1/t)$ does not exist.

11. Determine the behaviour of

(a) $\dfrac{x^2 + 3x + 2}{5x^2 + 2x + 1}$ as $x \to \infty$,

(b) $\dfrac{x + 7}{x - 7}$ as $x \to 7^{+}$ and as $x \to 7^{-}$,

(c) $\dfrac{x^3 + 4}{x^2 + 5}$ as $x \to \infty$,

(d) $\dfrac{x^2 + 3x - 4}{x - 3}$ as $x \to \infty$ and as $x \to -\infty$.

Chapter 8

Differentiation—Fundamentals

In the previous chapter, we studied continuous functions. Geometrically, a real functions is continuous on a interval if the graph is unbroken. Here, we consider *differentiable* functions. These are the continuous functions whose graphs are *smooth, i.e.* have a tangent at each point. The reader should note that not every *curve* is smooth. A circle is smooth since it has a tangent at each point (the tangent being perpendicular to the radius at the point). On the other hand, a square is not smooth since there is no tangent at any vertex.

The first section establishes an analytic characterisation of (the gradient of) the tangent to a graph in terms of a limit. Our knowledge of limits allows is to decide whether a graph has a tangent at a particular point, and find the tangent *when it exists*. This is known as *differentiation from first principles*.

Working from first principles is often laborious, even for simple functions. To extend the range of functions we can differentiate, we use the definition to prove rules which allow us to differentiate *combinations* of functions. These cover sums, products quotients and composites. These rules are very important and must be thoroughly mastered. Of course, to complete the calculation of a derivative, we must be able to differentiate the functions which make up the combination. Again, we use the definition to find the derivatives of standard functions. We are then in a position to differentiate any algebraic combination of standard functions.

In the final section we observe that, if f is a differentiable function, then its derivative, f', is another function. If this is itself differentiable, then we denote its derivative by f'', the *second derivative of f*. Of course, the process may continue giving *higher derivatives* (third, fourth,...). In later sections, we meet applications of the second derivative.

8.1 First Principles

Roughly speaking, a real function f is differentiable at $x \in \mathbb{R}$ if the curve $y = f(x)$ is smooth at the point $(x, f(x))$. Of the examples in Figure 8.1.1, only (a) is differentiable at x. The corner in (b) and the discontinuity in (c) are not allowed.

A distinguishing feature of Figure 8.1.1 (a) is that the curve $y = f(x)$ has a unique tangent at $(x, f(x))$. We use this idea as the basis of the formal definition of *differentiable*. The idea is to take a point B, on the curve, near $A(x, f(x))$ and consider what happens to the chord through A and B as B moves along the curve towards A. Two

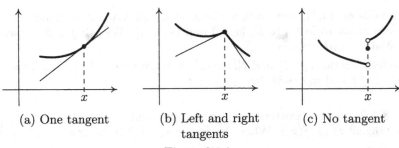

(a) One tangent   (b) Left and right   (c) No tangent
                             tangents

Figure 8.1.1

cases are shown in Figure 8.1.2. If the curve is smooth at A then, in the limit, the chord will become the tangent at A and *vice versa*.

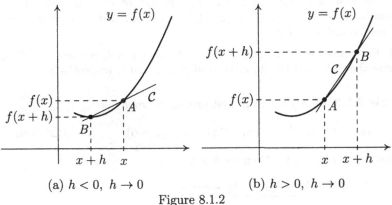

(a) $h < 0$, $h \to 0$ (b) $h > 0$, $h \to 0$

Figure 8.1.2

We now make things more precise.

Definitions 8.1.1 Let f be a real function defined on a neighbourhood N of $x \in \mathbb{R}$, let $h \neq 0$ with $x + h \in N$ and let $\mathcal{C}$ be the chord of the curve $y = f(x)$ which passes through the points $(x, f(x))$ and $(x + h, f(x + h))$. The gradient of $\mathcal{C}$ is

$$\frac{f(x + h) - f(x)}{h}.$$

This is called a *difference quotient*. If it approaches a limit as $h \to 0$ then f is said to be *differentiable at* x and the limit is called the *derivative of f at x*. It is denoted by

$$f'(x) \quad \text{or} \quad \frac{d}{dx} f(x) \quad \text{or} \quad \frac{df}{dx} \quad \text{or} \quad \frac{dy}{dx} \quad \text{or} \quad y'.$$

If the left [right] limit of the difference quotient exists then f is *left* [*right*] *differentiable at* x and the limit is the *left* [*right*] *derivative of f at x*, denoted by $f'_L(x)$ [$f'_R(x)$]. See Figure 8.1.2 (a) and (b).

If f is differentiable at every $x \in A$, we say that f is *differentiable on A*. If the domain of f contains $[a, b]$, we say that f is *differentiable on $[a, b]$* provided f is differentiable on (a, b), right differentiable at a and left differentiable at b. Similar definitions apply to $[a, b)$, $(a, b]$, $[a, \infty)$ and $(-\infty, a]$. We call f *differentiable* if it is differentiable on its domain.

The real functions f', f'_L and f'_R are called, respectively, the *derivative of f*, the *left derivative of f* and the *right derivative of f*.

Notes 8.1.2 (1) The existence of $f'(x)$ is equivalent to the curve $y = f(x)$ having a unique tangent at $(x, f(x))$. When this happens, $f'(x)$ is the gradient of the tangent.

(2) Either (or both) of $f'_L(x)$ and $f'_R(x)$ may exist without $f'(x)$ existing. We can think of the left [right] derivative, when it exists, as giving the gradient of the left [right] tangent to the curve $y = f(x)$ at the point $(x, f(x))$. See Figure 8.1.1 (b).

(3) The derivative $f'(x)$ exists if and only if $f'_L(x)$ *and* $f'_R(x)$ exist and are *equal*.

(4) An *essential* requirement for the existence of $f'(x)$ is that f is defined on a neighbourhood of x. However, $f'_L(x)$ $[f'_R(x)]$ may exist if f is defined only on a left [right] neighbourhood of x.

(5) For convenience, when f is differentiable on $[a, b]$ we may use $f'(a)$ and $f'(b)$ to denote the right and left derivatives of f at a and b, respectively.

Example 8.1.3 Find $f'(x)$ from first principles when

$$\text{(a)} \quad f(x) = x^4, \qquad \text{(b)} \quad f(x) = \sqrt{x} \ (x > 0), \qquad \text{(c)} \quad f(x) = \sin x.$$

Solution (a) Let $x \in \mathbb{R}$. Then for $h \neq 0$,

$$\begin{aligned}
\frac{f(x + h) - f(x)}{h} &= \frac{(x + h)^4 - x^4}{h} \\
&= \frac{1}{h}[x^4 + 4x^3 h + 6x^2 h^2 + 4xh^3 + h^4 - x^4] \\
&= 4x^3 + 6x^2 h + 4xh^2 + h^3 \\
&\to 4x^3 \quad \text{as} \quad h \to 0.
\end{aligned}$$

Thus f is differentiable on $\mathbb{R}$ with $f'(x) = 4x^3$.

(b) Let $x > 0$. Then, for $0 < |h| < x$,

$$\begin{aligned}
\frac{f(x + h) - f(x)}{h} &= \frac{\sqrt{x + h} - \sqrt{x}}{h} \\
&= \frac{\sqrt{x + h} - \sqrt{x}}{h} \cdot \frac{\sqrt{x + h} + \sqrt{x}}{\sqrt{x + h} + \sqrt{x}} \\
&= \frac{(x + h) - x}{h(\sqrt{x + h} + \sqrt{x})} \\
&= \frac{1}{\sqrt{x + h} + \sqrt{x}} \to \frac{1}{2\sqrt{x}} \quad \text{as} \quad h \to 0.
\end{aligned}$$

Thus f is differentiable on $(0, \infty)$ with $f'(x) = \dfrac{1}{2\sqrt{x}}$.

(c) Let $x \in \mathbb{R}$. Then, for $h \neq 0$,

$$\frac{f(x+h) - f(x)}{h} = \frac{\sin(x+h) - \sin x}{h}$$

$$= \frac{2\cos(x + h/2)\sin(h/2)}{h}$$

$$= \cos(x + h/2)\frac{\sin(h/2)}{h/2}$$

$$\to (\cos x).1 = \cos x \quad \text{as} \quad h \to 0.$$

Thus f is differentiable on $\mathbb{R}$ with $f'(x) = \cos x$. □

Remark In the solution of Example 8.1.3 (b) the condition $0 < |h| < x$ is imposed to ensure that $f(x + h)$ is defined. The essential requirement is that h should be *sufficiently small*. In situations like this, where we eventually let $h \to 0$, it is normally acceptable to begin simply with $h \neq 0$.

Lemma 8.1.4 *Let* $K \in \mathbb{R}$ *and let* $n \in \mathbb{N}$. *Then, for* $x \in \mathbb{R}$,

$$\text{(a)} \quad \frac{d}{dx}K = 0, \qquad \text{(b)} \quad \frac{d}{dx}x^n = nx^{n-1}.$$

Proof (a) Let f be the real function defined by $f(x) = K$. Let $x \in \mathbb{R}$. Then, for $h \neq 0$,

$$\frac{f(x+h) - f(x)}{h} = \frac{K - K}{h} = \frac{0}{h} \to 0 \quad \text{as} \quad h \to 0.$$

(b) Let f be the real function defined by $f(x) = x^n$. Let $x \in \mathbb{R}$. Then, for $h \neq 0$, (see Lemma 3.2.8)

$$\frac{(x+h)^n - x^n}{h} = \frac{1}{h}[(x+h) - x][(x+h)^{n-1} + (x+h)^{n-2}x + \cdots$$

$$\cdots + (x+h)x^{n-2} + x^{n-1}]$$

$$= (x+h)^{n-1} + (x+h)^{n-2}x + \cdots + (x+h)x^{n-2} + x^{n-1}$$

$$\to \underbrace{x^{n-1} + x^{n-2}x + \cdots + xx^{n-2} + x^{n-1}}_{n \text{ terms}} = nx^{n-1} \quad \text{as} \quad h \to 0. \quad □$$

Problem 8.1.5 Prove Lemma 8.1.4 using the binomial expansion of $(x+h)^n$ instead of the factorisation of $(x+h)^n - x^n$.

8.2 Properties of Derivatives

We saw, in Theorem 7.3.9, that the arithmetic and composition of functions can be used to build up new continuous functions from old. In this section we show that the same processes can be applied to differentiable functions with similar effect.

To begin with, however, we look at the relationship between continuity and differentiability.

Theorem 8.2.1 *Let f be a real function, differentiable at x. Then f is continuous at x.*

Proof First observe that, since f is differentiable at x, f is defined on a neighbourhood N (say) of x. Next, for $t \in N - \{x\}$, let $h = t - x$ so that $h \neq 0$. Then

$$f(x + h) = \frac{f(x + h) - f(x)}{h}.h + f(x)$$
$$\to f'(x).0 + f(x) = f(x) \quad \text{as} \quad h \to 0.$$

It follows from Note 7.2.5 (5) that $\lim_{t \to x} f(t)$ exists and equals $f(x)$. Hence f is continuous at x. $\qquad \square$

Hazard The converse of Theorem 8.2.1 is false as the next example shows.

Example 8.2.2 Let f be the real function defined by $f(x) = |x|$. Show that f is not differentiable at 0.

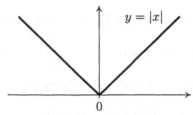

Figure 8.2.1

Solution We can see in Figure 8.2.1 that the curve $y = |x|$ is not smooth at $(0, 0)$. The gradient of the left tangent is -1 while the gradient of the right tangent is 1. We prove this as follows. For $h \neq 0$,

$$\frac{f(0 + h) - f(0)}{h} = \frac{|h| - |0|}{h} = \frac{|h|}{h}$$
$$= \begin{cases} -1 & \text{if } h < 0 \\ 1 & \text{if } h > 0 \end{cases} \quad \begin{array}{l} \to \quad -1 \quad \text{as } h \to 0^-, \\ \to \quad 1 \quad \text{as } h \to 0^+. \end{array}$$

Thus $f'_L(0) = -1$ and $f'_R(0) = 1$ so f cannot be differentiable at 0. $\qquad \square$

Theorem 8.2.3 *Let f and g be real functions, differentiable at x, and let $\lambda \in \mathbb{R}$. Then λf and $f + g$ are differentiable at x with*

$$(\lambda f)'(x) = \lambda f'(x) \quad and \quad (f + g)'(x) = f'(x) + g'(x).$$

Proof Since f and g are each defined on a neighbourhood of x, so are λf and $f + g$. See Note 2.2.3. For $h \neq 0$,

$$\frac{(\lambda f)(x + h) - (\lambda f)(x)}{h} = \frac{\lambda f(x + h) - \lambda f(x)}{h}$$
$$= \lambda.\frac{f(x + h) - f(x)}{h}$$
$$\to \lambda f'(x) \quad \text{as} \quad h \to 0$$

and

$$\frac{(f+g)(x+h) - (f+g)(x)}{h} = \frac{f(x+h) + g(x+h) - f(x) - g(x)}{h}$$

$$= \frac{f(x+h) - f(x)}{h} + \frac{g(x+h) - g(x)}{h}$$

$$\to f'(x) + g'(x) \quad \text{as} \quad h \to 0$$

which proves that λf and $f + g$ are differentiable at x and the derivatives are as stated.   □

Example 8.2.4 Find the equations of the tangent and normal to the curve

$$y = 2x^3 - 7x^2 + x + 5$$

at the point $(2, -5)$. Find, also, the point where the tangent meets the curve again.

Solution   Here we have

$$\frac{dy}{dx} = 2.3x^2 - 7.2x + 1 + 0 = -3 \quad \text{when} \quad x = 2.$$

So the tangent at $(2, -5)$ has equation

$$y + 5 = -3(x - 2), \qquad i.e. \ y = -3x + 1.$$

Since the normal at $(2, -5)$ is, by definition, perpendicular to the tangent, it has gradient $1/3$ and equation

$$y + 5 = \frac{1}{3}(x - 2), \qquad i.e. \ 3y = x - 17.$$

To find the points where the curve and the tangent meet, we must solve, simultaneously, the equations

$$y = 2x^3 - 7x^2 + x + 5 \qquad \text{and} \qquad y = -3x + 1.$$

Setting

$$y[\text{for the curve}] = y[\text{for the tangent}]$$

gives

$$2x^3 - 7x^2 + x + 5 = -3x + 1,$$

i.e.

$$2x^3 - 7x^2 + 4x + 4 = 0. \tag{8.2.1}$$

We know that the tangent meets the curve at $(2, -5)$ so $x = 2$ must satisfy equation (8.2.1) and hence $(x - 2)$ must be a factor of the left-hand-side. The complete factorisation gives

$$(x - 2)^2(2x + 1) = 0.$$

Thus the tangent meets the curve again where $x = -1/2$ and substituting in the equation of the tangent we get the point $(-1/2, 5/2)$.   □

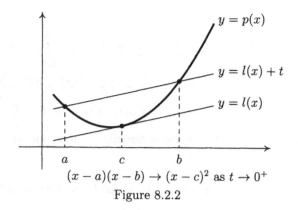

$$(x - a)(x - b) \to (x - c)^2 \text{ as } t \to 0^+$$

Figure 8.2.2

Note 8.2.5 In the solution of Example 8.2.4 it is not a coincidence that $(x - 2)^2$ is a factor of the left-hand-side of equation (8.2.1). In general, if $y = l(x)$ is the tangent to a polynomial curve $y = p(x)$ where $x = c$ then $(x - c)^2$ will be a factor of $p(x) - l(x)$. Figure 8.2.2 gives an indication of why one might expect this to hold.

Theorem 8.2.6 (The Chain Rule) *Let f and g be real functions with f differentiable at x and g differentiable at $f(x)$. Then $g \circ f$ is differentiable at x with*

$$(g \circ f)'(x) = g'(f(x))f'(x).$$

Proof [For a simpler proof of a special case see Note 8.2.7 (1).] By Theorem 8.2.1, f is continuous at x. Then since f is defined on a neighbourhood of x and g is defined on a neighbourhood of $f(x)$, $g \circ f$ is defined on a neighbourhood of x (see Note 7.2.5 (3)). Since f is defined on a neighbourhood of x, we can define

$$k(s) = f(x + s) - f(x)$$

for all s in a neighbourhood of 0. Then the continuity of f at x gives

$$\lim_{s \to 0} k(s) = 0. \tag{8.2.2}$$

Since g is defined on a neighbourhood of $f(x)$, we can define $\Delta(t)$, for all t in a neighbourhood of 0, by

$$\Delta(t) = \begin{cases} \dfrac{g(f(x) + t) - g(f(x))}{t} & \text{if } t \neq 0, \\ g'(f(x)) & \text{if } t = 0. \end{cases}$$

Then the differentiability of g at $f(x)$ gives

$$\lim_{t \to 0} \Delta(t) = g'(f(x)) = \Delta(0). \tag{8.2.3}$$

From equations (8.2.2) and (8.2.3) it follows, using Theorem 7.2.3, that

$$\lim_{s \to 0} \Delta(k(s)) = \Delta(0) = g'(f(x)).$$

Hence, for $h \neq 0$

$$\frac{(g \circ f)(x+h) - (g \circ f)(x)}{h} = \frac{g(f(x+h)) - g(f(x))}{h}$$

$$= \frac{g(f(x) + k(h)) - g(f(x))}{h}$$

$$= \frac{\Delta(k(h)).k(h)}{h}$$

$$= \Delta(k(h)) \frac{f(x+h) - f(x)}{h}$$

$$\rightarrow g'(f(x))f'(x) \quad as \quad h \rightarrow 0$$

which establishes the chain rule. $\qquad\qquad\qquad\qquad\qquad\qquad\qquad\qquad\qquad$ □

Notes 8.2.7 (1) In the proof of the chain rule, $\Delta(t)$ is defined specifically so that

$$g(f(x) + k(x)) - g(f(x)) = \Delta(k(h)).k(h)$$

whether or not $k(h) = 0$. If the function f has the property that $f(x+h) - f(x) \neq 0$ for all h in a punctured neighbourhood of 0 then $\Delta(t)$ is not needed and the proof can be simplified as follows.

As in the proof of the theorem, $g \circ f$ is defined on a neighbourhood of x. For $h \neq 0$, let $k = f(x+h) - f(x)$ so that, by the continuity of f at x, $k \rightarrow 0$ as $h \rightarrow 0$. Then

$$\frac{(g \circ f)(x+h) - (g \circ f)(x)}{h} = \frac{g(f(x+h)) - g(f(x))}{f(x+h) - f(x)} \cdot \frac{f(x+h) - f(x)}{h}$$

$$= \frac{g(f(x) + k) - g(f(x))}{k} \cdot \frac{f(x+h) - f(x)}{h}$$

$$\rightarrow g'(f(x))f'(x) \quad \text{as} \quad h \rightarrow 0.$$

(2) The chain rule can be extended (inductively) to three or more suitable functions. For example,

$$(h \circ g \circ f)'(x) = h'(g(f(x)))g'(f(x))f'(x).$$

To see this, let $g \circ f = k$ and begin by applying the chain rule to $h \circ k$.

(3) Writing $y = f(x)$ and $z = g(f(x)) = g(y)$ the chain rule becomes

$$\frac{dz}{dx} = \frac{dz}{dy}\frac{dy}{dx}$$

and if we write $w = h(g(f(x))) = h(g(y)) = h(z)$ then, using (2),

$$\frac{dw}{dx} = \frac{dw}{dz}\frac{dz}{dy}\frac{dy}{dx}.$$

Example 8.2.8 Find (a) $\dfrac{d}{dx}\sin(3x^2)$, (b) $\dfrac{d}{dx}\sin^4\sqrt{x}$.

Solution   (a) Let $y = 3x^2$ and $z = \sin y$. Then using Note 8.2.7 (3),

$$\frac{d}{dx}\sin(3x^2) = \frac{dz}{dx} = \frac{dz}{dy}\frac{dy}{dx} = \left(\frac{d}{dy}\sin y\right)\left(\frac{d}{dx}3x^2\right)$$
$$= (\cos y)(6x) = 6x\cos(3x^2).$$

(b)   Let $y = \sqrt{x}$, $z = \sin y$ and $w = z^4$. Then using Note 8.2.7 (3),

$$\frac{d}{dx}\sin^4\sqrt{x} = \frac{dw}{dx} = \frac{dw}{dz}\frac{dz}{dy}\frac{dy}{dx} = 4z^3\cos y\frac{1}{2\sqrt{x}} = \frac{2\sin^3\sqrt{x}\cos\sqrt{x}}{\sqrt{x}}. \qquad \square$$

Theorem 8.2.9 *Let f be a real bijection, continuous on a neighbourhood of $x = f^{-1}(y)$ and differentiable at x with $f'(x) \neq 0$. Then f^{-1} is differentiable at y with*

$$(f^{-1})'(y) = \frac{1}{f'(x)}. \tag{8.2.4}$$

Proof   By Corollary 7.3.19 (2), f^{-1} is defined on a neighbourhood of y. Since

$$\lim_{h\to 0}\frac{f(x+h) - f(x)}{h} = f'(x) \neq 0,$$

it follows, using Theorem 7.2.3 (2)(d), that

$$\lim_{h\to 0}\frac{h}{f(x+h) - f(x)} = \frac{1}{f'(x)}.$$

For $k \neq 0$, let $h = f^{-1}(y+k) - f^{-1}(y)$. Then, by Corollary 7.3.19 (2), f^{-1} is continuous at y so that $h \to 0$ as $k \to 0$. Further, $f^{-1}(y+k) = f^{-1}(y) + h = x + h$ so that $y + k = f(x+h)$. Hence

$$\frac{f^{-1}(y+k) - f^{-1}(y)}{k} = \frac{h}{(y+k) - y} = \frac{h}{f(x+h) - f(x)}$$
$$\to \frac{1}{f'(x)} \quad \text{as} \quad k \to 0.$$

Thus f^{-1} is differentiable at y and equation (8.2.4) holds. $\square$

Notes 8.2.10 (1) In Theorem 8.2.9, since $x = f^{-1}(y)$ and $y = f(x)$, equation (8.2.4) may be written

$$\frac{dx}{dy} = 1 \left/ \frac{dy}{dx}\right..$$

(2)   For all x in the domain of f, $(f^{-1} \circ f)(x) = x$. If we can differentiate the left-hand-side then, using the chain rule, $(f^{-1})'(f(x))f'(x) = 1$ which leads directly to equation (8.2.4). Note, however, that this does not prove that f^{-1} is differentiable.

(3)   The mirror property for inverse functions, Note 2.5.8, suggests, intuitively, that smoothness will carry over from f to f^{-1} and equation (8.2.4) can be compared with the fact that, when $m \neq 0$ the mirror image of the line $y - b = m(x - a)$ in the diagonal $y = x$ is the line $y - a = (1/m)(x - b)$.

To illustrate Theorem 8.2.9, consider the real bijection $f : (0, \infty) \to (0, \infty)$ defined by $f(x) = x^2$. Then $f^{-1} : (0, \infty) \to (0, \infty)$ is defined by $f^{-1}(x) = \sqrt{x}$ and, for $x \in (0, \infty)$,

$$(f^{-1})'(x) = \frac{1}{f'(y)} \quad \text{where} \quad y = f^{-1}(x)$$

$$= \frac{1}{2y} \quad \text{where} \quad y = \sqrt{x}$$

$$= \frac{1}{2\sqrt{x}}.$$

The next lemma takes this illustration a stage further.

Lemma 8.2.11 *Let $n \in \mathbb{N}$, $n \geq 2$. Then, for $x \in A$,*

$$\frac{d}{dx} x^{1/n} = \frac{1}{n} x^{(1/n)-1}$$

where $A = (0, \infty)$ in n is even and $A = \mathbb{R} - \{0\}$ if n is odd.

Proof Define $f : A \to A$ by $f(x) = x^n$. Then f is a bijection. For $x \in A$, $f^{-1}(x) = x^{1/n} = y$ (say), $f'(y) \neq 0$ and

$$\frac{dy}{dx} = 1 \bigg/ \frac{dx}{dy} = \left(\frac{d}{dy} y^n \right)^{-1} = \frac{1}{n} y^{1-n}$$

$$= \frac{1}{n} \left(x^{1/n} \right)^{(1-n)} = \frac{1}{n} x^{(1/n)-1}. \qquad \square$$

Theorem 8.2.12 (The Product, Quotient and Reciprocal Rules) *Let f and g be real functions, differentiable at x. Then fg is differentiable at x with*

$$(fg)'(x) = f'(x)g(x) + f(x)g'(x) \qquad (\text{product rule})$$

and, provided $g(x) \neq 0$, f/g is differentiable at x with

$$\left(\frac{f}{g} \right)' (x) = \frac{f'(x)g(x) - f(x)g'(x)}{g(x)^2} \qquad (\text{quotient rule})$$

and, in particular,

$$\left(\frac{1}{g} \right)' (x) = -\frac{g'(x)}{g(x)^2} \qquad (\text{reciprocal rule}).$$

Proof By Theorem 8.2.1, g is continuous at x and it follows that $g(x + h) \to g(x)$ as $h \to 0$.

By Note 2.2.3, since f and g are defined on neighbourhoods of x, so is fg. For $h \neq 0$,

$$\frac{(fg)(x+h) - (fg)(x)}{h} = \frac{1}{h}[f(x+h)g(x+h) - f(x)g(x)]$$

$$= \frac{1}{h}[f(x+h)g(x+h) - f(x)g(x+h) + f(x)g(x+h) - f(x)g(x)]$$

$$= \frac{f(x+h) - f(x)}{h}g(x+h) + f(x)\frac{g(x+h) - g(x)}{h}$$

$$\to f'(x)g(x) + f(x)g'(x) \text{ as } h \to 0$$

which establishes the product rule.

Now suppose that $g(x) \neq 0$. We consider $1/g$ first. By Theorem 8.2.1, g is continuous at x and hence, by Theorem 7.3.9 (1), $1/g$ is defined on a neighbourhood of x. For $h \neq 0$,

$$\frac{1}{h}\left[\frac{1}{g}(x+h) - \frac{1}{g}(x)\right] = \frac{1}{h}\left[\frac{1}{g(x+h)} - \frac{1}{g(x)}\right]$$

$$= \frac{1}{h}\left[\frac{g(x) - g(x+h)}{g(x+h)g(x)}\right]$$

$$= \frac{-1}{g(x+h)g(x)}\left[\frac{g(x+h) - g(x)}{h}\right]$$

$$\to \frac{-1}{g(x)g(x)}g'(x) = -\frac{g'(x)}{g(x)^2} \text{ as } h \to 0$$

which establishes the reciprocal rule.

Since $f/g = f.(1/g)$ we can deduce, from the product and reciprocal rules, that f/g is differentiable at x with

$$\left(\frac{f}{g}\right)'(x) = \left(f\frac{1}{g}\right)'(x) = f'(x)\left(\frac{1}{g}\right)(x) + f(x)\left(\frac{1}{g}\right)'(x)$$

$$= \frac{f'(x)}{g(x)} - \frac{f(x)g'(x)}{g(x)^2} = \frac{f'(x)g(x) - f(x)g'(x)}{g(x)^2}$$

which establishes the quotient rule.   □

Example 8.2.13 Differentiate, with respect to x,

$$\text{(a)} \quad \sin^3 x \sin 3x, \qquad \text{(b)} \quad \frac{x^2 - 3x + 1}{3x - 1}.$$

Solution (a) Using the product rule and the chain rule,

$$\frac{d}{dx}(\sin^3 x \sin 3x) = \left(\frac{d}{dx}\sin^3 x\right)\sin 3x + \sin^3 x\left(\frac{d}{dx}\sin 3x\right)$$

$$= (3\sin^2 x \cos x)\sin 3x + \sin^3 x(3\cos 3x)$$

$$= 3\sin^2 x(\cos x \sin 3x + \sin x \cos 3x)$$

$$= 3\sin^2 x \sin 4x.$$

(b) Using the quotient rule,

$$\frac{d}{dx}\left(\frac{x^2 - 3x + 1}{3x - 1}\right)$$

$$= \frac{\left(\frac{d}{dx}(x^2 - 3x + 1)\right)(3x - 1) - (x^2 - 3x + 1)\left(\frac{d}{dx}(3x - 1)\right)}{(3x - 1)^2}$$

$$= \frac{(2x - 3)(3x - 1) - (x^2 - 3x + 1)(3)}{(3x - 1)^2}$$

$$= \frac{3x^2 - 2x}{(3x - 1)^2} = \frac{x(3x - 2)}{(3x - 1)^2}.$$ □

Remark The rules for derivatives in Theorems 8.2.3, 8.2.6 and 8.2.12 can be applied even when the functions involved are not given explicitly. For example, if y is a differentiable function of x then applying the chain rule to y^5 and the product rule to $x^4 y$ gives

$$\frac{d}{dx}(y^5) = 5y^4 \frac{dy}{dx} \qquad \text{and} \qquad \frac{d}{dx}(x^4 y) = 4x^3 y + x^4 \frac{dy}{dx}.$$

Example 8.2.14 Given that y is a differentiable function of x and

$$xy + 1 = 2\sin(x + y), \tag{8.2.5}$$

find $\dfrac{dy}{dx}$ and the equation of the tangent to the curve defined by (8.2.5) at the point $(-1, 1)$.

Solution Differentiating both sides of equation (8.2.5), with respect to x, we get

$$1.y + x\frac{dy}{dx} + 0 = 2\cos(x + y)\left(1 + \frac{dy}{dx}\right),$$

i.e.

$$y + x\frac{dy}{dx} = 2\cos(x + y) + 2\frac{dy}{dx}\cos(x + y),$$

i.e.

$$\frac{dy}{dx}[x - 2\cos(x + y)] = 2\cos(x + y) - y.$$

Thus, provided $x - 2\cos(x + y) \neq 0$,

$$\frac{dy}{dx} = \frac{2\cos(x + y) - y}{x - 2\cos(x + y)}.$$

So when $x = -1$ and $y = 1$,

$$\frac{dy}{dx} = \frac{2\cos 0 - 1}{-1 - 2\cos 0} = \frac{2 - 1}{-1 - 2} = -\frac{1}{3}$$

and the equation of the tangent at $(-1, 1)$ is

$$y - 1 = -\frac{1}{3}(x + 1), \qquad\qquad \text{i.e. } x + 3y = 2.$$ □

Remark If f is a real function then the equation $y = f(x)$ is said to define y *explicitly*. On the other hand, an equation such as (8.2.5), is said to define y *implicitly* and differentiating both sides of such an equation is called *implicit differentiation*.

8.3   Some Standard Derivatives

In this section we investigate the derivatives of rational power functions, the trigonometric functions and the inverse trigonometric functions.

Theorem 8.3.1 *Let $q \in \mathbb{Q} - \{0\}$. Then, for $x \in D'(q)$,*

$$\frac{d}{dx}x^q = qx^{q-1},$$

where $D'(q)$ is defined as follows. Let $q = k/n$ where $k \in \mathbb{Z}$, $n \in \mathbb{N}$ and k and n have no common factor. Then

$$D'(q) = \begin{cases} \mathbb{R} & \text{if } q \geq 1 \text{ and } n \text{ is odd,} \\ \mathbb{R} - \{0\} & \text{if } q < 1 \text{ and } n \text{ is odd,} \\ (0,\infty) & \text{if } n \text{ is even.} \end{cases}$$

[Compare this with $D(q)$ defined in Note 1.5.8.]

Proof   Lemmas 8.1.4 and 8.2.11 cover the cases $q = n, 1/n$ $(n \in \mathbb{N})$.
Now suppose that $k > 0$. Using the lemmas and the chain rule, for $x \in D'(q)$,

$$\frac{d}{dx}x^q = \frac{d}{dx}(x^{1/n})^k = k(x^{1/n})^{k-1}\frac{1}{n}x^{(1/n)-1} = \frac{k}{n}x^{[(k/n)-(1/n)+(1/n)-1]} = qx^{q-1}.$$

Lastly, suppose that $q < 0$. Let $p = -q$. Then using the reciprocal rule, for $x \in D'(q)$,

$$\frac{d}{dx}x^q = \frac{d}{dx}\left(\frac{1}{x^p}\right) = -\frac{px^{p-1}}{x^{2p}} = -px^{-p-1} = qx^{q-1}. \qquad \square$$

Theorem 8.3.2 *The functions sine, cosine, tangent, cosecant, secant and cotangent are differentiable on their maximal domains with*

$$\frac{d}{dx}\sin x = \cos x, \qquad\qquad \frac{d}{dx}\operatorname{cosec} x = -\operatorname{cosec} x \cot x.$$

$$\frac{d}{dx}\cos x = -\sin x, \qquad\qquad \frac{d}{dx}\sec x = \sec x \tan x,$$

$$\frac{d}{dx}\tan x = \sec^2 x, \qquad\qquad \frac{d}{dx}\cot x = -\operatorname{cosec}^2 x.$$

Proof   From Example 8.1.3, for $x \in \mathbb{R}$, $\dfrac{d}{dx}\sin x = \cos x$.
Using the chain rule, for $x \in \mathbb{R}$,

$$\frac{d}{dx}\cos x = \frac{d}{dx}\sin(\tfrac{\pi}{2} - x) = \cos(\tfrac{\pi}{2} - x)(-1) = -\sin x.$$

Using the quotient rule, for $\cos x \neq 0$,

$$\frac{d}{dx}\tan x = \frac{d}{dx}\left(\frac{\sin x}{\cos x}\right) = \frac{\cos x \cos x - \sin x(-\sin x)}{\cos^2 x} = \frac{1}{\cos^2 x} = \sec^2 x.$$

Using the reciprocal rule, for $\sin x \neq 0$,

$$\frac{d}{dx}\operatorname{cosec} x = \frac{d}{dx}\left(\frac{1}{\sin x}\right) = -\frac{\cos x}{\sin^2 x} = -\frac{1}{\sin x}.\frac{\cos x}{\sin x} = -\operatorname{cosec} x \cot x.$$

Using the reciprocal rule, for $\cos x \neq 0$,

$$\frac{d}{dx}\sec x = \frac{d}{dx}\left(\frac{1}{\cos x}\right) = \frac{\sin x}{\cos^2 x} = \frac{1}{\cos x}.\frac{\sin x}{\cos x} = \sec x \tan x.$$

Using the quotient rule, for $\sin x \neq 0$,

$$\frac{d}{dx}\cot x = \frac{d}{dx}\left(\frac{\cos x}{\sin x}\right) = \frac{-\sin x \sin x - \cos x \cos x}{\sin^2 x} = \frac{-1}{\sin^2 x} = -\operatorname{cosec}^2 x. \qquad \square$$

Theorem 8.3.3 *The inverse sine and inverse cosine functions are differentiable on* $(-1, 1)$ *with*

$$\frac{d}{dx}\sin^{-1} x = \frac{1}{\sqrt{1-x^2}} \qquad and \qquad \frac{d}{dx}\cos^{-1} x = -\frac{1}{\sqrt{1-x^2}}.$$

The inverse tangent function is differentiable on $\mathbb{R}$ *with*

$$\frac{d}{dx}\tan^{-1} x = \frac{1}{x^2 + 1}.$$

Proof　From Definition 5.5.2, $\sin^{-1}$ is the inverse of the bijection $f : [-\frac{\pi}{2}, \frac{\pi}{2}] \to [-1, 1]$ defined by $f(y) = \sin y$. For $x \in (-1, 1)$, let $y = \sin^{-1} x$. Then $-\frac{\pi}{2} < y < \frac{\pi}{2}$ and $\sin y = x$ so that

$$\cos y = \sqrt{1 - \sin^2 y} = \sqrt{1 - x^2}.$$

Hence, by Theorem 8.2.9, since f is differentiable at y and $f'(y) = \cos y \neq 0$,

$$\frac{d}{dx}\sin^{-1} x = \frac{1}{f'(y)} = \frac{1}{\cos y} = \frac{1}{\sqrt{1 - x^2}}.$$

A proof similar to that for $\sin^{-1}$ can be applied to $\cos^{-1}$. This is left to the reader.

From Definition 5.5.4, $\tan^{-1}$ is the inverse of the bijection $h : (-\frac{\pi}{2}, \frac{\pi}{2}) \to \mathbb{R}$ defined by $f(y) = \tan y$. For $x \in \mathbb{R}$, let $y = \tan^{-1} x$. Then $\tan y = x$. Hence, by Theorem 8.2.9, since h is differentiable at y and $h'(y) = \sec^2 y = \tan^2 y + 1 \neq 0$,

$$\frac{d}{dx}\tan^{-1} x = \frac{1}{h'(y)} = \frac{1}{\tan^2 y + 1} = \frac{1}{x^2 + 1}. \qquad \square$$

The formulae for the derivatives in Theorem 8.3.3 can also be found by implicit differentiation. For example, differentiating both sides of the equation $\tan y = x$, with respect to x, gives

$$\sec^2 y \frac{dy}{dx} = 1, \qquad i.e. \quad \frac{dy}{dx} = \frac{1}{\sec^2 y} = \frac{1}{\tan^2 y + 1} = \frac{1}{x^2 + 1}.$$

Example 8.3.4 Differentiate, with respect to x,

$$\text{(a)} \quad \left(\frac{1+4x}{1-5x}\right)^{2/3}, \qquad \text{(b)} \quad \sin^{-1}\left(\frac{x}{a}\right) \quad (a>0).$$

Solution (a) Using the chain and quotient rules, for $x \in \mathbb{R} - \{-\frac{1}{4}, \frac{1}{5}\}$,

$$\frac{d}{dx}\left(\frac{1+4x}{1-5x}\right)^{2/3} = \frac{2}{3}\left(\frac{1+4x}{1-5x}\right)^{-1/3}\frac{4(1-5x)-(1+4x)(-5)}{(1-5x)^2}$$

$$= \frac{2(1+4x)^{-1/3}(9)}{3(1-5x)^{-1/3}(1-5x)^2} = \frac{6}{(1+4x)^{1/3}(1-5x)^{5/3}}.$$

(b) Using the chain rule, for $|x| < a$,

$$\frac{d}{dx}\sin^{-1}\left(\frac{x}{a}\right) = \frac{1}{\sqrt{1-\left(\frac{x}{a}\right)^2}}\frac{1}{a} = \frac{1}{\sqrt{a^2-x^2}}. \qquad \square$$

Example 8.3.5 Let f be the real function defined by

$$f(x) = \cos x (\sec 3x)^{1/3}.$$

For which x, in the range $0 \le x \le \pi$, is $f'(x) = 0$?

Solution We have

$$\frac{df}{dx} = (-\sin x)(\sec 3x)^{1/3} + \cos x.\frac{1}{3}(\sec 3x)^{-2/3}\sec 3x \tan 3x.3$$

$$= (\sec 3x)^{1/3}(\cos x \tan 3x - \sin x)$$

$$= \frac{\sin 3x \cos x - \cos 3x \sin x}{(\cos 3x)^{4/3}} = \frac{\sin 2x}{(\cos 3x)^{4/3}}$$

$$= 0 \quad \text{when } \sin 2x = 0 \text{ } and \text{ } \cos 3x \ne 0.$$

Hence, in the range $0 \le x \le \pi$, $f'(x) = 0$ when $x = 0$ or $x = \pi$. $\qquad \square$

Example 8.3.6 Find $\dfrac{dx}{dt}$ when $t = \tan\frac{1}{2}x$.

Solution Differentiating both sides of the equation $t = \tan\frac{1}{2}x$ with respect to t we get

$$1 = (\sec^2\tfrac{1}{2}x)(\tfrac{1}{2})\frac{dx}{dt}, \quad i.e. \quad \frac{dx}{dt} = \frac{2}{\sec^2\frac{1}{2}x} = \frac{2}{1+\tan^2\frac{1}{2}x} = \frac{2}{1+t^2}. \qquad \square$$

Remark The expression for $\dfrac{dx}{dt}$ in Example 8.3.6 is one of the *t-formulae*, which will be used in Section 16.3.

8.4　Higher Derivatives

Definitions 8.4.1 Let f be a real function. If f' exists and is itself differentiable then the resulting function is called the *second derivative of f* and is denoted by f'' or $f^{(2)}$. Its value at x is denoted by

$$f''(x) \quad \text{or} \quad f^{(2)}(x) \quad \text{or} \quad \frac{d^2 f}{dx^2} \quad \text{or} \quad \frac{d^2}{dx^2} f(x) \quad \text{or} \quad \frac{d^2 y}{dx^2} \quad \text{or} \quad y^{(2)} \quad \text{or} \quad y''$$

where $y = f(x)$.

More generally, for $n = 2, 3, 4, \ldots$, the nth *derivative of f*, when it exists, is denoted by $f^{(n)}$ and its value at x is denoted by

$$f^{(n)} \quad \text{or} \quad \frac{d^n f}{dx^n} \quad \text{or} \quad \frac{d^n}{dx^n} f(x) \quad \text{or} \quad \frac{d^n y}{dx^n} \quad \text{or} \quad y^{(n)}.$$

It will also be convenient, on occasions, to extend this notation and write $f^{(1)}$ for f' and $f^{(0)}$ for f.

Note 8.4.2 A more formal approach is to define higher derivatives recursively. Let $f^{(0)}$ be the function f and, for $n \in \mathbb{N}$, let $f^{(n)}$ be the derivative of $f^{(n-1)}$. See also Notes 4.1.8 (1).

Example 8.4.3 Find all the derivatives of

$$y = 4x^3 + 3x^2 + 2x + 1.$$

Solution　We have

$$\frac{dy}{dx} = 12x^2 + 6x + 2, \qquad \frac{d^2 y}{dx^2} = 24x + 6, \qquad \frac{d^3 y}{dx^3} = 24,$$

$$\frac{d^4 y}{dx^4} = 0 \quad \text{and} \quad \frac{d^n y}{dx^n} = 0 \quad \text{for } n = 5, 6, 7, \ldots. \qquad \square$$

Note 8.4.4 In general, if $p(x)$ is a polynomial of degree m then $\dfrac{d^n}{dx^n} p(x) = 0$ for $n > m$.

Example 8.4.5 Prove that, for all $n \in \mathbb{N}$,

$$\frac{d^n}{dx^n} \sin x = \sin(x + n\tfrac{\pi}{2}) \qquad (x \in \mathbb{R}). \tag{8.4.1}$$

Solution　We use induction. For $n \in \mathbb{N}$, let $P(n)$ be the statement (8.4.1). Since

$$\frac{d}{dx} \sin x = \cos x = \sin(x + (1)\tfrac{\pi}{2}) \qquad (x \in \mathbb{R})$$

it follows that $P(1)$ is true. Now let $k \in \mathbb{N}$ and assume that $P(k)$ is true. That is

$$\frac{d^k}{dx^k} \sin x = \sin(x + k\tfrac{\pi}{2}) \qquad (x \in \mathbb{R}).$$

We must deduce that $P(k+1)$ is true. That is

$$\frac{d^{k+1}}{dx^{k+1}}\sin x = \sin(x+(k+1)\tfrac{\pi}{2}) \qquad (x \in \mathbb{R}).$$

We have, for $x \in \mathbb{R}$,

$$\frac{d^{k+1}}{dx^{k+1}}\sin x = \frac{d}{dx}\left(\frac{d^k}{dx^k}\sin x\right) = \frac{d}{dx}\sin(x+k\tfrac{\pi}{2}) \qquad \text{(using } P(k))$$

$$= \cos(x+k\tfrac{\pi}{2}).1 \qquad \text{(by the chain rule)}$$

$$= \sin(x+k\tfrac{\pi}{2}+\tfrac{\pi}{2}) = \sin(x+(k+1)\tfrac{\pi}{2})$$

Thus it follows, by induction, that $P(n)$ is true for all $n \in \mathbb{N}$.   □

Problem 8.4.6 *Deduce* from Example 8.4.5 that, for all $n \in \mathbb{N}$,

$$\frac{d^n}{dx^n}\cos x = \cos(x+n\tfrac{\pi}{2}) \qquad (x \in \mathbb{R}).$$

Example 8.4.7 Given that y is a twice differentiable function of x and

$$x^2 + 4xy + 3y^2 = 7, \tag{8.4.2}$$

obtain expressions for y' and y'' in terms of x and y.

Solution Differentiating both sides of equation (8.4.2), with respect to x, we get

$$2x + 4(1.y + xy') + 6yy' = 0,$$

i.e.

$$x + 2(y + xy') + 3yy' = 0, \tag{8.4.3}$$

i.e.

$$y'(2x+3y) = -x - 2y. \tag{8.4.4}$$

Hence, provided $2x + 3y \neq 0$,

$$y' = -\frac{x+2y}{2x+3y}. \tag{8.4.5}$$

To find y'' we could differentiate equation (8.4.3), (8.4.4) or (8.4.5). Differentiating (8.4.4), with respect to x, we get

$$y''(2x+3y) + y'(2+3y') = -1 - 2y'.$$

Hence

$$y''(2x+3y) = -1 - 4y' - 3(y')^2 = -1 + 4\frac{x+2y}{2x+3y} - 3\frac{(x+2y)^2}{(2x+3y)^2}$$

$$= \frac{x^2 + 4xy + 3x^2}{(2x+3y)^2} = \frac{7}{(2x+3y)^2} \qquad \text{(using (8.4.2))}.$$

Hence, provided $2x + 3y \neq 0$,

$$y'' = \frac{7}{(2x+3y)^3}. \qquad □$$

Problem 8.4.8 The product rule for derivatives gives us the formula

$$(fg)'(x) = f'(x)g(x) + f(x)g'(x)$$

or, in an alternative notation and omitting the variable,

$$(fg)^{(1)} = f^{(1)}g^{(0)} + f^{(0)}g^{(1)}.$$

By differentiating both sides of this equation, and simplifying the right-hand-side, find a formula for $(fg)^{(2)}$. Continue this process to obtain formulae for $(fg)^{(3)}$ and $(fg)^{(4)}$. Can you now guess the formula for $(fg)^{(n)}$ $(n \in \mathbb{N})$?

8.X   Exercises

1. By arguing from first principles, find $f'(x)$ when $f(x)$ is

    (a) x^5,   (b) $x^3 + 2x$,   (c) $3x^2 + 2x + 1$,

    (d) $\dfrac{1}{x}$,   (e) $\dfrac{1}{3x-1}$,   (f) $\dfrac{1}{2-5x}$,

    (g) $\sqrt{x+2}$,   (h) $\sqrt{1-x}$,   (i) $\sqrt{x^2+1}$,

    (j) $\dfrac{1}{x^3}$,   (k) $\dfrac{1}{x^2+2x}$,   (l) $\dfrac{1}{\sqrt{1-x}}$,

    (m) $x^{2/3}$,   (n) $\sin(2x+1)$,   (o) $\cos x$.

2. Let f be a real function, continuous at 0. Prove that the real function g defined by $g(x) = xf(x)$ is differentiable at 0 with $g'(0) = f(0)$.

3. For each of the following curves, find the equations of the tangent and the normal at the given point and find any other point(s) where the tangent meets the curve.

    (a) $y = x^3 - x^2 - 15x - 17$ at $(-2, 1)$,

    (b) $y = x^4 - 2x^3 - 7x^2 + 22x - 13$ at $(2, 3)$,

    (c) $y = x^5 + 2x^4 - x^3 - 4x^2 - 4x + 3$ at $(-1, 5)$,

    (d) $y = 2x^3 - 12x^2 + 23x - 18$ at $(2, -4)$.

    In (d) find also the other points where the normal meets the curve.

4. Find the points on the curve $y = x^4 - 2x^3 - 2x^2 + 3x + 4$ where the tangent is parallel to the tangent at the point $(1, 4)$.

5. Differentiate with respect to x

(a) $(2x - 3)^5$,

(b) $(x^2 - 4)^6$,

(c) $(2x^2 + 5x - 1)^3$,

(d) $\sqrt{3x - 5}$,

(e) $\sqrt{1 - x^2}$,

(f) $(x^3 + 5)^{1/2}$,

(g) $(2x^2 + 5)(x^3 - 7)$,

(h) $(x - 5)^3(x^2 + 9)$,

(i) $(2x + 1)^4(3x - 4)^2$,

(j) $(4 - x)^3(1 + 2x)^5$,

(k) $(2x + 3)^{1/2}(3x - 1)^{1/2}$,

(l) $(6x^2 + 7x - 3)^{1/2}$,

(m) $\dfrac{1}{x^2 + 1}$,

(n) $\dfrac{4x - 9}{3x + 1}$,

(o) $\dfrac{2 - 5x}{3 + x}$,

(p) $\dfrac{x^2 - 2x - 1}{x^2 + 3x + 1}$,

(q) $\dfrac{2x + 1}{(x + 3)^2}$,

(r) $\left(\dfrac{9x + 7}{x - 6}\right)^3$,

(s) $\left(\dfrac{2x + 7}{3x + 1}\right)^{1/2}$,

(t) $\dfrac{\sqrt{x - 4}}{\sqrt{x + 4}}$,

(u) $\dfrac{1}{\sqrt{1 - x^2}}$,

(v) $\sqrt{2 + \sin^2 x}$,

(w) $\sin \sqrt{2 + x^2}$,

(x) $\sin 5x \sin^5 x$,

(y) $\dfrac{\sin\left(x + \frac{\pi}{2}\right)}{\sin\left(x - \frac{\pi}{2}\right)}$,

(z) $\sin(\sin(\sin x))$.

6. Find y' in terms of x and y when

(a) $x^2 + y^2 - 8x - 6y = 0$,

(b) $x^3 + y^3 = 5xy + 7$,

(c) $xy^2 - x^2y = \sin(x + y)$,

(d) $(x - y)^4 = 1 - \sin(xy)$.

7. Find the point P in the first quadrant where the curves

$$\frac{x^2}{49} + \frac{y^2}{36} = 1 \quad \text{and} \quad \frac{x^2}{9} - \frac{y^2}{4} = 1$$

intersect. Show that the tangents to these curves at P are perpendicular. The curves are said to intersect *orthogonally* at P.

8. Differentiate, with respect to x,

(a) $(x^2 + 2)^{3/2}$,

(b) $(3x^2 + 5x - 1)^{3/4}$,

(c) $\left(\dfrac{2x + 1}{7x - 3}\right)^{3/5}$,

(d) $(7 + 3x)^{2/3}(1 - 4x)^{1/3}$,

(e) $\sin(x^{2/3})$,

(f) $(\sin x)^{2/3}$,

(g) $\cos\sqrt{2x^2 + 3}$,

(h) $\sqrt{2\cos^2 x + 3}$,

(i) $\cos(\sin(x^2))$,

(j) $\sin(\cos(\tan x))$,

(k) $\tan^3 x$,

(l) $\tan x \sec x$,

(m) $\operatorname{cosec}^2\sqrt{x}$,

(n) $\sin^3 4x \cos^2 6x$,

(o) $(\sin 5x)^{4/3}\sin 2x$,

(p) $(\cos 4x)^{1/2}(\cos 6x)^{1/3}$,

(q) $\cos^7 x \sin 7x$,

(r) $\dfrac{\cos^7 x}{\sin 7x}$,

(s) $\dfrac{\sqrt{\cos 4x}}{\sin 5x}$,

(t) $\dfrac{\sec x}{x^2 + 1}$,

(u) $\dfrac{\sin(2x^2 + x + 1)}{\cos(2x^2 + x - 1)}$,

(v) $(\tan x + \sec x)^2$,

(w) $\sin^{-1}\sqrt{4x - 1}$,

(x) $\sin^{-1}\sqrt{\dfrac{1 - x^2}{1 + x^2}}$,

(y) $\cos^{-1}\left(\dfrac{x - 1}{x + 1}\right)$,

(z) $(1 - x)\tan^{-1}\sqrt{x}$.

9. Prove by induction that, for all $n \in \mathbb{N}$,

(a) $\dfrac{d^n}{dx^n}(x\sin x) = x\sin\left(x + \dfrac{n\pi}{2}\right) - n\cos\left(x + \dfrac{n\pi}{2}\right)$,

(b) $\dfrac{d^n}{dx^n}\left(\dfrac{1}{x + a}\right) = \dfrac{(-1)^n n!}{(x + a)^{n+1}}$ $(a \in \mathbb{R},\ x \neq -a)$.

Deduce from (b) that, for all $n \in \mathbb{N}$,

$$\dfrac{d^n}{dx^n}\left(\dfrac{2x}{x^2 - 1}\right) = (-1)^n n!\left[\dfrac{1}{(x - 1)^{n+1}} + \dfrac{1}{(x + 1)^{n+1}}\right] \qquad (x \neq \pm 1).$$

10. Given that y is a twice differentiable function of x and $x^2 + 3xy + y^2 = 1$, show that

$$\dfrac{d^2y}{dx^2} = \dfrac{10}{(3x + 2y)^3} \qquad (3x + 2y \neq 0).$$

11. Given that y is a twice differentiable function of x and $2y^2 - 3xy + x = 4$, show that

$$\dfrac{d^2y}{dx^2} = \dfrac{68}{(4y - 3x)^3} \qquad (4y \neq 3x).$$

Chapter 9

Differentiation—Applications

Derivatives are used to describe *dynamic* situations, *i.e.* situations where change in one quantity produces a change in another. Indeed, the difference quotient (see Definition 8.1.1) measures the *ratio* of the change in $f(x)$ to the change in x, and its limit is the derivative $f'(x)$. In this chapter, we develop theories which allow us to investigate such situations.

In many applications, it is important not merely to solve a problem, but to find the solution which is, in some sense, optimal. For example, there are infinitely many rectangles having a given perimeter, but the square is the choice which gives the maximum area. We will see that differentiation can be used to *locate* optimal (*i.e.* extremal) values. The reader should note that ideas of continuity show the existence of extreme values, but do *not* help us to find them.

Sections 9.3 and 9.4 develop the connection between differentiation and the theory of monotonic functions, *i.e.* functions which are increasing or decreasing on an interval. This is mainly used within mathematics. Along the way, we meet two powerful theorems—Rolle's theorem, which relates the zeros of f' to those of f, and the mean value theorem, which was once used to obtain approximations to function values which are hard to evaluate. The mean value theorem is a special case of Taylor's theorem which is introduced in Chapter 18.

The final section shows how differentiation can be used to study the *rate* at which functions change. The most obvious application is to the study of quantities which change in *time*. For example, velocity is the rate of change of position and acceleration is the rate of change of velocity. Notice that acceleration is thus a natural occurrence of a second derivative.

9.1 Critical Points

Definitions 9.1.1 Let f be a real function, differentiable at c. If $f'(c) = 0$ we call the point $(c, f(c))$ on the curve $y = f(x)$ a *critical point* or *stationary point* of the curve or of f.

Now suppose that f' is continuous at c, $f'(c) = 0$ and $f'(c) \neq 0$ for all x in a punctured neighbourhood of c. If $f'(x)$ changes sign at $x = c$, we have a *minimum turning point* or *maximum turning point*. See Figure 9.1.1.

If $f'(x)$ does not change sign at c, we have a *horizontal point of inflection* which can be *down-up* or *up-down*. See Figure 9.1.2. In Definition 10.2.10, a more general type

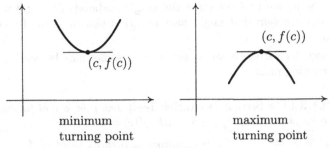

Figure 9.1.1

of point of inflection will be defined.

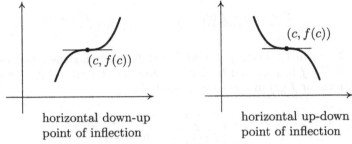

Figure 9.1.2

Example 9.1.2 Determine the nature of each critical point of the curve

$$y = \frac{1 - 5x + 10x^2 - 2x^3}{(x - 1)^5}.$$

Solution Using the quotient rule, we get

$$y' = \frac{4x^2(x - 6)}{(x - 1)^6}.$$

Hence y has critical points where $x = 0$ and $x = 6$. To determine their natures we draw a table of signs for y' in neighbourhoods of 0 and 6.

x	$\to$	0	$\to$	$\to$	6	$\to$
$4x^2$	+	0	+	+	+	+
$x - 6$	−	−	−	−	0	+
$(x - 1)^6$	+	+	+	+	+	+
y'	−	0	−	−	0	+
	$\searrow$	$\to$	$\searrow$	$\searrow$	$\to$	$\nearrow$

Thus y has a horizontal up-down point of inflection at $(0, -1)$ and a minimum turning point at $(6, -101/5^5)$. □

Hazard In the above table of signs the neighbourhoods of 0 and 6 are 'separated'. This avoids the problem that might have occurred had the sign of any factor changed between 0 and 6.

The second derivative, when it exists, can sometimes be used to determine the nature of a turning point.

Theorem 9.1.3 (The Second Derivative Test) *Let f be a real function, twice differentiable on a neighbourhood of $c \in \mathbb{R}$, with $f'(c) = 0$. Then*

(1) $f''(c) > 0 \;\Rightarrow\; (c, f(c))$ *is a minimum turning point of f,*

(2) $f''(c) < 0 \;\Rightarrow\; (c, f(c))$ *is a maximum turning point of f.*

Proof We prove (1). The proof of (2) is similar and is left to the reader.

Let $f''(c) > 0$. Then

$$\lim_{h \to 0} \frac{f'(c+h) - f'(c)}{h} > 0, \qquad i.e. \; \lim_{h \to 0} \frac{f'(c+h)}{h} > 0.$$

Hence, by Theorem 7.2.3 (1), $f'(c+h)/h > 0$ for all h in some punctured neighbourhood N of 0. So $f'(c+h)$ and h have the same sign for all $h \in N$. We can now draw a table of signs for $f'(x)$ in a neighbourhood of c.

x	$\to$	c	$\to$
$f'(x)$	$-$	0	$+$
	$\searrow$	$\longrightarrow$	$\nearrow$

It follows that $(c, f(c))$ is a minimum turning point of f.   □

To illustrate Theorem 9.1.3, consider

$$y = \frac{1 - 5x + 10x^2 - 2x^3}{(x-1)^5}.$$

We have seen that y has critical points where $x = 0$ and $x = 6$. Now

$$y'' = -\frac{12x(x^2 - 7x - 4)}{(x-1)^7}$$

so that $y'' = 0$ when $x = 0$ and $y'' = 120/5^7 > 0$ when $x = 6$. This confirms that the curve has a minimum turning point where $x = 6$ but gives no information about $x = 0$.

 Hazard For a critical point $(c, f(c))$, knowing that $f''(c) = 0$ tell us nothing about its nature. To see this, consider the curves $y = x^4$, $y = -x^4$, $y = x^3$ and $y = -x^3$ at $x = 0$. All satisfy $y' = y'' = 0$ at $x = 0$ but all have different shapes at the critical point $(0, 0)$. They are sketched in Figure 9.1.3.

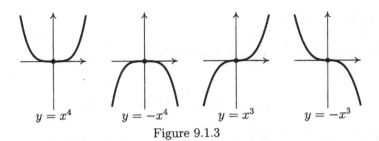

$$y = x^4 \qquad y = -x^4 \qquad y = x^3 \qquad y = -x^3$$

Figure 9.1.3

9.2   Local and Global Extrema

The word *extrema* is the plural of *extremum*. An *extremum* is a *minimum* or a *maximum*.

Definitions 9.2.1 Let f be a real function. Then f has a *local minimum* [*local maximum*] at c if f is defined on a neighbourhood N of c and

$$\text{for all } x \in N, \quad f(x) \geq f(c) \quad [f(x) \leq f(c)].$$

For example, the real function f defined by

$$f(x) = \begin{cases} x^3 - 3x - 1 & \text{if } x \leq 0, \\ x^3 - 3x + 1 & \text{if } x > 0 \end{cases}$$

has a local maximum at -1 and local minima at 0 and 1. See Figure 9.2.1

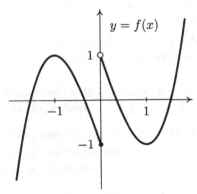

Figure 9.2.1

Theorem 9.2.2 (The Local Extrema Theorem) *Let f be a real function, differentiable at c, with a local extremum at c. Then $f'(c) = 0$.*

Proof We prove the case where f has a local minimum at c.

For x in some neighbourhood of c, $f(x) \geq f(c)$. So, by Note 1.2.4(3), for h in some neighbourhood of 0,

$$f(c + h) \geq f(c).$$

Hence, by Note 7.2.5 (4),

$$f'_L(c) = \lim_{h \to 0^-} \frac{f(c+h) - f(c)}{h} \leq 0$$

and

$$f'_R(c) = \lim_{h \to 0^+} \frac{f(c+h) - f(c)}{h} \geq 0.$$

Since $f'_L(c) = f'_R(c) = f'(c)$ it follows that $f'(c) = 0$ as required. $\qquad\square$

Hazard The converse of Theorem 9.2.2 is false. Consider horizontal points of inflection.

Definitions 9.2.3 Let f be a real function defined on a set A and let $c \in A$. Then f has a *global minimum* [*global maximum*] on A at c, and we call $f(c)$ the *global minimum value* [*global maximum value*] *of* f *on* A, if

$$\text{for all } x \in A, \quad f(x) \geq f(c) \quad [f(x) \leq f(c)].$$

The word 'global' is often omitted.

Notes 9.2.4 (1) It follows directly from Definition 9.2.3 that if a real function f has a global minimum [global maximum] on A at c and $c \in B \subseteq A$, then f has a global minimum [global maximum] on B at c.

(2) It follows directly from Definitions 9.2.1 and 9.2.3 that if a real function f has a global minimum [global maximum] on a neighbourhood of c at c then f has a local minimum [local maximum] at c.

(3) The extreme value theorem (Theorem 7.3.11) guarantees that a real function, continuous on $[a, b]$, has a global minimum value and a global maximum value on $[a, b]$.

Lemma 9.2.5 *Let f be a real function with a global minimum [global maximum] on A at c and let $c \in (a, b) \subseteq A$. Then f has a local minimum [local maximum] at c.*

Proof By Note 1.2.4(2), there is a neighbourhood N of c such that $N \subseteq (a, b) \subseteq A$. Then, by Note 9.2.4(1), f has a global minimum [global maximum] on N at c and hence, by Note 9.2.4(2), f has a local minimum [local maximum] at c. $\qquad\square$

Theorem 9.2.6 (The Global Extrema Theorem) *Let f be a real function, continuous on $[a, b]$ and differentiable on (a, b), with a global extremum on $[a, b]$ at c. Then either $c = a$, $c = b$ or $f'(c) = 0$.*

Proof Suppose that $c \neq a$ and $c \neq b$. Then $c \in (a, b)$ and Lemma 9.2.5 applies together with Theorem 9.2.2 to give the required result. $\qquad\square$

Example 9.2.7 Find the global minimum and maximum values on $[0, 9]$ of the real function f defined by

$$f(x) = x^3 - 9x^2 + 15x - 6.$$

Solution Since f is continuous on $[0,9]$, the minimum and maximum values on $[0,9]$ exist (see Note 9.2.4(3)). Also, f is differentiable on $(0,9)$ with

$$f'(x) = 3x^2 - 18x + 15 = 3(x-1)(x-5)$$
$$= 0 \text{ if and only if } x = 1 \text{ or } 5.$$

Hence, by the global extrema theorem, the required values occur where $x = 0, 1, 5$ or 9. Evaluating gives

$$f(0) = -6, \quad f(1) = 1, \quad f(5) = -31, \quad f(9) = 244.$$

So the minimum value is -31 and the maximum value is 244. $\square$

Example 9.2.8 A storage tank is to be made from a sheet of metal, 9 metres square, by cutting a square of side x metres from each corner and bending up the sides. See Figure 9.2.2. What is the maximum volume of the tank

(a) if the height of the tank must not exceed 3 metres?

(b) if the height must not exceed 3·5 metres and the base must not exceed 4 metres by 4 metres?

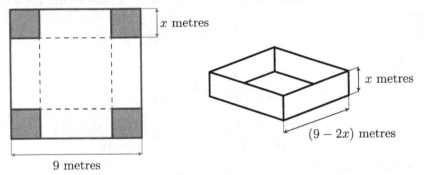

Figure 9.2.2

Solution Let the volume of the tank be V cubic metres. Then

$$V = (9 - 2x)^2 x = 4x^3 - 36x^2 + 81x$$

so that

$$\frac{dV}{dx} = 12x^2 - 72x + 81 = 3(2x - 3)(2x - 9)$$
$$= 0 \text{ when } x = 3/2 \text{ or } 9/2.$$

(a) In this case, $0 \le x \le 3$. As a function of x, V is continuous on $[0,3]$ and differentiable on $(0,3)$. So, by the global extrema theorem, the maximum volume occurs when $x = 0, 3/2$ or 3. Since

$$V(0) = 0, \quad V(3/2) = 54, \quad V(3) = 27,$$

the maximum volume of the tank is 54 cubic metres.

(b) In this case, $0 \le x \le 7/2$ and $0 \le (9 - 2x) \le 4$. This is equivalent to $5/2 \le x \le 7/2$. As a function of x, V is continuous on $[5/2, 7/2]$ and differentiable on $(5/2, 7/2)$. So, by the global extrema theorem, the maximum volume occurs when $x = 5/2$ or $7/2$. Since

$$V(5/2) = 40 \quad \text{and} \quad V(7/2) = 14,$$

the maximum volume of the tank is 40 cubic metres.   □

Example 9.2.9 Which point on the curve

$$y = \frac{2}{\sqrt{x^2 + 1}} \quad (x \ge 0)$$

is nearest to the origin O?

Solution See Figure 9.2.3. The point $(0, 2)$ lies on the curve and is distance 2 from O. So, in seeking the point on the curve nearest to O, we can ignore any point whose distance from O is greater than 2. In particular, we need not consider any point (x, y) with $x > 2$. Let $P(x, y)$ be a point on the curve with x in the range $0 \le x \le 2$. To minimise $|OP|$ it is enough to minimise $|OP|^2$ (see Example 2.6.6). Let $|OP|^2 = z$. Then

$$z = x^2 + y^2 = x^2 + \frac{4}{x^2 + 1}$$

so that

$$\frac{dz}{dx} = 2x - \frac{8x}{(x^2 + 1)^2} = 2x \left[\frac{(x^2 + 1)^2 - 4}{(x^2 + 1)^2} \right]$$

$$= \frac{2x(x^2 - 1)(x^2 + 3)}{(x^2 + 1)^2} = \frac{2x(x + 1)(x - 1)(x^2 + 3)}{(x^2 + 1)^2}.$$

Hence, in the range $0 < x < 2$,

$$\frac{dx}{dx} = 0 \text{ if and only if } x = 1.$$

As a function of x, z is continuous on $[0, 2]$ and differentiable on $(0, 2)$. So, by the global extrema theorem, the minimum distance occurs when $x = 0$, 1 or 2. Since

$$z(0) = 4, \quad z(1) = 3, \quad z(2) > 4,$$

we take $x = 1$. Hence, the point on the curve, nearest to O, is $(1, \sqrt{2})$.   □

9.3   The Mean Value Theorem

The mean value theorem is an extremely important result with a variety of applications. In particular, as we shall see in Chapter 18, it leads to power series representations of certain functions.

We begin with a special case.

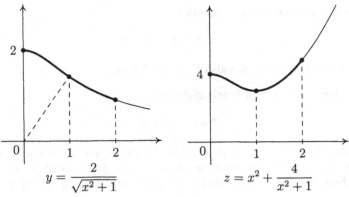

$$y = \frac{2}{\sqrt{x^2 + 1}} \qquad z = x^2 + \frac{4}{x^2 + 1}$$

Figure 9.2.3

Theorem 9.3.1 (Rolle's Theorem) *Let f be a real function, continuous on $[a, b]$ and differentiable on (a, b), with $f(a) = f(b)$. Then there exists $c \in (a, b)$ such that $f'(c) = 0$.*

[The situation is illustrated in Figure 9.3.1.]

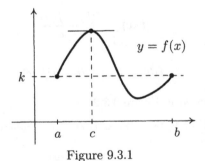

Figure 9.3.1

Proof By the extreme value theorem, there exist $c_1, c_2 \in [a, b]$ such that, for all $x \in [a, b]$,

$$f(c_1) \le f(x) \le f(c_2).$$

So f has global extrema on $[a, b]$ at c_1 and c_2. Let $f(a) = f(b) = k$.

Case 1. If $f(c_1) = f(c_2) = k$, then $f(x) = k$ for all $x \in [a, b]$ so that $f'(c) = 0$ for all $c \in (a, b)$.

Case 2. If $f(c) \ne k$ for $c = c_1$ or c_2, then $c \ne a$ and $c \ne b$ so that $c \in (a, b)$. It now follows from Lemma 9.2.5 that f has a local extremum at c and hence, by Theorem 9.2.2, that $f'(c) = 0$. □

Remark Rolle's theorem is sometimes stated with $f(a) = f(b) = 0$ replacing the weaker condition $f(a) = f(b)$.

Example 9.3.2 Show that the equation

$$x^3 + 2x - 7 = 0 \qquad (9.3.1)$$

has at most one real root. [See also Example 7.3.14.]

Solution Let f be the real function defined by

$$f(x) = x^3 + 2x - 7.$$

Suppose that equation (9.3.1) has two real roots a and b with $a < b$. Then f is continuous on $[a, b]$ and differentiable on (a, b) with $f(a) = f(b)$ $(= 0)$. So, by Rolle's theorem, there exists $c \in (a, b)$ such that $f'(c) = 0$. But, for all $x \in \mathbb{R}$,

$$f'(x) = 3x^2 + 2 \geq 2 > 0.$$

Thus we have a contradiction. Hence equation (9.3.1) has at most one real root, as required.   □

Theorem 9.3.3 (The Mean Value Theorem) *Let f be a real function, continuous on $[a, b]$ and differentiable on (a, b). Then there exists $c \in (a, b)$ such that*

$$f'(c) = \frac{f(b) - f(a)}{b - a},$$

i.e.

$$f(b) = f(a) + (b - a)f'(c).$$

[The situation is illustrated in Figure 9.3.2(a).]

Proof Define $g : [a, b] \to \mathbb{R}$ by $g(x) = f(x) - mx$, where the constant m is chosen so that $g(a) = g(b)$, *i.e.*

$$m = \frac{f(b) - f(a)}{b - a}. \qquad (9.3.2)$$

See Figure 9.3.2. Then, by Theorems 7.3.9 and 8.2.3, g is continuous on $[a, b]$ and differentiable on (a, b). So, by Rolle's theorem, there exists $c \in (a, b)$ such that $g'(c) = 0$, *i.e.* $f'(c) = m$. Using (9.3.2) the result follows immediately.   □

Note 9.3.4 With the notation of Theorem 9.3.3, let A, B and C be the points $(a, f(a))$, $(b, f(b))$ and $(c, f(c))$, respectively, on the curve $y = f(x)$. See Figure 9.3.2(a). Then $f'(c)$ is the gradient of the tangent to the curve at C and $[f(b) - b(a)]/(b - a)$ is the gradient of the chord AB. So the conclusion of the mean value theorem says that, for some C, the tangent at C is parallel to the chord AB.

Example 9.3.5 By comparing $\sqrt{39}$ with $\sqrt{36}$, show that

$$6 \cdot 24 < \sqrt{39} < 6 \cdot 25.$$

(a) (b)

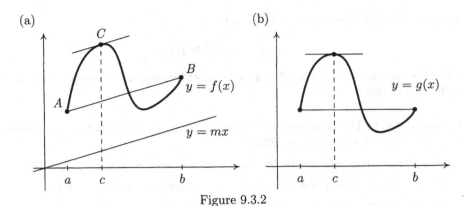

Figure 9.3.2

Solution Let f be the real function defined by $f(x) = \sqrt{x}$. Then f is continuous on $[36, 39]$ and differentiable on $(36, 39)$. So, by the mean value theorem, there exists $c \in (36, 39)$ such that

$$f(39) = f(36) + (39 - 36)f'(c),$$

i.e.

$$\sqrt{39} = 6 + \frac{3}{2\sqrt{c}}.$$

Since $36 < c < 39$ and f is strictly increasing on $[0, \infty)$ (see Example 2.6.6), it follows that

$$6 + \frac{3}{2\sqrt{39}} < \sqrt{39} < 6 + \frac{3}{2\sqrt{36}}. \tag{9.3.3}$$

The right-hand inequality of (9.3.3) gives

$$\sqrt{39} < \frac{25}{4}. \tag{9.3.4}$$

Using this in the left-hand inequality of (9.3.3) gives

$$6 + \frac{3}{2(25/4)} < \sqrt{39}, \qquad i.e. \quad \frac{156}{25} < \sqrt{39}. \tag{9.3.5}$$

Combining (9.3.4) and (9.3.5) gives

$$6{\cdot}24 < \sqrt{39} < 6{\cdot}25$$

as required. $\qquad \square$

9.4 More on Monotonic Functions

Recall part of Definitions 2.6.1. Let f be a real function whose domain contains the set A. Then f is *strictly increasing on* A if, for $s, t \in A$,

$$s < t \implies f(s) < f(t)$$

and f is *strictly decreasing on A* if, for $s, t \in A$,

$$s < t \implies f(s) > f(t).$$

When the function involved is differentiable, strict increase and decrease can be related to the sign of the derivative.

Theorem 9.4.1 (The Monotonicity Theorem) *Let f be a real function, continuous on $[a, b]$ and differentiable on (a, b). Then*

(1) $f'(x) = 0$ *for all* $x \in (a, b)$ $\implies$ f *is constant on* $[a, b]$,

(2) $f'(x) > 0$ *for all* $x \in (a, b)$ $\implies$ f *is strictly increasing on* $[a, b]$,

(3) $f'(x) < 0$ *for all* $x \in (a, b)$ $\implies$ f *is strictly decreasing on* $[a, b]$.

The variations given in Note 9.4.2 *are included in the theorem.*

Proof Let $s, t \in [a, b]$ with $s < t$. Then f is continuous on $[s, t]$ and differentiable on (s, t). So, by the mean value theorem, there exists $c \in (s, t) \subseteq (a, b)$ such that

$$f(t) - f(s) = (t - s)f'(c).$$

Since $(t - s) > 0$, we have

$$f'(x) \begin{cases} = 0 \\ > 0 \\ < 0 \end{cases} \text{ for all } x \in (a, b) \implies f'(c) \begin{cases} = 0 \\ > 0 \\ < 0 \end{cases}$$

$$\implies f(t) - f(s) \begin{cases} = 0 \\ > 0 \\ < 0 \end{cases} \implies f(s) \begin{cases} = f(t) \\ < f(t) \\ > f(t) \end{cases}$$

and (1), (2) and (3) follow immediately. $\square$

Note 9.4.2 The proof of Theorem 9.4.1 allows for variations in the statement of the theorem. Specifically, we can replace $[a, b]$ with I and (a, b) with J where I and J are given by a row of the following table.

I	J
$[a, b)$, $(a, b]$ or (a, b)	(a, b)
$[a, \infty)$ or (a, ∞)	(a, ∞)
$(-\infty, a]$ or $(-\infty, a)$	$(-\infty, a)$

Example 9.4.3 Determine the intervals on which the real function f defined by

$$f(x) = \frac{x}{(x + 1)^2}$$

is (a) strictly increasing, (b) strictly decreasing.

Solution Here,

$$f'(x) = \frac{1-x}{(x+1)^3}.$$

So a table of signs for $f'(x)$ is as follows.

x	$\to$	-1	$\to$	1	$\to$
$1-x$	$+$	$+$	$+$	0	$-$
$(x+1)^3$	$-$	0	$+$	$+$	$+$
$f'(x)$	$-$	?	$+$	0	$-$

From this we see that $f'(x) > 0$ when $-1 < x < 1$ and $f'(x) < 0$ when $x < -1$ or $x > 1$. Hence, by the monotonicity theorem, f is strictly increasing on $(-1, 1]$ and strictly decreasing on $(-\infty, -1)$ and on $[1, \infty)$.   □

Example 9.4.4 Prove that, for all $x \in (0, \frac{\pi}{2}]$,

$$\sin x < x. \tag{9.4.1}$$

Deduce that (9.4.1) holds for all $x > 0$.

Solution Define $f : [0, \frac{\pi}{2}] \to \mathbb{R}$ by

$$f(x) = x - \sin x.$$

Then f is continuous on $[0, \frac{\pi}{2}]$ and differentiable on $(0, \frac{\pi}{2})$ with

$$f'(x) = 1 - \cos x > 0 \qquad (0 < x < \tfrac{\pi}{2}).$$

So, by the monotonicity theorem, f is strictly increasing on $[0, \frac{\pi}{2}]$. Hence, for all $x \in (0, \frac{\pi}{2}]$, we have $0, x \in [0, \frac{\pi}{2}]$ and $0 < x$ so that

$$f(0) < f(x), \qquad i.e. \ 0 < \sin x - x, \qquad i.e. \ \sin x < x.$$

For $x > \frac{\pi}{2}$,

$$\sin x \le 1 < \tfrac{\pi}{2} < x.$$

Hence equation (9.4.1) holds for all $x > 0$.   □

Example 9.4.5 Show that, for all $x \in [-1, 1]$,

$$\sin^{-1} x + \cos^{-1} x = \tfrac{\pi}{2}.$$

[See Problem 5.5.6.]

Solution Define $f : [-1, 1] \to \mathbb{R}$ by

$$f(x) = \sin^{-1} x + \cos^{-1} x.$$

Then f is continuous on $[-1, 1]$ and differentiable on $(-1, 1)$ with

$$f'(x) = \frac{1}{\sqrt{1-x^2}} - \frac{1}{\sqrt{1-x^2}} = 0 \qquad (-1 < x < 1).$$

So, by the monotonicity theorem, f is constant on $[-1, 1]$. Hence, for all $x \in [-1, 1]$,

$$f(x) = f(0) = \sin^{-1} 0 + \cos^{-1} 0 = 0 + \tfrac{\pi}{2} = \tfrac{\pi}{2}$$

as required.   □

Hazard A real function whose domain D is not an interval may have zero derivative on the whole of D and yet not be constant on D. The next example illustrates this.

Example 9.4.6 Let f be the real function defined by

$$f(x) = \tan^{-1}\left(\frac{x-1}{x+1}\right) + \tan^{-1}\left(\frac{x+1}{x-1}\right).$$

Find the maximal domain D of f. Show that $f'(x) = 0$ for all $x \in D$. Sketch the graph of f.

Solution Since $\tan^{-1}$ is defined on $\mathbb{R}$, the only x for which $f'(x)$ is not defined are -1 and 1. Hence

$$D = \mathbb{R} - \{-1, 1\} = (-\infty, -1) \cup (-1, 1) \cup (1, \infty).$$

For $x \in D$,

$$f'(x) = \frac{1}{1 + \left(\dfrac{x-1}{x+1}\right)^2} \frac{(x+1)-(x-1)}{(x+1)^2} + \frac{1}{1 + \left(\dfrac{x+1}{x-1}\right)^2} \frac{(x-1)-(x+1)}{(x-1)^2}$$

$$= \frac{2}{(x+1)^2 + (x-1)^2} - \frac{2}{(x-1)^2 + (x+1)^2} = 0.$$

It follows from the monotonicity theorem that f is constant on each of the intervals $(-\infty, -1)$, $(-1, 1)$ and $(1, \infty)$. Since

$$f(0) = \tan^{-1}(-1) + \tan^{-1}(-1) = \left(-\frac{\pi}{4}\right) + \left(-\frac{\pi}{4}\right) = -\frac{\pi}{2}$$

and

$$f(2 + \sqrt{3}) = \tan^{-1}\frac{1}{\sqrt{3}} + \tan^{-1}\sqrt{3} = \frac{\pi}{6} + \frac{\pi}{3} = \frac{\pi}{2}$$

it follows that $f(x) = -\pi/2$ for all $x \in (-1, 1)$, $f(x) = \pi/2$ for all $x \in (1, \infty)$. Since f is an even function, $f(x) = \pi/2$ for all $x \in (-\infty, -1)$. Alternatively, evaluate $f(-2 - \sqrt{3})$.

The graph of f is sketched in Figure 9.4.1.   □

Problem 9.4.7 For the function f of Example 9.4.6, establish the constant value of f on $(1, \infty)$ by considering $\lim_{x \to \infty} f(x)$.

Corollary 9.4.8 (to the Monotonicity Theorem) *Let g and h be real functions, continuous on $[a, b]$ and differentiable on (a, b) with*

$$g'(x) = h'(x) \qquad (a < x < b).$$

Then there exists $K \in \mathbb{R}$ such that

$$g(x) = h(x) + K \qquad (a \leq x \leq b),$$

i.e. g and h differ by a constant on $[a, b]$.

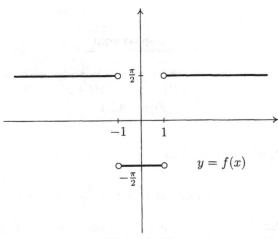

Figure 9.4.1

Proof Define $f : [a, b] \to \mathbb{R}$ by

$$f(x) = g(x) - h(x).$$

Then f is continuous on $[a, b]$ and differentiable on (a, b) with

$$f'(x) = g'(x) - h'(x) = 0 \qquad (a \leq x \leq b).$$

So, by the monotonicity theorem, f is constant on $[a, b]$ and the result follows immediately. $\qquad\qquad\qquad\qquad\qquad\qquad\qquad\qquad\qquad\qquad\qquad\qquad\qquad\qquad\square$

Remark Corollary 9.4.8 will be used in Section 14.3 to prove the fundamental theorem of calculus.

9.5   Rates of Change

If u is a differentiable function of s, we sometimes refer to $\dfrac{du}{ds}$ as the *rate of change of u with respect to s*. For example, if A is the area of a circle of radius r, so that $A = \pi r^2$, then $\dfrac{dA}{dr} = 2\pi r$ is the rate of change of the area with respect to the radius. If v is a differentiable function of time t then $\dfrac{dv}{dt}$, which is often written $\dot{v}$, is simply called the *rate of change of v*.

    The reason for the terminology can be understood by considering a particle moving on the x-axis. The rate of change of the position of the particle is its *velocity*. Suppose the x-coordinate of the particle at time t is $x(t)$ where x is a differentiable function of t. The average velocity between times t and $t + h$ $(h \neq 0)$ is

$$\frac{\text{displacement}}{\text{time taken}} = \frac{x(t + h) - x(t)}{h}.$$

See Figure 9.5.1. The limit of this quotient, as h approaches 0, is the velocity of the

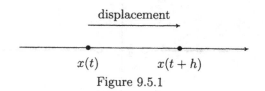

displacement

$x(t)$ $x(t + h)$

Figure 9.5.1

particle at time t. It is also $\dfrac{dx}{dt}$. Thus

$$\text{velocity at time } t \; = \; \frac{dx}{dt} = \dot{x}.$$

If x is a twice differentiable function of t then, taking the rate of change of the velocity of the particle to be its *acceleration* and arguing as above, we get

$$\text{acceleration at time } t \; = \; \frac{d}{dt}\left(\frac{dx}{dt}\right) = \frac{d^2x}{dt^2} = \ddot{x}.$$

Note the extended use of the 'dot notation' for differentiation with respect to t.

Example 9.5.1 At time t the x-coordinate of a particle moving on the x-axis is given by
$$x(t) = t^3 - 6t^2 + 9t + 2.$$
Find (a) the position and velocity of the particle when its acceleration is 0, (b) the acceleration when the velocity is 0.

Solution Here,
$$\dot{x}(t) = 3t^2 - 12t + 9 = 3(t - 1)(t - 3),$$
$$\ddot{x}(t) = 6t - 12 = 6(t - 2).$$

(a) The acceleration is 0 when $t = 2$. Then

$$\text{position} \; = \; x(2) = 4,$$
$$\text{velocity} \; = \; \dot{x}(2) = -3.$$

(b) The velocity is 0 when $t = 1$ or $t = 3$. When $t = 1$,

$$\text{acceleration} \; = \; \ddot{x}(1) = -6,$$

and when $t = 3$,

$$\text{acceleration} \; = \; \ddot{x}(3) = 6. \qquad\qquad \square$$

Example 9.5.2 An object is propelled, from ground level, vertically upwards with an initial velocity u. The deceleration of the object, due to gravity, is g (constant). Disregarding any other factors, when will the object return to earth?

Solution Let s, v and a be, respectively, the height, velocity and acceleration of the object at time t. Then

$$\dot{v}(t) = a(t) = -g.$$

Hence, using Corollary 9.4.8,

$$v(t) = -gt + K \qquad (K \text{ a constant}).$$

But $v(0) = u$. So $K = u$ giving

$$\dot{s}(t) = v(t) = -gt + u.$$

Hence, using Corollary 9.4.8,

$$s(t) = -\tfrac{1}{2}gt^2 + ut + C \qquad (C \text{ a constant}).$$

But $s(0) = 0$. So $C = 0$ giving

$$s(t) = ut - \tfrac{1}{2}gt^2 = t(u - \tfrac{1}{2}gt)$$

$$= 0 \text{ when } t = 0 \text{ or } 2u/g.$$

Thus the object returns to earth at time $t = 2u/g$.   □

In problems on *related rates*, one rate of change, $\dfrac{du}{dt}$ (say), is given and another, $\dfrac{dv}{dt}$ (say), has to be found. The first step is to relate u and v. This might be possible in one of several ways. For example, $u = \sqrt{1 - v^2}$ or $v = \sqrt{1 - u^2}$ or $u^2 + v^2 = 1$ or $u = \cos w$, $v = \sin w$ where w is a third function of t. Then differentiating with respect to t relates $\dfrac{du}{dt}$ and $\dfrac{dv}{dt}$.

Example 9.5.3 Water is running out of a conical funnel at the rate of 0·002 cubic metres per second. The radius of the top of the funnel is 0·25 metres and its height is 0·5 metres. At what rate is the water level changing when it is 0·3 metres from the top?

Solution Let r and h be, respectively, the radius and height of the surface of the water at time t. See Figure 9.5.2. Let V be the volume of water left in the tank at time t.

Here we are given $\dfrac{dV}{dt} = -0\cdot002$ and have to find $\dfrac{dh}{dt}$.

By similar triangles,

$$\frac{r}{h} = \frac{0\cdot25}{0\cdot5} = \frac{1}{2}, \qquad i.e. \ r = \frac{h}{2}.$$

Hence,

$$V = \frac{1}{3}\pi r^2 h = \frac{1}{3}\pi \frac{h^2}{4} h = \frac{\pi}{12}h^3$$

so that, differentiating with respect to t,

$$\frac{dV}{dt} = \frac{\pi}{12}3h^2\frac{dh}{dt} = \frac{\pi}{4}h^2\frac{dh}{dt}.$$

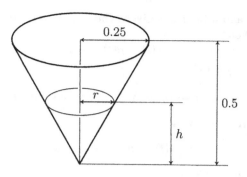

Figure 9.5.2

So, when $h = (0{\cdot}5 - 0{\cdot}3) = 0{\cdot}2$,

$$\frac{dh}{dt} = \frac{dV}{dt}\frac{4}{\pi}\frac{1}{h^2} = (-0{\cdot}002)\frac{4}{\pi}\frac{1}{0{\cdot}04} = -\frac{0{\cdot}2}{\pi},$$

i.e. the water level is dropping at $0{\cdot}2/\pi$ metres per second.   □

Example 9.5.4 A weight W is being raised vertically by a rope through a block B, 30 metres above a point A at ground level. The rope is 70 metres long and is held by a man M who walks away from A at $0{\cdot}75$ metres per second. How fast is the weight rising when it is 10 metres above A?

Solution We ignore the dimensions of the weight and the block (pulley), the thickness of the rope and the height of the man. Let x and y be the distances from A to M and W, respectively. See Figure 9.5.3.

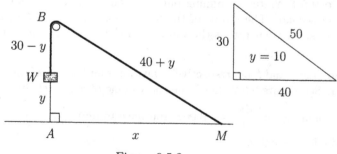

Figure 9.5.3

Here we are given $\dfrac{dx}{dt} = 0{\cdot}75$ and have to find $\dfrac{dy}{dt}$.

From triangle ABM, we get

$$(40 + y)^2 = x^2 + 30^2.$$

Differentiating with respect to time t gives

$$2(40 + y)\frac{dy}{dt} = 2x\frac{dx}{dt}.$$

Hence, when $y = 10$ so that $x = 40$ (see Figure 9.5.3),

$$\frac{dy}{dt} = \frac{x}{40 + y}\frac{dx}{dt} = \frac{40}{50}\frac{3}{4} = \frac{3}{5},$$

i.e. the weight is rising at 0·6 metres per second. □

Example 9.5.5 The volume of a spherical balloon is increasing at the rate of 0·05 cubic metres per second. How fast is the surface area increasing when the radius is 0·4 metres?

Solution Let V, S and r be the volume, surface area and radius, respectively, at time t.

Here we are given $\dfrac{dV}{dt} = 0·05$ and have to find $\dfrac{dS}{dt}$.

We have

$$V = \frac{4}{3}\pi r^3 \qquad \text{and} \qquad S = 4\pi r^2.$$

So, differentiating with respect to t,

$$\frac{dV}{dt} = 4\pi r^2\frac{dr}{dt} \qquad \text{and} \qquad \frac{dS}{dt} = 8\pi r\frac{dr}{dt}.$$

Hence, when $r = 0·4$,

$$\frac{dS}{dt} = 8\pi r\frac{1}{4\pi r^2}\frac{dV}{dt} = \frac{2}{r}\frac{dV}{dt} = \frac{2}{0·4}0·05 = 0·25,$$

i.e. the surface area is increasing at 0·25 square metres per second. □

9.6 L'Hôpital's Rule

Let f and g be the real functions defined by

$$f(x) = x^2 + 4x + 3 \qquad \text{and} \qquad g(x) = x^2 + 5x + 4,$$

respectively. In Example 7.2.6 (b), we showed that

$$\lim_{x \to -1}\frac{f(x)}{g(x)} = \lim_{x \to -1}\frac{(x+1)(x+3)}{(x+1)(x+4)} = \lim_{x \to -1}\frac{x+3}{x+4} = \frac{2}{3}. \tag{9.6.1}$$

If we now consider

$$f'(x) = 2x + 4 \qquad \text{and} \qquad g'(x) = 2x + 5,$$

we have

$$\lim_{x \to -1}\frac{f'(x)}{g'(x)} = \lim_{x \to -1}\frac{2x+4}{2x+5} = \frac{2}{3}, \tag{9.6.2}$$

and it is no coincidence that the limits in (9.6.1) and (9.6.2) are equal. The theorem which guarantees this is called L'Hôpital's rule. Together with its variations, it provides a means of evaluating certain otherwise awkward limits.

Theorem 9.6.1 (L'Hôpital's Rule) *Let f and g be real functions, continuous on a neighbourhood N of c and differentiable on $N - \{c\}$, with $f(c) = g(c) = 0$ and $g'(x) \neq 0$ $(x \in \mathbb{N} - \{c\})$. Let $L \in \mathbb{R}$. Then*

$$\lim_{x \to c} \frac{f'(x)}{g'(x)} = L \quad \Rightarrow \quad \lim_{x \to c} \frac{f(x)}{g(x)} = L.$$

Proof   We omit the proof here as it requires the formal definition of function limit. A proof is given in Appendix C, Theorem C.3.2.   □

Example 9.6.2   Evaluate   (a) $\displaystyle\lim_{x \to \pi} \frac{\sin(4x - \pi)}{\sin(5x - \pi)}$,   (b) $\displaystyle\lim_{x \to 1} \frac{x^3 - 3x + 2}{x^3 + x^2 - 5x + 3}$.

Solution   (a) Let f and g be the real functions defined by $f(x) = \sin(3x - \pi)$ and $g(x) = \sin(4x - \pi)$. Then f and g satisfy the requirements of L'Hôpital's rule and

$$\frac{f'(x)}{g'(x)} = \frac{4\cos(4x - \pi)}{5\cos(5x - \pi)} \to \frac{-4}{5} \quad \text{as } x \to \pi.$$

Hence, by L'Hôpital's rule, $\displaystyle\lim_{x \to \pi} \frac{f(x)}{g(x)} = -\frac{4}{5}$.

Alternatively, we could simply write

$$\lim_{x \to \pi} \frac{\sin(4x - \pi)}{\sin(5x - \pi)} = \lim_{x \to \pi} \frac{4\cos(4x - \pi)}{5\cos(5x - \pi)} = -\frac{4}{5}.$$

However, if we opt for this abbreviated solution, we must still be sure that the conditions of L'Hôpital's rule are satisfied and we must be aware that only when the final limit is evaluated can we assert the existence of the first limit.

   (b) Here we use L'Hôpital's rule twice. We have

$$\lim_{x \to 1} \frac{x^3 - 3x + 2}{x^3 + x^2 - 5x + 3} = \lim_{x \to 1} \frac{3x^2 - 3}{3x^2 + 2x - 5} = \lim_{x \to 1} \frac{6x}{6x + 2} = \frac{6}{6 + 2} = \frac{3}{4}. \qquad □$$

Note 9.6.3   The conclusion of L'Hôpital's rule, *i.e.*

$$\frac{f'(x)}{g'(x)} \to L \;\text{ as } x \to c \quad \Rightarrow \quad \frac{f(x)}{g(x)} \to L \;\text{ as } x \to c,$$

holds for suitably continuous and differentiable functions f and g if, instead of $f(c) = g(c) = 0$, we have $f(x) \to \pm\infty$ and $g(x) \to \pm\infty$ as $x \to c$. We can also replace $x \to c$ with $x \to c^{\pm}$ or $x \to \pm\infty$, and L with $\pm\infty$.

Example 9.6.4   Investigate the behaviour of

$$x^2 \sin(1/x) \quad \text{as } x \to \infty.$$

Solution   We have

$$x^2 \sin(1/x) = \frac{\sin(1/x)}{1/x^2}.$$

Let f and g be the real functions defined by $f(x) = \sin(1/x)$ and $g(x) = 1/x^2$. Then f and g satisfy the requirements of L'Hôpital's rule (modified by Note 9.6.3) and, making use of Example 7.4.10, we have

$$\frac{f'(x)}{g'(x)} = \frac{-(1/x^2)\cos(1/x)}{-(2/x^3)} = (x/2)\cos(1/x) \to \infty \text{ as } x \to \infty.$$

Hence,

$$x^2 \sin(1/x) \to \infty \text{ as } x \to \infty. \qquad \square$$

 Hazard Do not confuse the application of L'Hôpital's rule with the quotient rule for differentiation.

9.X   Exercises

1. Find the coordinates and nature of each critical point of the real function f defined by

    (a) $f(x) = 2x^3 - 3x^2 - 12x + 5$,

    (b) $f(x) = 3x^3 - 6x^2 + 4x + 1$,

    (c) $f(x) = x^4 - 8x^2 - 5$,

    (d) $f(x) = \dfrac{1}{(x-2)(x-6)}$,

    (e) $f(x) = \dfrac{4x^2 - 2x - 1}{x - 1}$,

    (f) $f(x) = \left(\dfrac{x}{2} - \dfrac{2}{x}\right)^3$.

2. Find the coordinates and nature of each critical point, in the given range, of the real function f defined by

    (a) $f(x) = \cos^2 x \sin 2x \qquad (0 < x < \pi)$,

    (b) $f(x) = \dfrac{\sin 3x}{\cos^3 x} \qquad (-\pi/2 < x < \pi/2)$,

    (c) $f(x) = \sin 2x - 2\sin x \qquad (-\pi < x < \pi)$.

3. Find the global minimum and maximum values, on the set A, of the real function f defined by

    (a) $f(x) = 10x - x^2, \qquad A = [0, 12]$,

    (b) $f(x) = x^3 - 4x^2 - 3x + 1, \qquad A = [-1, 1]$,

    (c) $f(x) = x - \sqrt{2}\sin x, \qquad A = [0, 2\pi]$.

    For (c) you may assume that $\pi < 4$.

4. Find the maximum value of $u^2 v^3$ when $u \geq 0$, $v \geq 0$ and $u + v = 75$.

5. Determine the maximum area of an isosceles triangle with perimeter 2.

6. A function $f : (0, \infty) \to \mathbb{R}$ is strictly decreasing on $(0, c]$ and strictly increasing on $[c, \infty)$. Prove that f has a global minimum on $(0, \infty)$ at c.

7. A cylindrical container, closed at both ends, is to have a volume of $\pi/2$ cubic metres. The cost per square metre of the material for the ends is twice that of the material for the curved surface. What dimensions for the container minimise the cost of materials?

8. Show that the equation $x^5 + 5x + 1 = 0$ has at most one real root.

9. Show that the equation $x^5 + 3x^2 - 1 = 0$ has at most one positive root.

10. Show that the equation $x^3 + 3x^2 + 6x - 1 = 0$ has at most one real root. Deduce that the equation $x^4 + 4x^3 + 12x^2 - 4x - 1 = 0$ has at most two real roots.

11. Show that, for all $x > 0$, $\tan^{-1} x < x$.

12. Show that, for all $x > 0$, $\cos x > 1 - x^2/2$. Deduce that

    (a) for all $x \in \mathbb{R}$, $\cos x \geq 1 - \dfrac{x^2}{2}$,   (b) for all $x > 0$, $\sin x > x - \dfrac{x^3}{6}$.

13. Use the mean value theorem to compare $\tan^{-1} x$ with $\tan^{-1} 0$ and show that, for all $x > 0$,

$$\tan^{-1} x > \frac{x}{x^2 + 1}.$$

14. Determine the intervals of strict increase and strict decrease for the real function f when $f(x)$ is

    (a) $2x^4 - x^2 + 1$,   (b) $\dfrac{x^2 - 2x + 5}{x - 1}$,   (c) $\dfrac{x^3}{(x^2 + 5)^4}$.

15. Determine the intervals of strict increase and strict decrease for the function $f : [0, \pi] \to \mathbb{R}$ defined by $f(x) = 2 \sin x + \cos 2x$.

16. The x-coordinate, at time t, of a particle moving on the x-axis is given by

$$x(t) = t^4 + 6t^3 - 3t^2 + 2t + 1.$$

    Find the velocity and acceleration of the particle when $t = 3$.

17. The coordinate, at time t, of a particle moving on the x-axis is given by

$$x(t) = 4t^3 - 5t^2 + 2t + 7.$$

Find the position and acceleration of the particle when its velocity is 0.

18. The volume of a sphere is increasing at the rate of 0·6 cubic metres per minute. At what rate is the surface area increasing when the radius is 1·5 metres?

19. A boat travelling north at 10 knots, will pass 1 nautical mile to the west of a lighthouse. How fast is the distance between the boat and the lighthouse decreasing when the boat is south west of the lighthouse? [1 knot = 1 nautical mile per hour.]

20. Sand is being poured, at the rate of 2 cubic metres per minute, on to level ground to form a conical heap whose semi-vertical angle remains constant at $\pi/3$. How fast will the height of the heap be increasing when the diameter of the base of the heap is 2 metres?

21. Use L'Hôpital's rule to evaluate

(a) $\displaystyle\lim_{x \to 2} \frac{x^2 + x - 6}{x^2 - x - 2}$,

(b) $\displaystyle\lim_{x \to 3} \frac{2x^3 - 7x^2 + 9}{3x^3 - 17x - 30}$,

(c) $\displaystyle\lim_{x \to \pi/2} \frac{1 - \sin x}{\cos x}$,

(d) $\displaystyle\lim_{x \to 0} \frac{\tan x + \tan 3x}{\tan 2x + \tan 4x}$,

(e) $\displaystyle\lim_{x \to 1} \frac{x^3 + 2x^2 - 7x + 4}{x^3 - 4x^2 + 5x - 2}$,

(f) $\displaystyle\lim_{x \to 0} \left(\frac{1}{\sin x} - \frac{1}{x} \right)$.

For (e) and (f), apply L'Hôpital's rule twice.

22. Use L'Hôpital's rule to investigate

(a) $\displaystyle\frac{\sin^2 x}{x - \tan x}$ as $x \to 0^+$,

(b) $x(\pi - 2\tan^{-1} x)$ as $x \to \infty$.

Chapter 10

Curve Sketching

It is often said that 'a picture is worth a thousand words'. In mathematics, it is often the case that the graph of a function is more useful than a table of values.

In the first two sections of this chapter, we develop techniques which allow us to sketch the graph of any *rational* function. The techniques include the location of critical points and points of inflection using differentiation. We will also look at what happens to the graph $y = f(x)$ as x (or y) becomes large. This involves limits involving ∞, as discussed in Section 7.4. Most of the ideas apply to general functions, and some of the examples consider such cases.

Many plane curves are *not* the graph of a function. If we observe that a function takes a unique value, $f(c)$, for a given c in its domain, we see that the graph of f crosses $x = c$ only once. It follows that, for example, the circle

$$\mathcal{C} : \ x^2 + y^2 = 1,$$

cannot be described as the graph of a function—it crosses $x = 0$ at $(0, 1)$ and $(0, -1)$.

In Section 10.3, we consider curves, like $\mathcal{C}$, defined by an *implicit* equation. Again using the ideas of differentiation, we can obtain sketches of such curves.

The circle $\mathcal{C}$ may also be defined *parametrically* as

$$\mathcal{C} : \ x = \cos t, \ y = \sin t \quad (t \in [0, 2\pi)).$$

In Section 10.4, we will see how differentiation can be used to locate the key features of a curve so defined, and to obtain a sketch.

The ideas of Sections 10.3 and 10.4 are used to obtain sketches of curves known as *conic sections*. These occur in mathematics as plane sections of a cone; they arise naturally in the theory of orbits of planets.

Finally, we introduce *polar coordinates* (r, θ)—an alternative way of describing the points of a plane. With this notation, the circle $\mathcal{C}$ is

$$r = 1.$$

Once more, differentiation can be used to obtain sketches of such polar curves.

10.1   Types of Curve

The simplest type of curve on the plane is the graph of a real function f with domain A, *i.e.* the set

$$\{(x, y) \in \mathbb{R}^2 : x \in A \text{ and } y = f(x)\}.$$

208

When we refer to 'the curve $y = f(x)$' we mean the graph of f with the assumption that the domain is maximal unless it is stated otherwise. The techniques for sketching graphs will be discussed first and then extended to certain types of implicit curves and to parametric and polar curves.

10.2 Graphs

Zeros and vertical asymptotes Given a polynomial $f(x)$, Theorem 6.4.2 states that $f(x)$ may be factorised into real linear and irreducible quadratic factors. As discussed in Section 3.4, there exist $\alpha_1, \alpha_2, \ldots, \alpha_p \in \mathbb{R}$, the distinct zeros of $f(x)$, and $m_1, m_2, \ldots, m_p \in \mathbb{N}$, the *multiplicities* of these zeros, such that we may write

$$f(x) = (x - \alpha_1)^{m_1} (x - \alpha_2)^{m_2} \ldots (x - \alpha_p)^{m_p} F(x), \qquad (10.2.1)$$

where $F(x)$ is a product of irreducible quadratic polynomials and so has no real zero.

Let γ be a zero of $f(x)$ of multiplicity k and choose a neighbourhood N of γ which contains no other zero. To examine the behaviour of $f(x)$ on N we isolate the factors involving γ and write

$$f(x) = (x - \gamma)^k R(x). \qquad (10.2.2)$$

For $x \in N$, $R(x)$ is non-zero and, since $R(x)$ is a polynomial, it is continuous and therefore of fixed sign on N. It follows that $f(x)$ will change sign at γ if and only if $(x - \gamma)^k$ changes sign, *i.e.* if and only if k is odd.

If we take a second polynomial $g(x)$, expressed in a similar way as

$$g(x) = (x - \beta_1)^{n_1} (x - \beta_2)^{n_2} \ldots (x - \beta_q)^{n_q} G(x), \qquad (10.2.3)$$

then the rational function $h(x) = f(x)/g(x)$ is

$$h(x) = \frac{(x - \alpha_1)^{m_1} (x - \alpha_2)^{m_2} \ldots (x - \alpha_p)^{m_p}}{(x - \beta_1)^{n_1} (x - \beta_2)^{n_2} \ldots (x - \beta_q)^{n_q}} H(x), \qquad (10.2.4)$$

where $H(x) = F(x)/G(x)$. Without loss of generality we may suppose that $\alpha_1, \ldots, \alpha_p$ and $\beta_1, \ldots, \beta_q$ are all distinct since otherwise we could cancel any common factors in $f(x)$ and $g(x)$.

The maximal domain of the rational function $H = F/G$ is $\mathbb{R}$ since G is non-zero on $\mathbb{R}$. Thus the maximal domain of h is $\mathbb{R} - \{\beta_1, \beta_2, \ldots, \beta_q\}$.

The zeros of $h(x)$ are $\alpha_1, \alpha_2, \ldots, \alpha_p$ and those of $1/h(x)$ are $\beta_1, \beta_2, \ldots, \beta_q$. According to Definition 7.4.5, this means that $h(x) \to \pm\infty$ as $x \to \beta_i^-$ and $x \to \beta_i^+$ for $i = 1, \ldots, q$ and the graph of h 'approaches' the vertical lines $x = \beta_1, x = \beta_2, \ldots, x = \beta_q$. These lines are the *vertical asymptotes* of the curve.

Since the sign of $h(x)$ is determined by the signs of $f(x)$ and $g(x)$, so $h(x)$ may only change sign at a zero of $f(x)$ or of $g(x)$. In other words, the possible changes of sign on the curve $y = h(x)$ occur at the zeros and vertical asymptotes. To investigate further such possible changes of sign and the approach to the vertical asymptotes, it is often best to construct a table of signs.

Example 10.2.1 Determine the signs of y and the approaches to the vertical asymptotes for the curves

$$\text{(a)} \quad y = \frac{x^2(x-3)(x-4)}{(x+1)(x-1)}, \qquad \text{(b)} \quad y = \frac{(3-x)^3(1+x^2)}{(1+x)^2}.$$

Draw sketches of the curves based on this information.

Solution (a) We draw up a table of signs

x	$\to$	-1	$\to$	0	$\to$	1	$\to$	3	$\to$	4	$\to$
x^2	$+$	$+$	$+$	0	$+$	$+$	$+$	$+$	$+$	$+$	$+$
$x-3$	$-$	$-$	$-$	$-$	$-$	$-$	$-$	0	$+$	$+$	$+$
$x-4$	$-$	$-$	$-$	$-$	$-$	$-$	$-$	$-$	$-$	0	$+$
$x+1$	$-$	0	$+$	$+$	$+$	$+$	$+$	$+$	$+$	$+$	$+$
$x-1$	$-$	$-$	$-$	$-$	$-$	0	$+$	$+$	$+$	$+$	$+$
y	$+$	$?$	$-$	0	$-$	$?$	$+$	0	$-$	0	$+$

from which we deduce the signs of y. In particular,

$$\begin{aligned} y \to \infty &\quad \text{as} \quad x \to -1^-, &\quad y \to -\infty &\quad \text{as} \quad x \to -1^+, \\ y \to -\infty &\quad \text{as} \quad x \to 1^-, &\quad y \to \infty &\quad \text{as} \quad x \to 1^+. \end{aligned}$$

See Figure 10.2.1 for a sketch of the curve based on the table of signs.

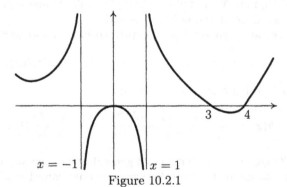

$x = -1$ $x = 1$

Figure 10.2.1

(b) In this case the table of signs is

x	$\to$	-1	$\to$	3	$\to$
$(3-x)^3$	$+$	$+$	$+$	0	$-$
$(1+x)^2$	$+$	0	$+$	$+$	$+$
y	$+$	$?$	$+$	0	$-$

and the sketch is shown in Figure 10.2.2. $\qquad\qquad\qquad\qquad\qquad\quad\square$

An essential feature of the graph of a function, that was not considered in Example 10.2.1, is the behaviour of the function for large values of the argument. If such a curve approaches a straight line as $x \to \pm\infty$ then we may sketch this behaviour quite precisely.

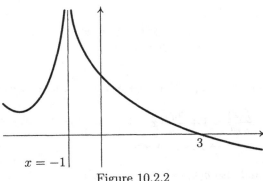

Figure 10.2.2

Non-vertical asymptotes

Definition 10.2.2 Let f be a real function such that for some $a, b \in \mathbb{R}$

$$\lim_{x \to \infty} [f(x) - (ax + b)] = 0.$$

The line $y = ax + b$ is said to be a *non-vertical asymptote as $x \to \infty$*.

If, in addition, there is some $K \in \mathbb{R}$ such that $f(x) - (ax+b) > 0 \ [< 0]$ for all $x > K$ then we say that the curve $y = f(x)$ *approaches its asymptote from above [below] as $x \to \infty$*. A non-vertical asymptote to a curve as $x \to -\infty$ is defined in a similar way.

To investigate further the way in which a graph approaches a non-vertical asymptote we introduce the following notation and establish some basic properties.

Notation 10.2.3 Let f and a be non-zero real functions and let $\alpha \in \mathbb{R}$ or $\alpha = \pm\infty$. We write $f(x) \sim a(x)$ as $x \to \alpha$ when

$$\lim_{x \to \alpha} \frac{f(x)}{a(x)} = 1.$$

For example, since $\sin x / x \to 1$ as $x \to 0$ (see Theorem 7.2.9), we write $\sin x \sim x$ as $x \to 0$.

Example 10.2.4 Show that $6x^3 - 2x + 1 \sim 6x^3$ as $x \to \pm\infty$.

Proof For $x \neq 0$,

$$\frac{6x^3 - 2x + 1}{6x^3} = 1 - \frac{1}{3x^2} + \frac{1}{6x^3} \to 1 \quad \text{as } x \to \pm\infty. \qquad \square$$

This result readily generalises to any polynomial and in turn to all rational functions.

Theorem 10.2.5 *Let $f(x)$ and $g(x)$ be non-zero polynomials. If $f(x)$ has order n and leading coefficient a, and $g(x)$ has order m and leading coefficient b then*

$$f(x) \sim ax^n, \quad g(x) \sim bx^m$$

and

$$\frac{f(x)}{g(x)} \sim \frac{a}{b}x^{n-m}$$

as $x \to \pm\infty$.

Proof Let $f(x) = ax^n + \alpha_{n-1}x^{n-1} + \cdots + \alpha_1 x + \alpha_0$ where $\alpha_0, \alpha_1, \ldots, \alpha_{n-1} \in \mathbb{R}$. Then

$$\frac{f(x)}{ax^n} = 1 + \frac{\alpha_{n-1}}{a}x^{-1} + \cdots + \frac{\alpha_1}{a}x^{1-n} + \frac{\alpha_0}{a}x^{-n}$$
$$\to 1 \text{ as } x \to \pm\infty.$$

Hence $f(x) \sim ax^n$ and also $g(x) \sim bx^m$.

Now consider

$$\frac{f(x)/g(x)}{ax^n/bx^m} = \frac{f(x)/ax^n}{g(x)/bx^m}$$
$$\to \frac{1}{1} = 1 \text{ as } x \to \pm\infty,$$

and so

$$\frac{f(x)}{g(x)} \sim \frac{a}{b}x^{n-m}. \qquad \square$$

Notes 10.2.6 (1) We say that a polynomial $f(x)$ is *dominated* by its leading term for large values of x.

(2) To show that the graph of f has non-vertical asymptote $y = \alpha x + \beta$ as $x \to \infty$ say, it is sufficient to find a function $a(x)$ such that $f(x) - (\alpha x + \beta) \sim a(x)$ as $x \to \infty$ which tends to zero as $x \to \infty$. If, in addition, $a(x)$ is positive [negative] for all sufficiently large x then we may conclude that the graph approaches the asymptote from above [below].

Example 10.2.7 Investigate the non-vertical asymptotes of

$$\text{(a)} \quad y = \frac{x^3 - 5x^2 - 2x + 24}{x^2 - 1}, \qquad \text{(b)} \quad y = \frac{2x^2 + 1}{x^4 + 3x^2 + 1}.$$

Proof (a) Using long division we get

$$y = x - 5 + \frac{-x + 19}{x^2 - 1}.$$

Therefore,

$$y - (x - 5) = \frac{-x + 19}{x^2 - 1}$$
$$\sim \frac{-x}{x^2} = -\frac{1}{x} \text{ as } x \to \pm\infty$$

using Theorem 10.2.5. Since $-1/x \to 0$ as $x \to \pm\infty$ the line $y = x - 5$ is a non-vertical asymptote both as $x \to \infty$ and as $x \to -\infty$. Furthermore, as $-1/x > 0$ for $x < 0$, the

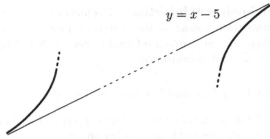

Figure 10.2.3

graph approaches the asymptote from above as $x \to -\infty$. Similarly, since $-1/x < 0$ for $x > 0$, it approaches from below as $x \to \infty$. As a shorthand for the above we write

$$y - (x - 5) \sim -1/x \to \begin{cases} 0^+ & \text{as } x \to -\infty, \\ 0^- & \text{as } x \to \infty. \end{cases}$$

The approach to the asymptote is illustrated in Figure 10.2.3.

(b) In this case long division is not needed and, using Theorem 10.2.5, we have

$$y = \frac{2x^2 + 1}{x^4 + 3x^2 + 1} \sim \frac{2x^2}{x^4} = \frac{2}{x^2} \quad \text{as } x \to \pm\infty.$$

Therefore $y - 0 \sim 2/x^2 \to 0^+$ as $x \to \pm\infty$ and so $y = 0$ is a non-vertical asymptote which the graph approaches from above as $x \to \pm\infty$. This is shown in Figure 10.2.4 □

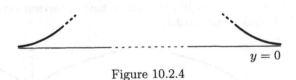

$y = 0$

Figure 10.2.4

Note 10.2.8 Consider a rational function in which the degree of the numerator exceeds that of the denominator by k.

(1) If $k < 0$ then $y = 0$ is an asymptote. This is a *horizontal asymptote*.

(2) If $k = 0$ then $y = b$, where $b \neq 0$, is a horizontal asymptote.

(3) If $k = 1$ then $y = ax + b$, where $a \neq 0$, is an asymptote. This is a *slant asymptote*.

(4) If $k \geq 2$ then there is no non-vertical asymptote.

Critical points and points of inflection   The location and nature of critical points on a curve is an important aspect of the sketch of a curve. The definitions of, and methods for studying, critical points are found in Section 9.1. Recall that the tangent to a curve at a critical point is horizontal.

Definitions 10.2.9 Let f be a real function and let $c \in \mathbb{R}$.

(1) We say that f has a *vertical tangent* at $(c, f(c))$ if f is continuous on a neighbourhood N of c, differentiable on $N - \{c\}$ and either $f'(x) \to \infty$ or $f'(x) \to -\infty$ as $x \to c$.

(2) We say that f has a *vertical left* (respectively *right*) *tangent* at $(c, f(c))$ if f is continuous on a left (right) neighbourhood N of c, differentiable on $N - \{c\}$ and either $f'(x) \to \infty$ or $f'(x) \to -\infty$ as $x \to c^-$ $(x \to c^+)$.

(3) We say that f has a *vertical cusp* at $(c, f(c))$ if it has a vertical left tangent and a vertical right tangent but not a vertical tangent.

(4) If a is the left (respectively right) end-point of the domain of f and it has a vertical left (right) tangent at $(a, f(a))$ then we also say that f has a vertical tangent at this point.

If f has a vertical tangent at $(a, f(a))$ we say that the gradient of the tangent is *infinite* at this point.

For example, the curve $y = \sqrt[3]{x}$ has a vertical tangent at $(0,0)$ since $y' = x^{-2/3}/3 \to \infty$ as $x \to 0$. By contrast, if $y = |\sqrt[3]{x}|$ then $y' = -x^{-2/3}/3$ for $x < 0$ and $y' = x^{-2/3}/3$ for $x > 0$ so that $y' \to -\infty$ as $x \to 0^-$ and $y' \to \infty$ as $x \to 0^+$. The origin is a vertical cusp in this case.

Consider also the curve $y = f(x) = \sqrt{x}$. Since the maximal domain of f is $[0, \infty)$ and $y' = x^{-1/2}/2 \to \infty$ as $x \to 0^+$, this curve have a vertical tangent at $(0,0)$ (see Definition 10.2.9 (4) and Figure 10.3.1).

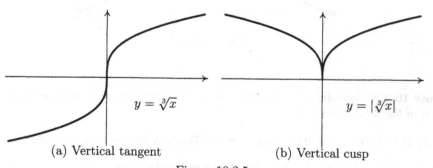

$$y = \sqrt[3]{x}$$

$$y = |\sqrt[3]{x}|$$

(a) Vertical tangent   (b) Vertical cusp

Figure 10.2.5

In Section 9.1 horizontal points of inflection were defined. We now extend this definition.

Definitions 10.2.10 Let f be a real function with continuous second derivative on an open interval I. If $f''(x) > 0$ $[< 0]$ for all $x \in I$ then f is said to be *concave up* [*down*] on I.

Let $(c, f(c))$ be a point on the curve $y = f(x)$ at which there is a tangent. Suppose there is a punctured left neighbourhood of c on which f is concave up [down] and a punctured right neighbourhood on which f is concave down [up]. Then $(c, f(c))$ is called an *up-down* [*down-up*] *point of inflection*. If in addition, the gradient of the tangent at $(c, f(c))$ is negative [zero, positive, infinite], we say that $(c, f(c))$ is a *decreasing* [*horizontal, increasing, vertical*] *point of inflection*.

We may use both sets of qualifiers together to refer to, for example, an increasing up-down point of inflection.

Notes 10.2.11 (1) As illustrated in Figure 10.2.6 (a) a curve that is concave up on an interval lies above any (non-vertical) tangent to the curve on this interval. Similarly, as shown in Figure 10.2.6 (b), a concave down curve lies below its tangents.

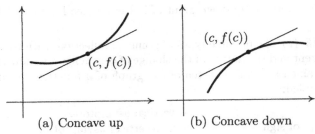

(a) Concave up   (b) Concave down

Figure 10.2.6

(2) The definition of a point of inflection $(c, f(c))$ requires that f be differentiable on a punctured neighbourhood of c but not necessarily at a itself. This allows the definition to cover vertical points of inflection.

Example 10.2.12 Determine the concavity and points of inflection on the curves

$$\text{(a)} \quad y = x^2 + 4\cos x, \ (x \in [0, \pi]), \qquad \text{(b)} \quad y = \sqrt[3]{x}.$$

Sketch each curve in the vicinity of the points of inflection.

Solution (a) We have

$$y' = 2x - 4\sin x,$$
$$y'' = 2 - 4\cos x = -4(\cos x - \tfrac{1}{2}).$$

Therefore y'' is continuous on $(0, \pi)$ and is zero in this interval if and only if $x = \frac{\pi}{3}$. Further,

$$y'' < 0 \ \text{ for } \ x < \frac{\pi}{3} \quad \text{and} \quad y'' > 0 \ \text{ for } \ x > \frac{\pi}{3}.$$

Also, the curve has a tangent with gradient $y'(\pi/3) = (2\pi - 6\sqrt{3})/3$ at $(\pi/3, \pi^2/9 + 2)$. Hence this is a point of inflection. The tangent has negative gradient and the shape changes from concave down to concave up, *i.e.* it is a decreasing down-up point of inflection as shown in Figure 10.2.7.

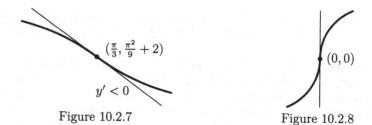

Figure 10.2.7   Figure 10.2.8

(b) Now

$$y' = \frac{1}{3x^{2/3}} \quad \text{and} \quad y'' = -\frac{2}{9x^{5/3}}.$$

At $x = 0$, y is defined and $y' \to \infty$ as $x \to 0$. Also,

$$y'' > 0 \quad \text{for} \quad x < 0 \quad \text{and} \quad y'' < 0 \quad \text{for} \quad x > 0.$$

and so $(0,0)$ is a vertical up-down point of inflection (see Figure 10.2.8). $\qquad\square$

Notes 10.2.13 (1) The possible types of points of inflection, with respect to the gradient of the tangent and the nature of the change of concavity, are shown in Figure 10.2.9.

(2) To obtained a detailed sketch of the graph of a function the following features should be included:

- zeros,
- changes of sign,
- intersection with the y-axis,
- critical points,
- vertical asymptotes,
- non-vertical asymptotes,
- intersection with horizontal asymptotes,
- points of inflection.

Example 10.2.14 Obtain a sketch of the curve

$$y = \frac{(3x + 1)(x + 3)}{2(x + 1)^2}.$$

Solution We investigate the eight features described in Note 10.2.13 (2).

Zeros: $x = -3, -\frac{1}{3}$.

Vertical asymptotes: $x = -1$.

Table of signs:

x	$\to$	-3	$\to$	-1	$\to$	$-\frac{1}{3}$	$\to$
$3x + 1$	$-$	$-$	$-$	$-$	$-$	0	$+$
$x + 3$	$-$	0	$+$	$+$	$+$	$+$	$+$
$(x + 1)^2$	$+$	$+$	$+$	0	$+$	$+$	$+$
y	$+$	0	$-$	$?$	$-$	0	$+$

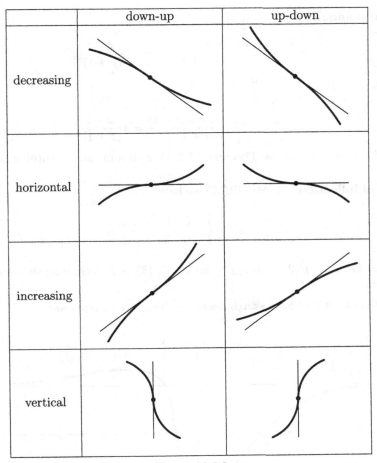

Figure 10.2.9

Non-vertical asymptotes: Using long division we have

$$y = \frac{3}{2} + \frac{2x}{(x+1)^2}$$

and so

$$y - \frac{3}{2} = \frac{2x}{x^2 + 2x + 1} \sim \frac{2}{x} \to \begin{cases} 0^- & \text{as } x \to -\infty, \\ 0^+ & \text{as } x \to \infty. \end{cases}$$

Therefore $y = 3/2$ is a non-vertical asymptote with approach from below (above) as $x \to -\infty$ (∞).

Intersection with y-axis: $y(0) = \frac{3}{2}$.

Intersection with horizontal asymptotes:

$$y = \frac{3}{2} \quad \Leftrightarrow \quad \frac{2x}{(x+1)^2} = 0 \quad \Leftrightarrow \quad x = 0.$$

Critical points:

$$y' = 2\frac{1 \times (x+1) - 2x}{(x+1)^3} = -2\frac{x-1}{(x+1)^3}, \tag{10.2.5}$$

$$= 0 \iff x = 1,$$

and

$$y'' = -2\frac{x+1 - 3(x-1)}{(x+1)^4} = 4\frac{x-2}{(x+1)^4}. \tag{10.2.6}$$

Since $y'' < 0$ at $x = 1$ so, by Theorem 9.1.3, $(1, 2)$ is a maximum turning point.

Points of inflection: From (10.2.6) we have

$$y'' = \begin{cases} < 0 & \text{for } x < 2, \\ > 0 & \text{for } x > 2, \end{cases}$$

and from (10.2.5), $y'(2) = -2/27$. So $(2, 35/18)$ is a decreasing down-up point of inflection.

See Figure 10.2.10 for a sketch based on the above information.   □

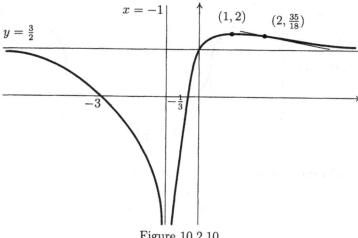

Figure 10.2.10

Note 10.2.15 We may deduce a rough sketch of the 'reciprocal' curve $y = 1/f(x)$ from a sketch of $y = f(x)$. The following table provides the rules to be followed in deducing the reciprocal curve.

$y = f(x)$	$y = 1/f(x)$
zero	vertical asymptote
vertical asymptote	zero
y positive (negative)	y positive (negative)
slant asymptote	'$y = 0$' asymptote
'$y = 0$' asymptote	$y \to \pm\infty$ as $x \to \pm\infty$
'$y = c \neq 0$' asymptote	'$y = 1/c$' asymptote
with approach from $\begin{cases} \text{above} \\ \text{below} \end{cases}$	with approach from $\begin{cases} \text{below} \\ \text{above} \end{cases}$
$\left.\begin{array}{l} \text{maximum} \\ \text{minimum} \end{array}\right\}$ point $(y \neq 0)$	$\left.\begin{array}{l} \text{minimum} \\ \text{maximum} \end{array}\right\}$ point

Example 10.2.16 Deduce from Example 10.2.14 a rough sketch of

$$y = \frac{2(x+1)^2}{(3x+1)(x+3)}.$$

Solution See Figure 10.2.11.   □

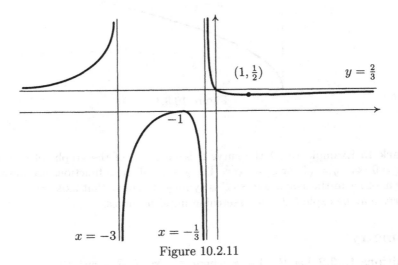

$(1, \tfrac{1}{2})$ $y = \tfrac{2}{3}$

-1

$x = -3$ $x = -\tfrac{1}{3}$

Figure 10.2.11

10.3 Implicit Curves

We may also sketch curves in the x, y-plane defined by an *implicit* relation $R(x, y) = 0$. Suppose that the function R has maximal domain A, which is a subset of $\mathbb{R}^2$, then the set

$$\mathcal{C} = \{(x, y) \in A : R(x, y) = 0\},$$

may be empty, finite or infinite. In the latter case $\mathcal{C}$ is called an *implicit curve*. We often write

$$\mathcal{C} : \ R(x, y) = 0$$

for short, with the assumption of the maximal domain for R.

By restricting to a subset of the plane it may be possible to consider part of an implicit curve as the graph of a function, and hence obtain a sketch of that part of the curve using the methods described in Section 10.2. This is the most important idea that we will use to sketch implicit curves.

Example 10.3.1 Sketch the part of the curve

$$\mathcal{C}: \; y^2 - x = 0$$

that lies in the first quadrant.

Solution Since $x \geq 0$ and $y \geq 0$,

$$(x, y) \in \mathcal{C} \;\Leftrightarrow\; y^2 = x \;\Leftrightarrow\; y = \sqrt{x}.$$

Thus the part of $\mathcal{C}$ that lies in the first quadrant is the graph of $x \mapsto \sqrt{x}$. The sketch is shown in Figure 10.3.1. $\qquad\qquad\qquad\qquad\qquad\qquad\qquad\qquad\qquad\qquad\qquad\quad$ $\square$

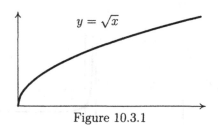

Figure 10.3.1

Remark In Example 10.3.1 the curve $\mathcal{C}$ is made up of the graphs of two functions: $y^2 - x = 0 \;\Leftrightarrow\; y = \sqrt{x}$ or $y = -\sqrt{x}$. The graphs of these functions are mirror images of one another in the x-axis and so $\mathcal{C}$ is symmetric about that axis. Such a symmetry property may be exploited when sketching implicit curves.

Symmetry

Definitions 10.3.2 Let $R : A \to \mathbb{R}$ where $A \subseteq \mathbb{R}^2$. If R is such that

$$R(x, y) = 0 \Rightarrow (x, -y) \in A \text{ and } R(x, -y) = 0,$$

then for each point (x, y) in the curve $\mathcal{C} : R(x, y) = 0$, its mirror image in the x-axis $(x, -y)$ also lies in the curve. We say that $\mathcal{C}$ is *symmetric about the x-axis* (*or $y = 0$*).
   Similarly, if

$$R(x, y) = 0 \Rightarrow (-x, y) \in A \text{ and } R(-x, y) = 0$$

then $\mathcal{C}$ is *symmetric about the y-axis* (*or $x = 0$*) and if

$$R(x, y) = 0 \Rightarrow (-x, -y) \in A \text{ and } R(-x, -y) = 0$$

then $\mathcal{C}$ is *symmetric about the origin*. See Figure 10.3.2.

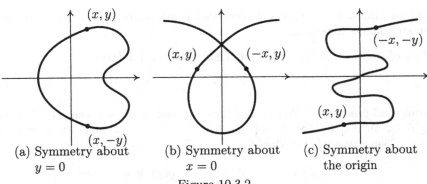

(a) Symmetry about
$y = 0$

(b) Symmetry about
$x = 0$

(c) Symmetry about
the origin

Figure 10.3.2

Remark In the special case that an implicit curve C is the graph of some function f, then symmetry of C about the y-axis [the origin] is equivalent to the statement that f is an even [odd] function. See Definition 2.2.8.

Example 10.3.3 Determine whether

(a) $C: \ x^3 + y^4 - 2y^2 = 0$,   (b) $D: \ \cos xy = x/y$,   (c) $\mathcal{E}: \ x^2y + \cos^{-1}(x-1) = 0$,

are symmetric about the x-axis, the y-axis or the origin.

Solution   (a) Let $f(x,y) = x^3 + y^4 - 2y^2$ then the maximal domain of f is $\mathbb{R}^2$. Suppose that $(a,b) \in \mathbb{R}^2$ then $(a,-b)$, $(-a,b)$ and $(-a,-b) \in \mathbb{R}^2$ also. If $f(a,b) = 0$ then

$$f(a,-b) = a^3 + (-b)^4 - 2(-b)^2 = f(a,b) = 0$$

and

$$f(-a,b) = f(-a,-b) = -a^3 + b^4 - 2b^2 \neq 0 \text{ (in general)},$$

so that C is symmetric about the x-axis only.

(b) Let $g(x,y) = \cos xy - x/y$ then the maximal domain of g is $A = \{(x,y) \in \mathbb{R}^2 : y \neq 0\}$. Suppose that $(a,b) \in A$ then so are $(a,-b)$, $(-a,b)$ and $(-a,-b)$. If $g(a,b) = 0$ then

$$g(a,-b) = g(-a,b) = \cos ab + a/b \neq 0 \text{ (in general)}$$

and

$$g(-a,-b) = \cos ab - a/b = g(a,b) = 0,$$

so that D is symmetric about the origin only.

(c) Let $h(x,y) = x^2y + \cos^{-1}(x-1)$ then the maximal domain of h is $B = \{(x,y) \in \mathbb{R}^2 : x \in [0,2]\} = [0,2] \times \mathbb{R}$ since the (maximal) domain of $\cos^{-1}$ is $[-1,1]$. If $(a,b) \in B$ then so is $(a,-b)$ but, in general, $(-a,b)$ and $(-a,-b) \notin B$. Thus we need only check for symmetry about the x-axis. If $h(a,b) = 0$ then

$$h(a,-b) = -a^2b + \cos^{-1}(a-1) \neq 0 \text{ (in general)}$$

and so $\mathcal{E}$ has none of the three types of symmetry.   □

Problem 10.3.4 Prove that any curve that is symmetric about the x- and y-axes is symmetric about the origin.

One of the commonest types of implicit curve has the form

$$C : \quad y^2 = f(x), \tag{10.3.1}$$

for some real function f. Points on C have x-coordinates satisfying $f(x) \geq 0$ because $y^2 \geq 0$ for all $y \in \mathbb{R}$. Let $A = \{x \in \operatorname{dom} f : f(x) \geq 0\}$. Since there is a function $g : A \to \mathbb{R}$ such that $f(x) = g^2(x)$ for all $x \in A$,

$$y^2 = f(x) \text{ for } x \in A \quad \Leftrightarrow \quad (y = g(x) \text{ or } y = -g(x)).$$

It follows that C consists of two pieces, the graph of g and the graph of $-g$. Note that the graph of $-g$ is the mirror image in the x-axis of the graph of g so that little extra work is required to obtain the full sketch of C once we have sketched the graph of g.

Example 10.3.5 Sketch the curve

$$C : \quad y^2 = x(x-3)^2.$$

Solution We see that $x(x-3)^2 \geq 0$ for $x \geq 0$ and so we first sketch

$$C' : \quad y = \sqrt{x}(x-3), \quad (x \geq 0).$$

Zeros: $x = 0, 3$.

Asymptotes: none.

Table of signs:

x	0	$\rightarrow$	3	$\rightarrow$
$\sqrt{x}$	0	+	+	+
$x - 3$	−	−	0	+
y	0	−	0	+

Critical points:

$$y' = \frac{1}{2\sqrt{x}}(x-3) + \sqrt{x} = \frac{3(x-1)}{2\sqrt{x}} = 0 \quad \Leftrightarrow \quad x = 1$$

and

$$y'' = \frac{3}{2}\left(\frac{1 \times \sqrt{x} - (x-1)/2\sqrt{x}}{x}\right) = \frac{3(x+1)}{4x\sqrt{x}} > 0 \text{ when } x = 1.$$

Therefore $(1, -2)$ is a minimum turning point.

Vertical tangents: $y' \to \infty$ as $x \to 0^+$ and so there is a vertical tangent at the origin.

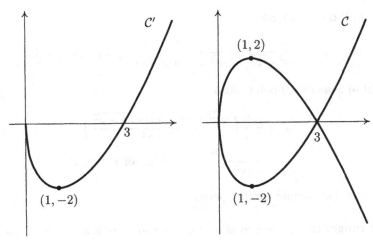

Figure 10.3.3

Points of inflection: $y'' > 0$ for all $x > 0$ therefore C' is concave up and has no points of inflection.

Sketches of C', based on the above information, and of C, obtained from C' using the symmetry about the x-axis, are shown in Figure 10.3.3.   □

Example 10.3.6 Let $a, b \in \mathbb{R}$ with $a, b > 0$. Sketch the curve

$$C: \frac{x^2}{a^2} + \frac{y^2}{b^2} = 1.$$

Solution Now if $(x, y) \in C$, $1 - x^2/a^2 = y^2/b^2 \geq 0$ and so $x^2 \leq a^2$, i.e. $x \in [-a, a]$ and similarly, $y \in [-b, b]$. We consider first the curve

$$y = b\sqrt{1 - \frac{x^2}{a^2}} \quad (x \in [-a, a]).$$

As well as having the symmetry about the x-axis of the general curve (10.3.1), this curve also has symmetry about the y-axis since $(x, y) \in C \Rightarrow (-x, y) \in C$. For this reason we consider

$$C': y = b\sqrt{1 - \frac{x^2}{a^2}} \quad (x \in [0, a]),$$

and then obtain the sketch of C by using the symmetry of this curve about the x- and y-axes.

Zeros: $x = a$.

Asymptotes: None.

Signs: $y > 0$ for all $x \in [0, a)$.

Critical points: We have

$$y' = -\frac{b}{a^2}\frac{x}{\sqrt{1 - x^2/a^2}} = -\frac{b}{a}\frac{x}{\sqrt{a^2 - x^2}} = 0 \iff x = 0,$$

so that $(0, b)$ is a critical point. Also

$$y'' = -\frac{b}{a}\left(\frac{\sqrt{a^2 - x^2} + x^2/\sqrt{a^2 - x^2}}{a^2 - x^2}\right)$$

$$= -\frac{ab}{(a^2 - x^2)^{3/2}} < 0 \text{ for all } x \in [0, a).$$

Thus $(0, b)$ is a maximum turning point.

Vertical tangents: $y' \to -\infty$ as $x \to a^-$ and so there is a vertical tangent at $(a, 0)$.

Points of inflection: $y'' < 0$ for all $x \in [0, a)$ and so there are no points of inflection and C' is concave down.

A sketch of C obtained using this information is shown in Figure 10.3.4.   □

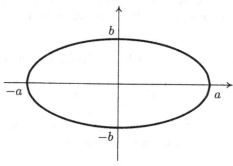

Figure 10.3.4

Remark The curve studied in Example 10.3.6, and shown in Figure 10.3.4, is an ellipse and is considered in more detail in Section 10.5.

10.4   Parametric Curves

Another way to describe a curve is by explicitly specifying the coordinates of all its points. For example, the set of points with coordinates $(\sin\theta, \cos\theta)$ for all $\theta \in [0, 2\pi)$ is, by Definitions 5.1.1, a circle with radius 1 and centre the origin. We say that this circle has *parametric equations*

$$x = \sin\theta, \quad y = \cos\theta \quad (\theta \in [0, 2\pi)),$$

and θ is the *parameter*.

As a second example, consider a particle moving in the plane and describing a curve. If the location of the particle has coordinates $(x(t), y(t))$ for all times t in an interval I, the curve has parametric equations

$$x = x(t), \quad y = y(t) \quad (t \in I).$$

Since we often think of the parameter as 'time', we use the notation $\dot{x}$, $\ddot{x}$ and so on, introduced in Section 9.5, to denote derivatives.

To sketch parametric curves it is useful to find the gradient of tangents to the curve and the concavity. The next result is used for this, and other, purposes.

Theorem 10.4.1 *Let x and y be real functions differentiable at some $t \in \mathbb{R}$. Then, if $\dot{x}(t) \neq 0$,*

$$\frac{dy}{dx} = \frac{\dot{y}}{\dot{x}} \tag{10.4.1}$$

and

$$\frac{d^2y}{dx^2} = \frac{\ddot{y}\dot{x} - \dot{y}\ddot{x}}{\dot{x}^3}. \tag{10.4.2}$$

Proof By the chain rule we have,

$$\frac{dy}{dx} = \frac{dy}{dt}\frac{dt}{dx}.$$

Also, by Theorem 8.2.9,

$$\frac{dt}{dx} = 1 \bigg/ \frac{dx}{dt}$$

and hence we get (10.4.1).

Differentiating again we get

$$\frac{d^2y}{dx^2} = \frac{d}{dx}\left(\frac{dy}{dx}\right) = \frac{d}{dt}\left(\frac{\dot{y}}{\dot{x}}\right)\frac{dt}{dx},$$
$$= \frac{\ddot{y}\dot{x} - \dot{y}\ddot{x}}{\dot{x}^3}$$

as required. $\qquad\square$

Example 10.4.2 Find an equation for the tangent to the curve

$$x = \frac{1}{1+t^3}, \quad y = \frac{t}{1+t^3} \quad (t \neq -1) \tag{10.4.3}$$

at the point where $t = 2$. Find, also, the coordinates of the other point where this tangent meets the curve.

Solution We have

$$\dot{x} = -\frac{3t^2}{(1+t^3)^2}, \quad \dot{y} = \frac{1-2t^3}{(1+t^3)^2}.$$

Hence

$$\frac{dy}{dx} = \frac{\dot{y}}{\dot{x}} = \frac{1-2t^3}{-3t^2} = \frac{-15}{-12} = \frac{5}{4} \text{ when } t = 2.$$

Also, $x(2) = 1/9$ and $y(2) = 2/9$. So the required tangent has equation

$$y - \frac{2}{9} = \frac{5}{4}\left(x - \frac{1}{9}\right),$$

i.e.

$$15x - 12y + 1 = 0. \tag{10.4.4}$$

To find where this tangent meets the curve (again) we substitute x and y from (10.4.3) into (10.4.4) to obtain

$$15\left(\frac{1}{1+t^3}\right) - 12\left(\frac{t}{1+t^3}\right) + 1 = 0,$$

i.e.

$$t^3 - 12t + 16 = 0. \tag{10.4.5}$$

Since the tangent touches the curve where $t = 2$, $(t - 2)^2$ is a factor of the left hand side of (10.4.5) (see Note 8.2.5). Hence (10.4.5) becomes

$$(t - 2)^2(t + 4) = 0.$$

So the tangent meets the curve again where $t = -4$, i.e. at the point $(-1/63, 4/63)$.   □

Example 10.4.3 Show that the curve

$$x = \cos t, \quad y = \sin t \quad (t \in [0, \pi])$$

has a maximum turning point at $(0, 1)$.

Solution Here

$$\frac{dy}{dx} = \frac{\cos t}{-\sin t} = -\cot t \quad (t \in (0, \pi)),$$
$$= 0 \quad \text{when} \quad t = \pi/2$$

and

$$\frac{d^2y}{dx^2} = \frac{1}{-\sin t}\frac{d}{dt}(-\cot t) = -\csc^3 t,$$
$$= -1 < 0 \text{ when } t = \pi/2.$$

Thus the curve has a critical point when $t = \pi/2$, i.e. at $(0, 1)$, and, by Theorem 9.1.3, it is a maximum turning point.   □

Note 10.4.4 The situation not covered by Theorem 10.4.1, when $\dot{x} = 0$, leads to a number of possibilities. Suppose that $\dot{x} = 0$ at a point on a parametric curve given by parameter $t = s$.

If $\dot{y}(s) \neq 0$ then, from (10.4.1), $\frac{dy}{dx} \to \pm\infty$ as $t \to s^{\pm}$ and so there is either a vertical tangent to the curve or there is a cusp. On the other hand, if $\dot{y}(s) = 0$ then the limit of $\frac{dy}{dx}$ may be zero, non-zero or $\pm\infty$. We will not make a systematic exploration

of all the possibilities but instead will illustrate by example how a sketch of the curve
in the vicinity of $(x(s), y(s))$ may be obtained.

Consider the parametric curve

$$C : \quad x = t^m + \alpha, \quad y = t^n + \beta \quad (t \in \mathbb{R}),$$

where $m, n \in \mathbb{N}$ and $\alpha, \beta \in \mathbb{R}$. We have $\dot{x} = mt^{m-1}$ and $\dot{y} = nt^{n-1}$ and hence, from
(10.4.1),

$$\frac{dy}{dx} = \frac{n}{m} t^{n-m}.$$

If $m > 1$ then $\dot{x} = 0$ for $t = 0$ and we consider the x, y, $\dot{x}$, $\dot{y}$ and $\dfrac{dy}{dx}$ near to $t = 0$. In
particular we determine in which quadrant of the plane, centred at (α, β), $(x(t), y(t))$
lies for $t < 0$ and $t > 0$ and the limit of $\dfrac{dy}{dx}$ as $t \to 0$.

For $m = 2$, $n = 1$ a table of values and resulting sketch is shown in Figure 10.4.1.

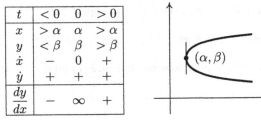

t	< 0	0	> 0
x	$> \alpha$	α	$> \alpha$
y	$< \beta$	β	$> \beta$
$\dot{x}$	$-$	0	$+$
$\dot{y}$	$+$	$+$	$+$
$\dfrac{dy}{dx}$	$-$	∞	$+$

Figure 10.4.1

Similarly, the case $m = 2$, $n = 3$ is shown in Figure 10.4.2.

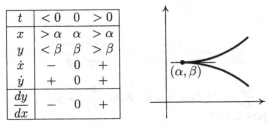

t	< 0	0	> 0
x	$> \alpha$	α	$> \alpha$
y	$< \beta$	β	$> \beta$
$\dot{x}$	$-$	0	$+$
$\dot{y}$	$+$	0	$+$
$\dfrac{dy}{dx}$	$-$	0	$+$

Figure 10.4.2

Example 10.4.5 As a circle rolls on a line, its original point of contact traces a curve
called a *cycloid*. Find the parametric equations of this curve and sketch it.

Solution Let the circle have radius a and let it roll on the x-axis. Fix the origin on
the plane at the original point of contact P. After the circle has turned through an
angle θ, let $(x(\theta), y(\theta))$ be the new coordinates of P and let Q be the new point of
contact. See Figure 10.4.3 (a)

The distance between Q and the original point of contact is the same as the arc
length PQ and so Q has coordinates $(a\theta, 0)$. As shown in Figure 10.4.3 (b), the horizontal and vertical displacement of P from Q are $a \sin \theta$ and $a - a \cos \theta$, respectively,

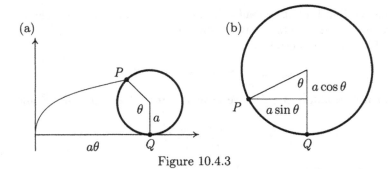

Figure 10.4.3

and hence the parametric equations are

$$x = a(\theta - \sin\theta), \quad y = a(1 - \cos\theta).$$

Note that since sin and cos are periodic with period 2π,

$$x(\theta + 2\pi) = x(\theta) + 2a\pi \quad \text{and} \quad y(\theta + 2\pi) = y(\theta). \tag{10.4.6}$$

For this reason it is sufficient to consider $\theta \in [0, 2\pi]$ and deduce the whole sketch using this symmetry.

First we determine the zeros of x and y:

$$x = 0 \iff \theta = 0, \quad y = 0 \iff \theta = 0, 2\pi.$$

Now, since

$$\dot{x} = a(1 - \cos\theta), \quad \dot{y} = a\sin\theta,$$

we have, for $\theta \neq 0, 2\pi$,

$$\frac{dy}{dx} = \frac{a\sin\theta}{a(1 - \cos\theta)} = \frac{2\sin\frac{1}{2}\theta\cos\frac{1}{2}\theta}{2\sin^2\frac{1}{2}\theta} = \cot\tfrac{1}{2}\theta.$$

Hence we see that $\dfrac{dy}{dx} = 0$ when $\theta = \pi$. Also, $\dfrac{dy}{dx} \to \infty$ as $\theta \to 0^+$ and $\dfrac{dy}{dx} \to -\infty$ as $\theta \to 2\pi^-$. See Figure 5.1.11.

We also have

$$\frac{d^2y}{dx^2} = \frac{1}{a(1 - \cos\theta)}\frac{d}{d\theta}\left(\cot\tfrac{1}{2}\theta\right) = \frac{1}{2a\sin^2\frac{1}{2}\theta}\left(-\tfrac{1}{2}\operatorname{cosec}^2\tfrac{1}{2}\theta\right),$$

$$= -\frac{1}{4a}\operatorname{cosec}^4\tfrac{1}{2}\theta < 0, \quad (\theta \neq 0, 2\pi).$$

Hence the curve is concave down for $\theta \in (0, 2\pi)$.

The important features on the curve are where $\theta = 0, \pi$ and 2π and we summarise the information about the curve at these points.

Table of values:

θ	0	π	2π
x	0	$a\pi$	$2a\pi$
y	0	$2a$	0
$\dfrac{dy}{dx}$	?	0	?

Using this information and the symmetry (10.4.6), we obtain a sketch of the cycloid (see Figure 10.4.4). Note that there are vertical cusps at $(0,0)$ and $(2a\pi, 0)$. $\qquad\square$

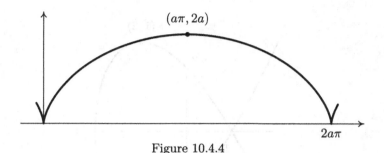

$(a\pi, 2a)$

$2a\pi$

Figure 10.4.4

Example 10.4.6 Sketch the curve

$$x = 2 - t^2, \quad y = t^3 - 3t \quad (t \in \mathbb{R}).$$

Solution We observe that

$$(x(t), y(t)) = (x(-t), -y(-t)) \quad \text{for all } t \in \mathbb{R}.$$

This means that the points with parameters t and $-t$ are mirror images of one another in the x-axis and so the curve is symmetric about this axis. Hence we will consider $t \geq 0$ in detail, and obtain the rest of the curve by means of this symmetry.

For $t \geq 0$ we have

$$x = 0 \iff t = \sqrt{2}, \qquad y = 0 \iff t = 0 \text{ or } t = \sqrt{3}.$$

Also,

$$\frac{dy}{dx} = \frac{3(t^2 - 1)}{-2t},$$

so that $\dfrac{dy}{dx} = 0$ for $t = 1$ and $\dfrac{dy}{dx} \to -\infty$ as $t \to 0^+$ and

$$\frac{d^2y}{dx^2} = \frac{1}{-2t} \frac{d}{dt}\left(-\frac{3}{2}\frac{t^2-1}{t}\right) = \frac{3}{4}\frac{t^2+1}{t^3} > 0,$$

so that the curve is concave up.

Table of values:

t	0	1	$\sqrt{2}$	$\sqrt{3}$
x	2	1	0	-1
y	0	-2	$-\sqrt{2}$	0
$\dfrac{dy}{dx}$	?	0	$-3\sqrt{2}/4$	$-\sqrt{3}$

A sketch is shown in Figure 10.4.5 in which the full curve is obtained by using the symmetry about the x-axis.   □

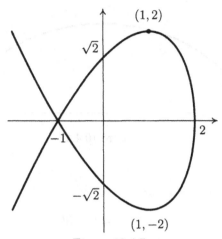

Figure 10.4.5

10.5   Conic Sections

A curve that is obtained as the cross-section of a circular cone is called a *conic section* or simply a *conic*. A conic can be one of three kinds: a parabola, an ellipse or a hyperbola.

Definitions 10.5.1 A *parabola* $\mathcal{P}$ has equation

$$y^2 = 4ax, \tag{10.5.1}$$

where $a \in \mathbb{R}$ with $a \neq 0$.

An *ellipse* $\mathcal{E}$ has equation

$$\frac{x^2}{a^2} + \frac{y^2}{b^2} = 1, \tag{10.5.2}$$

where $a, b \in \mathbb{R}$ with $a, b > 0$. The special case $a = b = r$ gives

$$x^2 + y^2 = r^2,$$

and is a circle of radius r.

A *hyperbola* $\mathcal{H}$ has equation

$$\frac{x^2}{a^2} - \frac{y^2}{b^2} = 1, \qquad (10.5.3)$$

where $a, b \in \mathbb{R}$ with $a, b > 0$.

According to the above definitions conic sections are examples of implicit curves. A sketch of $\mathcal{E}$ was obtained in Example 10.3.6 and shown in Figure 10.3.4. Figures 10.5.1 and 10.5.2 show $\mathcal{P}$, for $a > 0$, and $\mathcal{H}$, respectively.

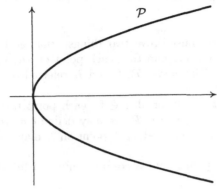

Figure 10.5.1

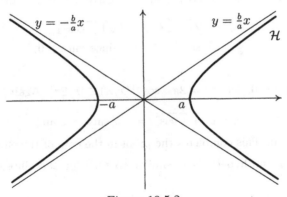

Figure 10.5.2

Remark The curve $xy = 1$ (for example) is also a hyperbola, although it does not have the form given in (10.5.3). As shown in Example 6.6.5, this curve is a rotation through $\pi/4$ of the standard hyperbola defined above.

Problem 10.5.2 Obtain the sketches of $\mathcal{P}$, where $a > 0$, and $\mathcal{H}$ by regarding them as curves of the type (10.3.1).

Theorem 10.5.3 *Parabola* $\mathcal{P}$ *(10.5.1), ellipse* $\mathcal{E}$ *(10.5.2) and hyperbola* $\mathcal{H}$ *(10.5.3) have parametric equations*

$$x = at^2, \quad y = 2at \quad (t \in \mathbb{R}), \tag{10.5.4}$$

$$x = a\cos t, \quad y = b\sin t \quad (t \in [0, 2\pi)) \tag{10.5.5}$$

and

$$x = a\sec t, \quad y = b\tan t \quad (t \in (-\pi/2, \pi/2) \cup (\pi/2, 3\pi/2)), \tag{10.5.6}$$

respectively.

Proof In each case we must prove two things: that each point given parametrically lies on the implicit curve and that each point on the implicit curve has such a parametric description. The curves $\mathcal{P}$, $\mathcal{E}$ and $\mathcal{H}$ referred to below are those given in Definitions 10.5.1.

Since $(2at)^2 - 4a(at^2) = 0$ for all $t \in \mathbb{R}$, each point $(at^2, 2at) \in \mathcal{P}$, *i.e.* satisfies $y^2 = 4ax$. Also, given any $(x, y) \in \mathcal{P}$, we may define $t = y/2a$ so that $y = 2at$ and $x = y^2/4a = at^2$. Hence, as required, each point on $\mathcal{P}$ may be written as $(at^2, 2at)$ for some $t \in \mathbb{R}$.

Since $(a\cos t)^2/a^2 + (b\sin t)^2/b^2 = \cos^2 t + \sin^2 t = 1$ for all $t \in [0, 2\pi)$, each point $(a\cos t, b\sin t) \in \mathcal{E}$.

As discussed in Example 10.3.6, for points $(x, y) \in \mathcal{E}$, $x \in [-a, a]$. We consider the cases $y \geq 0$ and $y \leq 0$ separately.

For $y \geq 0$ define $t = \cos^{-1}(x/a) \in [0, \pi]$. Clearly, $x = a\cos t$ and

$$y = b\sqrt{1 - x^2/a^2} = b\sqrt{1 - \cos^2 t},$$
$$= b|\sin t| = b\sin t, \text{ since } \sin t \geq 0,$$

as required.

Similarly, for $y \leq 0$, define $t = 2\pi - \cos^{-1}(x/a) \in [\pi, 2\pi]$. Again, $x = a\cos(2\pi - t) = a\cos t$ and

$$y = b\sqrt{1 - \cos^2 t} = -b|\sin t| = b\sin t,$$

since here $\sin t \leq 0$. This completes the proof in the case of the ellipse.

The proof for the hyperbola is similar to that for the ellipse and is left as an exercise. □

Many geometric properties of conics are most easily established by using the parametric representation of the curves.

Example 10.5.4 Let $U(au^2, 2au)$ and $V(av^2, 2av)$ be distinct points on the parabola $\mathcal{P} : y^2 = 4ax$. Find the equation of the tangent to $\mathcal{P}$ at U and the equation of the chord UV.

Deduce that UV cuts the x-axis at $S(a, 0)$ if and only if the tangents at U and V are perpendicular.

Solution From (10.5.4) the parametric equations for $\mathcal{P}$ are

$$x = at^2, \quad y = 2at \quad (t \in \mathbb{R}),$$

and so
$$\dot{x} = 2at, \quad \dot{y} = 2a,$$

giving
$$\frac{dy}{dx} = \frac{2a}{2at} = \frac{1}{t}, \text{ for } t \neq 0.$$

Hence the tangent at U has gradient $1/u$ and so its equation is

$$y - 2au = \frac{1}{u}(x - au^2), \quad i.e. \ uy = x + au^2.$$

The gradient of the chord UV is

$$\frac{2au - 2av}{au^2 - av^2} = \frac{2}{u + v}$$

and so its equation is

$$y - 2au = \frac{2}{u + v}(x - au^2), \quad i.e. \ (u + v)y = 2x + 2auv.$$

Now

$$S(a, 0) \text{ lies on } UV \ \Leftrightarrow \ (u + v) \times 0 = 2a + 2auv$$
$$\Leftrightarrow \ \frac{1}{u}\frac{1}{v} = -1$$
$$\Leftrightarrow \ \text{the product of the gradients of the}$$
$$\text{tangents at } U \text{ and } V \text{ is } -1$$
$$\Leftrightarrow \ \text{tangents at } U \text{ and } V \text{ are perpendicular,}$$

as required. $\qquad\qquad\qquad\qquad\qquad\qquad\qquad\qquad\qquad\qquad\qquad\qquad\Box$

10.6 Polar Curves

Let $P(x, y)$ be a point in the plane. Denote by r the distance $\sqrt{x^2 + y^2}$ of P from the origin O. If P is not the origin then $r > 0$ and we may measure, in an anti-clockwise direction, the angle $\theta \in [0, 2\pi)$ between the positive x-axis and OP (see Figure 6.2.2). The ordered pair (r, θ) gives the *polar coordinates* of P. Such coordinates were also discussed in relation to complex numbers in Section 6.2.

Conversely, any (r, θ) where $r \geq 0$ and $\theta \in \mathbb{R}$, determines the coordinates (x, y) of a point in the plane where, as shown in Theorem 6.2.4,

$$x = r\cos\theta, \quad y = r\sin\theta. \qquad\qquad\qquad (10.6.1)$$

If $r = 0$ then (r, θ), for any θ, determines the origin and θ may no longer be interpreted as an angle.

An equation of the form

$$r = r(\theta) \quad (r \geq 0, \ \theta \in I)$$

where $I \subseteq \mathbb{R}$ is called a *polar equation* and the set of points determined by $(r(\theta), \theta)$ for $\theta \in I$, forms a *polar curve*.

Example 10.6.1 Determine the polar equations of the curves

$$\text{(a) } x^2 + y^2 = 4, \quad \text{(b) } (x-1)^2 + y^2 = 1.$$

Solution (a) Using (10.6.1) we get

$$x^2 + y^2 - 4 = r^2 \cos^2 \theta + r^2 \sin^2 \theta - 4 = r^2 - 4 = 0,$$

since $\cos^2 \theta + \sin^2 \theta = 1$. As $r \geq 0$, the polar equation is

$$r = 2.$$

This is the circle centred at the origin with radius 2.

(b) Similarly,

$$(x-1)^2 + y^2 - 1 = r^2 - 2r \cos \theta = 0, \quad i.e. \ r = 0 \text{ or } r = 2 \cos \theta.$$

The polar equation is

$$r = 2 \cos \theta$$

since this has solution $(r, \theta) = (0, \pi/2)$ and so the possibility $r = 0$ is included automatically. This curve is the circle centred at $(1, 0)$ with radius 1. $\qquad \square$

Theorem 10.6.2 *Let* $r : I \to [0, \infty)$, *where* $I \subseteq \mathbb{R}$, *be differentiable at* $\theta \in I$ *and let* $\alpha \in \mathbb{R}$ *be such that*

$$\tan \alpha = r \left/ \frac{dr}{d\theta} \right. . \tag{10.6.2}$$

Then the gradient of the tangent to the polar curve

$$r = r(\theta) \quad (\theta \in I),$$

at the point with polar coordinates $(r(\theta), \theta)$ *is*

$$\frac{dy}{dx} = \tan(\theta + \alpha). \tag{10.6.3}$$

Proof From Theorem 10.4.1 and (10.6.1), we have

$$\frac{dy}{dx} = \frac{dy}{d\theta} \left/ \frac{dx}{d\theta} \right. = \frac{d}{d\theta}(r(\theta) \sin \theta) \left/ \frac{d}{d\theta}(r(\theta) \cos \theta) \right.$$

$$= \frac{\dfrac{dr}{d\theta} \sin \theta + r \cos \theta}{\dfrac{dr}{d\theta} \cos \theta - r \sin \theta} = \frac{\tan \theta + \left(r \left/ \dfrac{dr}{d\theta} \right. \right)}{1 - \tan \theta \left(r \left/ \dfrac{dr}{d\theta} \right. \right)}$$

$$= \frac{\tan \theta + \tan \alpha}{1 - \tan \theta \tan \alpha} = \tan(\theta + \alpha),$$

using Theorem 5.2.3 (5). $\qquad \square$

Note 10.6.3 Consider a point P on a polar curve, determined by polar angle θ, at which there is a tangent. Define $\beta \in [0, \pi)$ to be the angle between the radius and the tangent measured in an anti-clockwise direction. As shown in Figure 10.6.1, $\theta + \beta$ is the angle between the tangent and the x-axis (again measured anti-clockwise), and so the gradient of the tangent is

$$\frac{dy}{dx} = \tan(\theta + \beta). \tag{10.6.4}$$

By comparing (10.6.3) and (10.6.4) we see that $\tan(\theta + \beta) = \tan(\theta + \alpha)$, *i.e.* $\beta = \alpha + k\pi$ for some $k \in \mathbb{Z}$. This means that, by taking α satisfying (10.6.2), we may determine the angle β by taking an appropriate choice for $k \in \mathbb{Z}$. In this way the gradient of the tangent may be determined for any particular θ. Note that for any $k \in \mathbb{Z}$, $\tan(\theta + \alpha + k\pi)$ has the same value and so whatever choice of k is made, we obtain the same gradient for the tangent. This means that for any choice of α satisfying (10.6.2), the gradient is determined unambiguously.

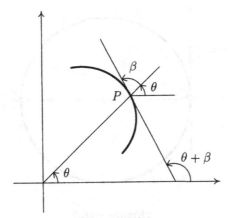

Figure 10.6.1

Example 10.6.4 Sketch the curve

$$r = 2\cos\theta \quad (\theta \in [-\pi/2, \pi/2]).$$

Solution Using Theorem 10.6.2

$$\tan\alpha = \frac{2\cos\theta}{-2\sin\theta} = -\frac{\sin(\pi/2 - \theta)}{\cos(\pi/2 - \theta)} = \tan(\theta - \pi/2).$$

Therefore we take $\alpha = \theta - \pi/2$ and, by Note 10.6.3, the gradient of the tangent at the point on the curve with polar angle θ is determined by the angle $\theta + \alpha = 2\theta - \pi/2$.

Observe that r is an even function of θ and so it is sufficient to consider non-negative θ. We may then complete the sketch by using the symmetry of the curve about the x-axis that this implies.

Table of values:

θ	0	$\frac{\pi}{6}$	$\frac{\pi}{4}$	$\frac{\pi}{3}$	$\frac{\pi}{2}$
r	2	$\sqrt{3}$	$\sqrt{2}$	1	0
$\theta - \frac{\pi}{2}$	$-\frac{\pi}{2}$	$-\frac{\pi}{3}$	$-\frac{\pi}{4}$	$-\frac{\pi}{6}$	0

From Example 10.6.1, the curve is the unit circle centred at $(1,0)$. This is illustrated in Figure 10.6.2.   □

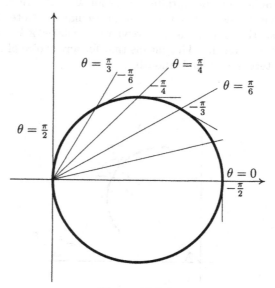

Figure 10.6.2

Example 10.6.5 Sketch the curve

$$r = 2(1 - \cos\theta) \quad (\theta \in [-\pi, \pi]),$$

called a *cardioid*.

Solution We have

$$\tan\alpha = \frac{2(1 - \cos\theta)}{2\sin\theta} = \frac{4\sin^2\frac{1}{2}\theta}{4\sin\frac{1}{2}\theta\cos\frac{1}{2}\theta} = \tan\tfrac{1}{2}\theta,$$

and so we take $\alpha = \frac{1}{2}\theta$ and the gradient of the tangent is determined by $\theta + \alpha = \frac{3}{2}\theta$. Again, r is an even function of θ and so we consider non-negative θ and then exploit the symmetry about the x-axis when we sketch the curve.

Table of values:

θ	0	$\frac{\pi}{3}$	$\frac{\pi}{2}$	$\frac{2\pi}{3}$	$\frac{5\pi}{6}$	π
r	0	1	2	3	$2 + \sqrt{3}$	4
$\frac{\theta}{2}$	0	$\frac{\pi}{6}$	$\frac{\pi}{4}$	$\frac{\pi}{3}$	$\frac{5\pi}{12}$	$\frac{\pi}{2}$

The curve is sketched in Figure 10.6.3.   □

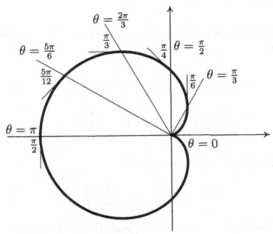

Figure 10.6.3

10.X   Exercises

1. Determine the sign of y and the approaches to the asymptotes for the curve $y = f(x)$ when $f(x)$ is

    (a) $y = \dfrac{x}{(x-2)(x-3)}$,   (b) $y = \dfrac{x^2(x+1)}{(x-3)(x-4)^2}$,   (c) $y = \dfrac{x(x+1)^2}{(x^2-4)}$.

    In each case, deduce a sketch of the curve $y = 1/f(x)$.

2. For each of the following curves, find the critical points and asymptotes and use this information sketch the curve.

    (a) $y = x + \dfrac{4}{x^2}$,   (b) $y = \dfrac{x^2}{x^3+1}$,

    (c) $y = \dfrac{(x-2)^2}{x(x-3)}$,   (d) $y = \dfrac{3x^2 + 2x - 2}{(x-1)(x^2-1)}$.

3. For each of the following curves, find the points of inflection, determine their type and sketch the curve near each such point.

   (a) $y = 3x - x^3$,   (b) $y = x^4 + 4x^3$,   (c) $y = \dfrac{x^2}{1-x}$,

   (d) $y = \dfrac{x}{x^4 + 3}$,   (e) $y = 2x + \cos x$ $(-\pi \le x \le \pi)$,

   (f) $y = x + 2\sin x$ $(-\pi/2 \le x \le 3\pi/2)$,

   (g) $y = 2x - \tan x$ $(-\pi/2 < x < \pi/2)$.

4. Sketch each of the following curves, showing critical points, asymptotes and points of inflection.

   (a) $y = \dfrac{x}{3x^2 + 1}$,   (b) $y = \dfrac{x}{(1+x)^2}$,   (c) $y = \dfrac{x^3}{x^2 - 3}$,

   (d) $y = \dfrac{64x}{(3-x)^4}$,   (e) $y = \dfrac{x^2}{2-x}$.

5. Find the critical points on the curve

   $$y = \frac{x^2 + 3}{x - 1}$$

   and sketch the curve. Deduce a rough sketch of the curve

   $$y = \frac{x - 1}{x^2 + 3}.$$

6. For the curve $y = (x - 1)^2/(x + 1)^3$, show that

   $$\frac{d^2y}{dx^2} = \frac{2(x^2 - 10x + 13)}{(x + 1)^5}.$$

   Find the nature of each critical point and point of inflection and sketch the curve. Deduce a rough sketch of $y = (x - 1)^4/(x + 1)^6$.

7. For each of the following curves, decide whether it is symmetric about the x-axis, the y-axis or the origin.

   (a) $x^2 + y^4 + y^2 = \cos x$,   (b) $x^2 + x = y^3 + \cos y$,

   (c) $(x - 2y)^2 = \cos(xy)$,   (d) $x^3 + \sin x = y^2$.

8. Sketch the curves

   (a) $16y^2 = x^2(2 - x^2)$,   (b) $y^2 = x^2 + x$.

9. Show that the critical points of the curve $x^2 + xy + y^2 = 27$ lie on the line $y + 2x = 0$. Find the coordinates of each such point.

10. Find the equation of the tangent at the point with parameter $t = 2$ on the curve

$$x = \frac{4t}{t^3 + 1}, \quad y = \frac{t^2}{t^3 + 1} \quad (t \neq -1).$$

Find the parameter of the other point at which this tangent meets the curve.

11. Sketch the curves

(a) $x = \sqrt{2}\cos\theta, \quad y = \frac{1}{4}\sin 2\theta \quad (0 \leq \theta \leq \pi/2),$

(b) $x = \cos^3 t, \quad y = \sin^3 t \quad (0 \leq t \leq 2\pi).$

12. Show that the tangent at the point (x', y') to the hyperbola

$$\frac{x^2}{a^2} - \frac{y^2}{b^2} = 1$$

has equation

$$\frac{xx'}{a^2} - \frac{yy'}{b^2} = 1.$$

13. Let P be the point $(at^2, 2at)$ on the parabola $y^2 = 4ax$ $(a \neq 0)$, let S be the point $(a, 0)$ and let U be the point on the line $x = -a$ such that PU is parallel to the x-axis. Show that (a) $|PS| = |PU|$, (b) the normal to the parabola at P makes equal angles with PS and PU.

14. Sketch the curves with polar equations

(a) $r = 1 + \cos\theta \quad (\theta \in [-\pi, \pi]),$

(b) $r = \sqrt{\sin 2\theta} \quad (\theta \in [0, \pi/2] \cup [\pi, 3\pi/2]),$

(c) $r = \theta/\pi \quad (\theta \in [0, 4\pi]).$

Chapter 11

Matrices and Linear Equations

In Chapter 1, we mentioned the fact that a pair (x, y) of real numbers can be used to represent a point in the plane. In Chapter 6, we used such a pair to represent a complex number. This is an example of an important aspect of mathematics—a *mathematical object* may be interpreted in several different ways. Once we have studied the mathematical objects, we have information which is available in every interpretation.

In this chapter we introduce a new mathematical object—the matrix. An $m \times n$ *matrix* is a rectangular array of mn numbers. We can think of the matrix as consisting of m *rows*, each having n entries, or as n *columns*, each having m entries. For example,

$$M = \begin{bmatrix} 2 & 3 & 4 \\ -1 & 6 & 0 \end{bmatrix}$$

is a 2×3 matrix. We regard a matrix as a single entity, to be manipulated as a whole.

We define *addition* on the set of $m \times n$ matrices. This operation has properties similar to the addition of real numbers. In particular, there is a *zero matrix* $O_{m \times n}$ in the set which behaves like 0 in $\mathbb{R}$, *i.e.*

$$A + O_{m \times n} = A = O_{m \times n} + A.$$

Under certain circumstances, we can define *multiplication* of matrices. Now some of the results are unfamiliar. For example, there are matrices A and B for which $AB \neq BA$.

Matrices have many uses in mathematics and its applications. Here we consider just one—the solution of (systems of) linear equations. Looking at such a system leads to the introduction of new operations on matrices. We develop a systematic method for finding all the solutions.

In the final sections, we return to the mathematical study of matrices. We consider the existence of matrix inverses. Along the way, we meet the concept of the *determinant* of a (square) matrix. This is an important topic in its own right and will be used in Chapter 13 in connection with the *vector product*.

11.1 Basic Definitions

Definitions 11.1.1 Let $\mathbb{F}$ be a number system. A *matrix defined over* $\mathbb{F}$ is a rectangular array of elements of $\mathbb{F}$. These elements are the *entries* of the matrix. If a matrix

A has m *rows* (horizontal lines of entries) and n *columns* (vertical lines), then A is *of type $m \times n$ or is an $m \times n$ bmatrix.*

The set of $m \times n$ matrices defined over $\mathbb{F}$ is denoted by $M_{m \times n}(\mathbb{F})$.

For example,

$$A = \begin{bmatrix} 1 & 2 & 3 \\ 4 & 5 & 6 \end{bmatrix} \text{ is a } 2 \times 3 \text{ matrix (defined over } \mathbb{R}),$$

$$B = \begin{bmatrix} 1 & 4 \\ 2 & 5 \\ 3 & 6 \end{bmatrix} \text{ is a } 3 \times 2 \text{ matrix (also defined over } \mathbb{R}).$$

We refer to individual entries of a matrix by their position in the array. The entry which occurs in the ith row and j^{th} column is the $(i, j)^{\text{th}}$ entry. Thus, with A and B as above, A has $(2, 1)^{\text{th}}$ entry 4, and B has $(3, 2)^{\text{th}}$ entry 6.

Notation 11.1.2 The $(i, j)^{\text{th}}$ entry of a matrix A is denoted by $(A)_{ij}$ or by a_{ij}. Thus, in our example B above, $(B)_{12} = 4$ and $b_{22} = 5$.

The general $m \times n$ matrix A has the form

$$A = \begin{bmatrix} a_{11} & a_{12} & \cdots & a_{1j} & \cdots & a_{1n} \\ a_{21} & a_{22} & \cdots & a_{2j} & \cdots & a_{2n} \\ \vdots & \vdots & & \vdots & & \vdots \\ a_{i1} & a_{i2} & \cdots & a_{ij} & \cdots & a_{in} \\ \vdots & \vdots & & \vdots & & \vdots \\ a_{m1} & a_{m2} & \cdots & a_{mj} & \cdots & a_{mn} \end{bmatrix} \leftarrow i^{\text{th}} \text{ row}$$

$$\uparrow$$
$$j^{\text{th}} \text{ column}$$

Definition 11.1.3 Matrices A and B are equal if and only if

(1) A and B are of the same type, say $m \times n$,

(2) for $1 \le i \le m$, $1 \le j \le n$, $(A)_{ij} = (B)_{ij}$.

Example 11.1.4 Determine the values of x and y for which

$$\begin{bmatrix} 3 & x+1 \\ y & 5 \end{bmatrix} = \begin{bmatrix} x-1 & 5 \\ x-3y & 5 \end{bmatrix}.$$

Solution Both matrices are 2×2, so they are equal if and only if the corresponding entries agree, *i.e.*

$$\begin{cases} 3 = x-1 \\ x+1 = 5 \\ y = x-3y \\ 5 = 5 \end{cases} \Leftrightarrow \begin{cases} x = 4 \\ x = 4 \\ 4y = x \end{cases} \Leftrightarrow \begin{cases} x = 4 \\ y = 1 \end{cases}. \qquad \square$$

11.2   Operations on Matrices

The arithmetic operations in $\mathbb{F}$ allow us to define operations on matrices.

Definition 11.2.1 Let $A, B \in M_{m \times n}(\mathbb{F})$, then $A+B$ is defined to be the $m \times n$ matrix with
$$(A + B)_{ij} = (A)_{ij} + (B)_{ij} \quad (1 \leq i \leq m, \ 1 \leq j \leq n).$$
Note that if A and B are not of the same type, then $A + B$ is *not defined*.

For example,
$$\begin{bmatrix} 1 & 2 & 3 \\ 4 & 5 & 6 \end{bmatrix} + \begin{bmatrix} 0 & -1 & 1 \\ 1 & 0 & -1 \end{bmatrix} = \begin{bmatrix} 1+0 & 2-1 & 3+1 \\ 4+1 & 5+0 & 6-1 \end{bmatrix} = \begin{bmatrix} 1 & 1 & 4 \\ 5 & 5 & 5 \end{bmatrix}$$
but
$$\begin{bmatrix} 1 & 2 & 3 \\ 4 & 5 & 6 \end{bmatrix} + \begin{bmatrix} 1 & 2 \\ 3 & 4 \end{bmatrix}$$
is not defined.

Remark If $A, B \in M_{m \times n}(\mathbb{F})$, then $A + B \in M_{m \times n}(\mathbb{F})$, *i.e.* $M_{m \times n}(\mathbb{F})$ is closed under addition.

Definitions 11.2.2 (1) For $m, n \in \mathbb{N}$, the *zero $m \times n$ matrix* $O_{m \times n}$ is the $m \times n$ matrix with all entries equal to 0. When the type is clear from the context, we just refer to 'the zero matrix'.

(2) If $A \in M_{m \times n}(\mathbb{F})$, then $-A$ is the $m \times n$ matrix with $(i, j)^{\text{th}}$ entry $-(A)_{ij}$, *i.e.* $(-A)_{ij} = -(A)_{ij}$, and is called the *negative of A*.

As usual, we define subtraction in terms of addition, so $A - B$ means $A + (-B)$.

Theorem 11.2.3 *If $A, B, C \in M_{m \times n}(\mathbb{F})$, then*
   (1) $A + B = B + A$,
   (2) $A + (B + C) = (A + B) + C$,
   (3) $A + O_{m \times n} = A$,
   (4) $A + (-A) = O_{m \times n}$.

Proof We prove (2) and (4), The others are left as an exercise to the reader. Note that, since addition does not change the type of matrix, all the sums are defined and the right hand side and left hand side of each equation are of the same size. Thus we need only check that the entries agree.

(2) Using Definition 11.2.1, we have
$$(A + (B + C))_{ij} = (A)_{ij} + (B + C)_{ij} = (A)_{ij} + (B)_{ij} + (C)_{ij}$$
$$= (A + B)_{ij} + (C)_{ij} = ((A + B) + C)_{ij},$$
as required.

(4) Using Definition 11.2.2, we have
$$(A + (-A))_{ij} = (A)_{ij} + (-A)_{ij} = (A)_{ij} - (A)_{ij} = 0,$$
as required.   □

Definition 11.2.4 If $A \in M_{m \times n}(\mathbb{F})$ and $\lambda \in \mathbb{F}$, then the *scalar multiple* λA is the $m \times n$ matrix with

$$(\lambda A)_{ij} = \lambda(A)_{ij}.$$

Example 11.2.5 If $A = \begin{bmatrix} 1 & 2 \\ 3 & 4 \end{bmatrix}$, $B = \begin{bmatrix} 1 & -1 \\ 0 & 1 \end{bmatrix}$, find $2A + 3B$.

Solution We have

$$2A + 3B = \begin{bmatrix} 2 \times 1 & 2 \times 2 \\ 2 \times 3 & 2 \times 4 \end{bmatrix} + \begin{bmatrix} 3 \times 1 & 3 \times (-1) \\ 3 \times 0 & 3 \times 1 \end{bmatrix}$$

$$= \begin{bmatrix} 2 & 4 \\ 6 & 8 \end{bmatrix} + \begin{bmatrix} 3 & -3 \\ 0 & 3 \end{bmatrix} = \begin{bmatrix} 5 & 1 \\ 6 & 11 \end{bmatrix}. \qquad \square$$

Theorem 11.2.6 *If $A, B \in M_{m \times n}(\mathbb{F})$ and $\lambda, \mu \in \mathbb{F}$, then*

(1) $\lambda(A + B) = \lambda A + \lambda B$,

(2) $\lambda(\mu A) = (\lambda \mu)A$,

(3) $\lambda A + \mu A = (\lambda + \mu)A$.

Proof Since addition and scalar multiplication do not change the type, the left and right hand sides of each equation have the same size.

(1) Using Definitions 11.2.4 and 11.2.1, we have

$$(\lambda(A + B))_{ij} = \lambda(A + B)_{ij} = \lambda(A)_{ij} + \lambda(B)_{ij}$$
$$= (\lambda A)_{ij} + (\lambda B)_{ij} = (\lambda A + \lambda B)_{ij},$$

as required.

Parts (2) and (3) are similar and are left to the reader. $\qquad \square$

Example 11.2.7 Let $A \in M_{m \times n}(\mathbb{F})$ and $\lambda \in \mathbb{F}$. Prove that

$$\lambda A = O_{m \times n} \iff \lambda = 0 \text{ or } A = O_{m \times n}.$$

Solution It is easy to see that if $\lambda = 0$ or $A = O_{m \times n}$, then every entry of λA is 0, i.e. $\lambda A = O_{m \times n}$. Thus, $\lambda = 0$ or $A = O_{m \times n} \Rightarrow \lambda A = O_{m \times n}$.

Now suppose that $\lambda A = O_{m \times n}$. Then

$$\lambda(A)_{ij} = 0 \quad (1 \leq i \leq m, 1 \leq j \leq n).$$

Thus either $\lambda = 0$ or *every* $(A)_{ij} = 0$, i.e. $A = O_{m \times n}$. Hence $\lambda A = O_{m \times n} \Rightarrow \lambda = 0$ or $A = O_{m \times n}$. $\qquad \square$

11.3 Matrix Multiplication

Definition 11.3.1 Let

$$R = \begin{bmatrix} r_1 & \cdots & r_n \end{bmatrix} \in M_{1 \times n}(\mathbb{F}) \quad \text{and} \quad C = \begin{bmatrix} c_1 \\ \vdots \\ c_n \end{bmatrix} \in M_{n \times 1}(\mathbb{F}).$$

Then the *dot product* $R \cdot C$ is defined by

$$R \cdot C = r_1 c_1 + r_2 c_2 + \cdots + r_n c_n.$$

Remarks (1) The dot product is defined *only* when R consists of a single row (R is a *row vector*), C of a single column (C is a *column vector*), and they have the same number of entries.

(2) Using the Σ notation, we may also write

$$R \cdot C = \sum_{j=1}^{n} r_j c_j.$$

A general $A \in M_{m \times n}(\mathbb{F})$ may be viewed as a collection of rows $R_1, \ldots, R_m$, *or* as a collection of columns $C_1, \ldots, C_n$. Here R_i denotes the ith row and C_j the jth column of A, *i.e.*

$$R_i = \begin{bmatrix} a_{i1} & a_{i2} & \cdots & a_{in} \end{bmatrix} \quad \text{and} \quad C_j = \begin{bmatrix} a_{1j} \\ a_{2j} \\ \vdots \\ a_{mj} \end{bmatrix}.$$

Note that $R_i \in M_{1 \times n}(\mathbb{F})$ and $C_i \in M_{m \times 1}(\mathbb{F})$.

Definition 11.3.2 Let $A \in M_{m \times n}(\mathbb{F})$ and $B \in M_{n \times p}(\mathbb{F})$. Suppose that A has rows $R_1, \ldots, R_m$ and that B has columns $C_1, \ldots, C_p$. Then we define the *matrix product* AB as the $m \times p$ matrix with

$$(AB)_{ij} = R_i \cdot C_j.$$

Remarks (1) The condition that the number of *columns* of A is equal to the number of *rows* of B is vital. It ensures that R_i and C_j have the same number of entries, so that $R_i \cdot C_j$ is defined. The product AB is not defined otherwise.

(2) It is probably best to think of AB as the matrix whose $(i, j)^{\text{th}}$ entry is the dot product of the ith row of A and the jth column of B. The size of the product is easily remembered using the mnemonic

$$(m \times \not{n}) \; (\not{n} \times p) \longrightarrow (m \times p).$$

Example 11.3.3 Let $A = \begin{bmatrix} 1 & 0 & -1 \\ 2 & 1 & 0 \end{bmatrix}$ and $B = \begin{bmatrix} 2 & 1 \\ 1 & 1 \\ 0 & 0 \end{bmatrix}$. Explain why only one of AB and BA exists and determine the one which does.

Solution Matrix A is 3×3 and B is 3×2. Thus AB exists and is of type 3×2 $((3 \times 3)\,(3 \times 2))$, but BA does not exist since the number of columns in B, 2, does not equal the number of rows in A, 3.

Now

$$AB = \begin{bmatrix} 1 & 0 & -1 \\ 2 & 1 & 0 \\ 1 & 1 & 1 \end{bmatrix} \begin{bmatrix} 2 & 1 \\ 1 & 1 \\ 0 & 0 \end{bmatrix}$$

$$= \begin{bmatrix} \begin{bmatrix} 1 & 0 & -1 \end{bmatrix} \cdot \begin{bmatrix} 2 \\ 1 \\ 0 \end{bmatrix} & \begin{bmatrix} 1 & 0 & -1 \end{bmatrix} \cdot \begin{bmatrix} 1 \\ 1 \\ 0 \end{bmatrix} \\[3mm] \begin{bmatrix} 2 & 1 & 0 \end{bmatrix} \cdot \begin{bmatrix} 2 \\ 1 \\ 0 \end{bmatrix} & \begin{bmatrix} 2 & 1 & 0 \end{bmatrix} \cdot \begin{bmatrix} 1 \\ 1 \\ 0 \end{bmatrix} \\[3mm] \begin{bmatrix} 1 & 1 & 1 \end{bmatrix} \cdot \begin{bmatrix} 2 \\ 1 \\ 0 \end{bmatrix} & \begin{bmatrix} 1 & 1 & 1 \end{bmatrix} \cdot \begin{bmatrix} 1 \\ 1 \\ 0 \end{bmatrix} \end{bmatrix}$$

$$= \begin{bmatrix} 1 \times 2 + 0 \times 1 + (-1) \times 0 & 1 \times 1 + 0 \times 1 + (-1) \times 0 \\ 2 \times 2 + 1 \times 1 + 0 \times 0 & 2 \times 1 + 1 \times 1 + 0 \times 0 \\ 1 \times 2 + 1 \times 1 + 1 \times 0 & 1 \times 1 + 1 \times 1 + 1 \times 0 \end{bmatrix}$$

$$= \begin{bmatrix} 2 & 1 \\ 5 & 3 \\ 3 & 2 \end{bmatrix}$$

is the required product. $\qquad\square$

As with addition, matrix multiplication has some properties similar to those of number systems. To prove these, we need a more convenient formula for the typical entry of a matrix product.

Lemma 11.3.4 *If $A \in M_{m \times n}(\mathbb{F})$ and $B \in M_{n \times p}(\mathbb{F})$, then the $(i, j)^{th}$ entry of AB can be expressed as*

$$(AB)_{ij} = \sum_{k=1}^{n} a_{ik} b_{kj}.$$

Proof From Definition 11.3.2,

$$(AB)_{ij} = R_i \cdot C_j,$$

where R_i is the ith row of A and C_j is the jth column of B. Thus

$$R_i = \begin{bmatrix} a_{i1} & \cdots & a_{in} \end{bmatrix} \quad \text{and} \quad C_j = \begin{bmatrix} b_{1j} \\ \vdots \\ b_{nj} \end{bmatrix}$$

(A has n columns and B has n rows), giving the required expression

$$R_i \cdot C_j = a_{i1} b_{1j} + \cdots + a_{in} b_{nj} = \sum_{k=1}^{n} a_{ik} b_{kj}. \qquad\square$$

Theorem 11.3.5 *Let* $A, D \in M_{m \times n}(\mathbb{F})$ *and* $B, C \in M_{n \times p}(\mathbb{F})$. *Then*

(1) $A(B + C) = AB + AC$,

(2) $(A + D)B = AB + DB$,

(3) $\lambda(AB) = (\lambda A)B$.

Proof In all cases, it is easy to check that all sums and products are defined, and that the matrices to be proved equal have the same type (namely $m \times p$).

(1) From Lemma 11.3.4 we have

$$(A(B + C))_{ij} = \sum_{k=1}^{n} a_{ik}(B + C)_{kj} = \sum_{k=1}^{n} a_{ik}(b_{kj} + c_{kj})$$

$$= \sum_{k=1}^{n} a_{ik}b_{kj} + \sum_{k=1}^{n} a_{ik}c_{kj} = (AB)_{ij} + (AC)_{ij}$$

$$= (AB + AC)_{ij},$$

as required. Parts (2) and (3) are similar and are left to the reader.   □

Theorem 11.3.6 *Let* $A \in M_{m \times n}(\mathbb{F})$, $B \in M_{n \times p}(\mathbb{F})$ *and* $C \in M_{p \times q}(\mathbb{F})$. *Then*

$$A(BC) = (AB)C.$$

Proof It is easy to see that all products are defined, and that $A(BC)$ and $(AB)C$ are both of type $m \times q$.

From Lemma 11.3.4,

$$(A(BC))_{ij} = \sum_{k=1}^{n} a_{ik}(BC)_{kj} = \sum_{k=1}^{n} a_{ik} \sum_{l=1}^{p} b_{kl}c_{lj} = \sum_{k=1}^{n}\sum_{l=1}^{p} a_{ik}(b_{kl}c_{lj})$$

$$= \sum_{k=1}^{n}\sum_{l=1}^{p}(a_{ik}b_{kl})c_{lj} = \sum_{l=1}^{p}\left(\sum_{k=1}^{n} a_{ik}b_{kl}\right)c_{lj} = \sum_{l=1}^{p}(AB)_{il}c_{lj}$$

$$= (AB)C)_{ij},$$

as required.   □

On the other hand, not all familiar looking properties hold for matrices, as we now see.

Example 11.3.7 Suppose that $A \in M_{m \times n}(\mathbb{F})$ and $B \in M_{n \times p}(\mathbb{F})$, so that AB is defined. Find the conditions on the types such that

(a) BA is defined,

(b) AB and BA are the same type.

(c) Give an example to show that, even when AB and BA have the same type, we need not have $AB = BA$.

Solution (a) For BA to be defined, the number of columns of B must equal the number of rows of A, *i.e.* $p = m$.

(b) From (a), we must have $p = m$. Then AB is of type $m \times m$ and BA is of type $n \times n$. Hence we require $m = n$.

(c) Let $A = \begin{bmatrix} 1 & 1 \\ 1 & 1 \end{bmatrix}$ and $B = \begin{bmatrix} 1 & 1 \\ -1 & -1 \end{bmatrix}$. Then, as the reader may verify,

$$AB = \begin{bmatrix} 0 & 0 \\ 0 & 0 \end{bmatrix} \quad \text{and} \quad BA = \begin{bmatrix} 2 & 2 \\ -2 & -2 \end{bmatrix}.$$

Clearly $AB \neq BA$ in this case. □

Definition 11.3.8 If A and B are matrices such that $AB = BA$, then we say that A and B *commute*.

Note 11.3.9 From Example 11.3.7, we see that if A and B commute then A and B must both be of type $n \times n$ for some $n \in \mathbb{N}$. The last part of this example shows that the converse of this statement is false.

Definition 11.3.10 A matrix A is *square* if it has an equal number of rows and columns, *i.e.* $A \in M_{n \times n}(\mathbb{F})$ for some $n \in \mathbb{N}$. We note that, if A is square then AA is defined. This is usually written A^2. As A^2 is also of the same type, we can form AA^2, written as A^3, and so on.

Example 11.3.11 Prove that, if $A, B \in M_{n \times n}(\mathbb{F})$, then

$$(A + B)(A - B) = A^2 - B^2 \iff A \text{ and } B \text{ commute.}$$

Solution It is clear that all sums and products are defined and are of type $n \times n$.

$$\begin{aligned} (A + B)(A - B) &= A(A - B) + B(A - B), && \text{(by Theorem 11.3.5 (2))} \\ &= AA - AB + BA - BB, && \text{(by Theorem 11.3.5 (1), (3))} \\ &= A^2 - AB + BA - B^2. \end{aligned}$$

Thus,

$$\begin{aligned} (A + B)(A - B) = A^2 - B^2 &\iff -AB + BA = O_{n \times n} \\ &\iff AB = BA \\ &\iff A \text{ and } B \text{ commute,} \end{aligned}$$

as required. □

Example 11.3.12 Let $C = \begin{bmatrix} 3 & 1 \\ -2 & 0 \end{bmatrix}$. Prove that, for $n \geq 1$,

$$C^n = \begin{bmatrix} 2^{n+1} - 1 & 2^n - 1 \\ -2^{n+1} + 2 & -2^n + 2 \end{bmatrix} \tag{11.3.1}$$

Proof We use a proof by induction since C^{n+1} is defined to be $C\,C^n$. As usual, we write $P(n)$ for the statement (11.3.1).

For $n = 1$, LHS $= C^1 = C$ and

$$RHS = \begin{bmatrix} 2^2 - 1 & 2^1 - 1 \\ -2^2 + 2 & -2^1 + 2 \end{bmatrix} = \begin{bmatrix} 3 & 1 \\ -2 & 0 \end{bmatrix} = C.$$

Hence $P(1)$ is true.

Now suppose that $P(k)$ holds for any $k \geq 1$, *i.e.*

$$C^k = \begin{bmatrix} 2^{k+1} - 1 & 2^k - 1 \\ -2^{k+1} + 2 & -2^k + 2 \end{bmatrix}.$$

Then

$$\text{LHS } P(k+1) = C^{k+1} = C\,C^k$$

$$= \begin{bmatrix} 3 & 1 \\ -2 & 0 \end{bmatrix} \begin{bmatrix} 2^{k+1} - 1 & 2^k - 1 \\ -2^{k+1} + 2 & -2^k + 2 \end{bmatrix}$$

$$= \begin{bmatrix} 3 \times 2^{k+1} - 3 - 2^{k+1} + 2 & 3 \times 2^k - 3 - 2^k + 2 \\ -2 \times 2^{k+1} + 2 & -2 \times 2^k + 2 \end{bmatrix}$$

$$= \begin{bmatrix} 2 \times 2^{k+1} - 1 & 2 \times 2^k - 1 \\ -2 \times 2^{k+1} + 2 & -2 \times 2^k + 2 \end{bmatrix}$$

$$= \begin{bmatrix} 2^{k+2} - 1 & 2^{k+1} - 1 \\ -2^{k+2} + 2 & -2^{k+1} + 2 \end{bmatrix}$$

$$= \text{RHS } P(k+1).$$

Thus $P(k) \Rightarrow P(k+1)$ for any $k \geq 1$ and hence, by induction, $P(n)$ is true for $n \geq 1$. $\square$

11.4   Further Properties of Multiplication

Theorem 11.4.1 *Let $A \in M_{m \times n}(\mathbb{F})$. Then*

(1) $AO_{n \times p} = O_{m \times p}$,

(2) $O_{q \times m} A = O_{q \times n}$.

Proof It is clear that the products exist and that, in each case, both sides are of the same type. The equalities follow since all the entries on the left hand side of both equations evaluate to 0. $\square$

This is analogous to the result $a \times 0 = 0$ in a number system. The other special case in number systems is that $a \times 1 = a$ for any a. We now introduce the matrix analogue of 1.

Definition 11.4.2 A *diagonal matrix* A is a square matrix in which entries not on the main diagonal are zero. (The *main diagonal* is the line of entries from the top left corner to the bottom right. It consists of entries with equal row and column indices.)

For example, $A = \begin{bmatrix} 1 & 0 \\ 0 & 2 \end{bmatrix}$ and $B = \begin{bmatrix} 4 & 0 & 0 \\ 0 & 0 & 0 \\ 0 & 0 & -1 \end{bmatrix}$ are diagonal matrices. Note from the second example that, as well as the off-diagonal entries being zero, some of the diagonal entries may also be zero.

Notation 11.4.3 To avoid writing large numbers of zeros, we will sometimes write the diagonal $n \times n$ matrix D as $\text{diag}(d_1, \ldots, d_n)$, where d_i denotes the ith entry down the diagonal. Thus

$$\text{diag}(d_1, \ldots, d_n) = \begin{bmatrix} d_1 & 0 & \cdots & 0 & 0 \\ 0 & d_2 & \cdots & 0 & 0 \\ \vdots & \vdots & \ddots & \vdots & \vdots \\ 0 & 0 & \cdots & d_{n-1} & 0 \\ 0 & 0 & \cdots & 0 & d_n \end{bmatrix}.$$

Definition 11.4.4 For $n \in \mathbb{N}$, the *identity matrix* I_n of size n is the $n \times n$ diagonal matrix with diagonal entries all equal to 1.

Thus, for example, $I_2 = \begin{bmatrix} 1 & 0 \\ 0 & 1 \end{bmatrix}$ and $I_3 = \begin{bmatrix} 1 & 0 & 0 \\ 0 & 1 & 0 \\ 0 & 0 & 1 \end{bmatrix}$. For convenience, we introduce the notation δ_{ij} for the $(i, j)^{\text{th}}$ entry of an identity matrix, so that

$$\delta_{ij} = \begin{cases} 1 & \text{if } i = j, \\ 0 & \text{if } i \neq j. \end{cases}$$

(This is sometimes known as the *Kronecker delta symbol*.)

Theorem 11.4.5 *For $A \in M_{m \times n}(\mathbb{F})$,*

(1) $AI_n = A$,

(2) $I_m A = A$.

Proof As usual, it is easy to see that the products on the left exist and are of the same type as A.

(1) Using Lemma 11.3.4, we have

$$(AI_n)_{ij} = \sum_{k=1}^{n} a_{ik} \delta_{kj}$$
$$= a_{i1} \delta_{1j} + \cdots + a_{ij} \delta_{jj} + \cdots + a_{in} \delta_{nj}$$
$$= a_{ij},$$

since $\delta_{ij} = 0$ when $i \neq j$, and $\delta_{jj} = 1$. Thus the entries of AI_n and A are the same and so $AI_n = A$.

(2) This is similar and is left as an exercise to the reader. $\qquad \square$

In some applications, it is useful to consider the matrix A^T formed from the matrix A by using the entries of the *rows* of A to make the *columns* of A^T. For example,

$$A = \begin{bmatrix} 1 & 2 & 3 \\ 4 & 5 & 6 \end{bmatrix}, \quad A^T = \begin{bmatrix} 1 & 4 \\ 2 & 5 \\ 3 & 6 \end{bmatrix}.$$

We call A^T the *transpose of A*. Although this description is easy to understand, it is difficult to use in calculation. We therefore make a more formal (but equivalent) definition.

Definition 11.4.6 If $A \in M_{m \times n}(\mathbb{F})$, then the *transpose of A* is the $n \times m$ matrix A^T with

$$(A^T)_{ij} = (A)_{ji}.$$

To see that this is equivalent to the above description, consider the jth row of A. This consists of the entries $a_{j1}, \ldots, a_{jn}$. These are used to form the jth column of A^T. Thus the $(i, j)^{\text{th}}$ entry of A^T (the ith entry in the jth column) is a_{ji}, as required. The result on the type of A^T is clear.

For example, as $\delta_{ji} = \delta_{ij}$, $I^T = I$.

Theorem 11.4.7 *Let* $A, B \in M_{m \times n}(\mathbb{F})$ *and* $C \in M_{n \times p}(\mathbb{F})$. *Then*

(1) $(A^T)^T = A$,

(2) $(A + B)^T = A^T + B^T$,

(3) $(AC)^T = C^T A^T$.

Proof (1) From its definition, A^T is of type $n \times m$ and hence $(A^T)^T$ is of type $m \times n$. So $(A^T)^T$ and A are of the same type. Also,

$$((A^T)^T)_{ij} = (A^T)_{ji}$$
$$= (A)_{ij}, \quad \text{(reversing the subscripts again).}$$

Thus $(A^T)^T = A$.

(2) This is similar to (but easier than) (3) and is left as an exercise to the reader.

(3) We first observe that AC is defined and is of type $m \times p$. Then $(AC)^T$ is of type $p \times m$. Also, C^T is of type $p \times n$, and A^T of type $n \times m$. Hence $C^T A^T$ is defined and is of type $p \times m$ (the same as $(AC)^T$). Finally,

$$((AC)^T)_{ij} = (AC)_{ji}$$
$$= \sum_{k=1}^{n} a_{jk} c_{ki},$$

and

$$(C^T A^T)_{ij} = \sum_{k=1}^{n} (C^T)_{ik} (A^T)_{kj}$$
$$= \sum_{k=1}^{n} c_{ki} a_{jk} = \sum_{k=1}^{n} a_{jk} c_{ki}.$$

Hence the $(i,j)^{\text{th}}$ entries agree, so the matrices are equal. □

 Hazard In Theorem 11.4.7,(3) note that the order of the factors is *reversed* after taking the transpose.

Corollary 11.4.8 *Let $n \in \mathbb{N}$. Then, provided the products exist,*

$$(A_1 \ldots A_n)^{\text{T}} = A_n^{\text{T}} \ldots A_1^{\text{T}}.$$

Proof This result may be proved by induction. It is left as an exercise. □

Definitions 11.4.9 Let A be a square matrix. Then

(1) A is *symmetric* if $A^{\text{T}} = A$,

(2) A is *skew-symmetric* if $A^{\text{T}} = -A$.

Remark We observe that if A^{T} is to equal A or $-A$, then A^{T} *must* have the same type as A. Hence the requirement that A is square.

For example,

$$A = \begin{bmatrix} 1 & 2 & 3 \\ 2 & 4 & 5 \\ 3 & 5 & 6 \end{bmatrix} \quad \text{is symmetric}$$

$$B = \begin{bmatrix} 0 & 2 & -3 \\ -2 & 0 & 4 \\ 3 & -4 & 0 \end{bmatrix} \quad \text{is skew-symmetric}$$

$$C = \begin{bmatrix} 1 & 2 \\ 3 & 4 \end{bmatrix} \quad \text{is neither,} \quad C^{\text{T}} = \begin{bmatrix} 1 & 3 \\ 2 & 4 \end{bmatrix} \neq \pm C.$$

From the first example it is clear that the symmetry referred to is a symmetry in the entries about the main diagonal. Any square matrix (in which the entries are given explicitly) may be recognised as symmetric by checking that all symmetrically placed entries agree.

Problem 11.4.10 Show that, if A is skew-symmetric, then all the main diagonal entries of A are zero. Describe the general appearance of a skew-symmetric matrix.

Example 11.4.11 Prove that, for any $C \in M_{m \times n}(\mathbb{F})$, $C^{\text{T}}C$ is symmetric.

Solution Since C^{T} is of type $n \times m$, $C^{\text{T}}C$ is defined and is of type $n \times n$, *i.e.* $C^{\text{T}}C$ is square. To show that it is symmetric, we must show that $C^{\text{T}}C$ is equal to its transpose. By Theorem 11.4.7 (3) and (1),

$$(C^{\text{T}}C)^{\text{T}} = C^{\text{T}}(C^{\text{T}})^{\text{T}}$$
$$= C^{\text{T}}C,$$

and the result follows. □

Example 11.4.12 Let S denote the set of skew-symmetric $n \times n$ matrices. Show that S is closed under matrix addition.

Solution We must show that, if $A, B \in S$, then $A + B \in S$.

Let $A, B \in S$. Then A and B are skew-symmetric and so

$$A^{\mathrm{T}} = -A, \quad B^{\mathrm{T}} = -B. \tag{11.4.1}$$

As A and B are both $n \times n$, $A + B$ is defined and of type $n \times n$. Also

$$\begin{aligned}
(A + B)^{\mathrm{T}} &= A^{\mathrm{T}} + B^{\mathrm{T}}, \quad \text{(by Theorem 11.4.7 (2))} \\
&= -A + (-B), \quad \text{from (11.4.1)} \\
&= -(A + B).
\end{aligned}$$

Hence $A + B$ is skew-symmetric, *i.e.* $A + B \in S$. $\qquad\square$

11.5   Linear Equations

To motivate topic we study later, we look at the applications of matrices to solving systems of linear equations.

Let S be a system of m linear equations in n unknowns, $x_1, \ldots, x_n$. Then S has the form

$$S : \begin{cases}
a_{11}x_1 + a_{12}x_2 + \cdots + a_{1n}x_n = h_1, \\
a_{21}x_1 + a_{22}x_2 + \cdots + a_{2n}x_n = h_2, \\
\quad\vdots \qquad\quad \vdots \qquad\qquad\quad \vdots \qquad \vdots \\
a_{m1}x_1 + a_{m2}x_2 + \cdots + a_{mn}x_n = h_m.
\end{cases}$$

The notation 'a_{ij}' for the coefficients is quite deliberate: it suggests matrix entries. Indeed, S can be written as

$$AX = H,$$

where A is the $m \times n$ matrix with $(i, j)^{\text{th}}$ entry a_{ij},

$$X = \begin{bmatrix} x_1 \\ \vdots \\ x_n \end{bmatrix} \quad \text{and} \quad H = \begin{bmatrix} h_1 \\ \vdots \\ h_m \end{bmatrix},$$

so that X is of type $n \times 1$, and H of type $m \times 1$. The matrix A is called the *matrix of coefficients* of S.

When manipulating equations, we must deal simultaneously with their left and right hand sides. To do this with matrix notation, we introduce the *augmented matrix* of S. This is written $[A|H]$ and is formed by adding H as an extra column to A. This is legitimate as both A and H have m rows. The result is of type $m \times (n + 1)$. As an example, consider the system

$$S : \begin{cases}
x_1 + x_2 - x_3 = 1, \\
x_1 + x_2 = 0, \\
x_1 - x_2 + 3x_3 = -1.
\end{cases}$$

Then S has augmented matrix

$$[A|H] = \begin{bmatrix} 1 & 1 & -1 & 1 \\ 1 & 1 & 0 & 0 \\ 1 & -1 & 3 & -1 \end{bmatrix}.$$

Remark If a variable is 'missing' from an equation, such as x_3 in the second equation of S, then the corresponding entry in the augmented matrix is 0.

The system S can be solved in a systematic way as follows:

$$S : \begin{cases} x_1 + x_2 - x_3 = 1, & (0.1) \\ x_1 + x_2 \phantom{{}- x_3} = 0, & (0.2) \\ x_1 - x_2 + 3x_3 = -1. & (0.3) \end{cases}$$

We eliminate x_1 from all but the first equation

$$S_1 : \begin{cases} x_1 + x_2 - x_3 = 1, & & (1.1) \\ \phantom{x_1 + x_2 -{}} x_3 = -1, & ((0.2)-(0.1)), & (1.2) \\ - 2x_2 + 4x_3 = -2, & ((0.3)-(0.1)). & (1.3) \end{cases}$$

We now interchange the second and third rows, so that x_2 appears in the second equation

$$S_2 : \begin{cases} x_1 + x_2 - x_3 = 1, & & (2.1) \\ - 2x_2 + 4x_3 = -2, & ((1.3)), & (2.2) \\ \phantom{x_1 - 2x_2 +{}} x_3 = -1, & ((1.2)). & (2.3) \end{cases}$$

We make the coefficient of x_2 in the second equation equal to 1

$$S_3 : \begin{cases} x_1 + x_2 - x_3 = 1, & & (3.1) \\ \phantom{x_1 +{}} x_2 - 2x_3 = 1, & (\text{divide } (2.2) \text{ by } -2), & (3.2) \\ \phantom{x_1 + x_2 -{}} x_3 = -1. & & (3.3) \end{cases}$$

We now eliminate x_2 from all but the second equation

$$S_4 : \begin{cases} x_1 + x_3 = 0, & ((3.1)-(3.2)), & (4.1) \\ x_2 - 2x_3 = 1, & & (4.2) \\ \phantom{x_1 + x_2 -{}} x_3 = -1, & (x_2 \text{ does not appear}). & (4.3) \end{cases}$$

Finally, we eliminate x_3 from all but the third equation

$$S_5 : \begin{cases} x_1 = 1, & ((4.1)-(4.3)), & (5.1) \\ x_2 = -1, & ((4.2)+2(4.3)), & (5.2) \\ \phantom{x_1 + x_2 -{}} x_3 = -1. & & (5.3) \end{cases}$$

Thus the solution set of S_5 (and so of S) is $\{(1, -1, -1)\}$.

We invite the reader to write down the augmented matrix of the system at each stage, and see how these matrices are related.

If you have completed the exercise, then you will have realised that there are three kinds of operation involved:

Type 1: Interchange row i and row j (denoted by $R_i \leftrightarrow R_j$).

Type 2: Multiply row i by a *non-zero* scalar λ (denoted by $R_i \to \lambda R_i$).

Type 3: Add a multiple of row j to row i, $(i \neq j)$. (denoted by $R_i \to R_i + \alpha R_j$).

These are called *elementary row operations*. We will use the abbreviation ERO for an elementary row operation.

Example 11.5.1 Determine the effect of the EROs $R_1 \to R_1 + 2R_2$ then $R_2 \leftrightarrow R_1$, then $R_1 \to -\frac{1}{2}R_1$ on the matrix

$$A = \begin{bmatrix} 0 & 1 & -1 \\ 2 & 0 & 2 \\ 1 & 0 & -1 \end{bmatrix}.$$

Solution Applying $R_1 \to R_1 + 2R_2$ to A, we get

$$B = \begin{bmatrix} 4 & 1 & 3 \\ 2 & 0 & 2 \\ 1 & 0 & -1 \end{bmatrix}.$$

Applying $R_2 \leftrightarrow R_1$ to B, we get

$$C = \begin{bmatrix} 2 & 0 & 2 \\ 4 & 1 & 3 \\ 1 & 0 & -1 \end{bmatrix}.$$

Applying $R_1 \to -\frac{1}{2}R_1$ to C, we get

$$\begin{bmatrix} -1 & 0 & -1 \\ 4 & 1 & 3 \\ 1 & 0 & -1 \end{bmatrix}. \qquad \square$$

Definition 11.5.2 Let $A, B \in M_{m \times n}(\mathbb{F})$. Then we say that A is *row equivalent to B* if there is a sequence of EROs which transforms A to B, and we write $A \sim B$.

Hence, from Example 11.5.1,

$$\begin{bmatrix} 0 & 1 & -1 \\ 2 & 0 & 2 \\ 1 & 0 & -1 \end{bmatrix} \sim \begin{bmatrix} -1 & 0 & -1 \\ 4 & 1 & 3 \\ 1 & 0 & -1 \end{bmatrix}.$$

We also note that the ERO $R_i \to R_i$ (type 2, with $\lambda = 1$) does not change the matrix A, so we always have $A \sim A$.

Remark If a matrix is the augmented matrix of a system of linear equations then any other matrix it is row equivalent to is the augmented matrix of an *equivalent* system of equations, *i.e.* a system having precisely the same solutions. As we saw earlier, the systematic solution of a system of linear equations is equivalent to applying EROs to the augmented matrix. The new matrix we obtained corresponds to a simpler system of equations which can be solved by inspection.

Example 11.5.3 Find all solutions of the systems of equations (in $x_1, \ldots, x_4$) which have augmented matrices row equivalent to

$$
\text{(a)} \begin{bmatrix} 1 & -1 & 0 & 0 & 1 \\ 0 & 0 & 1 & 0 & -1 \\ 0 & 0 & 0 & 1 & 1 \end{bmatrix}, \quad \text{(b)} \begin{bmatrix} 1 & 0 & 0 & 0 & 1 \\ 0 & 0 & 1 & 0 & -1 \\ 0 & 0 & 0 & 0 & 0 \end{bmatrix},
$$

$$
\text{(c)} \begin{bmatrix} 1 & 0 & 0 & 0 & 1 \\ 0 & 0 & 1 & 0 & -1 \\ 0 & 0 & 0 & 0 & 1 \end{bmatrix}.
$$

Proof (a) The system is

$$
\begin{cases} x_1 - x_2 & = 1, \\ \quad\quad\; x_3 & = -1, \\ \quad\quad\quad\; x_4 = 1. \end{cases}
$$

These equations determine, respectively, x_1, x_3 and x_4. There is no equation to determine x_2, so this remains unknown: it is called a *parameter* of the solution set.

The solution set is

$$
\{(1 + x_2, x_2, -1, 1) : x_2 \in \mathbb{R}\}.
$$

(b) Here, the system is

$$
\begin{cases} x_1 & = 1, \\ \;\; x_3 & = -1, \\ \;\; 0 = 0. \end{cases}
$$

Now we can determine x_1 and x_3. The third equation is true whatever values $x_1, \ldots, x_4$ have. The solution set is $\{(1, x_2, -1, x_4) : x_2, x_4 \in \mathbb{R}\}$, having two parameters.

(c) Here the system is

$$
\begin{cases} x_1 & = 1, \\ \;\; x_3 & = -1, \\ \;\; 0 = 1. \end{cases}
$$

The third equation cannot be satisfied whatever the values of the variables. Hence there are no solutions, *i.e.* the solution set is $\emptyset$. □

Remark From Example 11.5.3, we see that a system of linear equations may have no solutions, a unique solution or infinitely many solutions. In the last case, the solution may involve one or more parameters.

Definition 11.5.4 If R is a non-zero row of a matrix, then the leading entry of R is the left-most non-zero entry. A zero row has no leading entry.

In the example below, the leading entries are boxed

$$
A = \begin{bmatrix} \boxed{2} & 1 & 1 & 1 \\ 0 & 0 & 0 & 0 \\ 0 & \boxed{3} & 1 & 1 \end{bmatrix}.
$$

Note that the second row of A is a zero row, so has no leading entry.

The systems in Example 11.5.3 were particularly easy to solve because the augmented matrices were simple. The next definition characterises such matrices.

Definition 11.5.5 An m-rowed matrix E is a *reduced echelon matrix* if

(1) The zero rows of E (if any) lie below the non-zero rows.

(2) The leading entry of each non-zero row is 1.

(3) Let $i < j$. If the ith and j^{th} rows of E are non-zero, then the leading entry of the jth row lies to the right of that of the ith row.

(4) If a column of E contains the leading entry of some row, then all other entries in that column are 0.

A reduced echelon matrix E has the general form

$$E = \begin{bmatrix} 0 \dots 0 & 1 & * \dots * & 0 & * \dots * & 0 & * \dots * & 0 & * \dots * \\ 0 \dots 0 & 0 & 0 \dots 0 & 1 & * \dots * & 0 & * \dots * & 0 & * \dots * \\ 0 \dots 0 & 0 & 0 \dots 0 & 0 & 0 \dots 0 & 1 & * \dots * & 0 & * \dots * \\ \vdots & \vdots & \vdots \vdots & \vdots & \vdots \vdots & \vdots & \vdots \vdots & \vdots & \vdots \\ 0 \dots 0 & 0 & 0 \dots 0 & 0 & 0 \dots 0 & 0 & 0 \dots 0 & 1 & * \dots * \\ 0 \dots 0 & 0 & 0 \dots 0 & 0 & 0 \dots 0 & 0 & 0 \dots 0 & 0 & 0 \dots 0 \\ \vdots & \vdots & \vdots \vdots & \vdots & \vdots \vdots & \vdots & \vdots \vdots & \vdots & \vdots \\ 0 \dots 0 & 0 & 0 \dots 0 & 0 & 0 \dots 0 & 0 & 0 \dots 0 & 0 & 0 \dots 0 \end{bmatrix} \begin{matrix} \left. \begin{matrix} \\ \\ \\ \\ \\ \end{matrix} \right\} r \text{ rows} \\ \\ \left. \begin{matrix} \\ \\ \\ \end{matrix} \right\} (m-r) \text{ rows} \end{matrix} \quad (11.5.1)$$

where teh entries denoted $*$ can have any value.

Example 11.5.6 Determine which of the following are reduced echelon matrices.

$$E_1 = \begin{bmatrix} 1 & 2 & 3 & 4 \\ 0 & 1 & 2 & 3 \\ 0 & 0 & 0 & 1 \\ 0 & 0 & 0 & 0 \end{bmatrix}, \quad E_2 = \begin{bmatrix} 0 & 1 & 0 & 0 \\ 0 & 0 & -1 & 0 \\ 0 & 0 & 0 & 0 \\ 0 & 0 & 0 & 1 \end{bmatrix}, \quad E_3 = \begin{bmatrix} 1 & 0 & 3 & 4 \\ 0 & 1 & -6 & 8 \\ 0 & 0 & 0 & 0 \\ 0 & 0 & 0 & 0 \end{bmatrix}.$$

Find EROs which transform the others into reduced echelon form.

Solution Matrix E_1 fails because columns 2 and 4 contain leading entries (of rows 1 and 3, respectively), but also have other non-zero entries.

Matrix E_2 fails because the leading entry of row 2 is -1 (not 1), and also because the zero row 3 is above the non-zero row 4.

Matrix E_3 is a reduced echelon matrix.

Matrix E_1 may be transformed as follows

$$E_1 \sim \begin{bmatrix} 1 & 0 & -1 & -2 \\ 0 & 1 & 2 & 3 \\ 0 & 0 & 0 & 1 \\ 0 & 0 & 0 & 0 \end{bmatrix} \begin{matrix} R_1 \to R_1 - 2R_2 \\ \\ \\ \\ \end{matrix} \qquad \text{(to clear column 2)}$$

$$\sim \begin{bmatrix} 1 & 0 & -1 & 0 \\ 0 & 1 & 2 & 0 \\ 0 & 0 & 0 & 1 \\ 0 & 0 & 0 & 0 \end{bmatrix} \begin{matrix} R_1 \to R_1 + 2R_3 \\ R_2 \to R_2 - 3R_3 \\ \\ \\ \end{matrix} \qquad \text{(to clear column 4)},$$

which is a reduced echelon matrix.

Similarly,

$$E_2 \sim \begin{bmatrix} 0 & 1 & 0 & 0 \\ 0 & 0 & 1 & 0 \\ 0 & 0 & 0 & 0 \\ 0 & 0 & 0 & 1 \end{bmatrix} \begin{matrix} \\ R_2 \to -R_2 \\ \\ \\ \end{matrix} \qquad \text{(so that the leading entry is 1)}$$

$$\sim \begin{bmatrix} 0 & 1 & 0 & 0 \\ 0 & 0 & 1 & 0 \\ 0 & 0 & 0 & 1 \\ 0 & 0 & 0 & 0 \end{bmatrix} \begin{matrix} \\ \\ R_3 \leftrightarrow R_4 \\ \\ \end{matrix} \qquad \text{(to place the zero row at the bottom)},$$

is a reduced echelon matrix.　　　　　　　　　　　　　　　　　□

We now describe a general method which transforms any matrix to a reduced echelon matrix. The process is known as *row reduction*.

Strategy for row reduction　We deal with the matrix row-by-row, starting at the top and proceeding downwards one row at a time.

Suppose we have dealt with the first j rows. Then we apply the four steps below. In (1)–(3) only the remaining rows, *i.e.* those below the jth, are considered.

(1) Find a non-zero entry *as far left as possible* in the *remaining rows*. Suppose that this lies in the kth column.

(2) (If necessary) interchange the $(j + 1)^{\text{st}}$ with one of the rows with leading entry in the kth column. If possible choose a row in which the leading entry is 1 (see Note 11.5.11).

(3) (If necessary) scale the $(j + 1)^{\text{st}}$ row so that its leading entry is 1.

(4) Using type 3 EROs $R_i \to R_i + \alpha_i R_{j+1}$ (with suitable α_i), make *all* other entries in the kth column equal to zero.

The process halts when all rows have been dealt with, or no non-zero rows remain.

Example 11.5.7　Find a reduced echelon matrix which is row equivalent to

$$A = \begin{bmatrix} 1 & 1 & 1 & 1 \\ 2 & 2 & 1 & 3 \\ 3 & 3 & 2 & 4 \end{bmatrix}.$$

Solution We have

$$A \sim \begin{bmatrix} 1 & 1 & 1 & 1 \\ 0 & 0 & -1 & 1 \\ 0 & 0 & -1 & 1 \end{bmatrix} \begin{matrix} R_2 \to R_2 - 2R_1 \\ R_3 \to R_3 - 3R_1 \end{matrix}$$

$$\sim \begin{bmatrix} 1 & 1 & 1 & 1 \\ 0 & 0 & 1 & -1 \\ 0 & 0 & -1 & 1 \end{bmatrix} \begin{matrix} \\ R_2 \to -R_2 \\ \end{matrix}$$

$$\sim \begin{bmatrix} 1 & 1 & 0 & 2 \\ 0 & 0 & 1 & -1 \\ 0 & 0 & 0 & 0 \end{bmatrix} \begin{matrix} R_1 \to R_1 - R_2 \\ \\ R_3 \to R_3 + R_2 \end{matrix}$$

The last matrix is a reduced echelon matrix.   □

There are other strategies for row reduction, but it is a fact that *whatever* method is used, the reduced echelon matrix obtained will always be the same. A proof of this lies beyond the scope of this book.

We are therefore entitled to make the following definition.

Definition 11.5.8 Let $A \in M_{m \times n}(\mathbb{F})$. Then $E(A)$, the *reduced echelon form of A*, is *the* reduced echelon matrix satisfying

$$A \sim E(A).$$

Example 11.5.9 Find the solution set of the system of equations

$$\begin{cases} x + y + z = 1 \\ 2x + 2y + z = 3 \\ 3x + 3y + 2z = 4 \end{cases}$$

Solution The augmented matrix of the system is

$$\left[\begin{array}{ccc|c} 1 & 1 & 1 & 1 \\ 2 & 2 & 1 & 3 \\ 3 & 3 & 2 & 4 \end{array} \right].$$

By Example 11.5.7, the reduced echelon form is

$$\left[\begin{array}{ccc|c} 1 & 1 & 0 & 2 \\ 0 & 0 & 1 & -1 \\ 0 & 0 & 0 & 0 \end{array} \right].$$

This leads to the (simpler) system

$$\begin{cases} x + y \quad\;\; = 2 \\ \quad\quad\; z = -1 \\ \quad\quad\; 0 = 0 \end{cases}.$$

Hence we get the solution set $\{(2 - y, y, -1) : y \in \mathbb{R}\}$.   □

Example 11.5.10 Show that the system

$$\begin{cases} 2x + y + 8z = 0 \\ y + 2z = 3 \\ x - y + z = 1 \end{cases}$$

has no solutions.

Solution The augmented matrix is

$$\begin{bmatrix} 2 & 1 & 8 & | & 0 \\ 0 & 1 & 2 & | & 3 \\ 1 & -1 & 1 & | & 1 \end{bmatrix} \sim \begin{bmatrix} 1 & -1 & 1 & | & 1 \\ 0 & 1 & 2 & | & 3 \\ 2 & 1 & 8 & | & 0 \end{bmatrix} \quad R_1 \leftrightarrow R_3 \quad \text{(see Note 11.5.11)}$$

$$\sim \begin{bmatrix} 1 & -1 & 1 & | & 1 \\ 0 & 1 & 2 & | & 3 \\ 0 & 3 & 6 & | & -2 \end{bmatrix} \quad R_3 \to R_3 - 2R_1$$

$$\sim \begin{bmatrix} 1 & 0 & 3 & | & 4 \\ 0 & 1 & 2 & | & 3 \\ 0 & 0 & 0 & | & -11 \end{bmatrix} \quad \begin{matrix} R_1 \to R_1 + R_2 \\ \\ R_3 \to R_3 - 3R_2 \end{matrix} \quad .$$

We do not proceed further since the last row this matrix corresponds to the equation $0 = -11$. Clearly this equation (and hence the entire system) has no solution. $\square$

Note 11.5.11 The reader might have expected the first ERO to have been $R_1 \to \frac{1}{2}R_1$. This would have introduced fractions into the calculation. Interchanging two rows to get a 1 in the leading position is a useful way to avoid having to manipulate fractions.

In Example 11.5.9, we saw that a system of three equations reduced to a system of effectively two equations; the equation $0 = 0$ gives no information about the variables, so can be discarded. More generally, a system $AX = H$ of m equations may reduce to a system with fewer 'effective' equations. The number will be equal to the number of non-zero rows in the reduced echelon form of the augmented matrix for the system $E([A|H])$.

Definition 11.5.12 For a matrix A, the number of non-zero rows in $E(A)$ is the *rank* of A, written $r(A)$.

For example, the matrix A of Example 11.5.7 has two non-zero rows and so $r(A) = 2$.

Theorem 11.5.13 *Let S be the system of m equations in n variables*

$$S : AX = H,$$

with $A \in M_{m \times n}(\mathbb{F})$. Then $r([A|H]) \geq r(A)$, and, if

(1) $r([A|H]) > r(A)$, *then S has no solutions,*

(2) $r([A|H]) = r(A)$, *then S has a solution involving $(n - r(A))$ parameters.*

In particular, if $r([A|H]) = r(A) = n$, then S has a unique solution.

Proof Suppose that $E([A|H]) = [\widehat{A}|\widehat{H}]$, where $\widehat{A} \in M_{m \times n}(\mathbb{F})$ (as A is) and $\widehat{H}$ a single column. Since EROs act on the columns independently, the sequence of EROs which transform $[A|H]$ to $[\widehat{A}|\widehat{H}]$ will transform A into $\widehat{A}$. Also, from the definition of a reduced echelon matrix, as $[\widehat{A}|\widehat{H}]$ is such a matrix, so is $\widehat{A}$ (think of removing the final column from the general echelon matrix (11.5.1)). Thus $E(A) = \widehat{A}$. Finally, the removal of a column from $[\widehat{A}|\widehat{H}]$ cannot *increase* the number of non-zero rows, so $r([A|H]) \geq r(A)$.

(1) If $r([A|H]) > r(A)$, then $\widehat{H}$ must have the leading entry of *some* row in $[\widehat{A}|\widehat{H}]$. Since $\widehat{H}$ is the final column, this row must be $\begin{bmatrix} 0 & \cdots & 0 & 1 \end{bmatrix}$. This gives the equation $0 = 1$. Hence the system has *no* solutions.

(2) Suppose that $r([A|H]) = r(A) = r$. Then $[\widehat{A}|\widehat{H}]$ gives r non-trivial equations, with non-zero left hand side, and $m - r$ of the form $0 = 0$. The r non-trivial equations determine r of the n variables—those corresponding to the leading entry in each non-zero row. The other $n-r$ variables are unrestricted, so the solution has $n-r$ parameters.

For the final part, we apply (2) with $r = n$. Then the solution depends on $n - r = 0$ parameters, *i.e.* we have a unique solution. $\qquad \square$

Example 11.5.14 Find the value of a for which the system

$$\begin{cases} x + y + z + w = 2 \\ x + 2y + 3z + 2w = 6 \\ 2x + 3y + 4z + 3w = a \end{cases}$$

has solutions, and determine the solution set in this case.

Solution The augmented matrix is

$$[A|H] = \begin{bmatrix} 1 & 1 & 1 & 1 & 2 \\ 1 & 2 & 3 & 2 & 6 \\ 2 & 3 & 4 & 3 & a \end{bmatrix} \sim \begin{bmatrix} 1 & 1 & 1 & 1 & 2 \\ 0 & 1 & 2 & 1 & 4 \\ 0 & 1 & 2 & 1 & a-4 \end{bmatrix} \begin{matrix} \\ R_2 \to R_2 - R_1 \\ R_3 \to R_3 - 2R_1 \end{matrix}$$

$$\sim \begin{bmatrix} 1 & 0 & -1 & 0 & -2 \\ 0 & 1 & 2 & 1 & 4 \\ 0 & 0 & 0 & 0 & a-8 \end{bmatrix} \begin{matrix} R_1 \to R_1 - R_2 \\ \\ R_3 \to R_3 - R_2 \end{matrix}$$

From the first four columns (corresponding to A), $r(A) = 2$. Also,

$$r([A|H]) = \begin{cases} 2 & \text{if } a = 8, \\ 3 & \text{if } a \neq 8. \end{cases}$$

From Theorem 11.5.13, there are solutions only when $a = 8$.

When $a = 8$, the reduced echelon form yields the equations

$$\begin{cases} x - z = -2 \\ y + 2z + w = 4 \end{cases}$$

and hence the solution set is $\{(-2 + z, 4 - 2z - w, z, w) : z, w \in \mathbb{R}\}$. $\qquad \square$

11.6   Matrix Inverses

In a number system $\mathbb{F}$, the inverse of x is $y \in \mathbb{F}$ such that

$$xy = 1.$$

Since multiplication in $\mathbb{F}$ is commutative, we also have $yx = 1$. For matrix multiplication, we have to be more careful.

Definition 11.6.1 Let $A \in M_{n \times n}(\mathbb{F})$. If there exists a $P \in M_{n \times n}(\mathbb{F})$ such that

$$AP = PA = I_n,$$

then A is *invertible*, and P is an *inverse* of A.

Remarks (1) Since we need PA and AP to be of type $n \times n$, A *must* be $n \times n$ for Definition 11.6.1 to make sense.

(2) Since $O_{n \times n}P = PO_{n \times n} = O_{n \times n}(\neq I_n)$ for *any* $P \in M_{n \times n}(\mathbb{F})$, a square zero matrix is not invertible.

(3) As $I_n I_n = I_n$, I_n is invertible with inverse I_n.

Example 11.6.2 Determine which of the matrices

$$A = \begin{bmatrix} 2 & 1 \\ 0 & 1 \end{bmatrix}, \quad B = \begin{bmatrix} 1 & 0 \\ 2 & 0 \end{bmatrix}$$

is invertible.

Solution   Let $P = \begin{bmatrix} \alpha & \beta \\ \gamma & \delta \end{bmatrix}$. Then

$$AP = \begin{bmatrix} 2 & 1 \\ 0 & 1 \end{bmatrix} \begin{bmatrix} \alpha & \beta \\ \gamma & \delta \end{bmatrix} = \begin{bmatrix} 2\alpha + \gamma & 2\beta + \delta \\ \gamma & \delta \end{bmatrix}.$$

Thus

$$AP = I_2 \iff \begin{cases} 2\alpha & + \gamma & & = 1 \\ & 2\beta & + \delta & = 0 \\ & \gamma & & = 0 \\ & & \delta & = 1 \end{cases}.$$

It is easy to solve these equation without recourse to row reduction. We get $\gamma = 0$, $\delta = 1$, $\alpha = \frac{1}{2}$ and $\beta = -\frac{1}{2}$, giving

$$P = \begin{bmatrix} \frac{1}{2} & -\frac{1}{2} \\ 0 & 1 \end{bmatrix}.$$

To satisfy the definition we must also check PA:

$$PA = \begin{bmatrix} \frac{1}{2} & -\frac{1}{2} \\ 0 & 1 \end{bmatrix} \begin{bmatrix} 2 & 1 \\ 0 & 1 \end{bmatrix} = \begin{bmatrix} 1 & 0 \\ 0 & 1 \end{bmatrix} = I_2.$$

Hence A is invertible, with inverse $\begin{bmatrix} \frac{1}{2} & -\frac{1}{2} \\ 0 & 1 \end{bmatrix}$.

Let $Q = \begin{bmatrix} a & b \\ c & d \end{bmatrix}$. Then

$$QB = \begin{bmatrix} a & b \\ c & d \end{bmatrix} \begin{bmatrix} 1 & 0 \\ 2 & 0 \end{bmatrix} = \begin{bmatrix} a + 2b & 0 \\ c + 2d & 0 \end{bmatrix} \neq \begin{bmatrix} 1 & 0 \\ 0 & 1 \end{bmatrix} = I_2$$

for any $a, b, c, d \in \mathbb{R}$. Thus B is not invertible.   □

Theorem 11.6.3 *An invertible matrix has a unique inverse.*

Proof Suppose that P and Q are inverses of A. Then, from Definition 11.6.1, we have (in particular)

$$PA = I \text{ and } AQ = I.$$

Thus

$$P(AQ) = PI = P \tag{11.6.1}$$

and

$$(PA)Q = IQ = Q. \tag{11.6.2}$$

By Theorem 11.3.6, $P(AQ) = (PA)Q$, so, from (11.6.1) and (11.6.2), $P = Q$. Thus the inverse is unique.   □

After this result we are able to talk about *the* inverse of an invertible matrix A. We denote it by A^{-1} and we have

$$A^{-1}A = AA^{-1} = I. \tag{11.6.3}$$

Example 11.6.4 Suppose $A \in M_{n \times n}(\mathbb{F})$ and that $B \in M_{n \times p}(\mathbb{F})$ is non-zero and $AB = 0$. Prove that A is not invertible.

Solution Suppose that A is invertible, so A^{-1} exists.
As $AB = 0$, we can pre-multiply both sides by A^{-1} to get

$$A^{-1}(AB) = A^{-1}0 = 0.$$

Also, by Theorem 11.3.6,

$$A^{-1}(AB) = (A^{-1}A)B = IB = B.$$

Hence $B = 0$.
This contradicts the assumption that $B \neq 0$. Hence A is not invertible.   □

Theorem 11.6.5 *Let* $A = \begin{bmatrix} a & b \\ c & d \end{bmatrix}$. *Then*

$$A \text{ is invertible} \iff ad - bc \neq 0.$$

When $ad - bc \neq 0$, *the inverse is*

$$A^{-1} = \frac{1}{ad - bc} \begin{bmatrix} d & -b \\ -c & a \end{bmatrix}. \tag{11.6.4}$$

Proof Suppose that A is invertible. Therefore $A \neq 0$, *i.e.* a, b, c and d are not all zero and so $B = \begin{bmatrix} d & -b \\ -c & a \end{bmatrix} \neq 0$. Suppose that $ad - bc = 0$ then

$$AB = \begin{bmatrix} a & b \\ c & d \end{bmatrix} \begin{bmatrix} d & -b \\ -c & d \end{bmatrix} = \begin{bmatrix} ad - bc & 0 \\ 0 & ad - bc \end{bmatrix} = 0. \qquad (11.6.5)$$

From Example 11.6.4, since $B \neq 0$, A is not invertible. This is a contradiction and so $ad - bc \neq 0$.

Now suppose that $ad - bc \neq 0$, and let

$$C = \frac{1}{ad - bc} \begin{bmatrix} d & -b \\ -c & a \end{bmatrix}.$$

From (11.6.5), $AC = I_2$ and the reader may readily verify that $CA = I_2$. Hence A is invertible and $A^{-1} = C$, as required. $\qquad \square$

Remarks (1) The formula (11.6.4) is well worth remembering—it allows us to write down the inverse of any (invertible) 2×2 matrix. In Section 11.7, we will describe a method of finding the inverse of a general $n \times n$ invertible matrix. This method is rather cumbersome in the 2×2 case.

(2) Theorem 11.6.5 shows that the invertibility of a 2×2 matrix can be decided by evaluating a single number depending on the entries in the matrix. This is the *determinant* defined below.

Definition 11.6.6 Let $A = \begin{bmatrix} a & b \\ c & d \end{bmatrix}$. The *determinant* of A is $ad - bc$, and is denoted $\det(A)$ or $\begin{vmatrix} a & b \\ c & d \end{vmatrix}$.

For example,

$$\begin{vmatrix} 1 & 2 \\ 3 & 4 \end{vmatrix} = 1(4) - 2(3) = -2.$$

We *can* define the determinant of a general $n \times n$ matrix, and this is related to the problem of invertibility much as in Theorem 11.6.5. There is even a formula which gives the inverse (when it exists) analogous to (11.6.4). The definition of this general determinant, and the formula for the inverse, are rather complicated and will not be needed in this book.

We will, however, need the 3×3 determinant in Section 13.2.

Definition 11.6.7 Let

$$A = \begin{bmatrix} a & b & c \\ d & e & f \\ g & h & i \end{bmatrix}.$$

Then the *determinant* of A is

$$\begin{vmatrix} a & b & c \\ d & e & f \\ g & h & i \end{vmatrix} = a \begin{vmatrix} e & f \\ h & i \end{vmatrix} - b \begin{vmatrix} d & f \\ g & i \end{vmatrix} + c \begin{vmatrix} d & e \\ g & h \end{vmatrix}.$$

Remark It is best to remember this definition by observing the pattern.

(1) each term is the product of an entry from the first row and the 2×2 determinant taken from the *other* rows and columns.

(2) the second term has a minus sign. This is vital.

For example,

$$\begin{vmatrix} 1 & -2 & 3 \\ 4 & 5 & 6 \\ 1 & 1 & 0 \end{vmatrix} = 1 \begin{vmatrix} 5 & 6 \\ 1 & 0 \end{vmatrix} - (-2) \begin{vmatrix} 4 & 6 \\ 1 & 0 \end{vmatrix} + 3 \begin{vmatrix} 4 & 5 \\ 1 & 1 \end{vmatrix}$$

$$= 1(-6) - (-2)(-6) + 3(-1) = -21.$$

We finish this section with some algebraic results on matrix inverses. The proofs all use the same technique—to show that A is invertible, we find (somehow) a matrix P which satisfies $AP = PA = I$. In Section 11.7, we will show that $PA = I \Rightarrow AP = I$. After that it is sufficient to check just one of the equalities.

Theorem 11.6.8 *If A is invertible, then*

(1) A^{-1} *is invertible, with inverse A,*

(2) A^{T} *is invertible, with inverse $(A^{-1})^{\mathrm{T}}$.*

Proof As A is invertible, A^{-1} exists.

(1) From (11.6.3) we see that $P = A$ satisfies

$$A^{-1}P = PA^{-1} = I.$$

Hence A^{-1} is invertible with inverse $P = A$.

(2) By Theorem 11.4.7 (3), $(A^{-1}A)^{\mathrm{T}} = A^{\mathrm{T}}(A^{-1})^{\mathrm{T}}$ and $(AA^{-1})^{\mathrm{T}} = (A^{-1})^{\mathrm{T}}A^{\mathrm{T}}$. Also $I^{\mathrm{T}} = I$, so transposing each part of (11.6.3), we get

$$A^{\mathrm{T}}(A^{-1})^{\mathrm{T}} = (A^{-1})^{\mathrm{T}}A^{\mathrm{T}} = I.$$

Thus $Q = (A^{-1})^{\mathrm{T}}$ satisfies

$$A^{\mathrm{T}}Q = QA^{\mathrm{T}} = I.$$

Hence A^{T} is invertible with inverse $Q = (A^{-1})^{\mathrm{T}}$. $\square$

Theorem 11.6.9 *If $A, B \in M_{n \times n}(\mathbb{F})$ are invertible, then AB is invertible, with inverse $B^{-1}A^{-1}$.*

Proof Using Theorem 11.3.6,

$$(B^{-1}A^{-1})(AB) = B^{-1}(A^{-1}A)B$$
$$= B^{-1}IB$$
$$= B^{-1}B = I.$$

Similarly, $(AB)(B^{-1}A^{-1}) = I$. The result follows at once. $\square$

Hazard In Theorem 11.6.9 note that the order of the factors is *reversed* after taking the inverse.

Of course, Theorem 11.6.9 generalises to longer products.

Corollary 11.6.10 *Let $r \in \mathbb{N}$. If $A_1, \ldots, A_r$ are invertible, then the product $A_1 \ldots A_r$ (when defined) is invertible, with inverse $A_r^{-1} \ldots A_1^{-1}$.*

Proof It is left as an exercise to the reader to write down a formal proof of this result using principle of induction. □

Example 11.6.11 Prove that, if $A \in M_{n \times n}(\mathbb{F})$ is invertible, and $\lambda \in \mathbb{F}$ is non-zero, then λA is invertible. Find $(\lambda A)^{-1}$.

Solution Since A is invertible, A^{-1} exists and $AA^{-1} = A^{-1}A = I$. From Theorem 11.3.5 (3),

$$(\lambda A)\left(\frac{1}{\lambda}A^{-1}\right) = \lambda\frac{1}{\lambda}AA^{-1} = I,$$

and

$$\left(\frac{1}{\lambda}A^{-1}\right)(\lambda A) = \frac{1}{\lambda}\lambda A^{-1}A = I.$$

Hence, $P = (1/\lambda)A^{-1}$ satisfies

$$(\lambda A)P = P(\lambda A) = I,$$

so λA is invertible with inverse $P = (1/\lambda)A^{-1}$. □

Example 11.6.12 Give a counter-example to show that the implication

$$A, B \text{ invertible} \Rightarrow A + B \text{ invertible,}$$

is false.

Solution Take $A = I$, $B = -I$. Then A is invertible and, by Example 11.6.11 with $\lambda = -1$, B is invertible also. However, $A + B = O$ is not invertible. □

Let $f(x) = a_r x^r + \cdots + a_1 x + a_0 \in \mathbb{F}[x]$ and $A \in M_{n \times n}(\mathbb{F})$. Then the matrix powers A^k are defined and of type $n \times n$. We can therefore form the matrix $f(A)$ as follows:

$$f(A) = a_r A^r + \cdots + a_1 A + a_0 I_n.$$

Note that, to ensure that the matrix sum exists, we must replace the constant a_0 by $a_0 I_n$.

If $f(A) = O_{n \times n}$ and $a_0 \neq 0$, then A is invertible, and we can find A^{-1}, as the following example illustrates.

Example 11.6.13 Suppose that $A \in M_{n \times n}(\mathbb{F})$ satisfies

$$A^3 + A^2 - A - 2I = 0.$$

Show that A is invertible, and express A^{-1} as a polynomial in A.

Proof We isolate I on right hand side and obtain

$$\tfrac{1}{2}(A^3 + A^2 - A) = I. \tag{11.6.6}$$

By taking A out as a factor on the left or on the right we obtain

$$A(\tfrac{1}{2}(A^2 + A - I)) = (\tfrac{1}{2}(A^2 + A - I))A = I.$$

Hence A is invertible, with inverse $\tfrac{1}{2}(A^2 + A - I) = \tfrac{1}{2}A^2 + \tfrac{1}{2}A - \tfrac{1}{2}I$. $\qquad\square$

11.7 Finding Matrix Inverses

In this section, we describe a method for finding the inverse of an invertible $n \times n$ matrix. The key is the connection between EROs and matrix multiplication.

Definition 11.7.1 If θ is an ERO, then $_mM_\theta$ is the matrix obtained by applying the ERO θ to I_m.

    When the number of rows is clear from the context, we omit the subscript m, *i.e.* we write M_θ instead of $_mM_\theta$.

Example 11.7.2 Let θ_1 be $R_1 \leftrightarrow R_2$, θ_2 be $R_2 \to \lambda R_2$ and θ_3 be $R_2 \to R_2 + \alpha R_1$. Find $_2M_{\theta_1}$, $_2M_{\theta_2}$ and $_2M_{\theta_3}$.

Solution Applying each of the EROs θ_1, θ_2 and θ_3 to $I_2 = \begin{bmatrix} 1 & 0 \\ 0 & 1 \end{bmatrix}$, we get

$$_2M_{\theta_1} = \begin{bmatrix} 0 & 1 \\ 1 & 0 \end{bmatrix}, \quad _2M_{\theta_2} = \begin{bmatrix} 1 & 0 \\ 0 & \lambda \end{bmatrix}, \quad _2M_{\theta_3} = \begin{bmatrix} 1 & 0 \\ \alpha & 1 \end{bmatrix}. \qquad\square$$

Theorem 11.7.3 *Let $A \in M_{m\times n}(\mathbb{F})$, and let θ be an ERO. Then the matrix obtained by applying θ to A is (the product of matrices) $M_\theta A$.*

Proof As an example of the method of proof, we verify this result below for $m = n = 2$. The proof for general $m, n \in \mathbb{N}$ uses the same approach and is omitted. $\qquad\square$

    Let $A = \begin{bmatrix} a & b \\ c & d \end{bmatrix}$. We consider the three types of ERO in turn and make use of the results obtained in Example 11.7.2.

(1) $\theta = (R_1 \leftrightarrow R_2)$. Applying θ to A, we get $\begin{bmatrix} c & d \\ a & b \end{bmatrix}$ and

$$M_\theta A = \begin{bmatrix} 0 & 1 \\ 1 & 0 \end{bmatrix} \begin{bmatrix} a & b \\ c & d \end{bmatrix} = \begin{bmatrix} c & d \\ a & b \end{bmatrix},$$

as required.

(2) $\theta = (R_2 \to \lambda R_2)$. Applying θ to A, we get $\begin{bmatrix} a & b \\ \lambda c & \lambda d \end{bmatrix}$ and

$$M_\theta A = \begin{bmatrix} 1 & 0 \\ 0 & \lambda \end{bmatrix} \begin{bmatrix} a & b \\ c & d \end{bmatrix} = \begin{bmatrix} a & b \\ \lambda c & \lambda d \end{bmatrix},$$

as required. The case $\theta = (R_1 \to \lambda R_1)$ is similar.

(3) $\theta = (R_2 \to R_2 + \alpha R_1)$. Applying θ to A, we get $\begin{bmatrix} a & b \\ c + \alpha a & d + \alpha b \end{bmatrix}$ and

$$M_\theta A = \begin{bmatrix} 1 & 0 \\ \alpha & 1 \end{bmatrix} \begin{bmatrix} a & b \\ c & d \end{bmatrix} = \begin{bmatrix} a & b \\ c + \alpha a & d + \alpha b \end{bmatrix},$$

as required. The case $\theta = (R_1 \to R_1 + \alpha R_2)$ is similar.

Theorem 11.7.4 *If θ is an ERO, then M_θ is invertible.*

Proof We consider each type of ERO separately.

Case 1 $\theta = (R_i \leftrightarrow R_j)$. The ERO θ interchanges the ith and j^{th} rows and if we perform this twice, then any matrix is returned to its original form. Thus, in particular, $M_\theta(M_\theta I) = I$, and so $M_\theta M_\theta = I$. It follows that M_θ is invertible with inverse M_θ.

Case 2 $\theta = (R_i \to \lambda R_i)$ (where $\lambda \neq 0$). Let $\psi = (R_i \to (1/\lambda)R_i)$. Then if we perform θ and ψ, in either order, a matrix is returned to its original form. Hence $M_\theta M_\psi = M_\psi M_\theta = I$ and so M_θ is invertible, with inverse M_ψ.

Case 3 $\theta = (R_i \to R_i + \alpha R_j)$. Let $\psi = (R_i \to R_i - \alpha R_j)$. Again, if we perform θ and ψ, in either order, a matrix is returned to its original form. Hence $M_\theta M_\psi = M_\psi M_\theta = I$ and so M_θ is invertible, with inverse M_ψ. $\square$

We can combine these results to express the entire row reduction process in terms of matrix multiplication.

Theorem 11.7.5 *Let $A \in M_{m \times n}(\mathbb{F})$. There exists an invertible matrix $P \in M_{m \times m}$ such that*

$$PA = E(A),$$

the reduced echelon form of A.

Proof Suppose that θ_1, then θ_2, then $\ldots$, then θ_r transform A to $E(A)$. Then, applying Theorem 11.7.3 repeatedly, we see that A is transformed to $M_{\theta_1} A$, then to $M_{\theta_2}(M_{\theta_1} A) = M_{\theta_2} M_{\theta_1} A$, etc. Thus eventually we get

$$E(A) = M_{\theta_r} \cdots M_{\theta_2} M_{\theta_1} A.$$

Let $P = M_{\theta_r} \cdots M_{\theta_2} M_{\theta_1}$. Then P is invertible by Theorem 11.7.4 and Corollary 11.6.10, and

$$E(A) = PA,$$

as required. $\square$

Note 11.7.6 In the course of proving Theorem 11.7.5, we found that $P = M_{\theta_r} \cdots M_{\theta_1} = M_{\theta_r} \cdots M_{\theta_1} I_m$. Reversing the argument at the start of the proof, we see that P is obtained from I_m by applying ERO θ_1, then θ_2, then $\ldots$, then θ_r. Thus we observe that P is obtained from I_m by the same EROs, applied in the same order, as are used to obtain $E(A)$ from A. In practice, it is best to apply the EROs *simultaneously* to A and to I, rather than recording them as we reduce A to echelon form and *then* applying them to P. We combine these calculations by considering the matrix $[A|I_m]$, of type $m \times (n+m)$, whose rows are obtained by appending the rows of I_m to those of A. We carry out row reduction until we obtain $[E(A)|P]$, *i.e.* the matrix in which the part of the matrix originally occupied by A has been transformed into reduced echelon form. Then we obtain P such that $E(A) = PA$, from the final columns of $[E(A)|P]$.

Example 11.7.7 For the matrix

$$A = \begin{bmatrix} 0 & 1 & 1 \\ 1 & 0 & 1 \\ -1 & 2 & 2 \end{bmatrix},$$

find $E(A)$ and an invertible matrix P such that $E(A) = PA$.

Solution After Note 11.7.6, we consider the matrix $[A|I_3]$:

$$\begin{bmatrix} 0 & 1 & 1 & | & 1 & 0 & 0 \\ 1 & 0 & 1 & | & 0 & 1 & 0 \\ -1 & 2 & 2 & | & 0 & 0 & 1 \end{bmatrix} \sim \begin{bmatrix} 1 & 0 & 1 & | & 0 & 1 & 0 \\ 0 & 1 & 1 & | & 1 & 0 & 0 \\ -1 & 2 & 2 & | & 0 & 0 & 1 \end{bmatrix} \quad R_1 \leftrightarrow R_2$$

$$\sim \begin{bmatrix} 1 & 0 & 1 & | & 0 & 1 & 0 \\ 0 & 1 & 1 & | & 1 & 0 & 0 \\ 0 & 2 & 3 & | & 0 & 1 & 1 \end{bmatrix} \quad R_3 \to R_3 + R_1$$

$$\sim \begin{bmatrix} 1 & 0 & 1 & | & 0 & 1 & 0 \\ 0 & 1 & 1 & | & 1 & 0 & 0 \\ 0 & 0 & 1 & | & -2 & 1 & 1 \end{bmatrix} \quad R_3 \to R_3 - 2R_2$$

$$\sim \begin{bmatrix} 1 & 0 & 0 & | & 2 & 0 & -1 \\ 0 & 1 & 0 & | & 3 & -1 & -1 \\ 0 & 0 & 1 & | & -2 & 1 & 1 \end{bmatrix} \quad \begin{matrix} R_1 \to R_1 - R_3 \\ R_2 \to R_2 - R_3 \end{matrix} \, .$$

Hence

$$E(A) = \begin{bmatrix} 1 & 0 & 0 \\ 0 & 1 & 0 \\ 0 & 0 & 1 \end{bmatrix}, \quad \text{and} \quad P = \begin{bmatrix} 2 & 0 & -1 \\ 3 & -1 & -1 \\ -2 & 1 & 1 \end{bmatrix}. \qquad \square$$

Lemma 11.7.8 *Let A and P be square matrices of the same type with P invertible. Then*

$$PA = I \Rightarrow (A \text{ invertible with } A^{-1} = P \text{ and } P^{-1} = A).$$

Proof Since P is invertible, we have the inverse P^{-1} and so pre-multiplying both sides of $PA = I$ by P^{-1} gives

$$P^{-1}(PA) = P^{-1}I, \qquad \textit{i.e. } A = P^{-1}.$$

Post-multiplying each side of this equation by P we get

$$AP = P^{-1}P = I.$$

Hence we have $PA = AP = I$, and so A is invertible with $A^{-1} = P$. The result $P^{-1} = A$ follows from Theorem 11.6.8 (1) since $P = A^{-1}$. $\qquad \square$

In view of Theorem 11.7.5 and Lemma 11.7.8, if the reduced echelon form of (a square matrix) A is I then the invertible matrix P such that $PA = I$ is the inverse of A. This gives us a method for determining a matrix inverse when it exists. Before exploiting this further we will establish criteria for the existence of this inverse.

Lemma 11.7.9 *If A has a zero row, then A is not invertible.*

Proof Let A have its ith row a zero row.

If A is invertible. Then $AA^{-1} = I$, so that the $(i, i)^{\text{th}}$ entry of AA^{-1} is 1. However, the $(i, i)^{\text{th}}$ entry of AA^{-1} is the dot product of the ith row of A with the ith column of A^{-1}. Since the entries in the former are all zero, the $(i, i)^{\text{th}}$ entry is 0. This contradiction shows that A is not invertible.   □

Note 11.7.10 The echelon form of an $n \times n$ matrix A can take one of two forms. First, $E(A)$ may have at least one zero row. Otherwise, each of the n rows is non-zero and hence has a leading entry, which must be 1. All of these n 1's lie in a different one of the n columns. Finally, the requirement that the leading entry in the $(j + 1)^{\text{st}}$ row lies strictly to the right of the leading entry in the jth row, means that $E(A) = I_n$.

That is, the echelon form of a square matrix is either the identity matrix or a matrix with a zero row.

Theorem 11.7.11 *Let $A \in M_{n \times n}(\mathbb{F})$. Then*

$$A \text{ invertible } \Leftrightarrow r(A) = n.$$

Proof By Theorem 11.7.5, there is an invertible matrix P such that $E(A) = PA$.

Let $r(A) \neq n$, i.e. $r(A) < n$. Then $E(A)$ has at least one zero row. Suppose that A is invertible then, by Theorem 11.6.9, since P is invertible also, $PA = E(A)$ is invertible. This is a contradiction of Lemma 11.7.9 and so A is not invertible.

Now suppose that $r(A) = n$. Then $E(A)$ has no non-zero rows then, by Note 11.7.10, $E(A) = I$. We may now use Lemma 11.7.8 to deduce that A is invertible (with inverse P).   □

Note 11.7.12 We now have a method for determining whether a (square) matrix A is invertible, and of finding the inverse when it exists.

(1) Form $[A|I]$ and use row reduction to obtain $[E(A)|P]$.

(2) If $E(A) = I$ then A is invertible and the inverse is P.

(3) If $E(A) \neq I$ then A is not invertible.

Example 11.7.13 Show that $A = \begin{bmatrix} 1 & 2 & 1 \\ 2 & 5 & 2 \\ 2 & 4 & 1 \end{bmatrix}$ is invertible, and find its inverse. Hence find the solution of the system

$$\begin{cases} x + 2y + z = 3 \\ 2x + 5y + 2z = 4 \\ 2x + 4y + z = 5 \end{cases}.$$

Solution For the first part, we consider $[A|I]$:

$$\left[\begin{array}{ccc|ccc} 1 & 2 & 1 & 1 & 0 & 0 \\ 2 & 5 & 2 & 0 & 1 & 0 \\ 2 & 4 & 1 & 0 & 0 & 1 \end{array}\right] \sim \left[\begin{array}{ccc|ccc} 1 & 2 & 1 & 1 & 0 & 0 \\ 0 & 1 & 0 & -2 & 1 & 0 \\ 0 & 0 & -1 & -2 & 0 & 1 \end{array}\right] \begin{array}{l} \\ R_2 \to R_2 - 2R_1 \\ R_3 \to R_3 - 2R_1 \end{array}$$

$$\sim \left[\begin{array}{ccc|ccc} 1 & 0 & 1 & 5 & -2 & 0 \\ 0 & 1 & 0 & -2 & 1 & 0 \\ 0 & 0 & -1 & -2 & 0 & 1 \end{array}\right] \quad R_1 \to R_1 - 2R_2$$

$$\sim \left[\begin{array}{ccc|ccc} 1 & 0 & 1 & 5 & -2 & 0 \\ 0 & 1 & 0 & -2 & 1 & 0 \\ 0 & 0 & 1 & 2 & 0 & -1 \end{array}\right] \begin{array}{l} \\ \\ R_3 \to -R_3 \end{array}$$

$$\sim \left[\begin{array}{ccc|ccc} 1 & 0 & 0 & 3 & -2 & 1 \\ 0 & 1 & 0 & -2 & 1 & 0 \\ 0 & 0 & 1 & 2 & 0 & -1 \end{array}\right] \begin{array}{l} R_1 \to R_1 - R_3 \\ \\ \end{array}$$

Then A is invertible with $A^{-1} = \begin{bmatrix} 3 & -2 & 1 \\ -2 & 1 & 0 \\ 2 & 0 & -1 \end{bmatrix}$.

The system of equation is $AX = H$. Pre-multiplying by A^{-1}, we get

$$X = A^{-1}(AX) = A^{-1}H = \begin{bmatrix} 3 & -2 & 1 \\ -2 & 1 & 0 \\ 2 & 0 & -1 \end{bmatrix} \begin{bmatrix} 3 \\ 4 \\ 5 \end{bmatrix} = \begin{bmatrix} 6 \\ -2 \\ 1 \end{bmatrix}.$$

Hence the solution set is $\{(6, -2, 1)\}$. □

Example 11.7.14 Prove that the matrix $A = \begin{bmatrix} 1 & 2 & 3 \\ -1 & 0 & -4 \\ 1 & 6 & 1 \end{bmatrix}$ is not invertible and find the invertible matrix P such that $PA = E(A)$.

Solution We consider $[A|I]$,

$$\left[\begin{array}{ccc|ccc} 1 & 2 & 3 & 1 & 0 & 0 \\ -1 & 0 & -4 & 0 & 1 & 0 \\ 1 & 6 & 1 & 0 & 0 & 1 \end{array}\right] \sim \left[\begin{array}{ccc|ccc} 1 & 2 & 3 & 1 & 0 & 0 \\ 0 & 2 & -1 & 1 & 1 & 0 \\ 0 & 4 & -2 & -1 & 0 & 1 \end{array}\right] \begin{array}{l} \\ R_2 \to R_2 + R_1 \\ R_3 \to R_3 - R_1 \end{array}$$

$$\sim \left[\begin{array}{ccc|ccc} 1 & 2 & 3 & 1 & 0 & 0 \\ 0 & 1 & -\frac{1}{2} & \frac{1}{2} & \frac{1}{2} & 0 \\ 0 & 4 & -2 & -1 & 0 & 1 \end{array}\right] \begin{array}{l} \\ R_2 \to \frac{1}{2}R_2 \\ \\ \end{array}$$

$$\sim \left[\begin{array}{ccc|ccc} 1 & 0 & 4 & 0 & -1 & 0 \\ 0 & 1 & -\frac{1}{2} & \frac{1}{2} & \frac{1}{2} & 0 \\ 0 & 0 & 0 & -3 & -2 & 1 \end{array}\right] \begin{array}{l} R_1 \to R_1 - 2R_2 \\ \\ R_3 \to R_3 - 4R_2 \end{array}$$

Hence $E(A) = \begin{bmatrix} 1 & 0 & 4 \\ 0 & 1 & -\frac{1}{2} \\ 0 & 0 & 0 \end{bmatrix}$ has a zero row and so A is not invertible.

We have also found that $P = \begin{bmatrix} 0 & -1 & 0 \\ \frac{1}{2} & \frac{1}{2} & 0 \\ -3 & -2 & 1 \end{bmatrix}$ is such that $PA = E(A)$, which is invertible by construction. □

Finally, we justify the claim made earlier: that to show that A is invertible, it is enough to check that there is a matrix P with *either $PA = I$ or $AP = I$.*

Theorem 11.7.15 *Let $A, B \in M_{n \times n}(\mathbb{F})$ with $AB = I$. Then A and B are both invertible with $A^{-1} = B$ and $B^{-1} = A$.*

Proof Suppose that A is not invertible. Then, by Theorems 11.7.5 and 11.7.11, there exists an invertible matrix P such that $PA = E(A)$ has at least one zero row, the kth say.

We now consider PAB calculated in two different ways. First, $PAB = (PA)B$ has its kth row zero (see the proof of Lemma 11.7.9) and second, $PAB = P(AB) = PI = P$ which is invertible. This is a contradiction of Lemma 11.7.9. Hence A is invertible.

Now, since A is invertible, we may use Lemma 11.7.8 to deduce that B is invertible and that $A^{-1} = B$ and $B^{-1} = A$. $\qquad\qquad\square$

11.X   Exercises

1. Let
$$A = \begin{bmatrix} 1 & 2 & 3 \\ 4 & 5 & 6 \end{bmatrix}, \quad B = \begin{bmatrix} 1 & 1 & 1 \\ 2 & 1 & -1 \end{bmatrix}, \quad C = \begin{bmatrix} 1 & -1 & 0 \\ 0 & 1 & 2 \end{bmatrix},$$

$$X = \begin{bmatrix} 1 & 2 & 2 \\ -2 & -1 & 2 \\ -2 & 2 & 1 \end{bmatrix}, \quad Y = \begin{bmatrix} 1 & 2 \\ 3 & 4 \\ 5 & 6 \end{bmatrix}, \quad Z = \begin{bmatrix} 2 & -1 \\ 1 & 2 \end{bmatrix}.$$

   (a) Verify that $A + (B + C) = (A + B) + C$ and that $(A + B)^T = A^T + B^T$.

   (b) For each pair of matrices, check whether the product exists and, if it does, find it.

   (c) Verify that $(XY)Z = X(YZ)$, $(BY)^T = Y^T B^T$ and $Z(2A + 3B) = 2ZA + 3ZB$.

   (d) Find $X^T X$ and XX^T.

   (e) Show that $4Z^2 - Z^3 = 5Z$.

   (f) Check that $A^T A$, AA^T and $A^T Z^T Z A$ are symmetric.

2. Let $A = \begin{bmatrix} 4 & 0 \\ 0 & 1 \end{bmatrix}$ and $B = \begin{bmatrix} 0 & 0 \\ 1 & 0 \end{bmatrix}$. Find

   (a) the set of all matrices X such that $X^2 = A$,

   (b) the set of all matrices Y such that $Y^2 = B$.

3. Let $A = \begin{bmatrix} -1 & 3 \\ 1 & -1 \end{bmatrix}$. Prove that

   (a) there is no matrix X such that $AX - XA = \begin{bmatrix} 1 & 0 \\ 0 & 1 \end{bmatrix}$,

(b) there is exactly one matrix B of the form $\begin{bmatrix} 1 & a \\ 2 & b \end{bmatrix}$ such that A and B commute.

4.   Let $P = \begin{bmatrix} \cos\theta & -\sin\theta \\ \sin\theta & \cos\theta \end{bmatrix}$. Prove by induction that, for all $n \in \mathbb{N}$,

$$P^n = \begin{bmatrix} \cos n\theta & -\sin n\theta \\ \sin n\theta & \cos n\theta \end{bmatrix}.$$

5.   Let A be an $m \times m$ matrix such that $(A - 2I_m)^2 = O_{m\times m}$. Prove by induction that, for all $n \in \mathbb{N}$,
$$A^n = 2^{n-1}\big(nA - 2(n-1)I_m\big).$$

Deduce that, for all $n \in \mathbb{N}$,

$$\begin{bmatrix} 3 & 1 \\ -1 & 1 \end{bmatrix}^n = 2^{n-1}\begin{bmatrix} n+2 & n \\ -n & -n+2 \end{bmatrix}.$$

6.   Let $P, S \in M_{n\times n}(\mathbb{F})$. Prove that if S is symmetric then $P^T S P$ is symmetric.

7.   Let $P, Q \in M_{n\times n}(\mathbb{F})$. prove that if P and Q are symmetric then $PQ - QP$ is skew-symmetric.

8.   Let $A \in M_{n\times n}(\mathbb{R})$ with $AA^T = O_{n\times n}$. By considering the diagonal entries, prove that $A = O_{n\times n}$.

9.   A square matrix X is *idempotent* if $X^2 = X$.

   (a) Let $A, B \in M_{n\times n}(\mathbb{F})$ and let A, B and $A + B$ be idempotent. Show that $AB = -BA$. By considering ABA, show that $AB = O$.

   (b) Let $P, Q \in M_{n\times n}(\mathbb{F})$ and let $P + Q$ and $P - Q$ be idempotent. Show that $P^2 + Q^2 = P$. Deduce that Q^2 commutes with P.

10.   Find reduced echelon matrices corresponding to each of the following matrices.

(a) $\begin{bmatrix} 1 & 4 & 6 & 1 \\ 0 & 2 & 3 & 2 \\ -1 & 1 & 2 & 6 \end{bmatrix}$,   (b) $\begin{bmatrix} 1 & 2 & 3 & 4 \\ 2 & 4 & 5 & 6 \\ 3 & 6 & 7 & 11 \end{bmatrix}$,

(c) $\begin{bmatrix} 1 & 0 & -1 & 1 \\ 2 & 1 & 4 & -2 \\ 0 & -1 & -6 & 0 \end{bmatrix}$.

11. Find the general solutions of each of the following systems of equations.

(a) $\begin{cases} x + y + z + t = 1 \\ 2x + 3y + 4z + 4t = 2 \\ -x \quad\quad + z + t = -1 \end{cases}$,   (b) $\begin{cases} x + y + z = 1 \\ 2x + 3y + 4z = 2 \\ 3x + 4y - z = 7 \end{cases}$,

(c) $\begin{cases} x - y - 2z = 1 \\ 2x + y + z = 2 \\ 4x - y - 3z = 3 \end{cases}$,   (d) $\begin{cases} x - y - z = 2 \\ -x + 2y \quad + 3t = 1 \\ x \quad\quad + z = 5 \end{cases}$.

12. Given that the system of linear equations

$$\begin{cases} x + y + z - t = a \\ 2x - y + z + t = b \\ 4x + y + 3z - t = c \end{cases}$$

has solutions, find an equation connecting a, b and c. Find the general solution when $a = b = 1$ and $c = 3$.

13. Let S be the set of all 2×2 matrices over $\mathbb{R}$ of the form

$$\begin{bmatrix} x & 1-x \\ 1-x & x \end{bmatrix}$$

where $x \neq \frac{1}{2}$. Prove that

(a) each member of S is invertible,

(b) S is closed under matrix multiplication,

(c) the inverse of each element of S is also in S.

14. Let A and B be invertible matrices such that $A^2B = AB^2$. Show that $A = B$.

15. A matrix X is *orthogonal* if X is square and $XX^{\mathrm{T}} = X^{\mathrm{T}}X = I$. Prove that

(a) if X is orthogonal then X^{T} is orthogonal,

(b) if $A, B \in M_{n \times n}(\mathbb{F})$ with A and B orthogonal then AB is orthogonal,

(c) if $C \in M_{n \times n}(\mathbb{F})$ with $C+I$ and $C-I$ orthogonal then C is skew-symmetric.

16. Matrices $A, B \in M_{n \times n}(\mathbb{F})$ satisfy $AB = A$. Show that

(a) if $B \neq I$ then A is non-invertible,

(b) if $B^2 = O$ then $A = O$,

(c) if $B^3 = O$ then $A = O$.

17. Let $U, V \in M_{n \times n}(\mathbb{F})$ with $(UV)^k = I$ for some $k \in \mathbb{N}$. Use Theorem 11.7.15 to prove that U and V are invertible. Deduce that $(VU)^k = I$.

18.　Let $A, B \in M_{n \times n}(\mathbb{F})$ with $A + B = AB$. By considering $(I - A)(I - B)$, show that A and B commute.

19.　Let B be a square matrix such that $(B - I)^3 = O$.

(a) By expanding $(B - I)^3$, show that B is invertible and express B^{-1} in terms of B.

(b) Show also that $B + 3I$ is invertible and express its inverse in terms of B.

20.　Let $A, P \in M_{m \times m}(\mathbb{F})$ with P invertible. Use induction to prove that, for all $n \in \mathbb{N}$, $(P^{-1}AP)^n = P^{-1}A^nP$.

21.　In each of the following statements, A is a square matrix. For each statement, determine whether or not it is true. Give a proof of each true statement, and a counter-example for each false statement.

(a) If $A^2 = A$ and A is invertible then $A = I$.

(b) If $A^2 = A$ and A is non-invertible then $A = O$.

(c) If $A^2 = O$ then $A = O$.

(d) If $A^2 = O$ then A is non-invertible.

(e) If A is non-invertible then $A + A^{\mathrm{T}}$ is non-invertible.

(f) If A is invertible then $A + A^{\mathrm{T}}$ is invertible.

22.　For each of the following matrices, determine whether or not it is invertible and, if it is, find its inverse. Show the elementary row operations you use at each stage.

(a) $\begin{bmatrix} 3 & 4 \\ -1 & 1 \end{bmatrix}$,

(b) $\begin{bmatrix} 1 & 4 & 6 \\ 0 & 2 & 3 \\ -1 & 1 & 2 \end{bmatrix}$,

(c) $\begin{bmatrix} 1 & 2 & 3 \\ 5 & 5 & -10 \\ 2 & 3 & 1 \end{bmatrix}$,

(d) $\begin{bmatrix} 2 & 1 & 1 \\ 1 & -1 & 2 \\ 3 & 2 & -1 \end{bmatrix}$,

(e) $\begin{bmatrix} 1 & 3 & -2 \\ 2 & 4 & -3 \\ 3 & 2 & -1 \end{bmatrix}$,

(f) $\begin{bmatrix} 0 & 1 & 2 & 1 \\ -1 & 0 & 1 & -1 \\ -2 & -1 & 0 & -2 \\ -1 & 1 & 2 & 0 \end{bmatrix}$.

23.　Find the value of a for which

$$A = \begin{bmatrix} 1 & 2 & 3 \\ 2 & -1 & 5 \\ 2 & 4 & a \end{bmatrix}$$

is non-invertible. Find A^{-1} when $a = 1$.

Chapter 12

Vectors and Three Dimensional Geometry

In the real world we meet many things which can be described in terms of *magnitude*, *i.e.* a non-negative multiple of some unit. Some familiar examples are mass (*e.g.* 6.4 kilograms) and distance (*e.g.* 100 metres). These are known as *scalar quantities*.

Other things, known as *vector quantities*, have a *direction* as well as a magnitude. For example, the acceleration due to gravity at the Earth's surface has magnitude 9.81 metres per second per second and is directed towards the centre of the Earth. In this chapter, we introduce the mathematical objects which can be used to represent such quantities.

We define vectors in terms of *directed line segments* and give a geometrical definition of *addition* and *scalar multiplication* of vectors. We then show how a vector can be represented by a *triple of real numbers*.

This notation makes calculations involving vectors much more straightforward and we use this representation to study, amongst other things, the properties of lines in three-dimensional space.

12.1 Basic Properties of Vectors

Notation 12.1.1 We use bold face, lower-case roman letters (*e.g.* **u**, **v**) to denote vectors and upper-case roman letters (*e.g.* P, Q) to denote points in three-dimensional space.

Definitions 12.1.2 There is a unique vector **0** of magnitude 0 called the *zero vector*. The zero vector has no direction.

The magnitude of a vector **u** is written $|\mathbf{u}|$ and is non-negative. Furthermore, $|\mathbf{u}| = 0 \Leftrightarrow \mathbf{u} = \mathbf{0}$.

Given a pair of points P and Q we denote the *directed line segment* from P to Q by $\overrightarrow{PQ}$ and its length by $|PQ|$. If P and Q are distinct, $\overrightarrow{PQ}$ defines a direction and otherwise, if $Q = P$, $\overrightarrow{PP}$ is a trivial directed line segment of length 0 which does not define a direction.

We say that $\overrightarrow{AB}$ and $\overrightarrow{PQ}$ have the same direction if they are parallel with the same 'sense'. This is equivalent to saying that $ABQP$ forms a (convex) trapezium in which

275

the sides AB and PQ are parallel. See Figure 12.1.1. Note that $\overrightarrow{BA}$ does not have the same direction as $\overrightarrow{AB}$, the 'sense' is a vital part of the definition, and we say that $\overrightarrow{BA}$ has the *opposite* direction to $\overrightarrow{AB}$.

We write $\overrightarrow{PQ} \equiv \overrightarrow{AB}$ if $\overrightarrow{PQ}$ and $\overrightarrow{AB}$ have the same direction and $|PQ| = |AB|$.

Hazard In this chapter the notation $|\cdot|$ will be used to denote three distinct things: the magnitude of a vector, the length of a line segment and the modulus of a real number. It will always be clear from the context which of the three is meant.

Definitions 12.1.3 We say that a vector $\mathbf{u}$ is *represented by* a directed line segment $\overrightarrow{PQ}$ if the length of PQ equals the magnitude of $\mathbf{u}$ and the direction of $\mathbf{u}$, if $\mathbf{u} \neq \mathbf{0}$, is determined by $\overrightarrow{PQ}$. We call any such directed line segment a *representative* of $\mathbf{u}$. A vector has a representative starting at any given point, and $\overrightarrow{PQ}$ and $\overrightarrow{AB}$ both represent the same vector if and only if $\overrightarrow{AB} \equiv \overrightarrow{PQ}$.

If $\overrightarrow{PQ}$ represents $\mathbf{u} \neq \mathbf{0}$ then the vector represented by $\overrightarrow{QP}$ is called the *negative* of $\mathbf{u}$, denoted $-\mathbf{u}$, and is said to have *opposite* direction to $\mathbf{u}$. It follows that $-(-\mathbf{u}) = \mathbf{u}$ since $\overrightarrow{PQ}$ represents both vectors. The zero vector is represented by $\overrightarrow{PP}$ for each point P.

Note 12.1.4 We will assume the following property of a parallelogram $ABQP$. If lines AB and PQ are parallel and of equal length then AP and BQ are parallel and of equal length. More specifically, for directed line segments as shown in Figure 12.1.2, we have

$$\overrightarrow{AB} \equiv \overrightarrow{PQ} \;\Leftrightarrow\; \overrightarrow{AP} \equiv \overrightarrow{BQ}.$$

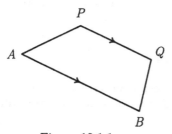
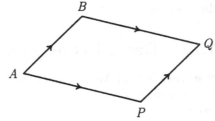

Figure 12.1.1   Figure 12.1.2

Consider vectors $\mathbf{u}$ and $\mathbf{v}$. We construct line segments to represent these vectors as in Figure 12.1.3. Take a starting point A and choose B so that $\overrightarrow{AB}$ represents $\mathbf{u}$, and then C so that $\overrightarrow{BC}$ represents $\mathbf{v}$. Now take a second starting point P and choose Q so that $\overrightarrow{PQ}$ represents $\mathbf{u}$, and then R so that $\overrightarrow{QR}$ represents $\mathbf{v}$.

Since $\overrightarrow{AB} \equiv \overrightarrow{PQ}$ and $\overrightarrow{BC} \equiv \overrightarrow{QR}$ it follows from the above property of parallelograms that $\overrightarrow{AP} \equiv \overrightarrow{BQ} \equiv \overrightarrow{CR}$. Thus, using the property once more, $\overrightarrow{AC} \equiv \overrightarrow{PR}$. This shows that $\overrightarrow{AC}$ and $\overrightarrow{PR}$ represent the same vector and hence that, if two sides of a triangle represent given vectors, then the vector represented by the third side is independent of the choice of this representation.

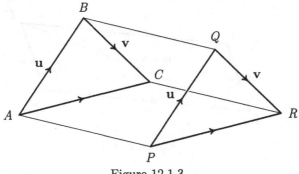

Figure 12.1.3

Definition 12.1.5 Let **u** and **v** be vectors and let $\overrightarrow{AB}$ and $\overrightarrow{BC}$ represent **u** and **v**, respectively. The *sum* of **u** and **v**, written **u** + **v**, is the vector represented by $\overrightarrow{AC}$ (see Figure 12.1.4). By Note 12.1.4 this definition is independent of the choice of the A.

The *difference* of vectors **u** and **v** is **u** − **v** = **u** + (−**v**), *i.e.* the sum of **u** and −**v**.

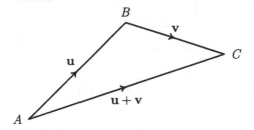

Figure 12.1.4

Theorem 12.1.6 *Let* **u**, **v** *and* **w** *be vectors*

   (1) **u** + **v** = **v** + **u**,

   (2) **u** + (**v** + **w**) = (**u** + **v**) + **w**,

   (3) **u** + **0** = **u** = **0** + **u**,

   (4) **u** + (−**u**) = **u** − **u** = **0**,

   (5) −(**u** + **v**) = −**u** − **v**.

Proof (1) Let vectors **u** and **v** be represented by the sides of the parallelogram as show in Figure 12.1.5. From $\triangle ABC$, $\overrightarrow{AC}$ represents **u** + **v** whereas from $\triangle ADC$, $\overrightarrow{AC}$ represents **v** + **u**. Hence **u** + **v** = **v** + **u**.

(2) As in Figure 12.1.6, let $\overrightarrow{AB}$, $\overrightarrow{BC}$ and $\overrightarrow{CD}$ represent vectors **u**, **v** and **w**, respectively. It follows that $\overrightarrow{AC}$ and $\overrightarrow{BD}$ represent **u** + **v** and **v** + **w**, respectively. Considering $\triangle ABD$ and $\triangle ACD$ we see that $\overrightarrow{AD}$ represents both **u** + (**v** + **w**) and (**u** + **v**) + **w**. Hence **u** + (**v** + **w**) = (**u** + **v**) + **w**.

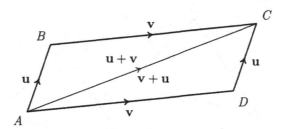

Figure 12.1.5

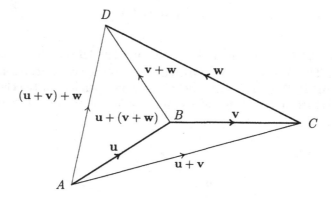

Figure 12.1.6

(3) In the definition of vector addition, identify the points B and C so that $\overrightarrow{AB}$ represents $\mathbf{u}$, $\overrightarrow{BB}$ represents $\mathbf{0}$ and $\overrightarrow{AB}$ represents $\mathbf{u} + \mathbf{0}$. Hence $\mathbf{u} + \mathbf{0} = \mathbf{u}$. The second equality follows from part (1).

(4) In the definition of vector addition identify points A and C so that $\overrightarrow{AB}$ represents $\mathbf{u}$, $\overrightarrow{BC}$ represents $-\mathbf{u}$ and $\overrightarrow{AC}$ represents $\mathbf{u} + (-\mathbf{u})$. As $\overrightarrow{AC}$ represents $\mathbf{0}$ also, the result follows.

(5) Let $\overrightarrow{AB}$ and $\overrightarrow{BC}$ represent $\mathbf{u}$ and $\mathbf{v}$ respectively so that $\overrightarrow{AC}$ represents $\mathbf{u} + \mathbf{v}$. Now $\overrightarrow{BA}$ represents $-\mathbf{u}$ and $\overrightarrow{CB}$ represents $-\mathbf{v}$ so that $\overrightarrow{CA}$ represents $-\mathbf{u} + (-\mathbf{v}) = -\mathbf{u} - \mathbf{v}$. Hence $\overrightarrow{CA}$ represents both $-(\mathbf{u} + \mathbf{v})$ and $-\mathbf{u} - \mathbf{v}$ and the result follows.   □

Remark The associativity property, Theorem 12.1.6 (2), allows us to write $\mathbf{a} + \mathbf{b} + \mathbf{c}$, and longer sums, without ambiguity.

Definitions 12.1.7 For $\alpha \in \mathbb{R}$, we define

$$\mathrm{sgn}(\alpha) = \begin{cases} -1 & \text{if } \alpha < 0, \\ 0 & \text{if } \alpha = 0, \\ 1 & \text{if } \alpha > 0. \end{cases}$$

For all $\alpha \in \mathbb{R}$, $|\alpha| = \mathrm{sgn}(\alpha)\alpha$.

Let **u** be a non-zero vector and let $\alpha \in \mathbb{R}$. We define a *scalar multiple* of **u** as follows, depending on the sign of α. We define $0\mathbf{u} = \mathbf{0}$ and otherwise, if $\alpha \neq 0$, $\alpha\mathbf{u}$ is the vector with magnitude $|\alpha|\|\mathbf{u}\|$ and the same [opposite] direction as **u** when $\alpha > 0$ [$\alpha < 0$]. For any $\alpha \in \mathbb{R}$, we define $\alpha\mathbf{0} = \mathbf{0}$.

See Figure 12.1.7 for some examples. If **u** is a vector and $\alpha \in \mathbb{R}$ is non-zero, we will sometimes write $\mathbf{u}/\alpha$ as shorthand for $(1/\alpha)\mathbf{u}$.

We say that vectors **u** and **v** are *parallel* if there exists $\alpha \in \mathbb{R}$ such that $\mathbf{u} = \alpha\mathbf{v}$ or $\mathbf{v} = \alpha\mathbf{u}$. If $\alpha > 0$ ($\alpha < 0$) then **u** and **v** have the *same* (*opposite*) direction. For any vector **u**, $0\mathbf{u} = \mathbf{0}$ and so **0** is parallel to *any* vector.

Hazard The zero vector is parallel to any vector.

If **u** and **v** are parallel then we sometimes write $\mathbf{u} \parallel \mathbf{v}$.

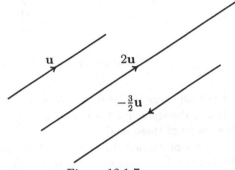

u   **2u**

$-\frac{3}{2}\mathbf{u}$

Figure 12.1.7

Theorem 12.1.8 *Let* **u** *and* **v** *be vectors and let* $\alpha, \beta \in \mathbb{R}$. *We have*

(1) $1\mathbf{u} = \mathbf{u}$ and $(-1)\mathbf{u} = -\mathbf{u}$,

(2) $0\mathbf{u} = \alpha\mathbf{0} = \mathbf{0}$,

(3) $\alpha(\beta\mathbf{u}) = (\alpha\beta)\mathbf{u}$.

(4) $(\alpha + \beta)\mathbf{u} = \alpha\mathbf{u} + \beta\mathbf{u}$,

(5) $\alpha(\mathbf{u} + \mathbf{v}) = \alpha\mathbf{u} + \alpha\mathbf{v}$,

Proof Property (2) is a restatement of part of the definition. The proofs of the other properties entail showing that the vectors on each side of the equations have the same magnitude and direction.

(1) From Definition 12.1.7, $1\mathbf{u}$ is the vector with magnitude $1 \times |\mathbf{u}| = |\mathbf{u}|$ and the same direction as **u**. As vectors $1\mathbf{u}$ and **u** have the same magnitude and direction they are equal. The proof of the other equality is similar and is left as an exercise to the reader.

(3) We take $\alpha, \beta \neq 0$ and $\mathbf{u} \neq \mathbf{0}$, since otherwise the equation is trivially satisfied. For the magnitude we have, using Definition 12.1.7 and Theorem 1.4.7,

$$|\alpha(\beta\mathbf{u})| = |\alpha||\beta\mathbf{u}| = |\alpha||\beta||\mathbf{u}| = |\alpha\beta||\mathbf{u}| = |(\alpha\beta)\mathbf{u}|.$$

The direction of $\beta\mathbf{u}$ is that of $\text{sgn}(\beta)\mathbf{u}$ and hence the direction of $\alpha(\beta\mathbf{u})$ is that of

$$\text{sgn}(\alpha)(\text{sgn}(\beta)\mathbf{u}) = \begin{cases} \mathbf{u} & \text{if } \alpha \text{ and } \beta \text{ have the same sign,} \\ -\mathbf{u} & \text{if } \alpha \text{ and } \beta \text{ have opposite sign} \end{cases}$$

(recall that $-(-\mathbf{u}) = \mathbf{u}$). The sign of $\alpha\beta$ is positive if α and β have the same sign and negative otherwise and so $(\alpha\beta)\mathbf{u}$ has the same direction as $\alpha(\beta\mathbf{u})$ in all cases.

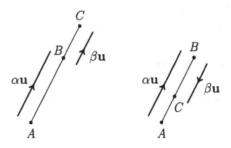

Figure 12.1.8

(4) The result is trivially true if $\alpha = 0$, $\beta = 0$ or $\mathbf{u} = \mathbf{0}$. If $\alpha + \beta = 0$ then the left hand side is $0\mathbf{u} = \mathbf{0}$ and the right hand side is $\alpha\mathbf{u} - \alpha\mathbf{u} = \mathbf{0}$, by Theorem 12.1.6 (4). We now suppose that none of these hold.

Let $\overrightarrow{AB}$ and $\overrightarrow{BC}$ represent $\alpha\mathbf{u}$ and $\beta\mathbf{u}$, respectively, so that $\overrightarrow{AC}$ represents $\alpha\mathbf{u}+\beta\mathbf{u}$. A, B and C are collinear (*i.e.* lie on a line) since $\alpha\mathbf{u}$ and $\beta\mathbf{u}$ are parallel. As the equation is symmetric in α and β we may suppose, without loss of generality, that $|\alpha| \geq |\beta|$, and so α and $\alpha + \beta$ have the same sign. Consequently the vector $(\alpha + \beta)\mathbf{u}$ has the same direction as $\alpha\mathbf{u}$.

With the above restriction there are just two possibilities as illustrated in the Figure 12.1.8. In each case it is clear that the vectors $\alpha\mathbf{u} + \beta\mathbf{u}$ and $\alpha\mathbf{u}$ have the same direction and hence the vectors on each side of the equality have the same direction.

The magnitude of $\alpha\mathbf{u} + \beta\mathbf{u}$ is the length of the line AC and we have

Case 1: $\text{sgn}(\alpha) = \text{sgn}(\beta)$.

$$|AC| = |AB| + |BC| = |\alpha||\mathbf{u}| + |\beta||\mathbf{u}| = \text{sgn}(\alpha)(\alpha + \beta)|\mathbf{u}|,$$

Case 2: $\text{sgn}(\alpha) \neq \text{sgn}(\beta)$.

$$|AC| = |AB| - |BC| = |\alpha||\mathbf{u}| - |\beta||\mathbf{u}| = \text{sgn}(\alpha)(\alpha + \beta)|\mathbf{u}|.$$

On the other hand,

$$|(\alpha + \beta)\mathbf{u}| = |\alpha + \beta||\mathbf{u}| = \text{sgn}(\alpha + \beta)(\alpha + \beta)|\mathbf{u}| = \text{sgn}(\alpha)(\alpha + \beta)|\mathbf{u}|,$$

and so in each case $|\alpha\mathbf{u} + \beta\mathbf{u}| = |(\alpha + \beta)\mathbf{u}|$.

(5) Again we only need deal with the nontrivial case in which $\alpha \neq 0$ and $\mathbf{u}$, $\mathbf{v}$ and $\mathbf{u} + \mathbf{v}$ are non-zero.

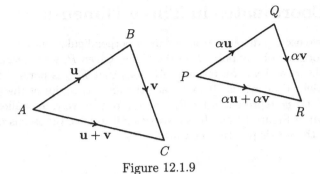

Figure 12.1.9

Case 1: $\alpha > 0$. Figure 12.1.9 shows similar triangles with directed line segments $\overrightarrow{AB}$, $\overrightarrow{BC}$ and $\overrightarrow{AC}$ representing $\mathbf{u}$, $\mathbf{v}$ and $\mathbf{u} + \mathbf{v}$, and $\overrightarrow{PQ}$, $\overrightarrow{QR}$ and $\overrightarrow{PR}$ representing $\alpha\mathbf{u}$, $\alpha\mathbf{v}$ and $\alpha\mathbf{u} + \alpha\mathbf{v}$, respectively. The ratio between the lengths of corresponding sides is

$$\frac{|PQ|}{|AB|} = \frac{|QR|}{|BC|} = |\alpha|$$

from which it is clear that

$$|PR| = |\alpha||AC|, \quad i.e. \ |\alpha\mathbf{u} + \alpha\mathbf{v}| = |\alpha||\mathbf{u} + \mathbf{v}| = |\alpha(\mathbf{u} + \mathbf{v})|.$$

From the similarity of $\triangle ABC$ and $\triangle PQR$, $\mathbf{u} + \mathbf{v}$ and $\alpha\mathbf{u} + \alpha\mathbf{v}$ are parallel and have the same sense.

Case 2: $\alpha = -\beta < 0$. We have

$$\begin{aligned}
\alpha(\mathbf{u} + \mathbf{v}) &= -\beta(\mathbf{u} + \mathbf{v}) \\
&= \beta(-(\mathbf{u} + \mathbf{v})) && \text{(using parts (1) and (3))} \\
&= \beta(-\mathbf{u} - \mathbf{v}) && \text{(using Theorem 12.1.6 (5))} \\
&= \beta(-\mathbf{u}) + \beta(-\mathbf{v}) && \text{(using Case 1, since } \beta > 0) \\
&= \alpha\mathbf{u} + \alpha\mathbf{v} && \text{(using parts (1) and (3) again).} \quad \square
\end{aligned}$$

Example 12.1.9 Let $\mathbf{u}$ and $\mathbf{v}$ be vectors and $a, b, c \in \mathbb{R}$. Write in simplest terms the vector

$$a(b\mathbf{u} - \mathbf{v}) + (b + c)\mathbf{v} + \mathbf{u}.$$

At each stage in the simplification, indicate the properties of addition and scalar multiplication that are used.

Solution Let I stand for Theorem 12.1.6 and II for Theorem 12.1.8. Then

$$\begin{aligned}
& a(b\mathbf{u} - \mathbf{v}) + (b + c)\mathbf{v} + \mathbf{u} \\
&= a(b\mathbf{u} + (-\mathbf{v})) + (b + c)\mathbf{v} + \mathbf{u} \\
&= a(b\mathbf{u}) + a(-\mathbf{v}) + (b + c)\mathbf{v} + \mathbf{u} \quad (\text{II} (5)) \\
&= (ab)\mathbf{u} - a\mathbf{v} + (b + c)\mathbf{v} + \mathbf{u} \quad\quad (\text{II} (3)) \\
&= (ab + 1)\mathbf{u} + (b + c - a)\mathbf{v} \quad\quad\quad (\text{I} (1), \text{II} (1),(4)). \quad \square
\end{aligned}$$

12.2   Coordinates in Three Dimensions

Recall that we use the distance from a pair of perpendicular lines, the x- and y-axes, to specify the location of any point P in the plane. Given P, we determine the points P_x and P_y on the x- and y-axes, respectively, such that PP_x is parallel to the y-axis and PP_y is parallel to the x-axis. If we regard the axes as copies of the real line, the real numbers x_P and y_P represented by P_x and P_y, respectively, are called the *coordinates of P* as shown in Figure 12.2.1. It is conventional to have the x-axis pointing from left to right and the y-axis pointing upwards.

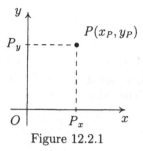

Figure 12.2.1

We extend this idea to (three dimensional) space by introducing a third axis, the z-axis, which is perpendicular to the x- and y-axes at the origin O. There are two possible choices for the direction of this axis; for the x- and y-axes as shown in Figure 12.2.1, the z-axis could either be directed *out of* the page (as shown in Figure 12.2.2) or *into* the page. The former choice gives a *right-handed system*, and the latter a *left-handed system* of coordinates. We will always use a right-handed system.

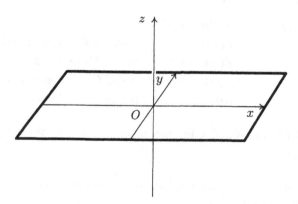

Figure 12.2.2

The plane containing the x- and y-axes is called the x, y-plane and we define the y, z- and z, x-planes similarly. See Figure 12.2.3

Take any point P in space and let P_{xy} be the point on the x, y-plane such that PP_{xy} is parallel to the z-axis. As described above, P_{xy} gives points P_x and P_y on the x- and y-axes respectively, which determine the (planar) coordinates x_P and y_P of P_{xy}. In a similar way we obtain P_{yz} and P_{zx} on the y, z- and z, x-planes, respectively, and

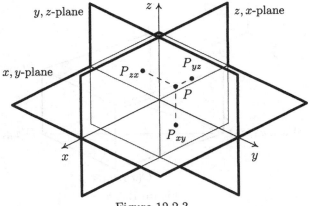

Figure 12.2.3

then P_z on the z-axis such that $P_{yz}P_z$ is parallel to the y-axis and $P_{zx}P_z$ is parallel to the x-axis. As illustrated in Figure 12.2.4, the eight points O, P, P_{xy}, P_{yz}, P_{zx}, P_x, P_y and P_z are the vertices of a cuboid, called a *coordinate brick*. The *coordinates* of P are the real numbers x_P, y_P and z_P represented by P_x, P_y and P_z, respectively.

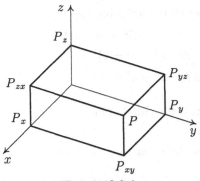

Figure 12.2.4

Theorem 12.2.1 *The distance between $A(x_A, y_A, z_A)$ and $B(x_B, y_B, z_B)$ is*

$$|AB| = \sqrt{(x_A - x_B)^2 + (y_A - y_B)^2 + (z_A - z_B)^2}$$

Proof See Figure 12.2.5.

By Pythagoras's Theorem in the right-angled triangles ABD and ACD we have

$$|AB|^2 = |AD|^2 + |BD|^2 = \left(|AC|^2 + |CD|^2\right) + |BD|^2$$
$$= (y_A - y_B)^2 + (x_A - x_B)^2 + (z_A - z_B)^2.$$

Since $|AB|$ is a distance it is non-negative, and so the result follows upon taking positive square roots. $\qquad\square$

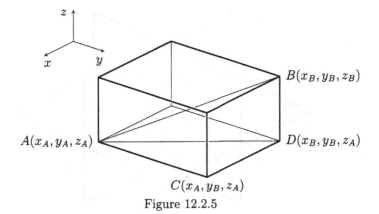

Figure 12.2.5

Corollary 12.2.2 *The distance of $P(x, y, z)$ from O is*

$$|OP| = \sqrt{x^2 + y^2 + z^2}.$$

Proof Taking $A = O$ and $B = P$ in Theorem 12.2.1 gives the result immediately. □

Example 12.2.3 Describe the following loci.

    (a) $S = \{(x, y, z) \in \mathbb{R}^3 : (x - a)^2 + (y - b)^2 + (z - c)^2 = r^2, \ (a, b, c \in \mathbb{R}, r \geq 0)\}$,

    (b) $C = \{(x, y, z) \in \mathbb{R}^3 : x^2 + y^2 = 1\}$,

    (c) $K = \{(x, y, z) \in \mathbb{R}^3 : x^2 + y^2 = z^2\}$.

Remark The word loci is the plural of locus, a collection of points which share a given property.

Solution　(a) Let A be the point (a, b, c). Then

$$\begin{aligned} P(x, y, z) \in S \ &\Leftrightarrow \ (x - a)^2 + (y - b)^2 + (z - c)^2 = r^2 \\ &\Leftrightarrow \ |AP|^2 = r^2 \\ &\Leftrightarrow \ |AP| = r, \text{ since } |AP| \geq 0 \text{ and } r \geq 0. \end{aligned}$$

Thus S is the set of points at a distance r from A. If $r = 0$ then S contains the single point A. If $r > 0$ then S is the surface of the sphere of radius r and centre A.

    (b) We first fix $z = c \in \mathbb{R}$,

$$\begin{aligned} P(x, y, z) \in C \ &\Leftrightarrow \ x^2 + y^2 = 1 \\ &\Leftrightarrow \ P \text{ lies on the unit circle centre } (0, 0, c) \text{ in plane } z = c. \end{aligned}$$

So, for each $c \in \mathbb{R}$, the cross-section through C by the plane $z = c$ is a unit circle and so C is surface of the (infinite) circular cylinder of radius 1 and *axis* the z-axis (see Figure 12.2.6).

    (c) If $z = 0$ then $P(x, y, z) \in K \ \Leftrightarrow \ x = y = 0$ while if $z = c \neq 0$ then

$$P(x, y, z) \in K \ \Leftrightarrow \ P \text{ lies on the circle centre } (0, 0, c), \text{ radius } |c| \text{ in plane } z = c.$$

Further, if $x = 0$ then

$$P(x, y, z) \in K \quad \Leftrightarrow \quad z = \pm y,$$

and so the cross-section of K with the y, z-plane is a pair of straight lines. Thus K is the surface of the circular cone with *vertex* O and *axis* the z-axis (see Figure 12.2.7). $\quad\square$

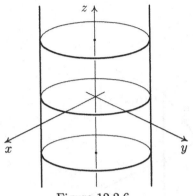

Figure 12.2.6

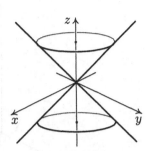

Figure 12.2.7

12.3 The Component Form of a Vector

Definition 12.3.1 A vector **u** is called a *unit vector* if $|\mathbf{u}| = 1$. The points $X(1, 0, 0)$, $Y(0, 1, 0)$ and $Z(0, 0, 1)$ are such that $|OX| = |OY| = |OZ| = 1$ and so $\overrightarrow{OX}$, $\overrightarrow{OY}$ and $\overrightarrow{OZ}$, represent unit vectors denoted **i**, **j** and **k**, respectively. These are called the (*standard*) *basis vectors* (see Figure 12.3.1).

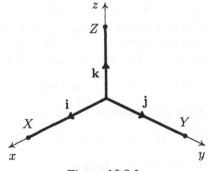

Figure 12.3.1

Theorem 12.3.2 *Every vector* **u** *may be expressed as a unique linear combination of basis vectors, i.e. we may write*

$$\mathbf{u} = \alpha\mathbf{i} + \beta\mathbf{j} + \gamma\mathbf{k},$$

for a unique choice of $\alpha, \beta, \gamma \in \mathbb{R}$.

Proof (see Figure 12.3.2). The point $U(\alpha, \beta, \gamma)$ such that $\overrightarrow{OU}$ represents **u** and the points $A(\alpha, 0, 0)$ and $B(\alpha, \beta, 0)$ are also shown. We will show that $\overrightarrow{OU}$ also represents $\alpha\mathbf{i} + \beta\mathbf{j} + \gamma\mathbf{k}$.

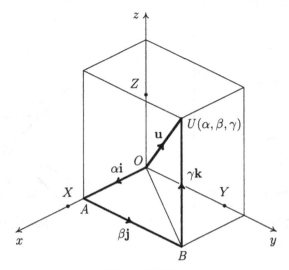

Figure 12.3.2

The points X, Y and Z are those appearing in the definition of the basis vectors.

Since OA is parallel to OX, $\overrightarrow{OA}$ represents a scalar multiple of **i**. The length of OA is $|\alpha|$ and $\overrightarrow{OA}$ has the direction of **i** if $\alpha > 0$, or of $-\mathbf{i}$ if $\alpha < 0$, *i.e.* the direction of $\operatorname{sgn}(\alpha)\mathbf{i}$. This implies that $\overrightarrow{OA}$ represents $\alpha\mathbf{i}$. In a similar way $\overrightarrow{AB}$ and $\overrightarrow{BU}$ represent $\beta\mathbf{j}$ and $\gamma\mathbf{k}$, respectively. From $\triangle OAB$ and $\triangle OBU$ we see that $\overrightarrow{OB}$ represents $\alpha\mathbf{i} + \beta\mathbf{j}$ and thus $\overrightarrow{OU}$ represents $(\alpha\mathbf{i} + \beta\mathbf{j}) + \gamma\mathbf{k} = \alpha\mathbf{i} + \beta\mathbf{j} + \gamma\mathbf{k}$.

To obtain a directed line segment of length 0 it is necessary and sufficient that $\alpha = \beta = \gamma = 0$. Thus the zero vector has a unique expression $\mathbf{0} = 0\mathbf{i} + 0\mathbf{j} + 0\mathbf{k}$. If a vector $\mathbf{u} = \alpha\mathbf{i} + \beta\mathbf{j} + \gamma\mathbf{k}$ may be also written as $\mathbf{u} = \alpha'\mathbf{i} + \beta'\mathbf{j} + \gamma'\mathbf{k}$ then

$$\mathbf{0} = \mathbf{u} + (-\mathbf{u}) = \mathbf{u} - \mathbf{u} = (\alpha - \alpha')\mathbf{i} + (\beta - \beta')\mathbf{j} + (\gamma - \gamma')\mathbf{k}.$$

Since the expression for the zero vector is unique so

$$\alpha' = \alpha, \ \ \beta' = \beta \ \text{and} \ \gamma' = \gamma,$$

and so the expression for *any* vector is unique.   □

Definition 12.3.3 For a vector $\mathbf{u} = u_x\mathbf{i} + u_y\mathbf{j} + u_z\mathbf{k}$ we call the ordered triple (u_x, u_y, u_z) the *components* of **u** and refer to u_x as the *x-component* or the *first component* of **u** and similarly for u_y and u_z. We will often use the notation $\mathbf{u} = (u_x, u_y, u_z)$ instead of the expression for **u** as a linear combination of unit vectors.

Corollary 12.3.4 *For the point* $U(\alpha, \beta, \gamma)$, $\overrightarrow{OU}$ *represents the vector* $\mathbf{u} = \alpha\mathbf{i} + \beta\mathbf{j} + \gamma\mathbf{k} = (\alpha, \beta, \gamma)$ *and* $|\mathbf{u}| = \sqrt{\alpha^2 + \beta^2 + \gamma^2}$.

Proof That $\overrightarrow{OU}$ represents $\mathbf{u}$ is evident from the proof of Theorem 12.3.2 and, by Corollary 12.2.2, $|\mathbf{u}| = |OU| = \sqrt{\alpha^2 + \beta^2 + \gamma^2}$. □

Corollary 12.3.5 *Given vectors* $\mathbf{u} = (u_1, u_2, u_3)$, $\mathbf{v} = (v_1, v_2, v_3)$ *and* $\alpha \in \mathbb{R}$ *we have*

(1) $\mathbf{u} = \mathbf{v} \Leftrightarrow u_k = v_k$ *for* $k = 1, 2, 3$,

(2) $\mathbf{u} + \mathbf{v} = (u_1 + v_1, u_2 + v_2, u_3 + v_3)$,

(3) $\alpha\mathbf{u} = (\alpha u_1, \alpha u_2, \alpha u_3)$.

Proof (1) This result follows immediately from the uniqueness of representation proved in Theorem 12.3.2.

(2) Using the properties of vector addition given in Theorem 12.1.6 we have

$$\mathbf{u} + \mathbf{v} = (u_1\mathbf{i} + u_2\mathbf{j} + u_3\mathbf{k}) + (v_1\mathbf{i} + v_2\mathbf{j} + v_3\mathbf{k})$$
$$= (u_1 + v_1)\mathbf{i} + (u_2 + v_2)\mathbf{j} + (u_3 + v_3)\mathbf{k}$$
$$= (u_1 + v_1, u_2 + v_2, u_3 + v_3)$$

(3) This is similar to (2) and is left as an exercise to the reader. □

Example 12.3.6 Find the unit vectors that have

(a) the same direction as, and

(b) opposite direction to

the vector $\mathbf{u} = (3, -1, 2\sqrt{2})$.

Solution Vectors parallel to $\mathbf{u}$ have the form $\hat{\mathbf{u}} = \alpha\mathbf{u}$ for some non-zero $\alpha \in \mathbb{R}$ where $\alpha > 0$ [$\alpha < 0$] gives a vector with the same direction as [opposite direction to] $\mathbf{u}$. For a unit vector we require that $|\hat{\mathbf{u}}| = 1$. Now

$$|\hat{\mathbf{u}}| = |\alpha||(3, -1, 2\sqrt{2})| = |\alpha|\sqrt{3^2 + (-1)^2 + (2\sqrt{2})^2} = 3\sqrt{2}|\alpha|,$$

and so $|\hat{\mathbf{u}}| = 1 \Rightarrow |\alpha| = 1/(3\sqrt{2})$ and so

(a) Taking $\alpha = 1/(3\sqrt{2})$ we get $\hat{\mathbf{u}} = (1/\sqrt{2}, -1/(3\sqrt{2}), 2/3)$ with the same direction as $\mathbf{u}$.

(b) Taking $\alpha = -1/(3\sqrt{2})$ we get $\hat{\mathbf{u}} = (-1/\sqrt{2}, 1/(3\sqrt{2}), -2/3)$ with opposite direction to $\mathbf{u}$. □

Definition 12.3.7 For a point $P(x, y, z)$ the vector $\mathbf{p} = x\mathbf{i} + y\mathbf{j} + z\mathbf{k} = (x, y, z)$ is called the *position vector* of P and is represented by $\overrightarrow{OP}$. We will often use the notation $\mathbf{p}$, $\mathbf{q}$ etc. for the position vectors of P, Q etc., respectively, without further explicit definition.

Theorem 12.3.8 *For any points P and Q the directed line segment $\overrightarrow{PQ}$ represents the vector* $\mathbf{q} - \mathbf{p}$.

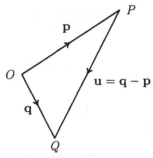

Figure 12.3.3

Proof Let $\overrightarrow{PQ}$ represent **u** and consider $\triangle OPQ$ as shown in Figure 12.3.3. From Definition 12.1.5 we have $\mathbf{q} = \mathbf{p} + \mathbf{u}$ and subtracting **p** from both sides give the required result.   □

Example 12.3.9 Consider the points $P(3, 4, -3)$ and $Q(-1, 9, 0)$. Determine the vector **u** which is represented by the directed line segment $\overrightarrow{PQ}$ and the point R such that $\overrightarrow{QR}$ also represents **u**.

Solution We have $\mathbf{p} = (3, 4, -3)$ and $\mathbf{q} = (-1, 9, 0)$ and hence

$$\mathbf{u} = (-1, 9, 0) - (3, 4, -3) = (-1 - 3, 9 - 4, 0 - (-3)) = (-4, 5, 3).$$

Let R be such that $\overrightarrow{QR} = \mathbf{r} - \mathbf{q} = \mathbf{u}$. Then

$$\mathbf{r} = \mathbf{u} + \mathbf{q} = (-4, 5, 3) + (-1, 9, 0) = (-4, 14, 3).$$

Thus the required point is $R(-5, 14, 3)$.   □

12.4   The Section Formula

If distinct points A, B and C are collinear (*i.e.* lie on a line) then the directed line segments for each pair of points are parallel. If $\overrightarrow{AC}$ and $\overrightarrow{CB}$ represent **u** and $\alpha\mathbf{u}$, for some $\alpha \in \mathbb{R}$, then $\overrightarrow{CB} \equiv \alpha\overrightarrow{AC}$ and we write $AC : CB = m : n$ if $m, n \in \mathbb{R}$, with $m \neq 0$, are such that $n/m = \alpha$. For example, if $\overrightarrow{CB} \equiv -2\overrightarrow{AC}$ then we have $AC : CB = -1 : 2$ or $AC : CB = 2 : -4$ or $AC : CB = -1/2 : 1$.

Example 12.4.1 Let collinear points A, B and C be such that $\overrightarrow{BC} \equiv 2\overrightarrow{AB}$. Determine $AC : CB$.

Solution Let **u** be the vector represented by $\overrightarrow{BC}$. Then $\overrightarrow{AB}$ represents $(1/2)\mathbf{u}$ and so $\overrightarrow{AC}$ represents $(1/2)\mathbf{u} + \mathbf{u} = (3/2)\mathbf{u}$. Thus $\overrightarrow{AC} \equiv (3/2)\overrightarrow{BC} \equiv -(3/2)\overrightarrow{CB}$ and so $AC : CB = -3 : 2$.   □

Figure 12.4.1

Remark The answer to Example 12.4.1 may also be found by inspection of Figure 12.4.1. Here the numbers 1 and 2 represent the *relative* lengths of the two line segments.

Theorem 12.4.2 (Section Formula) *Let the points A, B and C be collinear with A and B distinct. If $AC : CB = m : n$ then*

$$\mathbf{c} = \frac{m\mathbf{b} + n\mathbf{a}}{m + n}. \tag{12.4.1}$$

Figure 12.4.2

Proof See Figure 12.4.2. Since A and B are distinct it follows that $\overrightarrow{AC} \not\equiv -\overrightarrow{CB}$ and thus $m + n \neq 0$. If $n = 0$, so that B and C coincide, then the result is trivially true. Henceforth we take $n \neq 0$.

We have $\overrightarrow{AC} \equiv (m/n)\overrightarrow{CB}$ and $\overrightarrow{AC}$ and $\overrightarrow{CB}$ represent the vectors $\mathbf{c} - \mathbf{a}$ and $\mathbf{b} - \mathbf{c}$, respectively. Thus

$$\mathbf{c} - \mathbf{a} = \frac{m}{n}(\mathbf{b} - \mathbf{c}).$$

Solving for $\mathbf{c}$ we get

$$(m + n)\mathbf{c} = m\mathbf{b} + n\mathbf{a},$$

and hence (12.4.1) follows since $m + n \neq 0$.   □

Corollary 12.4.3 *The mid-point C of AB has position vector*

$$\mathbf{c} = \frac{\mathbf{a} + \mathbf{b}}{2}. \tag{12.4.2}$$

Proof We use Theorem 12.4.2 with $AC : CB = 1 : 1$ so that $m = n = 1$.   □

Example 12.4.4 Consider points $A(3, 0, 2)$, $B(-1, 1, 1)$ and $C(0, 1, 2)$. Let D be the mid-point of AB and E on BC (extended) such that $BE : EC = 2 : -1$. Determine the points D and E.

Solution See Figure 12.4.3. We have $\mathbf{a} = (3, 0, 2)$, $\mathbf{b} = (-1, 1, 1)$ and $\mathbf{c} = (0, 1, 2)$. Using (12.4.2) we get

$$\mathbf{d} = \frac{\mathbf{a} + \mathbf{b}}{2} = \frac{(3, 0, 2) + (-1, 1, 1)}{2}$$
$$= (1, 1/2, 3/2).$$

Hence D is the point $(1, 1/2, 3/2)$.

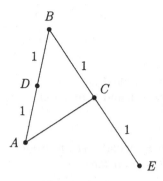

Figure 12.4.3

Using (12.4.1) we have

$$\mathbf{e} = \frac{2\mathbf{c} - \mathbf{b}}{2 - 1} = 2(0, 1, 2) - (-1, 1, 1)$$
$$= (1, 1, 3).$$

Hence E is the point $(1, 1, 3)$.   □

Example 12.4.5 Prove that $A(4, 0, 1)$, $B(2, 2, 1)$ and $C(1, 3, 1)$ are collinear and find m and n such $AB : BC = m : n$.

Solution We will show that $\overrightarrow{AB}$ and $\overrightarrow{BC}$ are parallel by proving that the vectors they represent are scalar multiples of each other. Using Theorem 12.3.8 we have

$$\overrightarrow{AB} = (2, 2, 1) - (4, 0, 1) = (-2, 2, 0)$$

and

$$\overrightarrow{BC} = (1, 3, 1) - (2, 2, 1) = (-1, 1, 0),$$

and so $\overrightarrow{AB} \equiv 2\overrightarrow{BC}$.

Hence A, B and C are collinear and $AB : BC = 2 : 1$.   □

Example 12.4.6 Let D, E and F lie on the sides BC, CA and AB, respectively, of $\triangle ABC$ so that $BD : DC = 1 : 4$, $CE : EA = 3 : 2$ and $AF : FB = 3 : 7$ and let $\overrightarrow{AD}$, $\overrightarrow{BE}$ and $\overrightarrow{CF}$ represent vectors $\mathbf{u}$, $\mathbf{v}$ and $\mathbf{w}$, respectively. Let K divide AB such that $AK : KB = 1 : 3$. Show that the vector represented by $\overrightarrow{CK}$ is parallel to $\mathbf{u} + \mathbf{v} + \mathbf{w}$ and find the ratio of the lengths of these vectors.

Solution Using (12.4.1) we have

$$\mathbf{d} = \frac{4\mathbf{b} + \mathbf{c}}{5}, \quad \mathbf{e} = \frac{2\mathbf{c} + 3\mathbf{a}}{5}, \quad \mathbf{f} = \frac{7\mathbf{a} + 3\mathbf{b}}{10} \quad \text{and} \quad \mathbf{k} = \frac{3\mathbf{a} + \mathbf{b}}{4}.$$

Therefore

$$\overrightarrow{CK} = \mathbf{k} - \mathbf{c} = \frac{3\mathbf{a} + \mathbf{b} - 4\mathbf{c}}{4}.$$

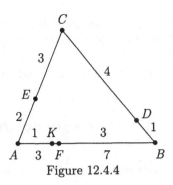

Figure 12.4.4

Also,

$$\mathbf{u} + \mathbf{v} + \mathbf{w} = (\mathbf{d} - \mathbf{a}) + (\mathbf{e} - \mathbf{b}) + (\mathbf{f} - \mathbf{c})$$

$$= \frac{(8\mathbf{b} + 2\mathbf{c} - 10\mathbf{a}) + (4\mathbf{c} + 6\mathbf{a} - 10\mathbf{b}) + (7\mathbf{a} + 3\mathbf{b} - 10\mathbf{c})}{10}$$

$$= \frac{(3\mathbf{a} + \mathbf{b} - 4\mathbf{c})}{10}$$

$$= \frac{2}{5}(\mathbf{k} - \mathbf{c}).$$

Thus $\mathbf{k} - \mathbf{c}$ and $\mathbf{u} + \mathbf{v} + \mathbf{w}$ are parallel and

$$\frac{|\mathbf{u} + \mathbf{v} + \mathbf{w}|}{|CK|} = \frac{2}{5}. \qquad \square$$

Definitions 12.4.7 A *median* of a triangle is a straight line joining one vertex to the mid-point of the opposite side.

Three or more lines are *concurrent* if they have a common point of intersection.

Theorem 12.4.8 *The medians of $\triangle ABC$ are concurrent at the point G with position vector*

$$\mathbf{g} = \frac{1}{3}(\mathbf{a} + \mathbf{b} + \mathbf{c}).$$

G is called the centroid *of $\triangle ABC$.*

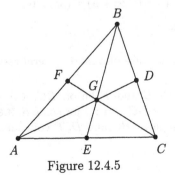

Figure 12.4.5

Proof Let D, E and F be the mid-points of BC, CA and AB, respectively, so that the lines AD, BE and CF are the medians of the triangle. We have

$$\mathbf{d} = \frac{1}{2}(\mathbf{b} + \mathbf{c}), \quad \mathbf{e} = \frac{1}{2}(\mathbf{c} + \mathbf{a}) \quad \text{and} \quad \mathbf{f} = \frac{1}{2}(\mathbf{a} + \mathbf{b}).$$

Consequently,

$$\overrightarrow{AD} = \mathbf{d} - \mathbf{a} = \frac{1}{2}(\mathbf{b} + \mathbf{c} - 2\mathbf{a})$$

and

$$\overrightarrow{AG} = \mathbf{g} - \mathbf{a} = \frac{1}{3}(\mathbf{b} + \mathbf{c} - 2\mathbf{a}).$$

Thus $\overrightarrow{AD}$ and $\overrightarrow{AG}$ are parallel and hence A, G and D are collinear. This means that G lies on the line AD.

Using the symmetry of the position vector of G with respect the $\mathbf{a}$, $\mathbf{b}$ and $\mathbf{c}$ we conclude that G also lies on the lines BE and CF. Hence the medians of the triangle are concurrent at G. □

Corollary 12.4.9 *The centroid of a triangle cuts each median in the ratio $2 : 1$ with the longer section adjoining the vertex.*

Proof From the proof of Theorem 12.4.8, using the notation established there, $\overrightarrow{AG} \equiv (2/3)\overrightarrow{AD}$. Hence $\overrightarrow{AG} \equiv 2\overrightarrow{GD}$ and similarly for each median. □

Example 12.4.10 Let D, E and F lie on BC, CA and AB, respectively, so that $AF : FB = CE : EA = BD : DC = 1 : 3$. Show that the centroids of triangles ABC and DEF coincide.

Solution Let G_1 and G_2 be the centroids of triangles ABC and DEF, respectively. Thus

$$\mathbf{g}_1 = \frac{1}{3}(\mathbf{a} + \mathbf{b} + \mathbf{c}) \quad \text{and} \quad \mathbf{g}_2 = \frac{1}{3}(\mathbf{d} + \mathbf{e} + \mathbf{f}).$$

Also,

$$\mathbf{d} = \frac{3\mathbf{b} + \mathbf{c}}{4}, \quad \mathbf{e} = \frac{3\mathbf{c} + \mathbf{a}}{4} \quad \text{and} \quad \mathbf{f} = \frac{3\mathbf{a} + \mathbf{b}}{4},$$

giving

$$\mathbf{g}_2 = \frac{1}{12}\left((3\mathbf{b} + \mathbf{c}) + (3\mathbf{c} + \mathbf{a}) + (3\mathbf{a} + \mathbf{b})\right) = \frac{1}{3}(\mathbf{a} + \mathbf{b} + \mathbf{c}) = \mathbf{g}_1.$$

Thus G_1 and G_2 coincide. □

This example is a special case of a more general result.

Problem 12.4.11 Let D, E and F lie on BC, CA and AB, respectively, so that $AF : FB = CE : EA = BD : DC = \alpha : \beta$ for some $\alpha, \beta \in \mathbb{R}$ such that $\alpha + \beta \neq 0$. Show that the centroids of triangles ABC and DEF coincide.

12.5   Lines in Three Dimensional Space

Given distinct points A and B, the set of points collinear with A and B form a line $\mathcal{L}$. A general point P on $\mathcal{L}$ will, by the section formula (12.4.1), have position vector

$$\mathbf{p} = \frac{\alpha\mathbf{a} + \beta\mathbf{b}}{\alpha + \beta}, \tag{12.5.1}$$

for some $\alpha, \beta \in \mathbb{R}$, where $\alpha + \beta \neq 0$. As a consequence of this we get the vector equation for a line.

Theorem 12.5.1 *The line $\mathcal{L}$ through points A and B has equation*

$$\mathbf{p} = (1 - t)\mathbf{a} + t\mathbf{b} \qquad (t \in \mathbb{R}). \tag{12.5.2}$$

Proof We consider (12.5.1) and define $t = \beta/(\alpha+\beta)$. It follows that $1-t = \alpha/(\alpha+\beta)$ and (12.5.2) is obtained. Since α and β are arbitrarily, t may take any real value. $\square$

Definition 12.5.2 For any pair of distinct points P and Q on a line $\mathcal{L}$, the vector $\mathbf{u}$ represented by $\overrightarrow{PQ}$ is called a *direction vector* for $\mathcal{L}$. If $\mathbf{u}$ is a direction vector for $\mathcal{L}$ then so is $\alpha\mathbf{u}$ for any $\alpha \neq 0$.

Corollary 12.5.3 *The line $\mathcal{L}$ through a point A with direction vector $\mathbf{u}$ has equation*

$$\mathbf{p} = \mathbf{a} + t\mathbf{u} \qquad (t \in \mathbb{R}). \tag{12.5.3}$$

Proof Let B be the point on $\mathcal{L}$ such that $\overrightarrow{AB}$ represents the direction vector $\mathbf{u}$. Thus $\mathbf{u} = \mathbf{b} - \mathbf{a}$ and (12.5.2) gives

$$\mathbf{p} = \mathbf{a} + t(\mathbf{b} - \mathbf{a}) = \mathbf{a} + t\mathbf{u}.$$ $\square$

Example 12.5.4 Find the equation of the line $\mathcal{L}$ through points $A(1, -2, 3)$ and $B(1, -1, 2)$. Determine also the points on $\mathcal{L}$ that are $2\sqrt{2}$ units from A.

Solution The equation of $\mathcal{L}$ is

$$\begin{aligned}
\mathbf{p} &= (1 - t)(1, -2, 3) + t(1, -1, 2) \\
&= (1, -2, 3) + t(0, 1, -1).
\end{aligned}$$

For a general point P, the distance of P from A is the magnitude of the vector $\mathbf{p} - \mathbf{a} = t(0, 1, -1)$, and so

$$|AP| = |t(0, 1, -1)| = |t|\sqrt{0^2 + 1^2 + (-1)^2} = |t|\sqrt{2}.$$

Thus $|AP| = 2\sqrt{2} \Leftrightarrow t = \pm 2$ and hence the required points have position vectors

$$\begin{aligned}
(1, -2, 3) + 2(0, 1, -1) &= (1, 0, 1), \\
(1, -2, 3) - 2(0, 1, -1) &= (1, -4, 5),
\end{aligned}$$

and so the required points are $(1, 0, 1)$ and $(1, -4, 5)$. $\square$

If we take the general point P on a line $\mathcal{L}$ to have position vector (x, y, z), the given direction vector to be $\mathbf{u} = (l, m, n)$ and given point to have position vector $\mathbf{a} = (\alpha, \beta, \gamma)$, then the equation of $\mathcal{L}$ (12.5.3), may be written as

$$(x, y, z) = (\alpha, \beta, \gamma) + t(l, m, n), \qquad (t \in \mathbb{R}).$$

This equation gives the coordinates of the general point on $\mathcal{L}$ as

$$(x, y, z) = (\alpha + tl, \beta + tm, \gamma + tn), \tag{12.5.4}$$

for some $t \in \mathbb{R}$. By writing the components of this equation separately we obtain equivalent forms for the equations of $\mathcal{L}$.

Parametric equations

$$x = \alpha + tl, \quad y = \beta + tm, \quad z = \gamma + tn \qquad (t \in \mathbb{R}).$$

If l, m and n are all non-zero we may eliminate the parameter t to obtain the

Symmetric equations

$$\frac{x - \alpha}{l} = \frac{y - \beta}{m} = \frac{z - \gamma}{n}.$$

When any of l, m or n vanishes we will still use this expression but with the understanding that a term of the form, for example,

$$\frac{x - \alpha}{0}$$

indicates that $x = \alpha + (0)t = \alpha$, *i.e.* all points on $\mathcal{L}$ have x-coordinate equal to α.

Given either the parametric or symmetric form for a line $\mathcal{L}$ we may determine a direction vector (l, m, n) and a point (α, β, γ) on $\mathcal{L}$ by inspection.

Example 12.5.5 Write in parametric and symmetric form the equations of the line $\mathcal{L}$ through $C(4, 5, -6)$ and parallel to the line through $A(1, -1, 3)$ and $B(2, -1, 2)$.

Solution As $\mathcal{L}$ is parallel to $\overrightarrow{AB}$, a direction vector for $\mathcal{L}$ is

$$\mathbf{b} - \mathbf{a} = (2, -1, 2) - (1, -1, 3) = (1, 0, -1).$$

Thus the parametric form of $\mathcal{L}$ is

$$x = 4 + t, \quad y = 5, \quad z = -6 - t \qquad (t \in \mathbb{R})$$

and the symmetric form is

$$\frac{x - 4}{1} = \frac{y - 5}{0} = \frac{z + 6}{-1}. \qquad \square$$

Distinct lines in a *plane* are either parallel and never intersect, or are non-parallel and have a unique point of intersection.

In contrast, a pair of distinct lines in *space* may be non-parallel but still have no point of intersection. If we randomly select two lines the most 'likely' occurrence is that they will be non-parallel and non-intersecting.

Example 12.5.6 Let $\mathcal{L}_1$ and $\mathcal{L}_2$ be the lines

$$\mathcal{L}_1: \quad \frac{x-1}{2} = \frac{y}{-1} = \frac{z+3}{1}, \qquad \mathcal{L}_2: \quad \frac{x-4}{1} = \frac{y+3}{1} = \frac{z+3}{2}.$$

Prove that $\mathcal{L}_1$ and $\mathcal{L}_2$ intersect and find their point of intersection.

Solution　In parametric form we have

$$\mathcal{L}_1: \quad \begin{cases} x &= 1 + 2t \\ y &= -t \\ z &= -3 + t \end{cases} \qquad \mathcal{L}_2: \quad \begin{cases} x &= 4 + s \\ y &= -3 + s \\ z &= -3 + 2s \end{cases} \qquad (s, t \in \mathbb{R}).$$

The lines $\mathcal{L}_1$ and $\mathcal{L}_2$ intersect if and only if there exist $s, t \in \mathbb{R}$ such that

$$1 + 2t = 4 + s, \tag{12.5.5}$$
$$-t = -3 + s, \tag{12.5.6}$$
$$-3 + t = -3 + 2s. \tag{12.5.7}$$

From (12.5.7) $t = 2s$ and substituting in (12.5.5) this gives $s = 1$ and hence $t = 2$. Finally, substituting these values we find that (12.5.6) is also satisfied. Thus $\mathcal{L}_1$ and $\mathcal{L}_2$ do intersect at the point on $\mathcal{L}_1$ with parameter $t = 2$. Hence the point of intersection is $(5, -2, -1)$.　　　□

Hazard　In Example 12.5.6 it is vital that we use *different* parameters s and t for the two lines.

Example 12.5.7 Determine the value of α for which the lines $\mathcal{L}_1$ and $\mathcal{L}_2$

$$\mathcal{L}_1: \quad \frac{x-\alpha}{1} = \frac{y-2}{-2} = \frac{z-1}{0}, \qquad \mathcal{L}_2: \quad \frac{x}{1} = \frac{y-3}{-1} = \frac{z+2}{2}.$$

intersect.

Solution　In parametric form we have

$$\mathcal{L}_1: \quad \begin{cases} x &= \alpha + t \\ y &= 2 - 2t \\ z &= 1 \end{cases} \qquad \mathcal{L}_2: \quad \begin{cases} x &= s \\ y &= 3 - s \\ z &= -2 + 2s \end{cases} \qquad (s, t \in \mathbb{R}).$$

$\mathcal{L}_1$ and $\mathcal{L}_2$ intersect if and only if there exist $s, t \in \mathbb{R}$ such that

$$\alpha + t = s, \tag{12.5.8}$$
$$2 - 2t = 3 - s, \tag{12.5.9}$$
$$1 = -2 + 2s. \tag{12.5.10}$$

From (12.5.10) we have $s = 3/2$ and substituting this value into (12.5.8) and (12.5.9) we get

$$\alpha + t = \frac{3}{2},$$
$$2 - 2t = \frac{3}{2}.$$

Eliminating t between this pair of equation gives

$$\alpha = \frac{5}{4},$$

and so $\mathcal{L}_1$ and $\mathcal{L}_2$ intersect if and only if $\alpha = 5/4$. $\square$

12.X   Exercises

1. Let $\mathbf{a} = (2, 1, -1)$, $\mathbf{b} = (-3, 1, 0)$ and $\mathbf{c} = (0, 1, -2)$. Find the components of the vectors

$$3\mathbf{a} + 2\mathbf{b} - 7\mathbf{c}, \qquad 4\mathbf{a} + 5\mathbf{b}, \qquad 3\mathbf{b} + 7\mathbf{c} - \mathbf{a}.$$

2. Let $\mathbf{a} = (1, 2, 0)$, $\mathbf{b} = (0, 1, 2)$ and $\mathbf{c} = (1, 0, -1)$. Find $\alpha, \beta, \gamma \in \mathbb{R}$ such that

$$(1, 1, 1) = \alpha \mathbf{a} + \beta \mathbf{b} + \gamma \mathbf{c}.$$

3. For which $\alpha \in \mathbb{R}$ does

$$\alpha(1, 1, 1) + (3, 1, 1)$$

have length $\sqrt{3}$?

4. Find $\alpha, \beta \in \mathbb{R}$ such that

$$\mathbf{u} = \alpha(1, 2, 1) + \beta(1, -1, 3)$$

is parallel to $(1, 5, -1)$. Do there exist $\alpha, \beta \in \mathbb{R}$ such that $\mathbf{u}$ is parallel to $(1, 5, -3)$?

5. Let the origin O be at the centre of a regular hexagon $ABCDEF$. Find the position vectors of C, D, E and F in terms of $\mathbf{a}$ and $\mathbf{b}$.

6. Which of the following sets of points are collinear?

     (a) $\{(1, -2, -4), (4, 1, 2), (5, 2, 4)\}$,

     (b) $\{(3, 2, 1), (5, 0, 2), (-1, 6, 0)\}$.

7. In $\triangle ABC$, G is the centroid, and D and E lie on BC and AD, respectively, such that $BD : DC = 1 : 3$ and $AE : ED = 4 : 1$. Prove that B, E and G are collinear and find $BE : EG$.

8. In $\triangle ABC$, D, E and F are the mid-points of BC, CA and AB, respectively, and P, Q and R are the centroids of $\triangle AFE$, $\triangle BDF$ and $\triangle CED$, respectively. Express the position vectors of D, E, F, P, Q and R in terms of $\mathbf{a}$, $\mathbf{b}$ and $\mathbf{c}$. Hence, prove that A, P and D are collinear and deduce that the opposite sides of the hexagon $PFQDRE$ are equal and parallel.

9. The points A, B, C, D, E, F and G are such that E is the mid-point of BC, F is the mid-point of CD, and G is the centroid of $\triangle ABC$. Write down the position vectors of E and F in terms of $\mathbf{a}$, $\mathbf{b}$, $\mathbf{c}$ and $\mathbf{d}$. Find the position vector of the point P on FG such that $FP : PG = 3 : 2$. Let Q be the mid-point of EF. Show that A, P and Q are collinear and find $AP : PQ$.

10. The points A, B and C are non-collinear and D, E and F are the points on AB, BC and CA, respectively, such that

$$AD : DB = 2 : 1, \quad BE : EC = 3 : 2, \quad CF : FA = 1 : \alpha.$$

Given that D, E and F are collinear, find α and the ratio $DE : EF$.

11. The points D and E lie on the lines AB and AC, respectively, such that

$$AD : DB = AE : EC = 1 : \lambda,$$

where $\lambda \neq 0$.

(a) Write down the position vectors of D and E in terms of $\mathbf{a}$, $\mathbf{b}$ and $\mathbf{c}$.

(b) Find the position vectors of F and G on BE and CD, respectively, such that

$$BF : FE = \alpha : 1 \quad \text{and} \quad CG : GD = \beta : 1.$$

(c) By choosing suitable α and β, show that BE and CD meet at the point P with position vector

$$\frac{1}{\lambda + 2}(\lambda \mathbf{a} + \mathbf{b} + \mathbf{c}).$$

(d) Show that AP (produced) bisects BC.

12. By considering $\triangle OAB$, where O is the origin, show that the vector

$$\frac{1}{|\mathbf{a}| + |\mathbf{b}|}(|\mathbf{b}|\mathbf{a} + |\mathbf{a}|\mathbf{b})$$

makes equal angles with non-zero vectors $\mathbf{a}$ and $\mathbf{b}$. See also Exercise 13.X.4.

13. Let P be the point (x, y, z). Write down the distance of P from the y, z-plane and from the x-axis.

14. Let A and B be the points $(-2, -5, 6)$ and $(2, 3, -2)$, respectively.

(a) Find the unit vectors parallel to AB.

(b) Write down equations for the line AB in vector form, in parametric form and in symmetric form.

(c) Find all the points on AB which are 6 units from A.

(d) Find all the points on AB which are 5 units from the y-axis (see Exercise 12.X.13).

15. Find the points where the line

$$\frac{x}{1} = \frac{y-3}{-2} = \frac{z+1}{3}$$

intersects the surface with equation $z = x^2 + y^2$.

16. A line $\mathcal{L}$ cuts the z-axis at $(0, 0, k)$, has direction vector $(1, 2, -6)$ and meets the line

$$\frac{x+7}{3} = \frac{y-7}{-1} = \frac{z-9}{-4}$$

at the point P. Find k and P.

17. Lines $\mathcal{L}_1$ and $\mathcal{L}_2$ have equations

$$\mathcal{L}_1: \quad \frac{x}{1} = \frac{y}{-1} = \frac{z+3}{1} \quad \text{and} \quad \mathcal{L}_2: \quad \frac{x-6}{2} = \frac{y+8}{-4} = \frac{z+5}{3}.$$

Prove that $\mathcal{L}_1$ and $\mathcal{L}_2$ do *not* intersect. Find the equations of a line parallel to $\mathcal{L}_2$ which does meet $\mathcal{L}_1$.

Chapter 13

Products of Vectors

In Chapter 12, we saw that vectors can be used to study points and lines in three dimensional space. Another fundamental concept in geometry that was not considered there is *angle*. In this chapter we define the angle between vectors. In connection with this we introduce the *scalar product* of vectors.

We also study a second product, the *vector product*. Both products have many properties which resemble those familiar properties of the product of real numbers. However the vector product is not commutative.

Whereas the scalar product is often used to compute angles, one purpose of the vector product is to construct a vector perpendicular to two given vectors. This property is used in solving problems related to planes and spheres.

We also study triple products in which the two types of product are combined and we consider both their mathematical properties and some of their applications.

13.1 Angles and the Scalar Product

Definitions 13.1.1 Suppose that $\mathbf{u}$ and $\mathbf{v}$ are non-zero vectors. The *angle between* $\mathbf{u}$ *and* $\mathbf{v}$ is the angle in $[0, \pi]$ between any representations of the vectors. The angle between a vector and the zero vector is not defined.

We say that non-zero vectors are *perpendicular* if the angle between them is $\pi/2$. If $\mathbf{u}$ and $\mathbf{v}$ are perpendicular then we sometimes write $\mathbf{u} \perp \mathbf{v}$.

Observe that the angle between $\mathbf{v}$ and $\mathbf{u}$ is the same as the angle between $\mathbf{u}$ and $\mathbf{v}$, *i.e.* no 'sense' is attached to the angle.

Remark Consider non-zero vectors $\mathbf{u}$ and $\mathbf{v}$. Let $\overrightarrow{AB}$ and $\overrightarrow{PQ}$ both represent $\mathbf{u}$, and let $\overrightarrow{AC}$ and $\overrightarrow{PR}$ both represent $\mathbf{v}$ as shown in Figure 13.1.1. Then $\overrightarrow{BC}$ and $\overrightarrow{QR}$ both represent $\mathbf{v} - \mathbf{u}$. It follows that $\triangle ABC$ and $\triangle PQR$ are congruent since their corresponding sides are of equal length. Hence $\angle BAC = \angle QPR = \theta \in [0, \pi]$.

This shows that the angle between the vectors $\mathbf{u}$ and $\mathbf{v}$ is independent of the choice of representation.

Theorem 13.1.2 *Let* $\mathbf{u}$ *and* $\mathbf{v}$ *be parallel, non-zero vectors. Then the angle between* $\mathbf{u}$ *and* $\mathbf{v}$ *is* 0 *if they have the same direction, or* π *if they have the opposite direction.*

Proof As $\mathbf{u}$ and $\mathbf{v}$ are parallel and non-zero vectors, $\mathbf{v} = \alpha \mathbf{u}$ for some non-zero $\alpha \in \mathbb{R}$.

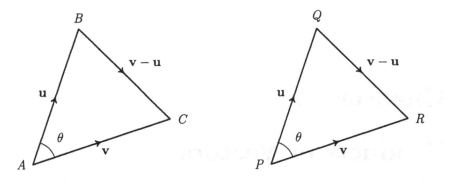

Figure 13.1.1

Further, **u** and **v** have the same [opposite] direction when $\alpha > 0$ $[\alpha < 0]$.

Suppose that $\overrightarrow{AB}$ and $\overrightarrow{AC}$ represent **u** and **v**, respectively. As **u** and **v** are parallel, then A, B and C lie on a line. If **u** and **v** have the same direction, then B and C are on the same side of A, so $\angle BAC = 0$. Otherwise A is between B and C, so $\angle BAC = \pi$.   □

Corollary 13.1.3 *Let θ be the angle between non-zero vectors **u** and **v** and let $\alpha, \beta \in \mathbb{R}$ be non-zero. Then the angle between α**u** and β**v** is θ, if α and β have the same sign, or $\pi - \theta$ if α and β have the opposite sign.*

Proof Since $|\alpha|, |\beta| > 0$, by Theorem 13.1.2, the vectors $|\alpha|$**u** and $|\beta|$**v** have the same directions as **u** and **v**, respectively. Thus the angle between $|\alpha|$**u** and $|\beta|$**v** is θ. Figure 13.1.2 shows the vectors $\pm|\alpha|$**u** and $\pm|\beta|$**v** with relevant angles.

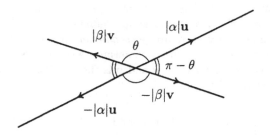

Figure 13.1.2

If α and β have the same sign, then $\alpha = |\alpha|$ and $\beta = |\beta|$ or $\alpha = -|\alpha|$ and $\beta = -|\beta|$. In either case, we see from Figure 13.1.2 that the angle between α**u** and β**v** is θ. When α and β have opposite sign, a similar argument shows that the angle between the vectors is $\pi - \theta$.   □

Definition 13.1.4 Let **u**, **v** be non-zero vectors. Let $\overrightarrow{AB}$ and $\overrightarrow{AC}$ represent **u** and **v**, respectively, and let B' be the point on AC (extended if necessary) with $\angle AB'B = \pi/2$. Then the *components of* $\mathbf{u}$ *relative to* $\mathbf{v}$ are $\mathbf{u}^{\parallel}$ and $\mathbf{u}^{\perp}$ represented by $\overrightarrow{AB'}$ and $\overrightarrow{B'B}$, respectively. See Figure 13.1.3.

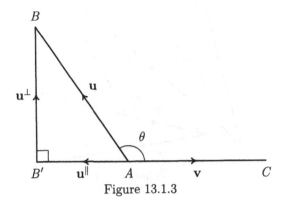

Figure 13.1.3

Note that $\mathbf{u}^{\parallel}$ is parallel to $\mathbf{v}$ and $\mathbf{u}^{\perp}$ is perpendicular to $\mathbf{v}$. If $\mathbf{u} = \mathbf{0}$ then we define $\mathbf{u}^{\parallel} = \mathbf{u}^{\perp} = \mathbf{0}$.

By considering $\triangle ABB'$, we have

$$\mathbf{u} = \mathbf{u}^{\parallel} + \mathbf{u}^{\perp}, \tag{13.1.1}$$

Remark From $\triangle ABB'$, we also see that

$$|\mathbf{u}^{\parallel}| = |\mathbf{u}||\cos\theta| \quad \text{and} \quad |\mathbf{u}^{\perp}| = |\mathbf{u}|\sin\theta. \tag{13.1.2}$$

Further, if we introduce the unit vector $\hat{\mathbf{v}} = (1/|\mathbf{v}|)\mathbf{v}$ in the same direction as $\mathbf{v}$, then $\mathbf{u}^{\parallel} = |\mathbf{u}|\cos\theta\,\hat{\mathbf{v}}$, *i.e.*

$$\mathbf{u}^{\parallel} = \frac{|\mathbf{u}|\cos\theta}{|\mathbf{v}|}\mathbf{v}. \tag{13.1.3}$$

Theorem 13.1.5 *Let $\mathbf{w}$ be a non-zero vector. Then for any vectors $\mathbf{u}$ and $\mathbf{v}$ we have,*

$$(\mathbf{u} + \mathbf{v})^{\parallel} = \mathbf{u}^{\parallel} + \mathbf{v}^{\parallel}, \tag{13.1.4}$$
$$(\mathbf{u} + \mathbf{v})^{\perp} = \mathbf{u}^{\perp} + \mathbf{v}^{\perp}, \tag{13.1.5}$$

where components are taken relative to $\mathbf{w}$.

Proof If either $\mathbf{u}$ or $\mathbf{v}$ is the zero vector then (13.1.4) and (13.1.5) are trivially true. Otherwise, consider Figure 13.1.4 in which $\overrightarrow{AB}$ represents $\mathbf{u}$, $\overrightarrow{BC}$ represents $\mathbf{v}$ and $\overrightarrow{AD}$ represents $\mathbf{w}$.

Let P and Q be such that BP and CQ are perpendicular to AD and let $\overrightarrow{PE}$ also represent $\mathbf{v}$. Since $BCEP$ is a parallelogram and BP is perpendicular to AD, CE is perpendicular to AD also. So two sides of $\triangle CQE$ are perpendicular to AD and hence so is the third, EQ.

It follows that $\overrightarrow{AP}$, $\overrightarrow{PQ}$ and $\overrightarrow{AQ}$ represent $\mathbf{u}^{\parallel}$, $\mathbf{v}^{\parallel}$ and $(\mathbf{u} + \mathbf{v})^{\parallel}$, respectively. Then (13.1.4) follows immediately.

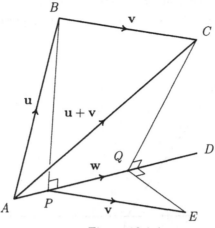

Figure 13.1.4

From (13.1.1) (for **u** and for **v**),

$$\mathbf{u} + \mathbf{v} = (\mathbf{u}^{\parallel} + \mathbf{u}^{\perp}) + (\mathbf{v}^{\parallel} + \mathbf{v}^{\perp}) \tag{13.1.6}$$

and (for (**u** + **v**)),

$$(\mathbf{u} + \mathbf{v}) = (\mathbf{u} + \mathbf{v})^{\parallel} + (\mathbf{u} + \mathbf{v})^{\perp}. \tag{13.1.7}$$

Equating the two expressions for **u** + **v** in (13.1.6) and (13.1.7), and using (13.1.4), we obtain (13.1.5).   □

Definition 13.1.6 The *scalar product* (or *dot product*) of non-zero vectors **u** and **v** is defined by

$$\mathbf{u} \cdot \mathbf{v} = |\mathbf{u}||\mathbf{v}| \cos \theta, \tag{13.1.8}$$

where θ is the angle between **u** and **v**. If **u** or **v** is the zero vector, then we define $\mathbf{u} \cdot \mathbf{v} = 0$.

 Hazard Although **u** and **v** are vectors, $\mathbf{u} \cdot \mathbf{v}$ is a *scalar*, *i.e.* $\mathbf{u} \cdot \mathbf{v} \in \mathbb{R}$. Thus the set of vectors is *not* closed under this type of multiplication.

Remarks (1) As the angle between vectors lies in $[0, \pi]$, and this is the codomain of $\cos^{-1}$, we may rearrange (13.1.8) as

$$\theta = \cos^{-1} \frac{\mathbf{u} \cdot \mathbf{v}}{|\mathbf{u}||\mathbf{v}|}. \tag{13.1.9}$$

(2) The scalar product of **u** and **v** is closely related to the component $\mathbf{u}^{\parallel}$ relative to **v**. Combining (13.1.3) and (13.1.8), we see that

$$\mathbf{u}^{\parallel} = \frac{1}{|\mathbf{v}|^2} (\mathbf{u} \cdot \mathbf{v}) \mathbf{v}. \tag{13.1.10}$$

Theorem 13.1.7 *Let* **u**, **v** *and* **w** *be vectors, and let* $\alpha \in \mathbb{R}$.

(1) $\mathbf{u} \cdot \mathbf{u} = |\mathbf{u}|^2$,

(2) $\mathbf{u} \cdot \mathbf{v} = \mathbf{v} \cdot \mathbf{u}$,

(3) $\mathbf{u} \cdot \mathbf{v} = 0 \iff \mathbf{u} = \mathbf{0},\ \mathbf{v} = \mathbf{0}$ or **u** and **v** are perpendicular,

(4) $(\alpha\mathbf{u}) \cdot \mathbf{v} = \alpha(\mathbf{u} \cdot \mathbf{v}) = \mathbf{u} \cdot (\alpha\mathbf{v})$,

(5) $(\mathbf{u} + \mathbf{v}) \cdot \mathbf{w} = \mathbf{u} \cdot \mathbf{w} + \mathbf{v} \cdot \mathbf{w}$.

Proof When **u** and **v** are non-zero, let θ be the angle between them.

(1) This is trivially true when $\mathbf{u} = \mathbf{0}$. Otherwise, since the angle between **u** and itself is 0, the result follows from (13.1.8).

(2) This is trivially true if either $\mathbf{u} = \mathbf{0}$ or $\mathbf{v} = \mathbf{0}$ since each side is 0. Otherwise, the result follows from (13.1.8) and the fact that the angle between **u** and **v** is the same as the angle between **v** and **u**.

(3) Suppose first that one of the conditions on the right hand side is satisfied. If $\mathbf{u} = \mathbf{0}$ or $\mathbf{v} = \mathbf{0}$, then $\mathbf{u} \cdot \mathbf{v} = 0$ by Definition 13.1.6. Otherwise, $\theta = \pi/2$, and (13.1.8) gives
$$\mathbf{u} \cdot \mathbf{v} = |\mathbf{u}||\mathbf{v}| \cos \frac{\pi}{2} = 0,$$
as $\cos(\pi/2) = 0$. Thus $\mathbf{u} \cdot \mathbf{v} = 0$ in any case.

Now suppose that $\mathbf{u} \cdot \mathbf{v} = 0$. Then, if **u** and **v** are non-zero, (13.1.8) shows that we must have $\cos \theta = 0$, *i.e.* $\theta = \pi/2$ since $\theta \in [0, \pi]$, and **u**, **v** are perpendicular. Otherwise, we must have $\mathbf{u} = \mathbf{0}$ or $\mathbf{v} = \mathbf{0}$.

(4) The result is clearly true if $\mathbf{u} = \mathbf{0}$ or $\mathbf{v} = \mathbf{0}$ or $\alpha = 0$ since all terms vanish. Otherwise, we can prove the first equality by considering separately the cases where $\alpha > 0$ and $\alpha < 0$.

Case 1: $\alpha > 0$. By Corollary 13.1.3 the angle between $\alpha\mathbf{u}$ and **v** is also θ. Hence
$$\begin{aligned}
(\alpha\mathbf{u}) \cdot \mathbf{v} &= |\alpha\mathbf{u}||\mathbf{v}| \cos \theta \\
&= |\alpha||\mathbf{u}||\mathbf{v}| \cos \theta = \alpha|\mathbf{u}||\mathbf{v}| \cos \theta \quad (\text{as } |\alpha| = \alpha) \\
&= \alpha(\mathbf{u} \cdot \mathbf{v}).
\end{aligned}$$

Case 2: $\alpha < 0$. Here the angles are $\pi - \theta$, and we have
$$\begin{aligned}
(\alpha\mathbf{u}) \cdot \mathbf{v} &= |\alpha\mathbf{u}||\mathbf{v}| \cos(\pi - \theta) \\
&= |\alpha||\mathbf{u}||\mathbf{v}|(-\cos \theta) = \alpha|\mathbf{u}||\mathbf{v}| \cos \theta \quad (\text{as } |\alpha| = -\alpha) \\
&= \alpha(\mathbf{u} \cdot \mathbf{v}).
\end{aligned}$$

Thus, in both cases,
$$(\alpha\mathbf{u}) \cdot \mathbf{v} = \alpha(\mathbf{u} \cdot \mathbf{v}). \tag{13.1.11}$$

Finally, using (13.1.11) and part (2), we have
$$\mathbf{u} \cdot (\alpha\mathbf{v}) = (\alpha\mathbf{v}) \cdot \mathbf{u} = \alpha(\mathbf{v} \cdot \mathbf{u}) = \alpha(\mathbf{u} \cdot \mathbf{v}).$$

(5) The result is true when $\mathbf{w} = \mathbf{0}$, since each side evaluates to zero. Otherwise, we take components relative to $\mathbf{w}$. Using (13.1.10), we have

$$(\mathbf{u} + \mathbf{v})^{\|} = \frac{1}{|\mathbf{w}|^2} ((\mathbf{u} + \mathbf{v}) \cdot \mathbf{w})\mathbf{w}.$$

and

$$\mathbf{u}^{\|} + \mathbf{v}^{\|} = \frac{1}{|\mathbf{w}|^2} (\mathbf{u} \cdot \mathbf{w})\mathbf{w} + \frac{1}{|\mathbf{w}|^2} (\mathbf{v} \cdot \mathbf{w})\mathbf{w}$$

From (13.1.4) these expression are equal, and so

$$\frac{1}{|\mathbf{w}|^2} ((\mathbf{u} + \mathbf{v}) \cdot \mathbf{w} - \mathbf{u} \cdot \mathbf{w} - \mathbf{v} \cdot \mathbf{w})\mathbf{w} = \mathbf{0}.$$

As $\mathbf{w} \neq \mathbf{0}$, the result follows.   □

Corollary 13.1.8 *The standard basis vectors satisfy*

$$\mathbf{i} \cdot \mathbf{i} = \mathbf{j} \cdot \mathbf{j} = \mathbf{k} \cdot \mathbf{k} = 1 \qquad (13.1.12)$$

and

$$\mathbf{i} \cdot \mathbf{j} = \mathbf{j} \cdot \mathbf{i} = \mathbf{i} \cdot \mathbf{k} = \mathbf{k} \cdot \mathbf{i} = \mathbf{j} \cdot \mathbf{k} = \mathbf{k} \cdot \mathbf{j} = 0. \qquad (13.1.13)$$

Proof Recall that $\mathbf{i}$, $\mathbf{j}$ and $\mathbf{k}$ are unit vectors, and are mutually perpendicular. Then (13.1.12) follows from Theorem 13.1.7 (1), and (13.1.13) follows from Theorem 13.1.7 (3).   □

Theorem 13.1.9 *Let* $\mathbf{u} = (u_x, u_y, u_z)$ *and* $\mathbf{v} = (v_x, v_y, v_z)$. *Then*

$$\mathbf{u} \cdot \mathbf{v} = u_x v_x + u_y v_y + u_z v_z. \qquad (13.1.14)$$

Proof We have

$$\mathbf{u} \cdot \mathbf{v} = (u_x \mathbf{i} + u_y \mathbf{j} + u_z \mathbf{k}) \cdot (v_x \mathbf{i} + v_y \mathbf{j} + v_z \mathbf{k}).$$

Using Theorem 13.1.7 (4) and (5) we may expand the brackets and then the result follows from (13.1.12) and (13.1.13).   □

Example 13.1.10 Verify the formula for the magnitude of a vector $\mathbf{u} = (u_x, u_y, u_z)$ using the scalar product.

Solution By Theorem 13.1.7 (1) $|\mathbf{u}|^2 = \mathbf{u} \cdot \mathbf{u}$ and so, since $|\mathbf{u}| \geq 0$ and using Theorem 13.1.9,

$$|\mathbf{u}| = \sqrt{u_x^2 + u_y^2 + u_z^2},$$

in agreement with Corollary 12.3.4.   □

Corollary 13.1.11 *The angle* θ *between non-zero vectors* $\mathbf{u} = (u_x, u_y, u_z)$ *and* $\mathbf{v} = (v_x, v_y, v_z)$ *is*

$$\theta = \cos^{-1} \left(\frac{u_x v_x + u_y v_y + u_z v_z}{\sqrt{u_x^2 + u_y^2 + u_z^2}\sqrt{v_x^2 + v_y^2 + v_z^2}} \right). \qquad (13.1.15)$$

Proof We simply write the expressions for the scalar product (13.1.14) and the magnitudes of the vectors in component form, Corollary 12.3.4, in (13.1.9). □

Remark In a more formal approach, taking vectors to be defined simply as triples (a, b, c) of real numbers, one could take (13.1.15) as a *definition* of the angle between vectors.

Example 13.1.12 Determine the angle between the vectors $(1, 1, \sqrt{2})$ and $(-1, 0, -\sqrt{2})$.

Solution Let θ be the angle we seek. Then

$$\cos\theta = \frac{(1, 1, \sqrt{2}) \cdot (-1, 0, -\sqrt{2})}{|(1, 1, \sqrt{2})||(-1, 0, -\sqrt{2})|} = \frac{-3}{\sqrt{4}\sqrt{3}} = -\frac{\sqrt{3}}{2}.$$

Thus $\theta = 5\pi/6$. □

Example 13.1.13 Prove that, for any $\triangle ABC$,

$$|BC|^2 = |AB|^2 + |AC|^2 - 2|AB||AC|\cos\angle BAC.$$

(This is the *cosine rule*.)

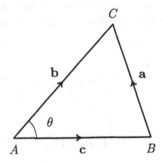

Figure 13.1.5

Solution Let $\overrightarrow{BC}$, $\overrightarrow{AC}$ and $\overrightarrow{AB}$ represent **a**, **b** and **c**, respectively, and let $\theta = \angle BAC$ be the angle between **b** and **c**, as shown in Figure 13.1.5. We have $\mathbf{a} = \mathbf{b} - \mathbf{c}$ and so

$$\mathbf{a} \cdot \mathbf{a} = (\mathbf{b} - \mathbf{c}) \cdot (\mathbf{b} - \mathbf{c}) = \mathbf{b} \cdot \mathbf{b} + \mathbf{c} \cdot \mathbf{c} - 2\mathbf{b} \cdot \mathbf{c}.$$

Using the Definition 13.1.6 and Theorem 13.1.7 (1) we get

$$|\mathbf{a}|^2 = |\mathbf{b}|^2 + |\mathbf{c}|^2 - 2|\mathbf{b}||\mathbf{c}|\cos\theta,$$

which is the required formula. □

13.2   Planes and the Vector Product

In introducing the scalar product in the previous section, we motivated its definition by discussing its geometrical applications. For the vector product which we consider next, one motivation comes from physics. If a charged particle moves through a magnetic field then the resulting force on the particle is directed perpendicular to both the direction of motion of the particle, and to the direction of the magnetic field. This is how the cathode ray tube in a television or CRT computer monitor works. Roughly speaking, the vector product of two vectors is a vector perpendicular to both the given vectors. To be definite we must fix the magnitude of this vector and also its sense since there are two choices for its direction (*cf.* the choice of coordinate systems in three dimensional space in Section 12.2). In order to specify this sense, it is necessary to discuss some properties of planes.

Definitions 13.2.1 Let $\mathbf{u}$, $\mathbf{v}$ and $\mathbf{w}$ be non-zero vectors. We say that $\mathbf{u}$, $\mathbf{v}$ and $\mathbf{w}$ are *coplanar* if there exist $\alpha, \beta, \gamma \in \mathbb{R}$, not all zero, such that

$$\alpha\mathbf{u} + \beta\mathbf{v} + \gamma\mathbf{w} = \mathbf{0}.$$

A set of more than three vectors is said to be coplanar if each subset of three vectors is coplanar.

Let $\mathbf{u}$ and $\mathbf{v}$ be non-parallel (and hence necessarily non-zero) vectors and let A be any given point. We call the set of all points P for which the vectors $\mathbf{u}$, $\mathbf{v}$ and $\mathbf{p} - \mathbf{a}$ are coplanar the *plane through A defined by $\mathbf{u}$ and $\mathbf{v}$* (see Figure 13.2.1).

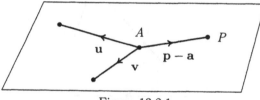

Figure 13.2.1

Three or more points $A_1, A_2, A_3, \ldots$ are said to be coplanar if, for all choices of i, j, the vectors represented by $\overrightarrow{A_i A_j}$ are coplanar.

Figure 13.2.2 shows four coplanar vectors and five coplanar points.

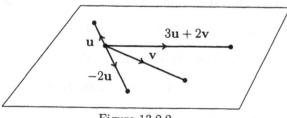

Figure 13.2.2

Note 13.2.2 Consider the non-zero vectors $\mathbf{u}$, $\alpha\mathbf{u}$ and $\mathbf{v}$. Since $(-\alpha)\mathbf{u}+1(\alpha\mathbf{u})+0\mathbf{v} = \mathbf{0}$, this set of vectors is coplanar. Thus any set of three non-zero vectors including a pair of parallel vectors is coplanar.

The above definition gives us the first of two vector equations of a plane.

Theorem 13.2.3 *The plane $\mathcal{P}$ through A defined by non-parallel vectors $\mathbf{u}$ and $\mathbf{v}$ has equation*

$$\mathbf{p} = \mathbf{a} + s\mathbf{u} + t\mathbf{v} \qquad (s, t \in \mathbb{R}). \tag{13.2.1}$$

Proof A point P lies on $\mathcal{P}$ if and only if

$$\alpha\mathbf{u} + \beta\mathbf{v} + \gamma(\mathbf{p} - \mathbf{a}) = \mathbf{0} \tag{13.2.2}$$

for some $\alpha, \beta, \gamma \in \mathbb{R}$, where $\gamma \neq 0$, since otherwise $\mathbf{u}$ and $\mathbf{v}$ would be parallel. Hence, (13.2.2) is equivalent to (13.2.1) with $s = -\alpha/\gamma$ and $t = -\beta/\gamma$. $\qquad \square$

Example 13.2.4 Determine the coordinates of all points on the plane $\mathcal{P}$ that contains the three points $A(1, -2, 3)$, $B(1, 1, 1)$ and $C(0, 1, -1)$. Show that the line

$$\mathcal{L}: \quad \frac{x-2}{-1} = \frac{y-1}{6} = \frac{z-3}{-6}$$

lies in $\mathcal{P}$.

Solution The vectors $\mathbf{b} - \mathbf{a} = (0, 3, -2)$ and $\mathbf{c} - \mathbf{a} = (-1, 3, -4)$ are non-parallel. The required plane is defined by the vectors $\mathbf{b} - \mathbf{a}$ and $\mathbf{c} - \mathbf{a}$ and contains the point A (or B or C).

Let $P(x, y, z)$ be a general point on the plane. Then, by Theorem 13.2.3,

$$(x, y, z) = (1, -2, 3) + s(0, 3, -2) + t(-1, 3, -4)$$
$$= (1 - t, -2 + 3s + 3t, 3 - 2s - 4t),$$

i.e. the coordinates of any point on $\mathcal{P}$ is $(1 - t, -2 + 3s + 3t, 3 - 2s - 4t)$ for some $s, t \in \mathbb{R}$.

Further, using (12.5.4), a general point on $\mathcal{L}$ has coordinates $(2 - u, 1 + 6u, 3 - 6u)$ for $u \in \mathbb{R}$. Hence $\mathcal{L}$ lies in $\mathcal{P}$ if and only if the system

$$\left\{ \begin{array}{rcl} 2 - u & = & 1 - t \\ 1 + 6u & = & -2 + 3s + 3t \\ 3 - 6u & = & 3 - 2s - 4t \end{array} \right. \quad i.e. \quad \left\{ \begin{array}{rcl} -u + t & = & -1 \\ 2u - s - t & = & -1 \\ -3u + s + 2t & = & 0 \end{array} \right.$$

is consistent. From the first two equations we get $t = u - 1$ and $s = 2u - t + 1 = u + 2$. Substituting into the final equation we see that it is also satisfied by these choices for s and t and hence $\mathcal{L}$ does lie in $\mathcal{P}$. $\qquad \square$

Definitions 13.2.5 Consider a plane $\mathcal{P}$ defined by non-parallel vectors $\mathbf{u}$ and $\mathbf{v}$. A vector $\mathbf{n}$ perpendicular to both $\mathbf{u}$ and $\mathbf{v}$ is called a *normal vector* to $\mathcal{P}$.

Remark Given a non-zero vector **n** one can always find (any number of) non-parallel vectors which are perpendicular to **n**.

The plane $\mathcal{P}$ containing a point A with normal vector **n** is the plane defined by two such vectors, **u** and **v** say, and passing through A. It consists of the set of points P such that $\overrightarrow{AP}$ represents a vector perpendicular to **n** (see Figure 13.2.3).

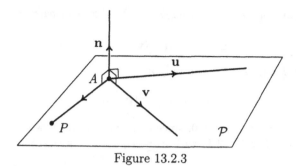

Figure 13.2.3

Theorem 13.2.6 *The plane* $\mathcal{P}$ *through* A *with normal vector* **n** *has equation*

$$(\mathbf{p} - \mathbf{a}) \cdot \mathbf{n} = 0. \tag{13.2.3}$$

Proof Consider a point P,

$$P \in \mathcal{P} \Leftrightarrow \mathbf{p} - \mathbf{a} \text{ is perpendicular to } \mathbf{n}, \text{ or } \mathbf{p} = \mathbf{a}$$
$$\Leftrightarrow (\mathbf{p} - \mathbf{a}) \cdot \mathbf{n} = 0.$$

The last equivalence follows from Theorem 13.1.7 (3), since $\mathbf{n} \neq \mathbf{0}$. $\square$

Corollary 13.2.7 *The equation of the plane* $\mathcal{P}$ *through point* $A(a, b, c)$ *with normal vector* $\mathbf{n} = (l, m, n)$ *is*

$$lx + my + nz = d, \tag{13.2.4}$$

where $d = la + mb + nc$.

Proof Let the general point on $\mathcal{P}$ be $P(x, y, z)$. Substituting the component forms of **p**, **a** and **n** into (13.2.3) we get

$$(x - a)l + (y - b)m + (z - c)n = 0,$$

which gives (13.2.4). $\square$

Definitions 13.2.8 Let **u** and **v** be non-parallel vectors and let θ be the angle between them. The *vector product* (or *cross product*) of **u** and **v** is the vector $\mathbf{u} \times \mathbf{v}$ with magnitude

$$|\mathbf{u} \times \mathbf{v}| = |\mathbf{u}||\mathbf{v}| \sin \theta, \tag{13.2.5}$$

which is perpendicular to **u** and **v** with the sense shown in Figure 13.2.4. A useful mnemonic for the orientation of $\mathbf{u} \times \mathbf{v}$ relative to **u** and **v** is the *right-hand screw rule*: if we turn a (right-handed) screw from **u** to **v** then it will move in the direction of $\mathbf{u} \times \mathbf{v}$.

If $\mathbf{u} = \mathbf{0}$ or $\mathbf{v} = \mathbf{0}$ or **u** and **v** are parallel then $\mathbf{u} \times \mathbf{v} = \mathbf{0}$.

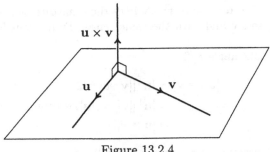

Figure 13.2.4

Notes 13.2.9 (1) In the case that **u** and **v** are non-parallel, **u** × **v** is perpendicular to both **u** and **v** and so it is a normal to the plane determined by **u** and **v**.

(2) Recall that **i**, **j** and **k** are mutually perpendicular unit vectors so that the sine of the angle between each pair is $\sin(\pi/2) = 1$. By comparing the direction of the vectors shown in Figures 12.2.2 and 13.2.4, and using (13.2.5), we get

$$\mathbf{i} \times \mathbf{j} = \mathbf{k}, \quad \mathbf{j} \times \mathbf{k} = \mathbf{i} \quad \text{and} \quad \mathbf{k} \times \mathbf{i} = \mathbf{j}. \tag{13.2.6}$$

Also, from the definition,

$$\mathbf{i} \times \mathbf{i} = \mathbf{j} \times \mathbf{j} = \mathbf{k} \times \mathbf{k} = \mathbf{0}. \tag{13.2.7}$$

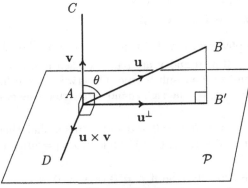

Figure 13.2.5

Theorem 13.2.10 *Let* **u** *and* **v** *be vectors, with* **v** *non-zero. Then, taking components relative to* **v**,

$$\mathbf{u} \times \mathbf{v} = \mathbf{u}^{\perp} \times \mathbf{v}.$$

Proof If **u** = **0**, then **u**$^{\perp}$ = **0**, and the result is trivially true.

Otherwise we have the situation shown in Figure 13.2.5. Here, $\mathcal{P}$ is the plane through A with normal vector **v**, and $\overrightarrow{AB}$ and $\overrightarrow{AC}$ represent **u** and **v**, respectively. As **u**$^{\perp}$ is perpendicular to **v**, its representation is $\overrightarrow{AB'}$, with B' on $\mathcal{P}$. Let θ be the angle between **u** and **v**.

From Figure 13.2.5, $\mathbf{u} \times \mathbf{v}$ and $\mathbf{u}^\perp \times \mathbf{v}$ have the same direction—normal to the plane containing $\mathbf{u}$, $\mathbf{u}^\perp$ and $\mathbf{v}$ and with the same sense. From triangle ABB', we see that $|\mathbf{u}^\perp| = |\mathbf{u}| \sin \theta$.

As $\mathbf{u}^\perp$ and $\mathbf{v}$ make angle $\pi/2$,

$$
\begin{aligned}
|\mathbf{u}^\perp \times \mathbf{v}| &= |\mathbf{u}^\perp||\mathbf{v}| \sin \pi/2 \\
&= |\mathbf{u}^\perp||\mathbf{v}| = (|\mathbf{u}| \sin \theta)|\mathbf{v}| \\
&= |\mathbf{u} \times \mathbf{v}|.
\end{aligned}
$$

Since lengths and directions agree, the result follows. $\square$

Note 13.2.11 Figure 13.2.5 and calculations in the proof of Theorem 13.2.10 lead to a useful description of $\mathbf{u} \times \mathbf{v}$ when it is non-zero. It is the vector obtained by

(a) rotating $\mathbf{u}^\perp$ by $\pi/2$ clockwise (viewed from the same side of $\mathcal{P}$ as $\mathbf{v}$), and then

(b) scaling by a factor of $|\mathbf{v}|$.

Theorem 13.2.12 *Let* $\mathbf{u}$, $\mathbf{v}$ *and* $\mathbf{w}$ *be vectors and let* $\alpha \in \mathbb{R}$.

(1) $\mathbf{u} \times \mathbf{u} = \mathbf{0}$,

(2) $\mathbf{u} \times \mathbf{v} = -(\mathbf{v} \times \mathbf{u})$,

(3) $\mathbf{u} \times \mathbf{v} = \mathbf{0} \Leftrightarrow \mathbf{u} = \mathbf{0}$, $\mathbf{v} = \mathbf{0}$ *or* $\mathbf{u}$ *and* $\mathbf{v}$ *are parallel*.

(4) $(\alpha \mathbf{u}) \times \mathbf{v} = \alpha(\mathbf{u} \times \mathbf{v}) = \mathbf{u} \times (\alpha \mathbf{v})$.

(5) $(\mathbf{u} + \mathbf{v}) \times \mathbf{w} = \mathbf{u} \times \mathbf{w} + \mathbf{v} \times \mathbf{w}$.

Proof (1) This is obvious from the definition since $\mathbf{u}$ is parallel to itself.

(2) If $\mathbf{u} = \mathbf{0}$, $\mathbf{v} = \mathbf{0}$ or $\mathbf{u}$ and $\mathbf{v}$ are parallel, then $\mathbf{u} \times \mathbf{v}$ and $\mathbf{v} \times \mathbf{u}$ are both equal to $\mathbf{0}$ by definition and so the result is true. Otherwise, $\mathbf{v} \times \mathbf{u}$ has the same magnitude, $|\mathbf{v}||\mathbf{u}| \sin \theta$, as $\mathbf{u} \times \mathbf{v}$ but, by the right-hand screw rule, the opposite direction. Hence the result follows.

(3) By definition, if $\mathbf{u} = \mathbf{0}$, $\mathbf{v} = \mathbf{0}$ or $\mathbf{u}$ and $\mathbf{v}$ are parallel, then $\mathbf{u} \times \mathbf{v} = \mathbf{0}$. Otherwise, the angle between $\mathbf{u}$ and $\mathbf{v}$ is $\theta \in (0, \pi)$. Thus $|\mathbf{u} \times \mathbf{v}| = |\mathbf{u}||\mathbf{v}| \sin \theta \neq 0$ and we get the required result.

(4) The result is trivially true if $\mathbf{u} = \mathbf{0}$ or $\mathbf{v} = \mathbf{0}$ or $\alpha = 0$. When none of these hold, the proof that the magnitudes of the vectors are equal follows exactly the same argument used in proving Theorem 13.1.7 (4) and is left as an exercise to the reader.

To show that the vectors all have the same direction, we observe first that $\alpha \mathbf{u}$ and $\alpha \mathbf{v}$ are both coplanar with $\mathbf{u}$ and $\mathbf{v}$ (see Note 13.2.2). If $\alpha > 0$ then $\alpha \mathbf{u}$ and $\alpha \mathbf{v}$ have the same direction as $\mathbf{u}$ and $\mathbf{v}$, respectively, and so $(\alpha \mathbf{u}) \times \mathbf{v}$ and $\mathbf{u} \times (\alpha \mathbf{v})$ both have the same direction as $\mathbf{u} \times \mathbf{v}$. When $\alpha < 0$, $\alpha \mathbf{u}$ and $\alpha \mathbf{v}$ have the opposite direction to $\mathbf{u}$ and $\mathbf{v}$, respectively, and, by the right-hand screw rule, $(\alpha \mathbf{u}) \times \mathbf{v}$ and $\mathbf{u} \times (\alpha \mathbf{v})$ have the opposite direction to $\mathbf{u} \times \mathbf{v}$. So in either case $(\alpha \mathbf{u}) \times \mathbf{v}$, $\alpha(\mathbf{u} \times \mathbf{v})$ and $\mathbf{u} \times (\alpha \mathbf{v})$ all have the same direction, that of $-(\mathbf{u} \times \mathbf{v})$. This completes the proof.

(5) If $\mathbf{w} = \mathbf{0}$, then all products are $\mathbf{0}$, so that the result holds.

Otherwise, we consider components relative to $\mathbf{w}$. Figure 13.2.6 shows a plane with normal $\mathbf{w}$, we imagine $\mathbf{w}$ coming out towards the reader. Let $\overrightarrow{AB}$, $\overrightarrow{AC}$ and $\overrightarrow{AD}$

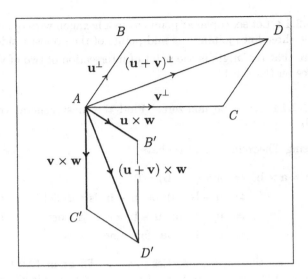

Figure 13.2.6

represent $\mathbf{u}^\perp$, $\mathbf{v}^\perp$ and $(\mathbf{u}+\mathbf{v})^\perp$, respectively. By Theorem 13.1.5 (2), $(\mathbf{u}+\mathbf{v})^\perp = \mathbf{u}^\perp + \mathbf{v}^\perp$ and so $ABCD$ is a parallelogram. Figure 13.2.7 show the relationship between the parallelogram $ABCD$ and the one formed by the vectors $\mathbf{u}$ and $\mathbf{v}$.

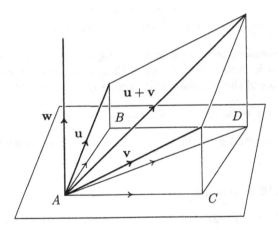

Figure 13.2.7

Rotating $ABCD$ through $\pi/2$ and scaling by a factor $|\mathbf{w}|$, we get another parallelogram $AB'D'C'$. Then, by Note 13.2.11, $\overrightarrow{AB'}$, $\overrightarrow{AC'}$ and $\overrightarrow{AD'}$ represent $\mathbf{u} \times \mathbf{w}$, $\mathbf{v} \times \mathbf{w}$ and $(\mathbf{u}+\mathbf{v}) \times \mathbf{w}$, respectively. As $AB'D'C'$ is a parallelogram, $(\mathbf{u}+\mathbf{v}) \times \mathbf{w} = \mathbf{u} \times \mathbf{w} + \mathbf{v} \times \mathbf{w}$ as required. $\square$

Problem 13.2.13 Let six coplanar points form a hexagon with opposite sides parallel. Prove that the three lines joining the mid-points of the opposite sides are concurrent. [Hint: Suppose that the origin is the point of intersection of two of the lines and prove that it also lies on the third.]

Example 13.2.14 Let $\mathbf{a}$ be non-zero. Find the most general vector $\mathbf{u}$ such that $\mathbf{a} \times \mathbf{u} = \mathbf{a} \times \mathbf{b}$.

Solution Using Theorem 13.2.12 we have

$$
\begin{aligned}
\mathbf{a} \times \mathbf{u} = \mathbf{a} \times \mathbf{b} \;&\Leftrightarrow\; \mathbf{a} \times (\mathbf{u} - \mathbf{b}) = 0 \\
&\Leftrightarrow\; \mathbf{u} - \mathbf{b} = 0 \;\text{ or }\; \mathbf{u} - \mathbf{b} \text{ is parallel to } \mathbf{a} \;(\text{as } \mathbf{a} \neq 0) \\
&\Leftrightarrow\; \mathbf{u} = \mathbf{b} \;\text{ or }\; \mathbf{u} - \mathbf{b} = \alpha \mathbf{a} \text{ for some } \alpha \neq 0 \\
&\Leftrightarrow\; \mathbf{u} = \mathbf{b} + \beta \mathbf{a} \text{ for some } \beta \in \mathbb{R}. \quad\quad\quad\square
\end{aligned}
$$

We will now use the basic properties given in Theorem 13.2.12 to determine the component form of the vector product. This is most conveniently expressed as a *determinant* (see Section 11.6).

Theorem 13.2.15 If $\mathbf{u} = (u_x, u_y, u_z)$ and $\mathbf{v} = (v_x, v_y, v_z)$ then

$$
\mathbf{u} \times \mathbf{v} = \begin{vmatrix} \mathbf{i} & \mathbf{j} & \mathbf{k} \\ u_x & u_y & u_z \\ v_x & v_y & v_z \end{vmatrix} \tag{13.2.8}
$$

$$
= (u_y v_z - u_z v_y)\mathbf{i} - (u_x v_z - u_z v_x)\mathbf{j} + (u_x v_y - u_y v_x)\mathbf{k}
$$

$$
= (u_y v_z - u_z v_y, -(u_x v_z - u_z v_x), u_x v_y - u_y v_x) \tag{13.2.9}
$$

Proof It is convenient to write $\mathbf{u}$ and $\mathbf{v}$ using the notation $\mathbf{u} = u_x \mathbf{i} + u_y \mathbf{j} + u_z \mathbf{k}$ and $\mathbf{u} = u_x \mathbf{i} + u_y \mathbf{j} + u_z \mathbf{k}$. Now, using Theorem 13.2.12 to expand the brackets and simplify and the properties of the standard unit vectors (13.2.6) and (13.2.7), we get

$$
\mathbf{u} \times \mathbf{v} = (u_y v_z - u_z v_y)\mathbf{i} - (u_x v_z - u_z v_x)\mathbf{j} + (u_x v_y - u_y v_x)\mathbf{k}.
$$

We leave it to the reader to check that this can also be written in the other forms given. $\quad\quad\quad\square$

Example 13.2.16 Determine

$$(a)\; (1, -3, -2) \times (0, 5, 1) \;\text{ and }\; (b)\; (2, -1, 1) \times (-4, 2, -2).$$

Solution (a) We have

$$
\begin{aligned}
(1, -3, -2) \times (0, 5, 1) &= \begin{vmatrix} \mathbf{i} & \mathbf{j} & \mathbf{k} \\ 1 & -3 & -2 \\ 0 & 5 & 1 \end{vmatrix} \\
&= \mathbf{i}\big(-3(1) - (-2)5\big) - \mathbf{j}\big(1(1) - (-2)0\big) + \mathbf{k}\big(1(5) - (-3)0\big) \\
&= (7, -1, 5).
\end{aligned}
$$

(b) We have

$$(2,-1,1) \times (-4,2,-2) = \begin{vmatrix} \mathbf{i} & \mathbf{j} & \mathbf{k} \\ 2 & -1 & 1 \\ -4 & 2 & -2 \end{vmatrix}$$
$$= (-1(-2) - (1)2, -(2(-2) - 1(-4)), 2(2) - (-1)(-4))$$
$$= (0,0,0).$$

Alternatively, we observe that $(-4,2,-2) = -2(2,-1,1)$ and so the vector product vanishes since the vectors are parallel. □

Example 13.2.17 Determine the plane $\mathcal{P}$ through the points $A(1,-2,3)$, $B(1,1,1)$ and $C(0,1,-1)$.

Solution A normal vector to $\mathcal{P}$ is $\mathbf{n} = (\mathbf{b} - \mathbf{a}) \times (\mathbf{c} - \mathbf{a})$, *i.e.*

$$\mathbf{n} = \begin{vmatrix} \mathbf{i} & \mathbf{j} & \mathbf{k} \\ 0 & 3 & -2 \\ -1 & 3 & -4 \end{vmatrix} = (-6,2,3).$$

Therefore, from Corollary 13.2.7, $\mathcal{P}$ has equation

$$-6x + 2y + 3z = -6(1) + 2(-2) + 3(3) = -1.$$ □

13.3 Spheres

We saw in Example 12.2.3 (1) that, for $r > 0$,

$$S = \{(x,y,z) : (x-a)^2 + (y-b)^2 + (z-c)^2 = r^2\},$$

is a sphere with centre $C(a,b,c)$ and radius r. We call

$$(x-a)^2 + (y-b)^2 + (z-c)^2 = r^2, \qquad (13.3.1)$$

the equation of the sphere.

More generally, given an equation

$$x^2 + y^2 + z^2 + 2\alpha x + 2\beta y + 2\gamma z + \delta = 0, \qquad (13.3.2)$$

we complete the squares in x, y and z to obtain

$$(x+\alpha)^2 + (y+\beta)^2 + (z+\gamma)^2 = \alpha^2 + \beta^2 + \gamma^2 - \delta,$$

which is the equation of a sphere if and only if $\alpha^2 + \beta^2 + \gamma^2 - \delta > 0$. In such a case, its centre is at $(-\alpha, -\beta, -\gamma)$ and its radius is $\sqrt{\alpha^2 + \beta^2 + \gamma^2 - \delta}$.

Definition 13.3.1 Given a point A on a sphere $\mathcal{S}$ with centre C the *tangent plane at A* is the plane through A with normal $\mathbf{c} - \mathbf{a}$ (see Figure 13.3.1).

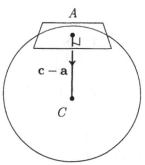

Figure 13.3.1

Example 13.3.2 Show that

$$S: \ x^2 + y^2 + z^2 - 2x - 4y - 6z - 13 = 0,$$

is the equation of a sphere and find the tangent plane at the point $P(4, 5, 6)$.

Solution Completing the squares we have

$$(x - 1)^2 + (y - 2)^2 + (z - 3)^2 = 1^2 + 2^2 + 3^2 + 13 = 27.$$

Thus S is a sphere with centre $C(1, 2, 3)$.

The tangent plane at P has normal vector $\mathbf{c} - \mathbf{p} = (1, 2, 3) - (4, 5, 6) = (-3, -3, -3)$. For simplicity, we take normal vector $(1, 1, 1)$ and so the tangent plane has equation

$$x + y + z = 4 + 5 + 6 = 15. \qquad \Box$$

Example 13.3.3 Find the two spheres of radius 6 which share the tangent plane

$$\mathcal{P}: \ x - 2y + 2z = 1$$

at $A(1, 0, 0)$.

Solution The centre C of such a sphere lies 6 units from A on the line $\mathcal{L}$ through A with direction vector normal to $\mathcal{P}$.

A general point on $\mathcal{L}$ is $\mathbf{p} = (1, 0, 0) + t(1, -2, 2) = (1 + t, -2t, 2t)$, for some $t \in \mathbb{R}$, since the normal $(1, -2, 2)$ for $\mathcal{P}$ is a direction vector for $\mathcal{L}$. Points 6 units from A are such that

$$(1 + t - 1)^2 + (-2t - 0)^2 + (2t - 0)^2 = 9t^2 = 36,$$

which gives $t = \pm 2$. It follows that the spheres have centres at $(1 \pm 2, \mp 4, \pm 4)$. Hence their equations are

$$(x - 3)^2 + (y + 4)^2 + (z - 4)^2 = 36$$

and

$$(x + 1)^2 + (y - 4)^2 + (z + 4)^2 = 36. \qquad \Box$$

Example 13.3.4 Find the equation of the smallest sphere which encloses both of the spheres

$$S_1 : x^2 + y^2 + z^2 - 2x - 2y - 6z + 7 = 0,$$
$$S_2 : x^2 + y^2 + z^2 + 2x - 2z - 7 = 0.$$

Solution　Completing the square for S_1 we get

$$(x-1)^2 + (y-1)^2 + (z-3)^2 = 1^2 + 1^2 + 3^2 - 7 = 4,$$

so that S_1 has centre $A(1,1,3)$ and radius 2. For S_2 we get

$$(x+1)^2 + (y-0)^2 + (z-1)^2 = 1^2 + 0^2 + (-1)^2 + 7 = 9,$$

so that S_2 has centre $B(-1,0,1)$ and radius 3.

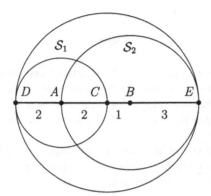

Figure 13.3.2

In Figure 13.3.2 we show a cross-section through the spheres S_1 and S_2 which passes through their centres A and B. The points C and D are the points of intersection of the line through A and B with S_1, with C lying between A and B.

Since $|AC| = 2 =$ radius of S_1 and $|AB| = \sqrt{(1+1)^2 + (1-0)^2 + (3-1)^2} = 3$, we see that A lies on S_2 as shown, with A and E being the points of intersection of the line with S_2. Since $|BE| = 3 =$ radius of S_2, the diameter of the smallest sphere enclosing S_1 and S_2 is $|DE| = 8$, *i.e.* its radius is 4.

Also, it is obvious that C is the mid-point of DE and hence is the centre of the enclosing sphere. Since $AC : CB = 2 : 1$, by the section formula (12.4.1) we have

$$\mathbf{c} = \frac{2\mathbf{b}+\mathbf{a}}{3} = \left(-\frac{1}{3}, \frac{1}{3}, \frac{5}{3}\right)$$

and so the required sphere has equation

$$\left(x+\frac{1}{3}\right)^2 + \left(y-\frac{1}{3}\right)^2 + \left(z-\frac{5}{3}\right)^2 = 16. \qquad \square$$

13.4   The Scalar Triple Product

As we have seen, two non-parallel vectors $\mathbf{v}$ and $\mathbf{w}$ determine a parallelogram $ABCD$ in which $\overrightarrow{AB}$ and $\overrightarrow{DC}$ represent $\mathbf{v}$ and $\overrightarrow{AD}$ and $\overrightarrow{BC}$ represent $\mathbf{w}$, as shown in Figure 13.4.1.

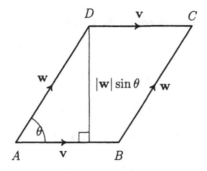

Figure 13.4.1

The area of this parallelogram is the length of one side multiplied by the perpendicular distance between this and the opposite side. If the angle between $\mathbf{v}$ and $\mathbf{w}$ is θ, then the distance between the sides representing $\mathbf{v}$ is $|\mathbf{w}|\sin\theta$. Hence the area is $|\mathbf{v}||\mathbf{w}|\sin\theta$. In other words, using (13.2.5), the area of any parallelogram defined by $\mathbf{v}$ and $\mathbf{w}$ is $|\mathbf{v}\times\mathbf{w}|$.

If we take another vector $\mathbf{u}$, which is not coplanar with $\mathbf{v}$ and $\mathbf{w}$, then the three vectors define a *parallelepiped*, the volume of which equals the area of a face (which is a parallelogram) multiplied by the perpendicular distance between this face and the opposite face.

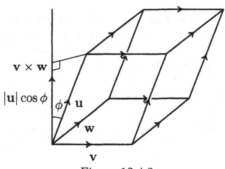

Figure 13.4.2

We first consider the case shown in Figure 13.4.2 in which the angle between $\mathbf{u}$ and $\mathbf{v}\times\mathbf{w}$ is $\phi\in[0,\pi/2)$. The distance between the pair of faces defined by $\mathbf{v}$ and $\mathbf{w}$ is $|\mathbf{u}|\cos\phi$ (as $\cos\phi>0$), and hence the volume of the parallelepiped is, by (13.1.8),

$$|\mathbf{u}||\mathbf{v}\times\mathbf{w}|\cos\phi=\mathbf{u}\cdot(\mathbf{v}\times\mathbf{w}).$$

When $\phi\in(\pi/2,\pi]$, so that $\cos\phi<0$, the volume of the parallelepiped is $-\mathbf{u}\cdot(\mathbf{v}\times\mathbf{w})$.

Definition 13.4.1 The *scalar triple product* of vectors $\mathbf{u}$, $\mathbf{v}$ and $\mathbf{w}$ is

$$[\mathbf{u}, \mathbf{v}, \mathbf{w}] = \mathbf{u} \cdot (\mathbf{v} \times \mathbf{w}) \quad (\in \mathbb{R}). \tag{13.4.1}$$

The ordered triple of vectors $(\mathbf{u}, \mathbf{v}, \mathbf{w})$ is said to be *right-handed* if $[\mathbf{u}, \mathbf{v}, \mathbf{w}] > 0$ and *left-handed* if $[\mathbf{u}, \mathbf{v}, \mathbf{w}] < 0$.

Theorem 13.4.2 *Let $\mathbf{u}$, $\mathbf{v}$ and $\mathbf{w}$ be non-zero. Then $[\mathbf{u}, \mathbf{v}, \mathbf{w}] = 0$ if and only if $\mathbf{u}$, $\mathbf{v}$ and $\mathbf{w}$ are coplanar.*

Proof In the following we use Theorem 13.1.7 (3) and Theorem 13.2.12 (3).

$$\mathbf{u} \cdot (\mathbf{v} \times \mathbf{w}) = 0 \;\Leftrightarrow\; \mathbf{v} \times \mathbf{w} = \mathbf{0} \;\;\text{or}\;\; (\mathbf{u} \text{ and } \mathbf{v} \times \mathbf{w} \text{ are perpendicular})$$
$$\Leftrightarrow\; (\mathbf{v} \text{ and } \mathbf{w} \text{ are parallel}) \;\;\text{or}\;\; (\mathbf{u}, \mathbf{v} \text{ and } \mathbf{w} \text{ are coplanar})$$
$$\Leftrightarrow\; \mathbf{u}, \mathbf{v} \text{ and } \mathbf{w} \text{ are coplanar} \;\;(\text{see Note 13.2.2}).$$

This completes the proof.   □

Theorem 13.4.3 *For any vectors $\mathbf{u}$, $\mathbf{v}$ and $\mathbf{w}$,*

$$[\mathbf{u}, \mathbf{v}, \mathbf{w}] = [\mathbf{v}, \mathbf{w}, \mathbf{u}] = [\mathbf{w}, \mathbf{u}, \mathbf{v}] =$$
$$-[\mathbf{u}, \mathbf{w}, \mathbf{v}] = -[\mathbf{v}, \mathbf{u}, \mathbf{w}] = -[\mathbf{w}, \mathbf{v}, \mathbf{u}]. \tag{13.4.2}$$

Proof If any of $\mathbf{u}$, $\mathbf{v}$ and $\mathbf{w}$ are zero or if they are coplanar, by Theorem 13.4.2, all of the triple product in (13.4.2) vanish and the result is true.

Otherwise, there are $3! = 6$ permutations of a set with 3 elements and so there are 6 distinct (ordered) triples involving the three vectors $\mathbf{u}$, $\mathbf{v}$ and $\mathbf{w}$. For each such triple, of general form $(\mathbf{a}, \mathbf{b}, \mathbf{c})$ where $\{\mathbf{a}, \mathbf{b}, \mathbf{c}\} = \{\mathbf{u}, \mathbf{v}, \mathbf{w}\}$, we may determine a parallelepiped, as described—for $(\mathbf{u}, \mathbf{v}, \mathbf{w})$—at the start of this section, with volume $V = \pm[\mathbf{a}, \mathbf{b}, \mathbf{c}]$. It has the positive sign if $(\mathbf{a}, \mathbf{b}, \mathbf{c})$ is right-handed, and the negative sign otherwise. It remains to determine which triples give the plus sign and which the negative.

We observe that, because the vectors are in the same 'cyclic order', $(\mathbf{v}, \mathbf{w}, \mathbf{u})$ and $(\mathbf{w}, \mathbf{u}, \mathbf{v})$ determine parallelepipeds with the same orientation as the one determined by $(\mathbf{u}, \mathbf{v}, \mathbf{w})$. On the other hand, $(\mathbf{u}, \mathbf{w}, \mathbf{v})$, $(\mathbf{v}, \mathbf{u}, \mathbf{w})$ and $(\mathbf{w}, \mathbf{v}, \mathbf{u})$ determine parallelepipeds which are the 'mirror image' of the first set. It follows that

$$\pm V = [\mathbf{u}, \mathbf{v}, \mathbf{w}] = [\mathbf{v}, \mathbf{w}, \mathbf{u}] = [\mathbf{w}, \mathbf{u}, \mathbf{v}]$$
$$= -[\mathbf{u}, \mathbf{w}, \mathbf{v}] = -[\mathbf{v}, \mathbf{u}, \mathbf{w}] = -[\mathbf{w}, \mathbf{v}, \mathbf{u}],$$

as required.   □

Remark As its proof suggests, Theorem 13.4.3 may be put more succinctly: a scalar triple product is unchanged if we reorder the vectors so that they maintain their 'cyclic order' (see Figure 13.4.3), and otherwise the sign changes.

Theorem 13.4.4 *If $\mathbf{u} = (u_x, u_y, u_z)$, $\mathbf{v} = (v_x, v_y, v_z)$ and $\mathbf{w} = (w_x, w_y, w_z)$ then*

$$[\mathbf{u}, \mathbf{v}, \mathbf{w}] = \begin{vmatrix} u_x & u_y & u_z \\ v_x & v_y & v_z \\ w_x & w_y & w_z \end{vmatrix}. \tag{13.4.3}$$

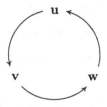

Figure 13.4.3

Proof This follows immediately from (13.2.8),

$$[\mathbf{u}, \mathbf{v}, \mathbf{w}] = \mathbf{u} \cdot (\mathbf{v} \times \mathbf{w})$$
$$= (u_x, u_y, u_z) \cdot (v_y w_z - v_z w_y, -(v_x w_z - v_z w_x), v_x w_y - v_y w_x)$$
$$= u_x(v_y w_z - v_z w_y) - u_y(v_x w_z - v_z w_x) + u_z(v_x w_y - v_y w_x)$$
$$= \begin{vmatrix} u_x & u_y & u_z \\ v_x & v_y & v_z \\ w_x & w_y & w_z \end{vmatrix}. \qquad \square$$

Corollary 13.4.5 *For a general* 3×3 *determinant* Δ *we have*

(1) $\Delta = 0$ if any row is zero,

(2) $\Delta = 0$ if any row is a linear combination of the other rows,

(3) If Δ' is a determinant obtained by reordering the rows of Δ then $\Delta' = \Delta$ if the reordering is cyclic and $\Delta' = -\Delta$ otherwise.

Proof This is simply a restatement of the properties of the scalar triple product, making use of Theorem 13.4.4. We consider three arbitrary vectors $\mathbf{u}$, $\mathbf{v}$ and $\mathbf{w}$ so that the determinant in (13.4.3) has arbitrary (real) entries.

(1) From Definition 13.4.1, $[\mathbf{u}, \mathbf{v}, \mathbf{w}] = 0$ if any of $\mathbf{u}$, $\mathbf{v}$ or $\mathbf{w}$ is zero.

(2) If, for example, $\mathbf{w} = \alpha \mathbf{u} + \beta \mathbf{v}$ then $\mathbf{u}$, $\mathbf{v}$ and $\mathbf{w}$ are coplanar and so, by Theorem 13.4.2, $[\mathbf{u}, \mathbf{v}, \mathbf{w}] = 0$.

(3) This follows directly from Theorem 13.4.3. $\qquad \square$

Example 13.4.6 (a) Show that $(\mathbf{i}, \mathbf{j}, \mathbf{k})$ is right-handed.

(b) Let $\mathbf{a} = (1, 0, 1)$, $\mathbf{b} = (2, -1, 3)$ and $\mathbf{c} = (-2, 0, 0)$. Show that $(\mathbf{a}, \mathbf{b}, \mathbf{c})$ is left-handed.

Solution (a) From Theorem 13.4.4,

$$[\mathbf{i}, \mathbf{j}, \mathbf{k}] = \begin{vmatrix} 1 & 0 & 0 \\ 0 & 1 & 0 \\ 0 & 0 & 1 \end{vmatrix} = 1 > 0$$

and so $(\mathbf{i}, \mathbf{j}, \mathbf{k})$ is right-handed.

(b) We make use of the properties of determinants obtained in Corollary 13.4.5 to simplify the calculation and obtain

$$[\mathbf{a}, \mathbf{b}, \mathbf{c}] = \begin{vmatrix} 1 & 0 & 1 \\ 2 & -1 & 3 \\ -2 & 0 & 0 \end{vmatrix} = \begin{vmatrix} -2 & 0 & 0 \\ 1 & 0 & 1 \\ 2 & -1 & 3 \end{vmatrix} = (-2) \begin{vmatrix} 0 & 1 \\ -1 & 3 \end{vmatrix} = -2.$$

Hence $(\mathbf{a}, \mathbf{b}, \mathbf{c})$ is left-handed. □

Remark It is no coincidence that the triple $(\mathbf{i}, \mathbf{j}, \mathbf{k})$ is right-handed. The definition of the term right-handed, for a triple of vectors, is made to be consistent with the right-handed coordinate system that was introduced in Section 12.2.

Example 13.4.7 Prove that the points $A(1, 2, 8)$, $B(2, 2, 7)$, $C(3, 1, 7)$ and $D(-1, 6, 6)$ are coplanar. Determine the equation of the plane on which they lie.

Solution Points A, B, C and D lie on a plane if and only if $\mathbf{b} - \mathbf{a}$, $\mathbf{c} - \mathbf{a}$ and $\mathbf{d} - \mathbf{a}$, represented by $\overrightarrow{AB}$, $\overrightarrow{AC}$ and $\overrightarrow{AD}$, respectively, are coplanar. Now

$$[\mathbf{b} - \mathbf{a}, \mathbf{c} - \mathbf{a}, \mathbf{d} - \mathbf{a}] = \begin{vmatrix} 1 & 0 & -1 \\ 2 & -1 & -1 \\ -2 & 4 & -2 \end{vmatrix}$$

$$= 1 \begin{vmatrix} -1 & -1 \\ 4 & -2 \end{vmatrix} + (-1) \begin{vmatrix} 2 & -1 \\ -2 & 4 \end{vmatrix} = 0,$$

and so, by Theorem 13.4.2, the vectors are coplanar. Hence the points are coplanar. A normal vector to the plane is, for example,

$$(\mathbf{b} - \mathbf{a}) \times (\mathbf{c} - \mathbf{a}) = \begin{vmatrix} \mathbf{i} & \mathbf{j} & \mathbf{k} \\ 1 & 0 & -1 \\ 2 & -1 & -1 \end{vmatrix} = (-1, -1, -1),$$

and so we take $\mathbf{n} = (1, 1, 1)$ for simplicity. Thus the equation of the plane is

$$x + y + z = 1 + 2 + 8 = 11.$$ □

13.5 The Vector Triple Product

Definition 13.5.1 The *vector triple product* of $\mathbf{u}$, $\mathbf{v}$ and $\mathbf{w}$ is $\mathbf{u} \times (\mathbf{v} \times \mathbf{w})$.

Hazard The vector triple product is *not* associative, *i.e.* in general

$$\mathbf{u} \times (\mathbf{v} \times \mathbf{w}) \neq (\mathbf{u} \times \mathbf{v}) \times \mathbf{w}.$$

To see why this should be so, we note that $(\mathbf{u} \times \mathbf{v}) \times \mathbf{w}$ is perpendicular to $\mathbf{u} \times \mathbf{v}$ which is normal to a plane determined by $\mathbf{u}$ and $\mathbf{v}$. So, $(\mathbf{u} \times \mathbf{v}) \times \mathbf{w}$ is coplanar with $\mathbf{u}$ and $\mathbf{v}$. By the same argument, $\mathbf{u} \times (\mathbf{v} \times \mathbf{w})$ is coplanar with $\mathbf{v}$ and $\mathbf{w}$. For this reason it is vital that we include the parentheses in a vector triple product to indicate which vector product should be performed first.

We now obtain a formula for the vector triple product which reflects the fact that $\mathbf{u} \times (\mathbf{v} \times \mathbf{w})$, as it is coplanar with $\mathbf{v}$ and $\mathbf{w}$, may be expressed as $\alpha \mathbf{v} + \beta \mathbf{w}$ for some $\alpha, \beta \in \mathbb{R}$.

Theorem 13.5.2 *For all vectors* $\mathbf{u}$, $\mathbf{v}$ *and* $\mathbf{w}$

$$\mathbf{u} \times (\mathbf{v} \times \mathbf{w}) = (\mathbf{u} \cdot \mathbf{w})\mathbf{v} - (\mathbf{u} \cdot \mathbf{v})\mathbf{w}. \tag{13.5.1}$$

Proof Let $\mathbf{u} = (u_x, u_y, u_z)$, $\mathbf{v} = (v_x, v_y, v_z)$ and $\mathbf{w} = (w_x, w_y, w_z)$ and let $\mathbf{v} \times \mathbf{w} = \mathbf{a} = (a_x, a_y, a_z)$. From (13.2.8) we have

$$(a_x, a_y, a_z) = (v_y w_z - v_z w_y, -(v_x w_z - v_z w_x), v_x w_y - v_y w_x).$$

The first component of $\mathbf{u} \times (\mathbf{v} \times \mathbf{w})$ is

$$\begin{aligned}
u_y a_z - u_z a_y &= u_y(v_x w_y - v_y w_x) + u_z(v_x w_z - v_z w_x) \\
&= (u_x w_x + u_y w_y + u_z w_z)v_x - (u_x v_x + u_y v_y + u_z v_z)w_x \\
&= (\mathbf{u} \cdot \mathbf{w})v_x - (\mathbf{u} \cdot \mathbf{v})w_x.
\end{aligned}$$

It is left as an exercise to the reader to verify that the second and third components are $(\mathbf{u} \cdot \mathbf{w})v_y - (\mathbf{u} \cdot \mathbf{v})w_y$ and $(\mathbf{u} \cdot \mathbf{w})v_z - (\mathbf{u} \cdot \mathbf{v})w_z$, respectively, and that the result follows. $\qquad\square$

Example 13.5.3 Express $(\mathbf{u} \times \mathbf{v}) \times \mathbf{w}$ as a linear combination of $\mathbf{u}$ and $\mathbf{v}$.

Solution Using Theorems 13.2.12 (2) and 13.5.2, we get

$$\begin{aligned}
(\mathbf{u} \times \mathbf{v}) \times \mathbf{w} &= -\mathbf{w} \times (\mathbf{u} \times \mathbf{v}) = -\big((\mathbf{w} \cdot \mathbf{v})\mathbf{u} - (\mathbf{w} \cdot \mathbf{u})\mathbf{v}\big) \\
&= -(\mathbf{w} \cdot \mathbf{v})\mathbf{u} + (\mathbf{w} \cdot \mathbf{u})\mathbf{v}. \qquad\square
\end{aligned}$$

Example 13.5.4 Let $\alpha = (\mathbf{a} \times \mathbf{b}) \cdot (\mathbf{a} \times \mathbf{c})$. Show that

$$\alpha = (\mathbf{c} \cdot \mathbf{b})|\mathbf{a}|^2 - (\mathbf{c} \cdot \mathbf{a})(\mathbf{a} \cdot \mathbf{b}).$$

Evaluate α when $\mathbf{a}$, $\mathbf{b}$ and $\mathbf{c}$ are unit vectors with $\mathbf{b}$ and $\mathbf{c}$ perpendicular, and the angles between $\mathbf{a}$ and $\mathbf{b}$, and between $\mathbf{a}$ and $\mathbf{c}$, are both $\pi/3$.

Solution Using Definition 13.4.1 and Theorems 13.4.3 and 13.5.2, we get

$$\begin{aligned}
(\mathbf{a} \times \mathbf{b}) \cdot (\mathbf{a} \times \mathbf{c}) &= [\mathbf{a} \times \mathbf{b}, \mathbf{a}, \mathbf{c}] = [\mathbf{a}, \mathbf{c}, \mathbf{a} \times \mathbf{b}] = \mathbf{a} \cdot (\mathbf{c} \times (\mathbf{a} \times \mathbf{b})) \\
&= \mathbf{a} \cdot ((\mathbf{c} \cdot \mathbf{b})\mathbf{a} - (\mathbf{c} \cdot \mathbf{a})\mathbf{b}) = (\mathbf{c} \cdot \mathbf{b})(\mathbf{a} \cdot \mathbf{a}) - (\mathbf{c} \cdot \mathbf{a})(\mathbf{a} \cdot \mathbf{b}) \\
&= (\mathbf{c} \cdot \mathbf{b})|\mathbf{a}|^2 - (\mathbf{c} \cdot \mathbf{a})(\mathbf{a} \cdot \mathbf{b}).
\end{aligned}$$

With the given conditions, we have $|\mathbf{a}| = |\mathbf{b}| = |\mathbf{c}| = 1$, $\mathbf{c} \cdot \mathbf{b} = 0$, $\mathbf{a} \cdot \mathbf{b} = \mathbf{c} \cdot \mathbf{a} = 1 \times 1 \times \cos \pi/3 = 1/2$ and so

$$\alpha = 0 - \left(\frac{1}{2}\right)^2 = -\frac{1}{4}. \qquad\square$$

13.6　Projections

Projection of a point on a line

Definitions 13.6.1 Consider a line $\mathcal{L}$ with direction vector $\mathbf{u}$ and let P be a point not on $\mathcal{L}$. The *projection* of P on $\mathcal{L}$ is the point Q such that $\mathbf{q} - \mathbf{p}$ is perpendicular to $\mathbf{u}$. If $P \in \mathcal{L}$ then P is its own projection. The *distance* from P to $\mathcal{L}$ is the length $|PQ|$. See Figure 13.6.1.

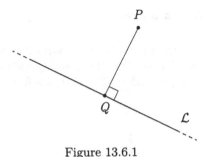

Figure 13.6.1

We refer to *the* projection of a point on a line. The use of the definite article is justified by the next result.

Theorem 13.6.2 *The projection of a point on a line is unique.*

Proof Suppose that Q and Q' are projections of a point P on a line $\mathcal{L}$ with direction vector $\mathbf{u}$. The vectors $\mathbf{q} - \mathbf{p}$ and $\mathbf{q}' - \mathbf{p}$ are both perpendicular to $\mathbf{u}$ and so

$$(\mathbf{q} - \mathbf{p}) \cdot \mathbf{u} = (\mathbf{q}' - \mathbf{p}) \cdot \mathbf{u} = 0,$$

giving,

$$(\mathbf{q} - \mathbf{q}') \cdot \mathbf{u} = 0. \tag{13.6.1}$$

Now suppose that Q and Q' are distinct so that $\mathbf{q} - \mathbf{q}' \neq \mathbf{0}$. Since Q and Q' lie on $\mathcal{L}$, $\mathbf{q} - \mathbf{q}' = \alpha\mathbf{u}$ for some non-zero $\alpha \in \mathbb{R}$ and (13.6.1) gives

$$\alpha\mathbf{u} \cdot \mathbf{u} = \alpha|\mathbf{u}|^2 = 0.$$

This is a contradiction, since both α and $\mathbf{u}$ are non-zero. Hence Q and Q' can not be distinct and the result follows.　　　　　　　　　　　　　　　　　　　　$\square$

Example 13.6.3 Find the projection of $A(1, -2, 3)$ on

$$\mathcal{L}: \frac{x - 4}{1} = \frac{y - 5}{1} = \frac{z + 6}{-1}$$

and the distance of A from $\mathcal{L}$.

Solution A general point B on $\mathcal{L}$ has position vector

$$\mathbf{b} = (4 + t, 5 + t, -6 - t),$$

for some $t \in \mathbb{R}$, and $\mathbf{u} = (1, 1, -1)$ is a direction vector for $\mathcal{L}$. If B is the projection of A on $\mathcal{L}$ then $(\mathbf{b} - \mathbf{a}) \cdot \mathbf{u} = 0$, *i.e.*

$$(3 + t, 7 + t, -9 - t) \cdot (1, 1, -1) = 19 + 3t = 0,$$

giving $t = -19/3$. Thus $\mathbf{b} = (4 - 19/3, 5 - 19/3, -6 + 19/3) = (-7/3, -4/3, 1/3)$.

The distance of A from $\mathcal{L}$ is

$$|AB| = \sqrt{(1 + 7/3)^2 + (-2 + 4/3)^2 + (3 - 1/3)^2} = \frac{2}{3}\sqrt{42}. \qquad \square$$

Projection of a point on a plane

Definitions 13.6.4 Consider a plane $\mathcal{P}$ with normal vector $\mathbf{n}$ and let P be a point not on $\mathcal{P}$. The *projection* of P on $\mathcal{P}$ is the point Q such that $\mathbf{q} - \mathbf{p}$ is parallel to $\mathbf{n}$. If $P \in \mathcal{P}$ then P is its own projection. The *distance* from P to $\mathcal{P}$ is the length $|PQ|$. See Figure 13.6.2.

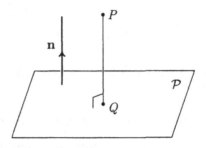

Figure 13.6.2

Example 13.6.5 Find the projection of $P(1, 2, 3)$ on the plane $\mathcal{P} : x + y + z = 15$ and determine the distance of P from $\mathcal{P}$.

Solution Let $Q(\alpha, \beta, \gamma)$ be the projection of P on $\mathcal{P}$. A normal vector for $\mathcal{P}$ is $\mathbf{n} = (1, 1, 1)$. Thus, since $\mathbf{q} - \mathbf{p}$ is normal to $\mathcal{P}$, for some $\lambda \in \mathbb{R}$,

$$(\alpha - 1, \beta - 2, \gamma - 3) = \lambda(1, 1, 1)$$

and, since $Q \in \mathcal{P}$,

$$\alpha + \beta + \gamma = (\lambda + 1) + (\lambda + 2) + (\lambda + 3) = 3(\lambda + 2) = 15.$$

So $\lambda = 3$ and $\mathbf{q} - \mathbf{p} = (3, 3, 3)$ giving $\mathbf{q} = (4, 5, 6)$.

The distance of P from $\mathcal{P}$ is

$$|PQ| = \sqrt{3^2 + 3^2 + 3^2} = 3\sqrt{3}. \qquad \square$$

Problem 13.6.6 Prove that the distance of $P(a, b, c)$ from $\mathcal{P} : lx + my + nz = d$ is

$$\frac{|la + mb + nc - d|}{\sqrt{l^2 + m^2 + n^2}}. \tag{13.6.2}$$

Projection of a line on a plane

Definition 13.6.7 The *projection* $\mathcal{L}'$ of a line $\mathcal{L}$ on a plane $\mathcal{P}$ is the line made up of the set of projections of the points in $\mathcal{L}$ on $\mathcal{P}$. See Figure 13.6.3.

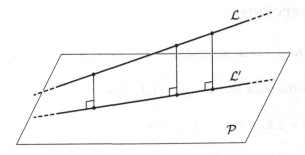

Figure 13.6.3

Remark The technique used in the next example may be generalized to show that the set of projections of the points on a line onto a plane do indeed form a line as Definition 13.6.7 requires.

Example 13.6.8 Find the projection $\mathcal{M}$ of the line

$$\mathcal{L}: \frac{x+2}{2} = \frac{y+1}{-1} = \frac{z}{3}$$

on the plane $\mathcal{P}: x + 2y + z = 2$.

Solution   A general point P on $\mathcal{L}$ has position vector $\mathbf{p} = (-2 + 2t, -1 - t, 3t)$, for some $t \in \mathbb{R}$. Let $Q(a, b, c)$ be the projection of P on $\mathcal{P}$, *i.e.* an arbitrary point on the projected line $\mathcal{M}$. Now

$$\mathbf{q} - \mathbf{p} = (a + 2 - 2t, b + 1 + t, c - 3t) = \alpha(1, 2, 1),$$

for some $\alpha \in \mathbb{R}$, since $\mathbf{q} - \mathbf{p}$ is parallel to a normal vector for $\mathcal{P}$. So

$$\mathbf{q} = (a, b, c) = (-2 + 2t + \alpha, -1 - t + 2\alpha, 3t + \alpha).$$

Further, since $\mathbf{q} \in \mathcal{P}$,

$$a + 2b + c = (-2 + 2t + \alpha) + 2(-1 - t + 2\alpha) + (3t + \alpha)$$
$$= -4 + 3t + 6\alpha = 2.$$

Thus $\alpha = 1 - t/2$, and so the vector equation of $\mathcal{L}'$ is

$$\mathbf{q} = (-1, 1, 1) + t(\tfrac{3}{2}, -2, \tfrac{5}{2}),$$

where $t \in \mathbb{R}$. For a simpler expression we can choose parameter $s = t/2$ so that this is written as

$$\mathbf{q} = (-1, 1, 1) + s(3, -4, 5),$$

giving the equation in symmetric form

$$\mathcal{M}: \frac{x+1}{3} = \frac{y-1}{-4} = \frac{z-1}{5}. \qquad \square$$

13.X   Exercises

1. Let **a** and **b** be unit vectors and let $\pi/3$ be the angle between them. Find $|2\mathbf{a}+\mathbf{b}|$.

2. Determine the angle between $\mathbf{u} = (1,1,2)$ and $\mathbf{v} = (1,0,1)$.

3. For which $\alpha \in \mathbb{R}$ is $(2, \alpha, 3)$ perpendicular to $(\alpha, -\alpha, 1)$?

4. Use scalar products to show that the vector $|\mathbf{b}|\mathbf{a} + |\mathbf{a}|\mathbf{b}$ makes equal angles with non-zero vectors **a** and **b**. See also Exercise 12.X.12.

5. Let $\mathbf{u} = 2\mathbf{i} + 2\mathbf{j} - \mathbf{k}$ and $\mathbf{v} = 2\mathbf{i} - 3\mathbf{j} + \mathbf{k}$. Find
    (a) $\mathbf{u} \cdot \mathbf{v}$,
    (b) the cosine of the angle between **u** and **v**,
    (c) $\alpha \in \mathbb{R}$ such that $2\mathbf{i} - 3\mathbf{j} + \alpha\mathbf{k}$ is perpendicular to **u**.

6. Find all the unit vectors **u** such that **u** is perpendicular to $\mathbf{i} - 2\mathbf{j}$ and the angle between **u** and $\mathbf{i} - \mathbf{k}$ is $\pi/4$.

7. Find the cosine of the acute angle between a body diagonal of a cube and a face diagonal at the same vertex.

8. A vector $\mathbf{v} = (x, y, z)$ makes angles α, β and γ with **i**, **j** and **k**, respectively. By considering $\mathbf{v} \cdot \mathbf{i}$, $\mathbf{v} \cdot \mathbf{j}$ and $\mathbf{v} \cdot \mathbf{k}$, show that
    $$\cos^2 \alpha + \cos^2 \beta + \cos^2 \gamma = 1.$$

9. For the vectors **u** and **v** of Exercise 13.X.5, find $\mathbf{u} \times \mathbf{v}$ and deduce the sine of the angle between **u** and **v**. Find two unit vectors perpendicular to both **u** and **v**.

10. (a) Let **u** and **v** be non-zero vectors such that $|\mathbf{u}| = |\mathbf{v}|$. Prove that $\mathbf{u} + \mathbf{v}$ is perpendicular to $\mathbf{u} - \mathbf{v}$.
    (b) It is given that A, B and C are not collinear and $\triangle ABC$ is not right-angled. Let O be the *circumcentre* of $\triangle ABC$, *i.e.* O is the centre of the unique circle through A, B and C. Let H be such that
    $$\overrightarrow{OH} = \overrightarrow{OA} + \overrightarrow{OB} + \overrightarrow{OC}.$$
    Use (a) to show that $\overrightarrow{AH} \perp \overrightarrow{BC}$. Deduce that the altitudes of $\triangle ABC$ meet at H, *i.e.* H is the *orthocentre* of $\triangle ABC$.

(c) Let G be the centroid of $\triangle ABC$. Show that G, H and O are collinear and find $GH : HO$.

11. Simplify $(\mathbf{a} + \mathbf{b}) \times (\mathbf{a} - \mathbf{b})$.

12. Find the centres and radii of the spheres

$$x^2 + y^2 + z^2 + 4x + 6y - 2z + 5 = 0$$

and

$$x^2 + y^2 + z^2 - 8x - 6y - 8z + 5 = 0.$$

Deduce that the spheres touch externally. Use the section formula to find their point of contact. Find the equation of their common tangent plane.

13. Find the radius of the circle in which the sphere

$$x^2 + y^2 + z^2 - 10x - 6y - 2z - 17 = 0$$

cuts the plane $x + y + z + 3 = 0$.

14. Find the equation of the sphere, with centre $(3, -3, 3)$, which touches the plane $x - 3y + 2z = 4$.

15. Find the equation of the sphere $\mathcal{S}$ which touches the x, y-plane at the origin and which passes through the point $(4, 2, 2)$. Find the equation of the other plane, parallel to the x, y-plane, which touches $\mathcal{S}$.

16. Let $\mathbf{a} = (-5, 4, 2)$, $\mathbf{b} = (-2, 1, 2)$ and $\mathbf{c} = (4, 1, -3)$. Calculate $\mathbf{a} \times \mathbf{b}$ and *deduce* the value of $[\mathbf{a}, \mathbf{b}, \mathbf{c}]$. Is $(\mathbf{a}, \mathbf{b}, \mathbf{c})$ right-handed or left-handed?

17. Prove that, for all vectors $\mathbf{u}$, $\mathbf{v}$, $\mathbf{w}$,

$$\mathbf{u} \times (\mathbf{v} \times \mathbf{w}) + \mathbf{v} \times (\mathbf{w} \times \mathbf{u}) + \mathbf{w} \times (\mathbf{u} \times \mathbf{v}) = \mathbf{0}.$$

18. Let $\mathcal{L}_1$, $\mathcal{L}_2$ be the lines with equations

$$\mathcal{L}_1 : \quad \frac{x-1}{2} = \frac{y}{-1} = \frac{z+3}{1} \quad \text{and} \quad \mathcal{L}_2 : \quad \frac{x-4}{1} = \frac{y+3}{1} = \frac{z+3}{2}.$$

(a) Prove that $\mathcal{L}_1$ and $\mathcal{L}_2$ intersect and find P, their point of intersection.

(b) Verify that Q, the point $(7, 0, 3)$, lies on $\mathcal{L}_2$.

(c) Find the angle between $\mathcal{L}_1$ and $\mathcal{L}_2$.

(d) Deduce from (b) and (c) that there is a point R on $\mathcal{L}_1$ such that $\triangle PQR$ is equilateral. Find R.

19. Find the equations of
    (a) the plane perpendicular to the line
    $$\frac{x-2}{3} = \frac{y+6}{2} = \frac{z-1}{4}$$
    and meeting it at $(2, -6, 1)$,

    (b) the plane through the origin, perpendicular to the line through the points $(2, 2, 5)$ and $(-1, 4, 3)$,

    (c) the plane which contains the point $(4, -1, 5)$ and is parallel to the plane $2x - 3y + z = 4$,

    (d) the plane which contains the points $(2, 5, 4)$ and $(1, 2, -3)$ and is parallel to the z-axis.

20. Find the equation of the plane $\mathcal{P}$ through the points $(-1, 1, 4)$, $(3, 0, 6)$ and $(-6, 2, 1)$. Find the projection of the point $(-1, 5, 0)$ on $\mathcal{P}$.

21. Find, in symmetric form, the equations of the line of intersection of the planes
    $$6x + y - z + 2 = 0 \quad \text{and} \quad 2x - y + 3z - 14 = 0.$$

22. Let $\mathcal{L}$ be the line
    $$\frac{x+1}{2} = \frac{y+3}{3} = \frac{z-2}{-1}$$
    and let $\mathcal{P}$ be the plane
    $$2x + y + 5z = 19.$$
    Find the equation of the plane $\mathcal{P}'$ which is perpendicular to $\mathcal{P}$ and contains $\mathcal{L}$. Find, in symmetric form, the equations of the line of intersection of $\mathcal{P}$ and $\mathcal{P}'$.

23. Find the equation of the plane which is parallel to the line
    $$\frac{x-2}{8} = \frac{y+3}{1} = \frac{z}{1}$$
    and contains the line
    $$\frac{x-6}{-4} = \frac{y+1}{3} = \frac{z-1}{2}.$$

24. Show that the line
    $$\mathcal{L}: \quad \frac{x-2}{1} = \frac{y+3}{-2} = \frac{z-1}{1}$$
    lies on the plane
    $$\mathcal{P}: \quad 3x + y - z = 2.$$
    Find
    (a) the equation of the plane which is perpendicular to $\mathcal{P}$ and contains $\mathcal{L}$,

(b) the equations of the line in $\mathcal{P}$ which meets $\mathcal{L}$ at right-angles at the point $(2, -3, 1)$.

25. Determine the points on the lines

$$\mathcal{L}_1: \quad \frac{x-2}{1} = \frac{y-2}{1} = \frac{z+1}{-2} \quad \text{and} \quad \mathcal{L}_2: \quad \frac{x+2}{1} = \frac{y-2}{0} = \frac{z}{1}$$

which are closest to the origin. Which of the lines passes closer to the origin?

26. Find the projection of the point $(2, 0, 3)$ on the plane $x - y = 4$.

27. Find the equations of the line which is the projection of the line

$$\frac{x-3}{1} = \frac{y+1}{2} = \frac{z}{-2}$$

on the plane $x - y + z = 1$.

Chapter 14

Integration—Fundamentals

In Chapters 8 and 9, we met one of the main branches of calculus—differential calculus. We now introduce the other branch—integral calculus.

In the first two sections, we consider two apparently unrelated problems. We then show that the two are intimately related by the *fundamental theorem of calculus*, a result which links differential and integral calculus.

The first problem considered is an inverse problem. We know from Chapter 8 that, if F is differentiable, then $f = F'$ is another function. If we are *given* f, can we *find* an F with $F' = f$? The function F is called an *antiderivative* of f. The most general antiderivative is the *indefinite integral* of f. Using our table of standard derivatives 'in reverse', we get some *standard integrals*.

The second problem is a more practical one. Suppose that the function f is positive on the interval $[a, b]$. The curve $y = f(x)$, the x-axis and the lines $x = a$ and $x = b$ bound a finite region. The area is the *definite integral of f over $[a, b]$*. The values can be obtained by splitting the area into a number of strips and adding the (approximate) areas of these strips. The sum is a *Riemann sum*. The definite integral is the limit of such sums. We will extend the definition to functions which may take non-positive values.

In Section 14.3, we show that if F is an antiderivative of f, then the definite integral over $[a, b]$ is given by $F(b) - F(a)$. This relates the two types of integral.

In the final section, we consider *improper integrals*. These generalise the definite integral to cases where the range of integration is infinite or the function f approaches infinity at one end of the range. This topic uses limits involving infinity.

14.1   Indefinite Integrals

Definitions 14.1.1 Let F be a differentiable real function with $F' = f$. Then F is an *antiderivative* of f. By Corollary 9.4.8, any other antiderivative of f differs from F by a constant. This is expressed by writing

$$\int f(x)\, dx = F(x) + C.$$

Here, $\displaystyle\int f(x)\, dx$ is the *general antiderivative* or *indefinite integral of f with respect to x*. We call $f(x)$ the *integrand* and C the *constant of integration*.

Some *standard integrals* are listed below. These are included in Appendix E.

$$\int x^q \, dx = \frac{x^{q+1}}{q+1} + C \quad (q \in \mathbb{Q} - \{-1\}),$$

$$\int \sin x \, dx = -\cos x + C, \qquad \int \cos x \, dx = \sin x + C,$$

$$\int \sec^2 x \, dx = \tan x + C, \qquad \int \operatorname{cosec}^2 x \, dx = -\cot x + C,$$

$$\int \sec x \tan x \, dx = \sec x + C, \qquad \int \operatorname{cosec} x \cot x \, dx = -\operatorname{cosec} x + C,$$

$$\int \frac{1}{\sqrt{a^2 - x^2}} \, dx = \sin^{-1} \frac{x}{a} + C \quad (a > 0),$$

$$\int \frac{1}{x^2 + a^2} \, dx = \frac{1}{a} \tan^{-1} \frac{x}{a} + C \quad (a \neq 0).$$

Each of the above formulae can be verified by differentiating the right-hand side. The same is true of the following properties.

Properties 14.1.2 Let F and G be differentiable real functions with $F' = f$ and $G' = g$. Let $\lambda, p, q \in \mathbb{R}$ with $p \neq 0$. Then

(1) $\displaystyle \int \lambda f(x) \, dx = \lambda F(x) + C,$

(2) $\displaystyle \int [f(x) + g(x)] \, dx = F(x) + G(x) + C,$

(3) $\displaystyle \int f(px + q) \, dx = \frac{1}{p} F(px + q) + C.$

Example 14.1.3 Find $\displaystyle \int f(x) \, dx$ when $f(x)$ is

(a) $3x - 1,$ (b) $x^2 + 8x - 3,$ (c) $\cos(5x + 2),$ (d) $\dfrac{1}{(6x + 7)^2},$

(e) $\dfrac{1}{4x^2 + 9},$ (f) $\dfrac{1}{\sqrt{7 - 6x - x^2}},$ (g) $\dfrac{5x - 3}{\sqrt{5x - 7}}.$

Solution We make use of Properties 14.1.2 together with standard integrals.

(a) Using Properties (1) and (2),

$$\int (3x - 1) \, dx = 3 \int x \, dx - \int 1 \, dx = \frac{3x^2}{2} - x + C.$$

Alternatively, using Property 14.1.2 (3),

$$\int (3x - 1) = \frac{1}{3} \frac{(3x - 1)^2}{2} + C = \frac{3x^2}{2} - x + \frac{1}{6} + C.$$

Both answers are correct. Remember that C is an *arbitrary* constant so that C and $(1/6) + C$ are effectively the same.

(b) Using Property (2) then Property (1) we have

$$\int (x^2 + 8x - 3)\, dx = \int x^2\, dx + \int 8x\, dx + \int (-3)\, dx$$

$$= \int x^2\, dx + 8 \int x\, dx - 3 \int 1\, dx$$

$$= \frac{x^3}{3} + 8\frac{x^2}{2} - 3x + C = \frac{x^3}{3} + 4x^2 - 3x + C.$$

(c) Property (3) gives

$$\int \cos(5x + 2)\, dx = \frac{1}{5}\sin(5x + 2) + C.$$

(d) Again using Property (3),

$$\int \frac{1}{(6x + 7)^2}\, dx = \int (6x + 7)^{-2}\, dx = \frac{1}{6}\frac{(6x + 7)^{-1}}{-1} + C$$

$$= \frac{-1}{6(6x + 7)} + C.$$

(e) Using Property (1),

$$\int \frac{1}{4x^2 + 9}\, dx = \frac{1}{4}\int \frac{1}{x^2 + (3/2)^2}\, dx = \frac{1}{4}\left[\frac{1}{(3/2)}\tan^{-1}\frac{x}{(3/2)} + K\right]$$

$$= \frac{1}{6}\tan^{-1}\frac{2x}{3} + C,$$

where $C = K/4$. Alternatively, using Property (3),

$$\int \frac{1}{4x^2 + 9}\, dx = \int \frac{1}{(2x)^2 + 3^2}\, dx = \frac{1}{2}\left(\frac{1}{3}\tan^{-1}\frac{2x}{3}\right) + C$$

$$= \frac{1}{6}\tan^{-1}\frac{2x}{3} + C.$$

(f) We first complete the square of the quadratic and then apply Property (3).

$$\int \frac{1}{\sqrt{7 - 6x - x^2}}\, dx = \int \frac{1}{\sqrt{16 - (x + 3)^2}}\, dx = \int \frac{1}{\sqrt{4^2 - (x + 3)^2}}\, dx$$

$$= \sin^{-1}\frac{x + 3}{4} + C.$$

(g) We have

$$\int \frac{5x - 3}{\sqrt{5x - 7}}\, dx = \int \frac{(5x - 7) + 4}{\sqrt{5x - 7}}\, dx = \int \left[\sqrt{5x - 7} + \frac{4}{\sqrt{5x - 7}}\right] dx$$

$$= \int (5x - 7)^{1/2}\, dx + 4\int (5x - 7)^{-1/2}\, dx$$

$$= \frac{1}{5}\frac{(5x - 7)^{3/2}}{(3/2)} + \frac{4}{5}\frac{(5x - 7)^{1/2}}{(1/2)} + C$$

$$= \frac{2}{5}(5x - 7)^{1/2}\left(\frac{5x - 7}{3} + 4\right) + C = \frac{2}{3}\sqrt{5x - 7}\,(x + 1) + C,$$

having used Properties (1), (2) and (3).   □

14.2   Definite Integrals

Let f be a real function, continuous on $[a,b]$. Roughly speaking, the *definite integral of f over $[a,b]$*, denoted by

$$\int_a^b f(x)\,dx,$$

is the area of the finite region bounded by the curve $y = f(x)$, the x-axis and the lines $x = a$ and $x = b$ where the areas of regions above and below the x-axis are regarded as positive and negative, respectively. Note, however, that in figures where areas of bounded regions are indicated, the areas will be given their positive values. Thus, for example, Figure 14.2.1 illustrates that

$$\int_0^2 (3-x)\,dx = 4 \qquad \text{and} \qquad \int_2^5 (3-x)\,dx = (1/2) + (-2) = -3/2.$$

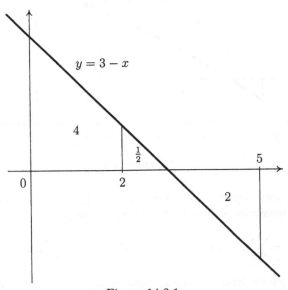

Figure 14.2.1

In making things more precise we begin with continuous functions and later extend the definition to certain functions with discontinuities.

Definitions 14.2.1 Let f be a real function, continuous on $[a,b]$, let $n \in \mathbb{N}$ and let $x_0, x_1, \ldots, x_n \in \mathbb{R}$, with

$$a = x_0 < x_1 < \cdots < x_n = b$$

We call $\mathcal{P} = (x_0, x_1, \ldots, x_n)$ a *partition* of $[a,b]$. The *size* of $\mathcal{P}$, written $|\mathcal{P}|$, is the length of the longest of the subintervals $[x_{r-1}, x_r]$ $(r = 1, \ldots, n)$, *i.e.*

$$|\mathcal{P}| = \max\{|x_r - x_{r-1}| : r = 1, \ldots, n\}.$$

For each $r = 1, \ldots, n$, let $c_r \in [x_{r-1}, x_r]$. Then

$$S = \sum_{r=1}^{n} (x_r - x_{r-1}) f(c_r)$$

is called a *Riemann sum* for $\mathcal{P}$. When $|\mathcal{P}|$ is 'small', the Riemann sum gives an approximation to the area of the finite region bounded by the curve $y = f(x)$, the x-axis and the lines $x = a$ and $x = b$, with the convention, already mentioned, that areas of regions above and below the x-axis are positive and negative, respectively. See Figure 14.2.2.

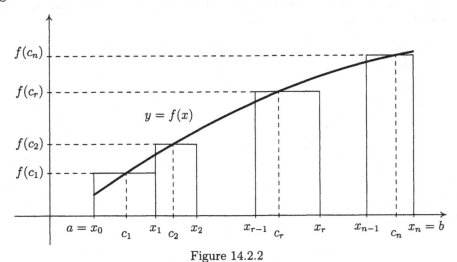

Figure 14.2.2

By the extreme value theorem, for each $r = 1, \ldots, n$, there exist m_r and M_r, the minimum and maximum values, respectively, of f in $[x_{r-1}, x_r]$. We call

$$L = \sum_{r=1}^{n} (x_r - x_{r-1}) m_r \qquad \text{and} \qquad U = \sum_{r=1}^{n} (x_r - x_{r-1}) M_r$$

the *lower* and *upper* Riemann sums, respectively. See Figure 14.2.3. Observe that, since $m_r \le f(c_r) \le M_r$ $(r = 1, \ldots, n)$,

$$L \le S \le U. \tag{14.2.1}$$

It may be proved that if we choose partitions of $[a, b]$ so that the sizes tend to 0 then the corresponding lower and upper Riemann sums tend to the same limit. We denote this limit by

$$\int_a^b f(x) \, dx$$

and call it the *Riemann* or *definite integral of f over $[a, b]$*. It is also referred to as the *area under the curve $y = f(x)$ from a to b*.

 Hazard The phrase 'area under the curve' is used even when the region in question lies wholly or partly *above* the curve. See Figure 14.2.1

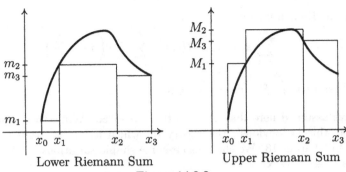

Lower Riemann Sum   Upper Riemann Sum

Figure 14.2.3

Notes 14.2.2 (1) Limits of Riemann sums behave in the same way as function limits. Where appropriate, we shall apply the properties of function limits given in Section 7.2 to limits of Riemann sums.

(2) The definition of Riemann integral assumes that the lower and upper Riemann sums tend to the same limit. A proof of this fact is beyond the scope of this book. However, having made the assumption, inequalities (14.2.1) together with the sandwich principle mean that for each partition we can take *any* Riemann sum and, provided the sizes of the partitions tend to 0, the limit of the Riemann sums will be the Riemann integral.

(3) When evaluating Riemann integrals from first principles, *i.e.* using Riemann sums, we often proceed as follows.

With the notation of Definition 14.2.1, let $\Delta = b - a$ and, for each $n \in \mathbb{N}$, choose the partition $\mathcal{P}_n$ which divides the interval $[a, b]$ into n subintervals each of length Δ/n. So

$$\mathcal{P}_n = (x_0, x_1, \ldots, x_n)$$

where,

$$x_r = a + r\frac{\Delta}{n} \quad (r = 0, 1, \ldots, n).$$

See Figure 14.2.4. Then $|\mathcal{P}_n| = \Delta/n \to 0$ as $n \to \infty$ (see below).

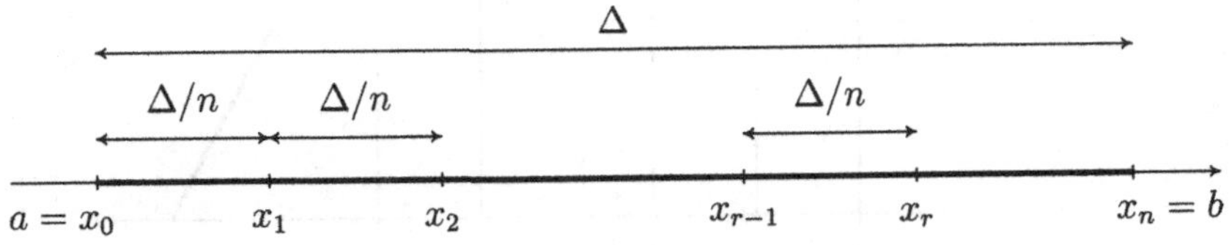

$$\Delta$$

$\Delta/n \quad \Delta/n \qquad\qquad \Delta/n$

$a = x_0 \qquad x_1 \qquad x_2 \qquad\qquad x_{r-1} \qquad x_r \qquad\qquad x_n = b$

Figure 14.2.4

Next we form the Riemann sum which evaluates the function at the right end-point of each subinterval, *i.e.* we take

$$c_r = x_r = a + r\frac{\Delta}{n} \quad (r = 1, \ldots, n).$$

This gives the Riemann sum

$$\sum_{r=1}^{n}(x_r - x_{r-1})f(c_r) = \sum_{r=1}^{n}\frac{\Delta}{n}f\left(a + r\frac{\Delta}{n}\right)$$

which will tend to $\displaystyle\int_a^b f(x)\,dx$ as $n \to \infty$.

We have assumed here that $\Delta/n \to 0$ as $n \to \infty$. While the reader may readily accept that this is 'obvious', the theory on which a formal proof can be based is contained in Chapter 18. We shall also assume simple variations on this result such as $2 + 1/n \to 2$ as $n \to \infty$.

(4)　In the expression $\displaystyle\int_a^b f(x)\,dx$, x is a 'dummy' variable. Any symbol will do. Thus

$$\int_a^b f(x)\,dx = \int_a^b f(t)\,dt = \int_a^b f(\theta)\,d\theta = \cdots.$$

Example 14.2.3 By arguing from first principles, evaluate

$$\text{(a) } \int_0^1 (1 - x^2)\,dx, \qquad \text{(b) } \int_1^3 (2x - 5)\,dx.$$

Solution　We use the notation of Definition 14.2.1 and the method given in Note 14.2.2 (3).

(a)　The solution is illustrated in Figure 14.2.5

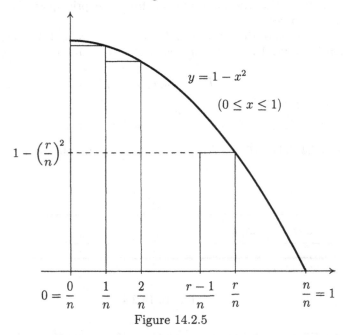

$$y = 1 - x^2$$
$$(0 \leq x \leq 1)$$

Figure 14.2.5

For $n \in \mathbb{N}$, let $\mathcal{P}_n$ be the partition

$$\left(\frac{0}{n}, \frac{1}{n}, \frac{2}{n}, \ldots, \frac{n}{n}\right),$$

i.e. $x_r = r/n$ $(r = 0, 1, \ldots, n)$. Then $|\mathcal{P}_n| = 1/n \to 0$ as $n \to \infty$.

Taking right end-points, so that $c_r = x_r = r/n$ $(r = 1, \ldots, n)$, we get the (lower) Riemann sum

$$\sum_{r=1}^{n}(x_r - x_{r-1})f(c_r) = \sum_{r=1}^{n}\frac{b-a}{n}[1-c_r^2] = \sum_{r=1}^{n}\frac{1}{n}\left[1-\left(\frac{r}{n}\right)^2\right]$$

$$= \frac{1}{n}\sum_{r=1}^{n}1 - \frac{1}{n^3}\sum_{r=1}^{n}r^2$$

$$= \frac{1}{n}n - \frac{1}{n^3}\frac{1}{6}(n+1)(2n+1) \quad \text{(see Theorem 4.1.5 (2))}$$

$$= 1 - \frac{1}{6}\left(1+\frac{1}{n}\right)\left(2+\frac{1}{n}\right)$$

$$\to 1 - \frac{1}{6}(1)(2) = \frac{2}{3} \quad \text{as } n \to \infty.$$

Hence

$$\int_0^1 (1-x^2)\,dx = \frac{2}{3}.$$

(b) For $n \in \mathbb{N}$, let $\mathcal{P}_n$ be the partition

$$\left(1+\frac{0}{n}, 1+\frac{2}{n}, 1+\frac{4}{n}, \ldots, 1+\frac{2n}{n}\right)$$

which divides $[1,3]$ into n subintervals each of length $2/n$. Then $|\mathcal{P}_n| = 2/n \to 0$ as $n \to \infty$. Taking right end-points we get the (upper) Riemann sum

$$\sum_{r=1}^{n}(x_r - x_{r-1})f(c_r) = \sum_{r=1}^{n}\frac{b-a}{n}[2c_r - 5] = \sum_{r=1}^{n}\frac{2}{n}\left[2\left(1+\frac{2r}{n}\right)-5\right]$$

$$= \sum_{r=1}^{n}\left(-\frac{6}{n}+\frac{8r}{n^2}\right) = -\frac{6}{n}\sum_{r=1}^{n}1 + \frac{8}{n^2}\sum_{r=1}^{n}r$$

$$= -\frac{6}{n}n + \frac{8}{n^2}\frac{n}{2}(n+1) \quad \text{(see Theorem 4.1.5 (1))}$$

$$= -6 + 4\left(1+\frac{1}{n}\right)$$

$$\to -6 + 4(1) = -2 \quad \text{as } n \to \infty.$$

Hence

$$\int_1^3 (2x-5)\,dx = -2. \qquad \square$$

Problem 14.2.4 Show that

$$\sum_{r=1}^{n}2\sin r\theta\sin\tfrac{1}{2}\theta = \cos\tfrac{1}{2}\theta - \cos(n+\tfrac{1}{2})\theta.$$

Deduce, by arguing from first principles, that $\displaystyle\int_0^{\pi/2}\sin x\,dx = 1$.

In the introduction to this section and in Definition 14.2.1 we referred to areas of regions bounded by continuous curves. Formally, we use definite integrals to define the areas of such regions.

Definition 14.2.5 Let f and g be real functions, continuous on $[a, b]$, with $f(x) \geq g(x)$ for all $x \in [a, b]$. Let $\mathcal{R}$ be the finite region bounded by the curves $y = f(x)$, $y = g(x)$ and the lines $x = a$ and $x = b$, *i.e.*

$$\mathcal{R} = \{(x, y) \in \mathbb{R}^2 : a \leq x \leq b,\ g(x) \leq y \leq f(x)\}.$$

Then we define the *area* of $\mathcal{R}$ to be

$$\int_a^b f(x)\,dx - \int_a^b g(x)\,dx.$$

In particular, when $g(x) = 0$ $(a \leq x \leq b)$ so that $f(x) \geq 0$ $(a \leq x \leq b)$, the area of $\mathcal{R}$ is $\int_a^b f(x)\,dx$.

Figure 14.2.6 gives an intuitive view of the definition. With $A, \ldots, D$ denoting areas as shown, we have

(a)   area $\mathcal{R} = (A + B) - B = A$,

(b)   area $\mathcal{R} = (A + B + C) - (B + C - D) = A + D$.

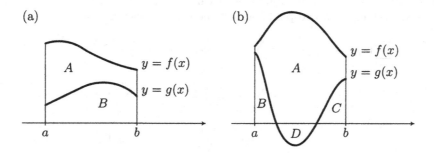

Figure 14.2.6

 Hazard In Definition 14.2.5, the convention whereby regions above and below the x-axis have positive and negative areas, respectively, does *not* apply to the area of $\mathcal{R}$.

We shall postpone examples on areas until after the fundamental theorem of calculus in Section 14.3. For the present, we leave the reader with a problem which relies on an earlier evaluation.

Problem 14.2.6 Evaluate $\int_0^1 \sqrt{1 - x}\,dx$ by regarding it as an area and comparing it with $\int_0^1 (1 - x^2)\,dx$.

Definitions 14.2.7 The method used in Definition 14.2.1 to define the Riemann integral of a real function, continuous on $[a, b]$, can be applied to certain functions with discontinuities. Any function to which the method applies is said to be *Riemann integrable over* $[a, b]$.

Let f be a real function, defined on $[a, b]$, and suppose we can find a partition $(x_0, x_1, \ldots, x_n)$ of $[a, b]$ such that, for each $r = 1, \ldots, n$ there exists a real function f_r, continuous on $[x_{r-1}, x_r]$ with $f(x) = f_r(x)$ for all $x \in (x_{r-1}, x_r)$. Then f is said to be *piecewise continuous on* $[a, b]$. It can be shown that f is Riemann integrable over $[a, b]$ with

$$\int_a^b f(x)\, dx = \int_{x_0}^{x_1} f_1(x)\, dx + \int_{x_1}^{x_2} f_2(x)\, dx + \cdots + \int_{x_{n-1}}^{x_n} f_n(x)\, dx.$$

To illustrate the definition of piecewise continuous, consider $f : [-1, 2] \to \mathbb{R}$ defined by

$$f(x) = \begin{cases} 3/2 & \text{if } -1 \le x \le 0, \\ (2 - x)^2 & \text{if } 0 < x \le 2. \end{cases}$$

We can take $f_1 : [-1, 0] \to \mathbb{R}$ defined by $f_1(x) = 3/2$ and $f_2 : [0, 2] \to \mathbb{R}$ defined by $f_2(x) = (2 - x)^2$. This shows that f is piecewise continuous on $[-1, 2]$ and

$$\int_{-1}^2 f(x)\, dx = \int_{-1}^0 (3/2)\, dx + \int_0^2 (2 - x)^2\, dx.$$

On the other hand, $g : [-1, 2] \to \mathbb{R}$ defined by

$$g(x) = \begin{cases} 3/2 & \text{if } -1 \le x \le 0, \\ (2 - x)/x & \text{if } 0 < x \le 2, \end{cases}$$

can be shown to be neither piecewise continuous on $[-1, 2]$ nor Riemann integrable over $[-1, 2]$.

The graphs of f and g are sketched in Figure 14.2.7.

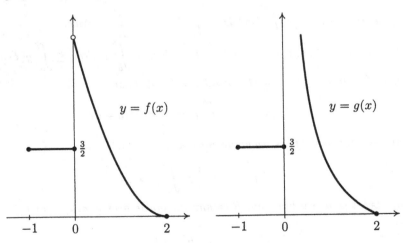

Figure 14.2.7

Notes 14.2.8 (1) If f is a real function, Riemann integrable over $[a, b]$, and $a \leq c < d \leq b$ then f is Riemann integrable over $[c, d]$.

(2) It follows directly from the Definition 14.2.7 that a real function f is piecewise continuous on $[a, b]$ if and only if there is a partition $(x_0, x_1, \ldots, x_n)$ of $[a, b]$ such that, for each $r = 1, \ldots, n$, f is continuous on (x_{r-1}, x_r) and the right and left limits of f at x_{r-1} and x_r, respectively, both exist.

For convenience, we further extend the definition of definite integral as follows.

Definitions 14.2.9 For all suitable real functions f, and $a, b \in \mathbb{R}$ with $a < b$, we define

$$\int_a^a f(x)\,dx = 0 \qquad \text{and} \qquad \int_b^a f(x)\,dx = -\int_a^b f(x)\,dx.$$

Theorem 14.2.10 (Properties of Definite Integrals) *Let* $K, \lambda \in \mathbb{R}$ *and let* f *and* g *be real functions, Riemann integrable over* $[a, b]$.

(1) *The real (constant) function* $x \mapsto K$ *is Riemann integrable over* $[a, b]$ *with*

$$\int_a^b K\,dx = K(b - a).$$

In particular,

$$\int_a^b 0\,dx = 0 \qquad \text{and} \qquad \int_a^b 1\,dx = b - a.$$

(2) *The functions* λf *and* $f + g$ *are Riemann integrable over* $[a, b]$ *with*

(a) $\displaystyle\int_a^b \lambda f(x)\,dx = \lambda \int_a^b f(x)\,dx,$

(b) $\displaystyle\int_a^b [f(x) + g(x)]\,dx = \int_a^b f(x)\,dx + \int_a^b g(x)\,dx.$

(3) *For* $c \in [a, b]$, $\displaystyle\int_a^b f(x)\,dx = \int_a^c f(x)\,dx + \int_c^b f(x)\,dx.$

(4) *Let* $f(x) \leq g(x)$ *for all* $x \in (a, b)$. *Then* $\displaystyle\int_a^b f(x)\,dx \leq \int_a^b g(x)\,dx.$

In particular, if $m \leq f(x) \leq M$ *for all* $x \in (a, b)$ *then*

$$m(b - a) \leq \int_a^b f(x)\,dx \leq M(b - a).$$

(5) *The function* $|f|$ *is Riemann integrable over* $[a, b]$ *and*

$$\left| \int_a^b f(x)\,dx \right| \leq \int_a^b |f(x)|\,dx.$$

(6) *Let* h *be a real function, Riemann integrable over* $[-k, k]$. *Then*

(a) h *odd* $\Rightarrow \displaystyle\int_{-k}^k h(x)\,dx = 0.$

(b) h *even* $\Rightarrow \displaystyle\int_{-k}^k h(x)\,dx = 2 \int_0^k h(x)\,dx,$

Proof We shall prove Properties (1) and (2) only. Properties (3) to (6) are illustrated in Figure 14.2.8 (a) to (d), respectively.

Let $\mathcal{P} = (x_0, x_1, \ldots, x_n)$ be a partition of $[a, b]$, let $c_r \in [x_{r-1}, x_r]$ $(r = 1, \ldots, n)$ and let $\Delta_r = x_r - x_{r-1}$ $(r = 1, \ldots, n)$. Using Theorem 7.2.2, we have

$$\sum_{r=1}^{n} \Delta_r K = K \sum_{r=1}^{n} \Delta_r = K(b - a) \to K(b - a) \text{ as } |\mathcal{P}| \to 0,$$

and using Property (2) of Theorem 7.2.3, we have

$$\sum_{r=1}^{n} \Delta_r \lambda f(c_r) = \lambda \sum_{r=1}^{n} \Delta_r f(c_r) \to \lambda \int_a^b f(x)\, dx \text{ as } |\mathcal{P}| \to 0$$

and

$$\sum_{r=1}^{n} \Delta_r [f(c_r) + g(c_r)] = \sum_{r=1}^{n} \Delta_r f(c_r) + \sum_{r=1}^{n} \Delta_r g(c_r)$$

$$\to \int_a^b f(x)\, dx + \int_a^b g(x)\, dx \text{ as } |\mathcal{P}| \to 0.$$

In particular, these limits hold for lower and upper Riemann sums so that the functions $x \mapsto K$, λf and $f + g$ are Riemann integrable, and Properties (1) and (2) follow immediately. $\quad\square$

Corollary 14.2.11 *Let f and g be real functions, continuous on $[a, b]$, with $f(x) \geq g(x)$ for all $x \in [a, b]$. Let $\mathcal{R}$ be the finite region bounded by the curves $y = f(x)$, $y = g(x)$ and the lines $x = a$ and $x = b$. Then*

$$\text{area } \mathcal{R} = \int_a^b [f(x) - g(x)]\, dx.$$

Proof This follows directly from Definition 14.2.5 and Theorem 14.2.10 (2). $\quad\square$

Notes 14.2.12 (1) Properties (1) to (3) of Theorem 14.2.10 remain true when $a = b$ and also when $a > b$ provided we replace $[a, b]$ with $[b, a]$.

(2) Property (3) of Theorem 14.2.10 remains true when c lies outside $[a, b]$ provided f is Riemann integrable on an interval containing a, b and c. For example, if $a < b < c$ and f is Riemann integrable over $[a, c]$ then, by Property (3),

$$\int_a^c f(x)\, dx = \int_a^b f(x)\, dx + \int_b^c f(x)\, dx$$

so that

$$\int_a^b f(x)\, dx = \int_a^c f(x)\, dx - \int_b^c f(x)\, dx = \int_a^c f(x)\, dx + \int_c^b f(x)\, dx.$$

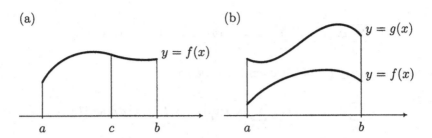

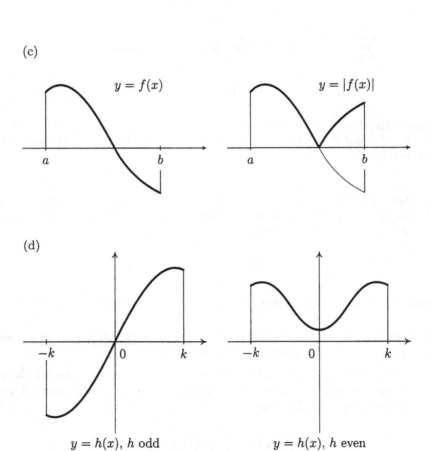

Figure 14.2.8

Example 14.2.13 Evaluate $\mathcal{I} = \int_{-2}^{2} [3 - 2\sin x]\, dx$.

Solution The properties referred to are those of Theorem 14.2.10. We have

$$\mathcal{I} = \int_{-2}^{2} 3\, dx - 2 \int_{-2}^{2} \sin x\, dx \qquad \text{(by Property (2))}$$

$$= 2 \int_{0}^{2} 3\, dx - 2(0) \qquad \text{(by Property (6))}$$

$$= 2(3)(2 - 0) + 0 = 12 \qquad \text{(by Property (1))}. \qquad \square$$

Example 14.2.14 Use the fact that $1 - x^2 \le \sqrt{1 - x^2} \le 1$ $(0 \le x \le 1)$ to show that

$$2/3 \le \int_{0}^{1} \sqrt{1 - x^2}\, dx \le 1.$$

Solution Using Property (4) of Theorem 14.2.10 we have

$$\int_{0}^{1} (1 - x^2)\, dx \le \int_{0}^{1} \sqrt{1 - x^2}\, dx \le \int_{0}^{1} 1\, dx.$$

Hence, using Example 14.2.3 and Property (1) of Theorem 14.2.10,

$$2/3 \le \int_{0}^{1} \sqrt{1 - x^2}\, dx \le 1$$

as required. $\qquad \square$

14.3 The Fundamental Theorem of Calculus

Our aim in this section is to relate indefinite and definite integrals. From a practical point of view this provides us with a means of using antiderivatives to evaluate Riemann integrals thereby avoiding the need for arguing from first principles.

We give the result in two parts (A and B). Together they form the *fundamental theorem of calculus*.

Theorem 14.3.1 (Fundamental Theorem of Calculus—A) *Let f be a real function, continuous on $[a, b]$ and let $c \in [a, b]$. Define $F : [a, b] \to \mathbb{R}$ by*

$$F(x) = \int_{c}^{x} f(t)\, dt.$$

Then F is differentiable on $[a, b]$ with $F' = f$, i.e. on $[a, b]$, and F is an antiderivative of f.

Proof The properties referred to are those of Theorem 14.2.10.

Let $x, x + h \in [a, b]$ with $h \ne 0$. We shall deal with the case where $h > 0$ and $x \in [a, b)$. The case $h < 0$ and $x \in (a, b]$ is similar. Let $I = [x, x + h]$. By Property (3) and Note 14.2.12 (2),

$$\int_{x}^{x+h} f(t)\, dt = \int_{x}^{c} f(t)\, dt + \int_{c}^{x+h} f(t)\, dt = \int_{c}^{x+h} f(t)\, dt - \int_{c}^{x} f(t)\, dt,$$

i.e.

$$\int_x^{x+h} f(t)\, dt = F(x+h) - F(x). \tag{14.3.1}$$

By the extreme value theorem, there exist $\xi, \eta \in I$ such that, for all $t \in I$,

$$f(\xi) \leq f(t) \leq f(\eta).$$

So, by Property (4),

$$\int_x^{x+h} f(\xi)\, dt \leq \int_x^{x+h} f(t)\, dt \leq \int_x^{x+h} f(\eta)\, dt$$

and hence, by Property (1),

$$f(\xi)\, h \leq \int_x^{x+h} f(t)\, dt \leq f(\eta)\, h.$$

Dividing by h and using (14.3.1) gives

$$f(\xi) \leq \frac{F(x+h) - F(x)}{h} \leq f(\eta).$$

Since f is continuous at x, $f(\xi) \to f(x)$ and $f(\eta) \to f(x)$ as $h \to 0$. So, by the sandwich principle,

$$\frac{F(x+h) - F(x)}{h} \to f(x) \text{ as } h \to 0$$

and it follows that $F'(x)$ exists and equals $f(x)$ as required.   □

Note 14.3.2 With the notation of Theorem 14.3.1,

$$\frac{d}{dx}\left[\int_c^x f(t)\, dt\right] = f(x).$$

Example 14.3.3 Find $\dfrac{d}{dx}\displaystyle\int_x^{2x} \sqrt{1+t^2}\, dt$.

Solution Let f be the real function defined by $f(x) = \displaystyle\int_0^x \sqrt{1+t^2}\, dt$. Then, using Theorem 14.2.10 (3) and Note 14.2.12 (2), for $x \in \mathbb{R}$,

$$\int_x^{2x} \sqrt{1+t^2}\, dt = \int_0^{2x} \sqrt{1+t^2}\, dt - \int_0^x \sqrt{1+t^2}\, dt = f(2x) - f(x).$$

By the fundamental theorem of calculus (A), f is differentiable on $\mathbb{R}$ with $f'(x) = \sqrt{1+x^2}$. Hence,

$$\frac{d}{dx}\int_x^{2x} \sqrt{1+t^2}\, dt = 2f'(2x) - f'(x) = 2\sqrt{1+4x^2} - \sqrt{1+x^2}.$$ □

Notation 14.3.4 Let G be any real function defined on $\{a, b\}$. Then we write

$$\left[G(x)\right]_a^b = G(b) - G(a).$$

Theorem 14.3.5 (Fundamental Theorem of Calculus—B) *Let f be a real function, continuous on $[a, b]$ and let G be an antiderivative of f on $[a, b]$. Then*

$$\int_a^b f(x)\, dx = [G(x)]_a^b.$$

Proof Define $F : [a, b] \to \mathbb{R}$ by

$$F(x) = \int_a^x f(t)\, dt.$$

Then, by Theorem 14.3.1, $F' = f$ on $[a, b]$. Hence, F and G are real functions, continuous on $[a, b]$ and differentiable on (a, b) with

$$F'(x) = G'(x) \qquad (a < x < b).$$

So, by Corollary 9.4.8, there exists $C \in \mathbb{R}$ such that

$$F(x) = G(x) + C \qquad (a \le x \le b).$$

Putting $x = a$ gives

$$F(a) = G(a) + C, \qquad \text{i.e. } C = 0 - G(a) = -G(a),$$

since $F(a) = 0$ (see Definition 14.2.9). Then putting $x = b$ gives

$$F(b) = G(b) + C, \qquad \text{i.e. } \int_a^b f(t)dt = G(b) - G(a)$$

as required. $\qquad\qquad\qquad\qquad\qquad\qquad\qquad\qquad\qquad\qquad\qquad\qquad\qquad$ $\square$

Example 14.3.6 Evaluate

$$\text{(a) } \int_0^1 (1 - x^2)\, dx, \qquad \text{(b) } \int_0^{\pi/6} \sin 3\theta \, d\theta, \qquad \text{(c) } \int_{-1}^1 \frac{1}{\sqrt{2 - x^2}}\, dx.$$

Solution We apply Theorem 14.3.5. The antiderivatives are found using standard integrals and Properties 14.1.2. We have

$$\text{(a) } \int_0^1 (1 - x^2)\, dx = \left[x - \frac{x^3}{3} \right]_0^1 = \left(1 - \frac{1}{3} \right) - \left(0 - \frac{0}{3} \right) = \frac{2}{3},$$

$$\text{(b) } \int_0^{\pi/6} \sin 3\theta \, d\theta = \left[\frac{1}{3}(-\cos 3\theta) \right]_0^{\pi/6} = \left(-\frac{\cos(\pi/2)}{3} \right) - \left(-\frac{\cos 0}{3} \right)$$

$$= 0 - \left(-\frac{1}{3} \right) = \frac{1}{3},$$

$$\text{(c) } \int_{-1}^1 \frac{dx}{\sqrt{2 - x^2}} = \left[\sin^{-1} \frac{x}{\sqrt{2}} \right]_{-1}^1 = \left(\sin^{-1} \frac{1}{\sqrt{2}} \right) - \left(\sin^{-1} \left(-\frac{1}{\sqrt{2}} \right) \right)$$

$$= \frac{\pi}{4} - \left(-\frac{\pi}{4} \right) = \frac{\pi}{2}. \qquad\qquad\qquad\qquad\qquad \square$$

Remark It is common practice to write, for example,

$$\int_0^1 \frac{dx}{x^2+4} \quad \text{for} \quad \int_0^1 \frac{1}{x^2+4}\,dx, \qquad \int_1^3 \frac{5x\,dx}{7x+1} \quad \text{for} \quad \int_1^3 \frac{5x}{7x+1}\,dx.$$

Indefinite integrals are treated similarly.

Example 14.3.7 Find the area of the finite region bounded by the curves

$$y = x^3 - 9x \quad \text{and} \quad y = 9 - x^2.$$

Solution Let $y_1 = x^3 - 9x$ and $y_2 = 9 - x^2$. Then

$$y_1 = y_2 \;\Leftrightarrow\; x^3 - 9x = 9 - x^2 \;\Leftrightarrow\; x^3 + x^2 - 9x - 9 = 0$$
$$\Leftrightarrow\; (x+1)(x^2-9) = 0 \;\Leftrightarrow\; x = -1,\; x = -3 \text{ or } x = 3.$$

Hence, the curves intersect at the points $(-3,0)$, $(-1,8)$ and $(3,0)$. The finite region bounded by the curves is sketched in Figure 14.3.1.

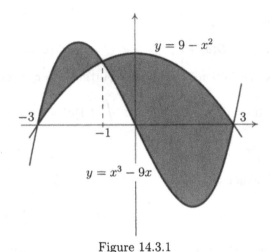

Figure 14.3.1

For $x \in [-3,-1]$, $y_1 \geq y_2$ and for $x \in [-1,3]$, $y_2 \leq y_1$. So, using Corollary 14.2.11, the required area is

$$\int_{-3}^{-1} (y_1 - y_2)\,dx + \int_{-1}^{3} (y_2 - y_1)\,dx$$

$$= \int_{-3}^{-1} (x^3 + x^2 - 9x - 9)\,dx + \int_{-1}^{3} (-x^3 - x^2 + 9x + 9)\,dx$$

$$= \left[\frac{x^4}{4} + \frac{x^3}{3} - \frac{9x^2}{2} - 9x \right]_{-3}^{-1} + \left[-\frac{x^4}{4} - \frac{x^3}{3} + \frac{9x^2}{2} + 9x \right]_{-1}^{3}$$

$$= \frac{53}{12} - \left(-\frac{9}{4} \right) + \frac{153}{4} - \left(-\frac{53}{12} \right) = \frac{148}{3}. \qquad \square$$

Example 14.3.8 Find the area of the finite region bounded by the curves

$$y^2 - y = x \quad \text{and} \quad y = \frac{x}{3}.$$

Solution The curves intersect at the points $(0,0)$ and $(12,4)$. The finite region bounded by the curves is sketched in Figure 14.3.2.

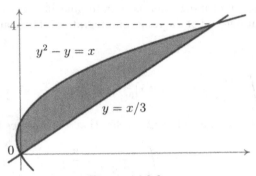

Figure 14.3.2

Here we can interchange the roles of x and y and find the area by integrating with respect to y from 0 to 4. Let $x_1 = y^2 - y$ and $x_2 = 3y$. Then the required area is

$$\int_0^4 (x_2 - x_1)\, dy = \int_0^4 (4y - y^2)\, dy = \left[2y^2 - (y^3/3)\right]_0^4 = 32/3. \qquad \square$$

Example 14.3.9 Let $a > 0$. Verify by differentiation that

$$\int \sqrt{a^2 - x^2}\, dx = \frac{x}{2}\sqrt{a^2 - x^2} + \frac{a^2}{2}\sin^{-1}\frac{x}{a} + C.$$

Hence show that (the region enclosed by) a circle of radius a has area πa^2.

Solution The verification of the indefinite integral is left to the reader.

We shall assume that all circles with radius a have the same area. Consider the circle, centre the origin, radius a, *i.e.* the circle $x^2 + y^2 = a^2$. We further assume that the part of this circle in the first quadrant encloses a quarter of its area. Hence, the area of the circle is

$$4\int_0^a y\, dx = 4\int_0^a \sqrt{a^2 - x^2}\, dx = 4\left[\frac{x}{2}\sqrt{a^2 - x^2} + \frac{a^2}{2}\sin^{-1}\frac{x}{a}\right]_0^a$$

$$= 4\frac{a^2}{2}\sin^{-1}(1) = 4\frac{a^2}{2}\frac{\pi}{2} = \pi a^2$$

as required. $\qquad \square$

Problem 14.3.10 Let $a < 0 < b$ and let $c > 0$. Let A, B and C be the points $(a,0)$, $(b,0)$ and $(0,c)$, respectively, in the x,y-plane. Show that (the region enclosed by) $\triangle ABC$ has area $(b-a)c/2$.

14.4　Improper Integrals

Consider the integrals

$$\mathcal{I} = \int_0^\infty \frac{dx}{x^2 + 1}, \qquad \mathcal{J} = \int_0^1 \frac{dx}{\sqrt{x}}.$$

In the first, the range of integration is infinite, and in the second, the integrand approaches infinity at one end of the range of integration. These are examples of *improper integrals*.

For $u > 0$,

$$\int_0^u \frac{dx}{x^2 + 1} = \left[\tan^{-1} x\right]_0^u = \tan^{-1} u \to \frac{\pi}{2} \quad \text{as} \quad u \to \infty.$$

Since the limit exists and equals $\pi/2$, we say that $\mathcal{I}$ *converges* to $\pi/2$ and write

$$\int_0^\infty \frac{dx}{x^2 + 1} = \frac{\pi}{2}.$$

Similarly, for $u \in (0, 1)$,

$$\int_u^1 \frac{dx}{\sqrt{x}} = \left[2\sqrt{x}\right]_u^1 = 2 - 2\sqrt{u} \to 2 \quad \text{as} \quad u \to 0^+,$$

so we say that $\mathcal{J}$ converges to 2 and write

$$\int_0^1 \frac{dx}{\sqrt{x}} = 2.$$

These evaluations are illustrated in Figure 14.4.1.

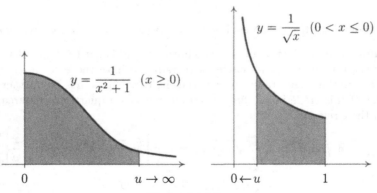

$$y = \frac{1}{\sqrt{x}} \quad (0 < x \le 0)$$

$$y = \frac{1}{x^2 + 1} \quad (x \ge 0)$$

$0 \qquad u \to \infty \qquad 0 \leftarrow u \qquad 1$

Figure 14.4.1

Definitions 14.4.1 Throughout, f is a real function and $a, b, L, M, N \in \mathbb{R}$.

(1)　Let f be Riemann integrable over $[a, u]$ for all $u > a$. Then the *improper integral* $\int_a^\infty f(x)\, dx$ *converges* to L if $\int_a^u f(x)\, dx \to L$ as $u \to \infty$.

(2) Let f be Riemann integrable over $[u, a]$ for all $u < a$. Then the *improper integral* $\int_{-\infty}^{a} f(x)\, dx$ *converges* to L if $\int_{u}^{a} f(x)\, dx \to L$ as $u \to -\infty$.

(3) Let f be Riemann integrable over $[a, u]$ for all $u \in (a, b)$, and let $f(u) \to \pm\infty$ as $u \to b^-$. Then the *improper integral* $\int_{a}^{b} f(x)\, dx$ *converges* to L if $\int_{a}^{u} f(x)\, dx \to L$ as $u \to b^-$.

(4) Let f be Riemann integrable over $[u, b]$ for all $u \in (a, b)$, and let $f(u) \to \pm\infty$ as $u \to a^+$. Then the *improper integral* $\int_{a}^{b} f(x)\, dx$ *converges* to L if $\int_{u}^{b} f(x)\, dx \to L$ as $u \to a^+$.

(5) Let $c \in (\alpha, \beta)$, where $\alpha \in \mathbb{R}$ or $\alpha = -\infty$ and $\beta \in \mathbb{R}$ or $\beta = \infty$. Let $\mathcal{I} = \int_{\alpha}^{c} f(x)\, dx$ and $\mathcal{J} = \int_{c}^{\beta} f(x)\, dx$ be improper integrals of types (1)–(4), above. Then the *improper integral* $\int_{\alpha}^{\beta} f(x)\, dx$ *converges* to L if $\mathcal{I}$ converges to M and $\mathcal{J}$ converges to N with $M + N = L$.

Notes 14.4.2 (1) An improper integral which converges to L is also said to *exist* and equal L or have *value* L, and we write, for example, $\int_{a}^{\infty} f(x)\, dx = L$.

Values of improper integrals must be finite, *i.e.* must belong to $\mathbb{R}$.

(2) In Definition 14.4.1 (5), suppose that the 'problems' occur at α and β, and f is continuous with an antiderivative F on (α, β). Then, to investigate the convergence of $\int_{\alpha}^{\beta} f(x)\, dx$ we consider, for $u \in (\alpha, c)$,

$$\int_{u}^{c} f(x)\, dx = F(c) - F(u) \quad \text{as} \quad u \to \alpha^+$$

and, for $v \in (c, \beta)$.

$$\int_{c}^{v} f(x)\, dx = F(v) - F(c) \quad \text{as} \quad v \to \beta^-.$$

It follows that the convergence of $\int_{\alpha}^{\beta} f(x)\, dx$ and its value, when it exists, is independent of c. Indeed, $F(c)$ need not be found. The 'short cut' is to consider $F(v) - F(u)$ as $u \to \alpha^+$ and $v \to \beta^-$. This is demonstrated in Example 14.4.5.

(3) Successive applications of Definition 14.4.1 (5) allow us to define improper integrals where the range of integration contains a finite number of points c for which $f(u) \to \pm\infty$ as $u \to c^-$ or $u \to c^+$.

Example 14.4.3 Investigate the convergence of the improper integrals

$$\text{(a)} \ \mathcal{I} = \int_{1}^{\infty} \frac{dx}{\sqrt{x}}, \qquad \text{(b)} \ \mathcal{J} = \int_{0}^{4} \frac{dx}{(4 - x)^{1/4}}.$$

Solution (a) For $u > 1$,

$$\int_{1}^{u} \frac{dx}{\sqrt{x}} = \left[2\sqrt{x} \right]_{1}^{u} = 2\sqrt{u} - 2 \to \infty \quad \text{as} \quad u \to \infty.$$

Hence, $\mathcal{I}$ does not converge (or does not exist).

(b) For $u \in (0, 4)$,

$$\int_0^u \frac{dx}{(4-x)^{1/4}} = \left[-\frac{4}{3}(4-x)^{3/4} \right]_0^u = -\frac{4}{3}\left((4-u)^{3/4} - 4^{3/4} \right)$$

$$\rightarrow -\frac{4}{3}(-4^{3/4}) = \frac{8\sqrt{2}}{3} \quad \text{as} \quad u \rightarrow 4^-.$$

Hence, $\mathcal{J}$ converges to $8\sqrt{2}/3$, *i.e.* $\int_0^4 \frac{dx}{(4-x)^{1/4}} = \frac{8\sqrt{2}}{3}.$ □

Example 14.4.4 Investigate the convergence of the improper integrals

(a) $\mathcal{I} = \int_0^\infty \frac{dx}{\sqrt{x}}$,   (b) $\mathcal{J} = \int_0^9 \frac{dx}{(x-1)^{2/3}}$,   (c) $\mathcal{K} = \int_0^\infty \frac{dx}{(x-1)^2}$.

Solution Each of these improper integrals has more than one 'problem'.

(a) Here, the range of integration is infinite and $1/\sqrt{x} \rightarrow \infty$ as $x \rightarrow 0^+$. So we write

$$\mathcal{I} = \int_0^1 \frac{dx}{\sqrt{x}} + \int_1^\infty \frac{dx}{\sqrt{x}} = \mathcal{I}_1 + \mathcal{I}_2 \quad \text{(respectively)}.$$

There is nothing significant about dividing the range at 1 (see Note 14.4.2 (2)). Then, for $\mathcal{I}$ to converge, both $\mathcal{I}_1$ and $\mathcal{I}_2$ must converge. But we have already shown, in Example 14.4.3, that $\mathcal{I}_2$ does not converge. Hence, $\mathcal{I}$ does not converge.

(b) Since, $1/(x-1)^{2/3} \rightarrow -\infty$ as $x \rightarrow 1^-$ and $1/(x-1)^{2/3} \rightarrow \infty$ as $x \rightarrow 1^+$, we write

$$\mathcal{J} = \int_0^1 \frac{dx}{(x-1)^{2/3}} + \int_1^9 \frac{dx}{(x-1)^{2/3}} = \mathcal{J}_1 + \mathcal{J}_2 \quad \text{(respectively)}.$$

Then, for $u \in (0, 1)$,

$$\int_0^u \frac{dx}{(x-1)^{2/3}} = \left[3(x-1)^{1/3} \right]_0^u = 3(u-1)^{1/3} - (-3) \rightarrow 3 \quad \text{as} \quad u \rightarrow 1^-$$

and, for $v \in (1, 9)$,

$$\int_v^9 \frac{dx}{(x-1)^{2/3}} = \left[3(x-1)^{1/3} \right]_v^9 = 6 - 3(v-1)^{1/3} \rightarrow 6 \quad \text{as} \quad u \rightarrow 1^+.$$

Hence $\mathcal{J}_1$ converges to 3 and $\mathcal{J}_2$ converges to 6 so that $\mathcal{J}$ converges to $3 + 6 = 9$.

(c) Here there are three problems which we isolate by writing

$$\mathcal{K} = \int_0^1 \frac{dx}{(x-1)^2} + \int_1^2 \frac{dx}{(x-1)^2} + \int_2^\infty \frac{dx}{(x-1)^2}$$

$$= \mathcal{K}_1 + \mathcal{K}_2 + \mathcal{K}_3 \quad \text{(respectively)}.$$

Then for $\mathcal{K}$ to converge, all three of $\mathcal{K}_1$, $\mathcal{K}_2$ and $\mathcal{K}_3$ must converge. For $u \in (0, 1)$,

$$\int_0^u \frac{dx}{(x-1)^2} = \left[\frac{-1}{x-1} \right]_0^u = \frac{-1}{u-1} - 1 \rightarrow \infty \quad \text{as} \quad u \rightarrow 1^-.$$

So $\mathcal{K}_1$ does not converge and, therefore, $\mathcal{K}$ does not converge.   □

Example 14.4.5 Show that the improper integral

$$\mathcal{I} = \int_{-\infty}^{\infty} \frac{dx}{x^2 + 2x + 2}$$

converges and find its value.

Solution We make use of the 'short cut' mentioned in Note 14.4.2 (2).

For $u < v$,

$$\int_u^v \frac{dx}{x^2 + 2x + 2} = \int_u^v \frac{dx}{(x+1)^2 + 1} = \left[\tan^{-1}(x+1) \right]_u^v$$

$$= \tan^{-1}(v+1) - \tan^{-1}(u+1)$$

$$\to (\pi/2) - (-\pi/2) = \pi \quad \text{as} \quad u \to -\infty \quad \text{and} \quad v \to \infty.$$

Hence, $\mathcal{I}$ converges to π.

Hazard When investigating an improper integral with 'multiple' problems, it is essential to deal with them independently. We achieved this in Example 14.4.5, without isolating the problems, by letting $u \to -\infty$ and $v \to \infty$. Do *not* be tempted to consider, for example,

$$\int_{-u}^u f(x)\,dx \quad \text{as} \quad u \to \infty.$$

By way of further examples illustrating this hazard, the reader is left to verify that none of the improper integrals

$$\int_{-\infty}^{\infty} x\,dx, \qquad \int_0^{\infty} \left(1 - \frac{1}{x^2}\right) dx, \qquad \int_{-1}^1 \frac{dx}{x^3}$$

converges but, for $u > 0$, $v > 1$ and $w \in (0,1)$,

$$\int_{-u}^u x\,dx, \qquad \int_{1/v}^v \left(1 - \frac{1}{x^2}\right) dx, \qquad \int_{-1}^{-w} \frac{dx}{x^3} + \int_w^1 \frac{dx}{x^3}$$

all equal 0.

14.X Exercises

1. Find $\displaystyle \int f(x)\,dx$ when $f(x)$ is

   (a) x^3,

   (b) $\dfrac{1}{x^2}$,

   (c) $\sqrt{x}$,

   (d) $\dfrac{1}{\sqrt{x}}$,

   (e) $(2x + 7)^5$,

   (f) $(5x - 4)^{-3}$,

   (g) $(3 - 2x)^{-2/3}$,

   (h) $4x^3 + 3x^2 + 2x + 1$,

(i) $\dfrac{(4x^3 - 3x + 2)}{(2x + 1)^2}$,

(j) $\sin 3x$,

(k) $\cos(2x + 1)$,

(l) $\sec^2(2 - x)$,

(m) $(x - \sec x)(x + \sec x)$,

(n) $\sec 3x(\sec 3x + \tan 3x)$,

(o) $\dfrac{1}{\sqrt{9 - x^2}}$,

(p) $\dfrac{1}{\sqrt{9 - 4x^2}}$,

(q) $\dfrac{1}{x^2 + 3}$,

(r) $\dfrac{1}{4x^2 + 3}$,

(s) $\dfrac{1}{(x - 1)^2 + 9}$,

(t) $\dfrac{1}{x^2 - 2x + 10}$.

2. Find $\dfrac{d}{dx} \displaystyle\int_0^{x^2} \sqrt{1 + t^2}\, dt$.

3. Let f be a real function, continuous on $\mathbb{R}$, and let $a, b \in \mathbb{R}$ with $a < b$. By applying the mean value theorem to the real function $x \mapsto \displaystyle\int_0^x f(t)\, dt$, show that there exists $c \in (a, b)$ such that

$$\int_a^b f(t)\, dt = f(c)(b - a).$$

4. By arguing from first principles, *i.e.* using Riemann sums, evaluate

    (a) $\displaystyle\int_0^1 (3x - 2)\, dx$,   (b) $\displaystyle\int_0^2 x^2\, dx$,   (c) $\displaystyle\int_1^4 (2x - x^2)\, dx$.

5. Evaluate

    (a) $\displaystyle\int_0^2 (3x + 1)\, dx$,   (b) $\displaystyle\int_{-1}^3 (5 - 4x)\, dx$,

    (c) $\displaystyle\int_1^3 (x^2 - 5x + 1)\, dx$,   (d) $\displaystyle\int_{-1}^2 x^2(2x^3 - 1)\, dx$,

    (e) $\displaystyle\int_1^2 (2x - 3)^7\, dx$,   (f) $\displaystyle\int_{2/3}^3 (3x - 1)^{-2/3}\, dx$,

    (g) $\displaystyle\int_0^1 \sqrt{2x + 3}\, dx$,   (h) $\displaystyle\int_0^1 \dfrac{dx}{\sqrt{2x + 3}}$,

    (i) $\displaystyle\int_0^\pi \sin(\theta/3)\, d\theta$,   (j) $\displaystyle\int_0^{\pi/6} \cos^2 \theta\, d\theta$,

    (k) $\displaystyle\int_0^{\pi/4} \tan^2 \theta\, d\theta$,   (l) $\displaystyle\int_{-1}^2 |x|\, dx$,

(m) $\displaystyle\int_0^4 |x - 1|\, dx,$ (n) $\displaystyle\int_0^\pi |\cos x|\, dx,$

(o) $\displaystyle\int_0^1 (2 - x^2)^{-1/2}\, dx,$ (p) $\displaystyle\int_0^{1/2} (4x^2 + 1)^{-1}\, dx.$

6.   Make use of odd and even functions to evaluate

     (a) $\displaystyle\int_{-\pi/2}^{\pi/2} \sin^3 x\, dx,$ (b) $\displaystyle\int_{-1}^1 (x^2 + \tan x)\, dx,$ (c) $\displaystyle\int_{-\pi}^\pi (x + 1) \cos x\, dx.$

7.   Given that, for all $x \geq 0,$ $x - \dfrac{x^3}{6} \leq \sin x \leq x - \dfrac{x^3}{6} + \dfrac{x^5}{120},$ show that

$$\frac{17}{18} \leq \int_0^1 \frac{\sin x}{x}\, dx \leq \frac{1703}{1800}$$

     (assuming that the integrand takes the value 1 at 0).

8.   Find the minimum and maximum values of $x^4(1 - x)^4$ when $x \in [0, 1]$. Hence show that

$$0 \leq \int_0^1 \frac{x^4(1 - x)^4}{x^2 + 1}\, dx \leq \frac{1}{256}.$$

     Use long division of polynomials to evaluate the integral.

9.   For each of the following pairs of curves, determine where they intersect, indicate their relative positions on a rough sketch and find the area of the finite region they enclose.

         (a) $y = x^2 - 4x - 4,$ $y = 5 + 2x - 2x^2,$

         (b) $y = x^3 - 3x^2 + 3x,$ $y = x,$

         (c) $y^2 - 2y = x,$ $x = 3.$

10.   By interpreting the integrals as the areas, evaluate

         (a) $\displaystyle\int_1^2 \sqrt{4 - x^2}\, dx,$ (b) $\displaystyle\int_{-1}^1 \sqrt{2 - x^2}\, dx.$

     Assume the usual formulae for areas of circular segments and triangles.

11.   Determine which of the following improper integrals converge and evaluate those that do.

         (a) $\displaystyle\int_0^1 \frac{dx}{x\sqrt{x}},$ (b) $\displaystyle\int_1^\infty \frac{dx}{x\sqrt{x}},$ (c) $\displaystyle\int_0^\infty \frac{dx}{x\sqrt{x}},$

         (d) $\displaystyle\int_0^3 \frac{dx}{(x - 3)^{2/3}},$ (e) $\displaystyle\int_0^\infty \frac{dx}{(x + 3)^{2/3}},$ (f) $\displaystyle\int_0^2 \frac{dx}{\sqrt{2x - x^2}}.$

Chapter 15

Logarithms and Exponentials

We expect that the reader has already met the logarithmic and exponential functions. The usual way to introduce the logarithmic function is to say that $\log_a b = c$, where $b = a^c$. In Chapter 1, we defined a^c only where c is *rational*. The most sensible way to define a^c for *general* c is to use logarithms. We have a circular definition!

In this chapter, we give a definition of logarithm in terms of a definite integral. We establish the usual properties of log, including its derivative. It turns out that $\log : (0, \infty) \to \mathbb{R}$ is bijective, and so has an inverse. This is the *exponential function* exp. Its properties are derived from those of log. The exponential function occurs in many applications, such as population growth and radioactive decay.

We are then in a position to give a proper definition of a^x $(a > 0)$ as $\exp(x \log a)$.

In the final sections, we introduce the *hyperbolic functions*. These are combinations of $x \mapsto \exp(x)$ and $x \mapsto \exp(-x)$. These functions and their inverses occur in many applications of mathematics, so that it is important to be familiar with them. In many ways their properties resemble those of the trigonometric functions and their names (sinh, cosh, ...) remind us of this fact. The point to watch is that, often, the signs in the formulae are different from those in the trigonometric analogues.

By the end of the chapter, the reader should be confident that the logarithmic and exponential functions can be formally defined in a logical way, be reminded of their properties, and be able to work confidently with the hyperbolic functions.

15.1   The Logarithmic Function

Definition 15.1.1 The function $\log : (0, \infty) \to \mathbb{R}$ defined by

$$\log x = \int_1^x \frac{1}{t}\, dt$$

is called the *logarithmic function* or *log function*.

For example, where A and B denote the areas indicated in Figure 15.1.1,

$$\log(1/2) = -A \qquad \text{and} \qquad \log 2 = B.$$

 Hazard The notation $\log_e x$ or $\ln x$ is sometimes used instead of $\log x$, while $\log x$ itself is used, particularly on electronic calculators, to denote $\log_{10} x$, the *logarithm of x to the base* 10. See Definition 15.3.10.

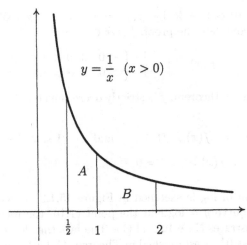

$$y = \frac{1}{x} \quad (x > 0)$$

Figure 15.1.1

Properties 15.1.2 These properties of the log function follow from the definition.

(1) $\log 1 = 0$ and $\log x \begin{cases} < 0 & \text{if } 0 < x < 1, \\ > 0 & \text{if } x > 1. \end{cases}$

(2) By the fundamental theorem of calculus (A), log is differentiable, and therefore continuous, on $(0, \infty)$ and

$$\frac{d}{dx} \log x = \frac{1}{x} > 0 \quad (x > 0).$$

So, by the monotonicity theorem, log is strictly increasing on $(0, \infty)$.

(3) Differentiating again,

$$\frac{d^2}{dx^2} \log x = -\frac{1}{x^2} < 0 \quad (x > 0).$$

So log is concave-down on $(0, \infty)$. See Definition 10.2.10.

Example 15.1.3 Differentiate $\log \sqrt{x^2 + 3}$ with respect to x.

Solution Using the chain rule, we have

$$\frac{d}{dx} \log \sqrt{x^2 + 3} = \frac{1}{\sqrt{x^2 + 3}} \frac{1}{2\sqrt{x^2 + 3}} 2x = \frac{x}{x^2 + 3}.$$

See also Example 15.1.7.   □

Theorem 15.1.4 *For all* $x \in (0, \infty)$,

$$\log x \leq x - 1.$$

Proof Define $f : (0, \infty) \to \mathbb{R}$ by $f(x) = x - 1 - \log x$. We show that $f(x) \geq 0$ $(x > 0)$. This will complete the proof. For $x \in (0, \infty)$,

$$f'(x) = 1 - \frac{1}{x} \begin{cases} < 0 & \text{if } 0 < x < 1, \\ > 0 & \text{if } x > 1. \end{cases}$$

So, by the monotonicity theorem, f is strictly decreasing on $(0, 1]$ and strictly increasing on $[1, \infty)$. Hence,

$$0 < x < 1 \implies f(x) > f(1) \qquad \text{and} \qquad 1 < x \implies f(1) < f(x).$$

So, for all $x \in (0, \infty)$, $f(x) \geq f(1) = 0$. $\square$

Remark The graph of log is sketched in Figure 15.1.2. Since the line $y = x - 1$ is the tangent to the curve $y = \log x$ at the point $(1, 0)$, the relationship established in Theorem 15.1.4 illustrates Note 10.2.11 (1). The fact that $\log x \to \infty$ as $x \to \infty$ and $\log x \to -\infty$ as $x \to 0^+$ is established in Theorem 15.1.10. The number e is explained in Definition 15.1.14.

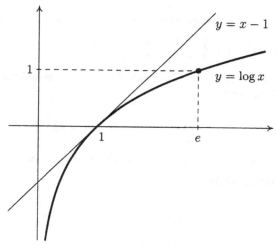

Figure 15.1.2 : Logarithmic Function

Theorem 15.1.5 *Let* $a, b \in (0, \infty)$ *and let* $q \in \mathbb{Q}$. *Then*

(1) $\log(ab) = \log a + \log b$,

(2) $\log(a^q) = q \log a$,

(3) $\log \dfrac{1}{b} = -\log b$,

(4) $\log \dfrac{a}{b} = \log a - \log b$.

Proof (1) For $x \in (0, \infty)$,

$$\frac{d}{dx} \log(ax) = \frac{1}{ax} a = \frac{1}{x} = \frac{d}{dx} \log x.$$

So, by Corollary 9.4.8, there is a constant C such that

$$\log(ax) = \log x + C \quad (x > 0).$$

Putting $x = 1$ gives

$$\log a = \log 1 + C, \qquad i.e. \ C = \log a.$$

Then putting $x = b$ gives (1).

(2) For $x \in (0, \infty)$,

$$\frac{d}{dx} \log(x^q) = \frac{1}{x^q} qx^{q-1} = \frac{q}{x} = \frac{d}{dx}(q \log x).$$

So, by Corollary 9.4.8, there is a constant C such that

$$\log(x^q) = q \log x + C \quad (x > 0).$$

Putting $x = 1$ gives $C = 0$. Then putting $x = a$ gives (2).

(3) Using (2),

$$\log \frac{1}{b} = \log(b^{-1}) = (-1) \log b = -\log b.$$

(4) Using (1) and (3),

$$\log \frac{a}{b} = \log \left(a \frac{1}{b} \right) = \log a + \log \frac{1}{b} = \log a - \log b. \qquad \square$$

Example 15.1.6 Simplify $E = \dfrac{\log 24 + \log 6 - 2 \log 3}{\log 24 - \log 6}$.

Solution Using Theorem 15.1.5, we have

$$E = \frac{\log[(24)(6)] - \log(3^2)}{\log(24/6)} = \frac{\log 144 - \log 9}{\log 4}$$

$$= \frac{\log(144/9)}{\log 4} = \frac{\log 16}{\log 4}$$

$$= \frac{\log(4^2)}{\log 4} = \frac{2 \log 4}{\log 4} = 2. \qquad \square$$

Example 15.1.7 Differentiate $\log \sqrt{x^2 + 3}$ with respect to x.

Solution Before applying the chain rule we make use of Theorem 15.1.5 (2).

$$\frac{d}{dx} \left[\log \sqrt{x^2 + 3} \right] = \frac{d}{dx} \left[\frac{1}{2} \log(x^2 + 3) \right] = \frac{1}{2} \frac{1}{(x^2 + 3)} 2x = \frac{x}{x^2 + 3}.$$

Compare this with the solution of Example 15.1.3. $\qquad \square$

The properties of the log function in Theorem 15.1.5 can be used to simplify the differentiation of products and quotients. The process is called *logarithmic differentiation*.

Example 15.1.8 Find $\dfrac{dy}{dx}$ when $y = \dfrac{x^4 \tan^{-1} x}{x^2 + 1}$ $(x > 0)$.

Solution Since $x > 0$ it follows that $x^4 > 0$, $\tan^{-1} x > 0$, $x^2 + 1 > 0$ and $y > 0$. We begin by 'taking logs'. Using Theorem 15.1.5, this gives

$$\log y = \log \frac{x^4 \tan^{-1} x}{x^2 + 1} = 4 \log x + \log(\tan^{-1} x) - \log(x^2 + 1).$$

Then differentiating implicitly, with respect to x,

$$\frac{1}{y} \frac{dy}{dx} = \frac{4}{x} + \frac{1}{\tan^{-1} x} \frac{1}{(x^2 + 1)} - \frac{1}{x^2 + 1} 2x,$$

so that

$$\frac{dy}{dx} = \frac{x^4 \tan^{-1} x}{x^2 + 1} \left[\frac{4}{x} + \frac{1}{(x^2 + 1) \tan^{-1} x} - \frac{2x}{x^2 + 1} \right]. \qquad \square$$

Problem 15.1.9 By writing

$$-y = \frac{(-x)^4(- \tan^{-1} x)}{x^2 + 1},$$

show that the expression for $\dfrac{dy}{dx}$, found in Example 15.1.8, holds for $x < 0$.

Theorem 15.1.10 (1) $\log x \to \infty$ *as* $x \to \infty$.

(2) $\log x \to -\infty$ *as* $x \to 0^+$.

Proof (1) We compare $\log$ with $f : (2, \infty) \to \mathbb{R}$, defined by

$$f(x) = n \log 2 \quad \text{where } n \in \mathbb{N} \text{ is such that } 2^n \le x < 2^{n+1}.$$

The graph of f is sketched in Figure 15.1.3. We shall assume that $f(x) \to \infty$ as $x \to \infty$. For $x \in (2, \infty)$, let $2^n \le x < 2^{n+1}$ where $n \in \mathbb{N}$. Then

$$f(x) = n \log 2 = \log(2^n) \le \log x$$

and result follows from the version of the sandwich principle stated in Note 7.4.11.

(2) As $x \to 0^+$, $1/x \to \infty$. Hence, with $x > 0$,

$$\log x = - \log \frac{1}{x} \to -\infty \quad \text{as} \quad x \to 0^+. \qquad \square$$

Corollary 15.1.11 (1) $\dfrac{\log x}{x} \to 0$ *as* $x \to \infty$.

(2) $x \log x \to 0$ *as* $x \to 0^+$.

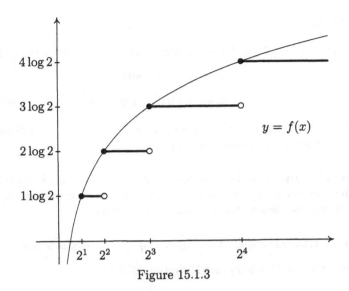

Figure 15.1.3

Proof (1) For $x > 1$,

$$0 < \frac{\log x}{x} = \frac{2\log \sqrt{x}}{x} < \frac{2\sqrt{x}}{x} = \frac{2}{\sqrt{x}},$$

and $\lim_{x \to \infty} 2/\sqrt{x} = \lim_{t \to 0^+} 2\sqrt{t} = 0$, since the real function $t \mapsto \sqrt{t}$ is right continuous at 0. Hence, the result follows by the sandwich principle (see Note 7.4.2).

(2) As $x \to 0^+$, $1/x \to \infty$. Hence, with $x > 0$,

$$x\log x = \frac{-\log(1/x)}{(1/x)} \to -0 = 0 \quad \text{as} \quad x \to 0^+. \qquad \square$$

Theorem 15.1.12 *The function* $\log : (0, \infty) \to \mathbb{R}$ *is a bijection.*

Proof Since the log function is strictly increasing on $(0, \infty)$, it is injective (see Theorem 2.6.3).

To see that log is surjective, let $t \in \mathbb{R}$. Then, by Theorem 15.1.10, there exist $s_1, s_2 \in (0, \infty)$ such that $\log s_1 < t < \log s_2$. Hence, by the intermediate value theorem, there exists s between s_1 and s_2, and therefore in $(0, \infty)$, such that $\log s = t$. $\qquad \square$

Example 15.1.13 Solve the equation $3\log(x + 5) = 2\log(7 - x)$.

Solution For any solution, we must have $x + 5 > 0$ and $7 - x > 0$, *i.e.* we must have $-5 < x < 7$. Provided this is true, Theorem 15.1.5 (2) and Theorem 15.1.12 (injectiveness), give

$$3\log(x + 5) = 2\log(7 - x) \Leftrightarrow \log[(x + 5)^3] = \log[(7 - x)^2]$$
$$\Leftrightarrow (x + 5)^3 = (7 - x)^2.$$

Now, for $x \in \mathbb{R}$,

$$(x+5)^3 = (7-x)^2 \Leftrightarrow x^3 + 15x^2 + 75x + 125 = x^2 - 14x + 49$$

$$\Leftrightarrow x^3 + 14x^2 + 89x + 76 = 0$$

$$\Leftrightarrow (x+1)(x^2 + 13x + 76) = 0 \Leftrightarrow x = -1,$$

since $x^2 + 13x + 76$ is irreducible. Hence, since $-5 < -1 < 7$, it follows that $x = -1$ is the only solution of the given equation. $\square$

Definition 15.1.14 Since $\log : (0, \infty) \to \mathbb{R}$ is a bijection (Theorem 15.1.12) and $1 \in \mathbb{R}$, there exists a unique element of $(0, \infty)$, which we denote by e, such that $\log e = 1$. It may be shown that $e = 2{\cdot}718 \ldots$ is irrational.

Theorem 15.1.15 *Let* $q \in \mathbb{Q}$. *Then* $\log x = q$ *if and only if* $x = e^q$.

Proof By Theorem 15.1.5 (2) and Definition 15.1.14,

$$\log(e^q) = q \log e = q.$$

Hence,

$$x = e^q \Rightarrow \log x = q$$

and, using the injectiveness of log (see Theorem 15.1.12),

$$\log x = q = \log(e^q) \Rightarrow x = e^q. \qquad \square$$

Example 15.1.16 Sketch the curve $y = \dfrac{\log x}{x}$.

Solution The curve is sketched in Figure 15.1.4.

First, observe that y is defined if and only if $x > 0$. For $x > 0$, we have the following.

$$y = 0 \Leftrightarrow \log x = 0 \Leftrightarrow x = 1.$$

$$y' = \frac{(1/x)x - \log x}{x^2} = \frac{1 - \log x}{x^2}$$

$$= 0 \Leftrightarrow \log x = 1 \Leftrightarrow x = e.$$

$$y'' = \frac{-(1/x)x^2 - (1 - \log x)2x}{x^4} = \frac{-3 + 2\log x}{x^3}$$

$$= 0 \Leftrightarrow \log x = \frac{3}{2} \Leftrightarrow x = e^{3/2}.$$

Tables of signs for y, y' and y'' are shown below.

x	$\to$	1	$\to$
y	$-$	0	$+$
y'		1	

x	$\to$	e	$\to$
y'	$+$	0	$-$
	$\nearrow$	$\longrightarrow$	$\searrow$

x	$\to$	$e^{3/2}$	$\to$
y''	$-$	0	$+$
y'		$-$	

Thus, the curve crosses the x-axis where $x = 1$ and $y'(1) = 1$, there is a maximum turning point at (e, e^{-1}) and a decreasing down-up point of inflection at $(e^{3/2}, 3e^{-3/2}/2)$. The concavity is down for $x < e^{3/2}$ and up for $x > e^{3/2}$.

Since $1/x \to \infty$ as $x \to 0^+$ and, by Theorem 15.1.10, $\log x \to -\infty$ as $x \to 0^+$, it follows (see Note 7.4.7) that

$$y = \frac{1}{x} \log x \to -\infty \text{ as } x \to 0^+.$$

From Corollary 15.1.11 we have

$$y = \frac{\log x}{x} \to 0 \text{ as } x \to \infty. \qquad \square$$

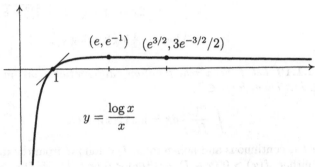

Figure 15.1.4

Note 15.1.17 The log function provides us with the following *standard integrals*.

$$\int \frac{1}{x} dx = \log |x| + C,$$

$$\int \frac{1}{\sqrt{x^2 + b}} dx = \log |x + \sqrt{x^2 + b}| + C \quad (b \neq 0),$$

$$\int \tan x \, dx = -\log |\cos x| + C, \qquad \int \cot x \, dx = \log |\sin x| + C,$$

$$\int \sec x \, dx = \log |\sec x + \tan x| + C,$$

$$\int \operatorname{cosec} x \, dx = -\log |\operatorname{cosec} x + \cot x| + C.$$

All these can be verified by differentiation. For example, with $b \neq 0$,

$$\frac{d}{dx} \log[\pm(x + \sqrt{x^2 + b})] = \frac{1}{\pm(x + \sqrt{x^2 + b})} \left[\pm \left(1 + \frac{2x}{2\sqrt{x^2 + b}} \right) \right]$$

$$= \frac{1}{x + \sqrt{x^2 + b}} \frac{\sqrt{x^2 + b} + x}{\sqrt{x^2 + b}} = \frac{1}{\sqrt{x^2 + b}}.$$

Example 15.1.18 Find/evaluate

$$\text{(a)} \quad \int \frac{5x+1}{5x+2}\,dx, \qquad \text{(b)} \quad \int_0^1 \frac{dx}{\sqrt{4x^2-4x+9}}.$$

Solution (a) Note that $5x+1 = (5x+2) - 1$. Hence,

$$\int \frac{5x+1}{5x+2}\,dx = \int \left[1 - \frac{1}{5x+2}\right]dx = x - \tfrac{1}{5}\log|5x+2| + C.$$

(b) Here we begin by completing the square of the quadratic.

$$\int_0^1 \frac{dx}{\sqrt{4x^2-4x+9}} = \int_0^1 \frac{dx}{\sqrt{(2x-1)^2+8}}$$

$$= \tfrac{1}{2}\left[\log\left|(2x-1) + \sqrt{(2x-1)^2+8}\right|\right]_0^1$$

$$= \tfrac{1}{2}\log 4 - \tfrac{1}{2}\log 2 = \log\sqrt{2} \qquad \square$$

Theorem 15.1.19 *Let f be a real function, differentiable on an interval I, with $f(x) \neq 0$ ($x \in I$). Then, for $x \in I$,*

$$\int \frac{f'(x)}{f(x)}\,dx = \log|f(x)| + C.$$

Proof Since f is continuous and non-zero on I, it follows from the intermediate value theorem that either $f(x) > 0$ ($x \in I$) or $f(x) < 0$ ($x \in I$). If $f(x) > 0$ ($x \in I$) then,

$$\frac{d}{dx}\log|f(x)|\,dx = \frac{d}{dx}\log f(x) = \frac{1}{f(x)}\,f'(x) = \frac{f'(x)}{f(x)}$$

and if $f(x) < 0$ ($x \in I$) then,

$$\frac{d}{dx}\log|f(x)|\,dx = \frac{d}{dx}\log[-f(x)] = \frac{1}{-f(x)}[-f'(x)] = \frac{f'(x)}{f(x)}. \qquad \square$$

Example 15.1.20 Find

$$\text{(a)} \quad \int \frac{x+1}{x^2+2x+5}\,dx, \qquad \text{(b)} \quad \int \frac{dx}{x\log x}, \qquad \text{(c)} \quad \int \tan x\,dx.$$

Solution In each case we make use of Theorem 15.1.19

(a) We have

$$\int \frac{x+1}{x^2+2x+5}\,dx = \frac{1}{2}\int \frac{2x+2}{x^2+2x+5}\,dx = \tfrac{1}{2}\log|x^2+2x+5| + C.$$

The modulus sign could be omitted here since $x^2 + 2x + 5 > 0$ ($x \in \mathbb{R}$).

(b) $\displaystyle \int \frac{dx}{x\log x} = \int \frac{(1/x)}{\log x}\,dx = \log|\log x| + C.$

(c) $\displaystyle \int \tan x\,dx = \int \frac{\sin x}{\cos x}\,dx = -\int \frac{-\sin x}{\cos x}\,dx = -\log|\cos x| + C. \qquad \square$

15.2 The Exponential Function

Recall that the function $\log : (0, \infty) \to \mathbb{R}$ is a bijection (Theorem 15.1.12).

Definition 15.2.1 The *exponential function* is the inverse of the logarithmic function. It is denoted by exp. Thus $\exp : \mathbb{R} \to (0, \infty)$ is defined by

$$\exp x = y \iff \log y = x.$$

In view of Theorem 15.1.15, for $q \in \mathbb{Q}$, $\exp q = e^q$. More generally, for any $x \in \mathbb{R}$ we write $\exp x = e^x$. This 'power' notation is justified in Section 15.3. To begin with we shall state results using both forms.

Using the mirror property for inverse functions we can deduce the graph of exp from the graph of log. The graph of *exp* is shown in Figure 15.2.1.

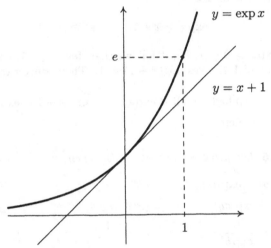

Figure 15.2.1 : Exponential Function

Properties 15.2.2 These properties of the exponential function follow from the definition of exp and properties of log.

(1) For $x \in (0, \infty)$, $\exp(\log x) = e^{\log x} = x$.
     For $x \in \mathbb{R}$, $\log(\exp x) = \log(e^x) = x$.

(2) $\exp(0) = e^0 = 1$ and $\exp(1) = e^1 = e$.

(3) For all $x \in \mathbb{R}$, $\exp x = e^x > 0$. Hence the exponential function has no real zero.

(4) By Theorem 8.2.9, exp is differentiable, and therefore continuous, on $\mathbb{R}$ and

$$\frac{d}{dx}\exp x = \frac{1}{1/\exp x} = \exp x, \quad \frac{d}{dx}e^x = e^x \quad (x \in \mathbb{R}).$$

Thus, in view of (3), exp is strictly increasing and concave-up on $\mathbb{R}$.

(5) We have the following *standard integral*.

$$\int \exp x \, dx = \exp x + C, \qquad \int e^x \, dx = e^x + C.$$

Example 15.2.3 Find   (a) $\dfrac{d}{dx} e^{\tan x}$,   (c) $\displaystyle\int e^{3x-5} \, dx$.

Solution   (a)   Using the chain rule, $\dfrac{d}{dx} e^{\tan x} = e^{\tan x} \sec^2 x$.

    (b)   Using Property 14.1.2 (3), $\displaystyle\int e^{3x-5} \, dx = \frac{1}{3} e^{3x-5} + C$.   □

Theorem 15.2.4 *For all $x \in \mathbb{R}$,*

$$\exp x \geq x + 1, \qquad e^x \geq x + 1.$$

Proof   Since $\exp x > 0$ $(x \in \mathbb{R})$, it follows that, for $x \leq -1$, $\exp x > 0 \geq x + 1$. For $x > -1$, Theorem 15.1.4, gives $\log(x + 1) < x$. Then, since $\exp$ is strictly increasing (on $\mathbb{R}$),

$$\exp[\log(x + 1)] < \exp x, \qquad i.e. \ x + 1 < \exp x,$$

and the proof is complete.   □

Theorem 15.2.5 *Let $a, b \in \mathbb{R}$ and let $q \in \mathbb{Q}$. Then*

(1) $\exp(a + b) = \exp a \exp b$, $e^{a+b} = e^a e^b$,

(2) $(\exp a)^q = \exp(aq)$, $(e^a)^q = e^{aq}$,

(3) $\exp(-b) = \dfrac{1}{\exp b}$, $e^{-b} = \dfrac{1}{e^b}$,

(4) $\exp(a - b) = \dfrac{\exp a}{\exp b}$, $e^{a-b} = \dfrac{e^a}{e^b}$.

Proof   (1)   We have

$$\log(e^{a+b}) = a + b = \log(e^a) + \log(e^b) = \log(e^a e^b)$$

and the injectiveness of log gives (1).

    (2)   We have

$$\log[(e^a)^q] = q \log(e^a) = qa = \log(e^{aq})$$

and the injectiveness of log gives (2).

    (3)   Using (2), $e^{-b} = e^{b(-1)} = (e^b)^{-1} = 1/e^b$.

    (4)   Using (1) and (3), $e^{a-b} = e^{a+(-b)} = e^a e^{-b} = e^a/e^b$.   □

Example 15.2.6 Differentiate $\dfrac{e^x - e^{-x}}{e^x + e^{-x}}$ with respect to x.

Solution Using the quotient rule, the chain rule and Theorem 15.2.5 (1),

$$\frac{d}{dx}\left(\frac{e^x - e^{-x}}{e^x + e^{-x}}\right) = \frac{(e^x + e^{-x})(e^x + e^{-x}) - (e^x - e^{-x})(e^x - e^{-x})}{(e^x + e^{-x})^2}$$

$$= \frac{(e^{2x} + 2 + e^{-2x}) - (e^{2x} - 2 + e^{-2x})}{(e^x + e^{-x})^2} = \frac{4}{(e^x + e^{-x})^2}. \qquad \square$$

Theorem 15.2.7 (1) $\exp x = e^x \to \infty$ *as* $x \to \infty$.

(2) $\exp x = e^x \to 0$ *as* $x \to -\infty$.

Proof Since $\exp x > x$ $(x \in \mathbb{R})$ and $x \to \infty$ as $x \to \infty$, (1) follows from the version of the sandwich principle stated in Note 7.4.11. For (2),

$$\exp x \to \infty \text{ as } x \to \infty \Rightarrow \exp(-x) = 1/\exp x \to 0 \text{ as } x \to \infty$$

$$\Rightarrow \exp u \to 0 \text{ as } u \to -\infty,$$

since $u = -x \to -\infty$ as $x \to \infty$. $\qquad \square$

Corollary 15.2.8 $\dfrac{\exp x}{x} = \dfrac{e^x}{x} \to \infty$ *as* $x \to \infty$.

Proof For $x > 0$,

$$\frac{\exp x}{x} = \frac{\exp x}{\exp(\log x)} = \exp(x - \log x) = \exp\left[x\left(1 - \frac{\log x}{x}\right)\right],$$

and the result follows from Theorem 15.2.7 (1), since $x(1 - \log x/x) \to \infty$ as $x \to \infty$. $\qquad \square$

Example 15.2.9 Show that $x^2 e^{-x} \to 0$ as $x \to \infty$.

Solution It is enough to show that $e^x/x^2 \to \infty$ as $x \to \infty$ (see Definition 7.4.5). Arguing as in the proof of Corollary 15.2.8, for $x > 0$,

$$\frac{\exp x}{x^2} = \frac{\exp x}{\exp(2\log x)} = \exp(x - 2\log x) = \exp\left[x\left(1 - 2\frac{\log x}{x}\right)\right],$$

and the result follows from Theorem 15.2.7, since $x(1 - 2\log x/x) \to \infty$ as $x \to \infty$. $\qquad \square$

Example 15.2.10 Sketch the curve $y = (x^2 + 1)e^{-x}$.

Solution The curve is sketched in Figure 15.2.2.

First, observe that y is defined for all $x \in \mathbb{R}$. We have the following.

$y > 0$ for all $x \in \mathbb{R}$.

$$y' = 2xe^{-x} + (x^2 + 1)(-e^{-x}) = -e^{-x}(x - 1)^2 \begin{cases} \leq 0 & \text{for all } x \in \mathbb{R}, \\ = 0 & \Leftrightarrow x = 1. \end{cases}$$

$$y'' = e^{-x}(x - 1)^2 - e^{-x}2(x - 1) = e^{-x}(x - 1)(x - 3)$$

$$= 0 \Leftrightarrow x = 1 \text{ or } x = 3.$$

A table of signs for y' and y'' is shown below.

x	$\to$	1	$\to$	3	$\to$
y''	+	0	−	0	+
y'		0		−	

Thus, the curve lies above the x-axis, there is a horizontal up-down point of inflection at $(1, 2/e)$ and a decreasing down-up point of inflection at $(3, 10/e^3)$. The concavity is down for $1 < x < 3$ and up for $x < 1$ and for $x > 3$.

Using Example 15.2.9 and properties referred to in Note 7.4.11,

$$y = x^2 e^{-x} + e^{-x} \to \begin{cases} 0 & \text{as } x \to \infty, \\ \infty & \text{as } x \to -\infty. \end{cases}$$

Note, in addition, that $y(0) = 1$ and $y'(0) = -1$. $\qquad\qquad\square$

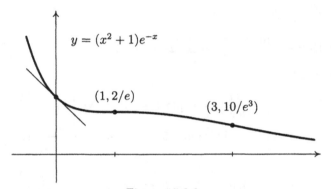

$$y = (x^2 + 1)e^{-x}$$

$$(1, 2/e)$$

$$(3, 10/e^3)$$

Figure 15.2.2

Example 15.2.11 (Newton's Law of Cooling) A metal rod, at a temperature of T degrees, cools at a rate proportional to $T - 20$ (20 degrees being the surrounding temperature which remains constant). The temperature of the rod is 100 degrees at time $t = 0$ and 60 degrees at $t = 1$.

(a) When has the temperature fallen to 30 degrees?

(b) Assuming that $T > 20$, show that $T = 20 + Ae^{-Bt}$ where A and B are positive constants. Find A and B.

Solution For some constant K,

$$\frac{dT}{dt} = K(T - 20), \qquad \text{i.e.} \quad \frac{\dfrac{dT}{dt}}{T - 20} = K.$$

Integrating, with respect to t, gives

$$\log |T - 20| = Kt + C. \tag{15.2.1}$$

(a) Putting $t = 0$ and $T = 100$ in (15.2.1) gives $C = \log 80$. Then putting $t = 1$ and $T = 60$ gives

$$K = \log 40 - \log 80 = \log(1/2) = -\log 2.$$

So (15.2.1) becomes

$$\log|T - 20| = -t\log 2 + \log 80$$

and $T = 30$ when

$$t\log 2 = \log 80 - \log 10 = \log 8 = 3\log 2, \quad i.e. \ t = 3.$$

(b) Assuming that $T > 20$, equation (15.2.1) becomes

$$\log(T - 20) = Kt + C, \quad i.e. \ T - 20 = e^{Kt+C} = e^{Kt}e^C.$$

Hence, $T = 20 + Ae^{-Bt}$, where $A = e^C = 80$ and $B = -K = \log 2$. $\qquad\square$

Remark The falling temperature of the rod (in Example 15.2.11) is an example of *exponential decay*.

15.3   Real Powers

In Section 1.5 we defined x^q where $q \in \mathbb{Q}$ and $x \in D(q)$. The set $D(q)$ depends on q (see Note 1.5.8) but in all cases it contains $(0, \infty)$.

Recall (Theorem 15.1.5 (2)) that, for $x \in (0, \infty)$ and $q \in \mathbb{Q}$, $\log(x^q) = q\log x$, so that

$$x^q = \exp(q\log x). \qquad (15.3.1)$$

The right-hand side of this equation makes sense if q is any real number. This leads us to the following definition of *real powers* of positive real numbers.

Definition 15.3.1 Let $r \in \mathbb{R}$ and $x \in (0, \infty)$. Then we define the rth *power* of x by

$$x^r = \exp(r\log x).$$

In view of equation (15.3.1), when $r \in \mathbb{Q}$, this definition agrees with the definitions in Section 1.5.

The justification for this definition is that the index laws, proved for rational powers in Section 1.5, remain true for real powers. Before proving these laws we extend Theorem 15.1.5 (2).

Theorem 15.3.2 *Let $r \in \mathbb{R}$ and $x \in (0, \infty)$. Then* $\log(x^r) = r\log x$.

Proof We use Definition 15.3.1 and the fact that exp is the inverse of log. Thus

$$\log(x^r) = \log[\exp(r\log x)] = r\log x. \qquad\square$$

Theorem 15.3.3 (Index Laws for Real Powers) *Let $r, s \in \mathbb{R}$ and $x, y \in (0, \infty)$ Then*

$$(1) \ x^r x^s = x^{r+s}, \qquad (2) \ (x^r)^s = x^{rs}, \qquad (3) \ x^r y^r = (xy)^r.$$

Proof Using Definition 15.3.1, Theorem 15.2.5 and, for (2), Theorem 15.3.2, we have

(1)　$x^r x^s = \exp(r \log x) \exp(s \log x) = \exp(r \log x + s \log x)$
$$= \exp[(r + s) \log x] = x^{r+s},$$

(2)　$(x^r)^s = \exp[s \log(x^r)] = \exp(rs \log x) = x^{rs},$

(3)　$x^r y^r = \exp(r \log x) \exp(r \log y) = \exp(r \log x + r \log y)$
$$= \exp[r(\log x + \log y)] = \exp[r \log(xy)] = (xy)^r. \qquad \square$$

Corollary 15.3.4 *Let* $r, s \in \mathbb{R}$ *and* $x, y \in (0, \infty)$. *Then*

$$(1)\ \frac{1}{x^s} = x^{-s}, \qquad (2)\ \frac{x^r}{x^s} = x^{r-s}, \qquad (3)\ \frac{x^r}{y^r} = \left(\frac{x}{y}\right)^r.$$

Proof This is similar to the proof of Corollary 1.5.10. $\qquad \square$

Remark As for rational powers, we shall assume the 'obvious' extensions to the index laws. For example,

$$x^r x^s x^t = x^{r+s+t}, \qquad ((x^r)^s)^t = x^{rst}, \qquad x^r y^r / z^r = (xy/z)^r,$$

where $r, s, t \in \mathbb{R}$ and $x, y, z \in (0, \infty)$.

The next result extends Theorem 8.3.1.

Theorem 15.3.5 *Let* $r \in \mathbb{R} - \{0\}$ *and* $a > 0$. *Then the real functions* $x \mapsto x^r$ *and* $x \mapsto a^x$ *are (continuous and) differentiable on* $(0, \infty)$ *and* $\mathbb{R}$, *respectively, and*

(1)　*For* $x \in (0, \infty)$,　$\dfrac{d}{dx} x^r = r x^{r-1}$,

(2)　*For* $x \in \mathbb{R}$,　$\dfrac{d}{dx} a^x = a^x \log a$.

Proof The differentiability of the functions follows from Theorem 8.2.6 since log and exp are differentiable.

(1)　For $x \in (0, \infty)$,　$\dfrac{d}{dx} x^r = \dfrac{d}{dx} e^{r \log x} = e^{r \log x} \left(\dfrac{r}{x}\right) = x^r \dfrac{r}{x} = r x^{r-1}$.

(2)　For $x \in \mathbb{R}$,　$\dfrac{d}{dx} a^x = \dfrac{d}{dx} e^{x \log a} = e^{x \log a} \log a = a^x \log a$. $\qquad \square$

Notes 15.3.6 (1) For $a > 0$, the real function $x \mapsto a^x$ is called a *general exponential function*, the function $x \mapsto e^x$ being a special case.

(2) Theorem 15.3.5 gives us two *standard integrals*. Let $r \in \mathbb{R} - \{-1\}$ and $a \in (0, \infty)$. Then

$$\int x^r \, dx = \frac{x^{r+1}}{r + 1} + C \quad \text{and} \quad \int a^x \, dx = \frac{a^x}{\log a} + C.$$

The next theorem generalises Corollaries 15.1.11 and 15.2.8.

Theorem 15.3.7 *Let* $r \in (0, \infty)$. *Then*

(1) $x^r \to \infty$ *as* $x \to \infty$,

(2) $\dfrac{\log x}{x^r} \to 0$ *as* $x \to \infty$,

(3) $x^r \log x \to 0$ *as* $x \to 0^+$,

(4) $\dfrac{e^x}{x^r} \to \infty$ *as* $x \to \infty$.

Proof (1) Since $\log x \to \infty$ as $x \to \infty$ and $e^y \to \infty$ as $y \to \infty$,

$$x^r = e^{r \log x} \to \infty \quad \text{as} \quad x \to \infty.$$

(2) Since $x^r \to \infty$ as $x \to \infty$ and $\log y / y \to 0$ as $y \to \infty$,

$$\frac{\log x}{x^r} = \frac{1}{r} \frac{\log(x^r)}{x^r} \to 0 \quad \text{as} \quad x \to \infty.$$

(3) Since $1/x \to \infty$ as $x \to 0^+$ and $\log y / y^r \to 0$ as $y \to \infty$,

$$x^r \log x = \frac{-\log(1/x)}{(1/x)^r} \to 0 \quad \text{as} \quad x \to 0^+.$$

(4) Since $x - r \log x = x(1 - r \log x / x) \to \infty$ as $x \to \infty$ and $e^y \to \infty$ as $y \to \infty$,

$$\frac{e^x}{x^r} = \frac{e^x}{e^{r \log x}} = e^{x - r \log x} \to \infty \quad \text{as} \quad x \to \infty. \qquad \square$$

Remark Theorem 15.3.7 (2) and (4) can be interpreted as saying that, as x increases towards ∞, $\log x$ increases more slowly than any positive power of x and $\exp x$ increases more rapidly than any positive power of x.

Example 15.3.8 Differentiate, with respect to x,

$$\text{(a)} \ \ x^x, \qquad \text{(b)} \ \ (x^2 + 1)^{\sin x}.$$

Solution We use Definition 15.3.1 and the chain rule.

(a) $\quad \dfrac{d}{dx} x^x = \dfrac{d}{dx} e^{x \log x} = e^{x \log x} \left(x \dfrac{1}{x} + 1. \log x \right) = x^x (1 + \log x).$

(b) $\quad \dfrac{d}{dx} (x^2 + 1)^{\sin x} = \dfrac{d}{dx} e^{\sin x \log(x^2 + 1)}$

$$= e^{\sin x \log(x^2 + 1)} \left[\sin x \frac{2x}{x^2 + 1} + \cos x \log(x^2 + 1) \right]$$

$$= (x^2 + 1)^{\sin x} \left[\frac{2x \sin x}{x^2 + 1} + \cos x \log(x^2 + 1) \right] \qquad \square$$

Example 15.3.9 Sketch the curve $y = x^x$ $(x > 0)$.

Solution The curve is sketched in Figure 15.3.1.

For $x > 0$,

$$y = e^{x \log x} > 0, \qquad y' = e^{x \log x}(1 + \log x) = 0 \iff x = 1/e$$

and

$$y'' = e^{x \log x}[(1/x) + (1 + \log x)^2] > 0.$$

Hence the curve lies above the x-axis, there is a minimum turning point at $(1/e, 1/e^{1/e})$ and the curve is concave-up for all $x > 0$.

Also

$$y \to \begin{cases} e^0 = 1 & \text{as } x \to 0^+, \\ \infty & \text{as } x \to \infty \end{cases} \quad \text{and} \quad y' \to -\infty \text{ as } x \to 0^+.$$

Note, in addition, that $y(1) = y'(1) = 1$.　　　　　　　　　　　　　□

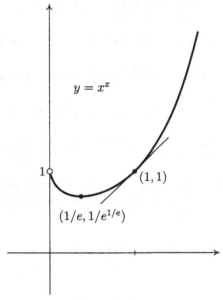

$y = x^x$

$(1, 1)$

$(1/e, 1/e^{1/e})$

Figure 15.3.1

We end this section with a definition which we shall not use! It may help to explain how *our* definition of $\log x$ corresponds to any definition the reader has met previously.

Definition 15.3.10 Let $a, b \in (0, \infty)$. Then we define the *logarithm of a to the base b*, written $\log_b a$, to be the real number c such that $a = b^c$. The number c exists and is unique since

$$a = b^c \iff a = e^{c \log b} \iff \log a = c \log b \iff c = \log a / \log b.$$

Thus, for $x \in (0, \infty)$, $\log x = \log_e x$. Logarithms to the base e are called *natural* logarithms. The notation $\ln x$ is sometimes used instead of $\log x$ or $\log_e x$.

15.4   Hyperbolic Functions

These functions are simple arithmetic combinations of the real function $x \mapsto e^x$ and $x \mapsto e^{-x}$. The notation is suggestive of the trigonometric functions and in many ways the hyperbolic functions behave like their trigonometric counterparts.

Definitions 15.4.1 The *hyperbolic cosine, sine, tangent, secant, cosecant* and *cotangent* functions are denoted by cosh, sinh, tanh, sech, cosech and coth, respectively. They are the real functions defined by

$$\cosh x = \frac{e^x + e^{-x}}{2}, \qquad \sinh x = \frac{e^x - e^{-x}}{2}, \qquad \tanh x = \frac{\sinh x}{\cosh x},$$

$$\operatorname{sech} x = \frac{1}{\cosh x}, \qquad \operatorname{cosech} x = \frac{1}{\sinh x}, \qquad \coth x = \frac{\cosh x}{\sinh x}.$$

The functions cosh, sinh, tanh and sech have maximal domain $\mathbb{R}$ while cosech and coth have maximal domain $\mathbb{R} - \{0\}$.

Note 15.4.2 It follows immediately from the definitions that, for $x \in \mathbb{R}$,

$$\cosh(-x) = \cosh x \qquad \text{and} \qquad \sinh(-x) = -\sinh x.$$

Thus the functions cosh and sinh are even and odd, respectively. It then follows easily that sech is even while cosech, tanh and coth are odd.

Theorem 15.4.3 (1) *For all $x \in \mathbb{R}$,* $\cosh x \geq 1,$
$$\cosh x > \sinh x.$$

(2) $\tanh x \to \begin{cases} 1 & \text{as } x \to \infty, \\ -1 & \text{as } x \to -\infty. \end{cases}$

So the lines $y = 1$ and $y = -1$ are horizontal asymptotes to the curve $y = \tanh x$.

Proof (1) For $w > 0$,
$$0 \leq (w^{1/2} - w^{-1/2})^2 = w - 2 + w^{-1}.$$

Hence, $w + w^{-1} \geq 2$ and $\cosh x \geq 1$ follows by putting $w = e^x$.
For the second inequality, observe that $\cosh x - \sinh x = e^{-x} > 0$.

(2) Since $e^{-u} \to 0$ as $u \to \infty$ and $e^v \to 0$ as $v \to -\infty$,

$$\tanh x = \frac{e^x - e^{-x}}{e^x + e^{-x}} = \begin{cases} \dfrac{1 - e^{-2x}}{1 + e^{-2x}} \to 1 & \text{as } x \to \infty, \\[2mm] \dfrac{e^{2x} - 1}{e^{2x} + 1} \to -1 & \text{as } x \to -\infty. \end{cases} \qquad \square$$

The graphs of cosh and sinh are easily deduced from that of exp. In Figure 15.4.1, the mid-point B of the vertical line segment AC traces out the curves $y = \cosh x$ and $y = \sinh x$ as A and C traverse the curves $y = e^x$ and $y = \pm e^{-x}$, respectively. The graphs of all six hyperbolic functions are shown in Figures 15.4.2, 15.4.3 and 15.4.4.

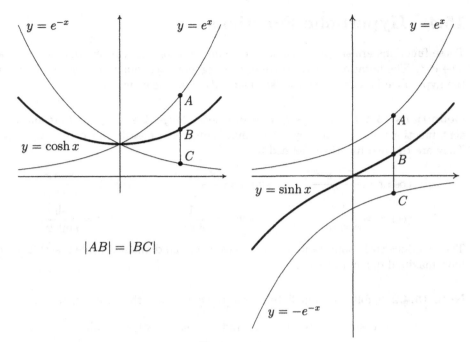

$$|AB| = |BC|$$

Figure 15.4.1

Example 15.4.4 Evaluate $\operatorname{sech}\big(\log(\sqrt{2}+1)\big)$.

Solution First, observe that $(\sqrt{2}-1)(\sqrt{2}+1) = 1$ so that

$$-\log(\sqrt{2}+1) = \log\big(1/(\sqrt{2}+1)\big) = \log(\sqrt{2}-1).$$

Hence,

$$\operatorname{sech}\big(\log(\sqrt{2}+1)\big) = \frac{2}{e^{\log(\sqrt{2}+1)} + e^{-\log(\sqrt{2}+1)}} = \frac{2}{e^{\log(\sqrt{2}+1)} + e^{\log(\sqrt{2}-1)}}$$

$$= \frac{2}{(\sqrt{2}+1) + (\sqrt{2}-1)} = \frac{1}{\sqrt{2}} \qquad \square$$

Example 15.4.5 Solve the equation $3\sinh x - \cosh x = 1$.

Solution We have

$$
\begin{aligned}
3\sinh x - \cosh x = 1 \;&\Leftrightarrow\; 3[\tfrac{1}{2}(e^x - e^{-x})] - \tfrac{1}{2}(e^x + e^{-x}) = 1\\
&\Leftrightarrow\; e^x - 2e^{-x} - 1 = 0\\
&\Leftrightarrow\; e^{2x} - e^x - 2 = 0 \quad \text{(multiplying by } e^x \neq 0)\\
&\Leftrightarrow\; (e^x)^2 - e^x - 2 = 0\\
&\Leftrightarrow\; e^x = -1 \text{ or } e^x = 2 \quad \text{(solving } t^2 - t - 2 = 0)\\
&\Leftrightarrow\; e^x = 2 \quad \text{(since } e^x > 0 \text{ so that } e^x \neq -1)\\
&\Leftrightarrow\; x = \log 2.
\end{aligned}
$$

Thus, $x = \log 2$ is the only solution of the given equation. $\qquad \square$

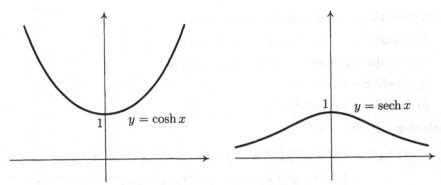

Figure 15.4.2 : Hyperbolic Cosine and Secant

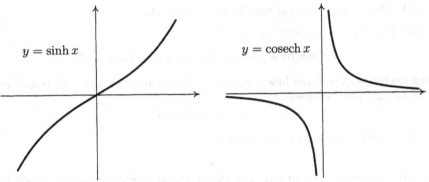

Figure 15.4.3 : Hyperbolic Sine and Cosecant

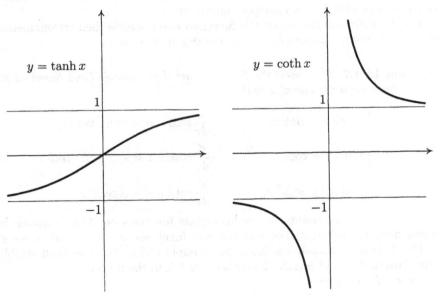

Figure 15.4.4 : Hyperbolic Tangent and Cotangent

Example 15.4.6 Show that, for all $x, y \in \mathbb{R}$,

(a) $\cosh(x+y) = \cosh x \cosh y + \sinh x \sinh y$,

(b) $\sinh(x+y) = \sinh x \cosh y + \cosh x \sinh y$,

(c) $\cosh^2 x - \sinh^2 x = 1$,

(d) $\operatorname{sech}^2 x = 1 - \tanh^2 x$.

Solution (a) We have

$$\text{RHS}\,(a) = \tfrac{1}{2}[e^x + e^{-x}]\tfrac{1}{2}[e^y + e^{-y}] + \tfrac{1}{2}[e^x - e^{-x}]\tfrac{1}{2}[e^y - e^{-y}]$$

$$= \tfrac{1}{4}[e^{x+y} + e^{x-y} + e^{-x+y} + e^{-x-y} + e^{x+y} - e^{x-y} - e^{-x+y} + e^{-x-y}]$$

$$= \tfrac{1}{4}[2e^{x+y} + 2e^{-(x+y)}] = \tfrac{1}{2}[e^{x+y} + e^{-(x+y)}] = \text{LHS}\,(a).$$

(b) This is similar to (a) and is left to the reader.

(c) Putting $y = -x$ in (a) gives

$$\cosh(0) = \cosh x \cosh(-x) + \sinh x \sinh(-x).$$

Since $\cosh(0) = 1$ and the functions cosh and sinh are even and odd, respectively (see Note 15.4.2), this becomes

$$1 = \cosh^2 x - \sinh^2 x.$$

(d) Divide both sides of (c) by $\cosh x$. $\qquad\square$

 Hazard As can be seen in Example 15.4.6, the hyperbolic functions satisfy identities similar to trigonometric identities *but* there are differences of sign. As a rough guide, the identities are 'the same' except that the sign is changed whenever $\sinh a \sinh b$ occurs *or is implied* by, for example, $\tanh a \tanh b$.

The derivatives of the hyperbolic functions also resemble their trigonometric counterparts but there is no simple rule for the sign differences.

Theorem 15.4.7 *The hyperbolic functions are differentiable (and therefore continuous) on their maximal domains with*

$$\frac{d}{dx}\cosh x = \sinh x, \qquad\qquad \frac{d}{dx}\operatorname{sech} x = -\operatorname{sech} x \tanh x,$$

$$\frac{d}{dx}\sinh x = \cosh x, \qquad\qquad \frac{d}{dx}\operatorname{cosech} x = -\operatorname{cosech} x \coth x,$$

$$\frac{d}{dx}\tanh x = \operatorname{sech}^2 x, \qquad\qquad \frac{d}{dx}\coth x = -\operatorname{cosech}^2 x.$$

Proof The differentiability of the hyperbolic functions on their maximal domains follows from the differentiability of the real functions $x \mapsto e^x$ and $x \mapsto e^{-x}$ on $\mathbb{R}$. The derivative of tanh was found in Example 15.2.6. Here we shall establish the derivatives of sinh and cosech. The others are left to the reader.

For $x \in \mathbb{R}$, we have

$$\frac{d}{dx}\sinh x = \frac{d}{dx}\left(\tfrac{1}{2}[e^x - e^{-x}]\right) = \tfrac{1}{2}[e^x - e^{-x}(-1)] = \tfrac{1}{2}[e^x + e^{-x}] = \cosh x$$

and for $x \neq 0$, using the reciprocal rule,

$$\frac{d}{dx}\operatorname{cosech} x = \frac{d}{dx}\left(\frac{1}{\sinh x}\right) = -\frac{1}{\sinh^2 x}\frac{d}{dx}\sinh x = -\frac{1}{\sinh^2 x}\cosh x$$

$$= -\left(\frac{1}{\sinh x}\right)\left(\frac{\cosh x}{\sinh x}\right) = -\operatorname{cosech} x \coth x. \qquad \square$$

Example 15.4.8 Differentiate $\cosh^5 3x \sinh^3 5x$ with respect to x.

Solution Using the product and chain rules,

$$\frac{d}{dx}\cosh^5 3x \sinh^3 5x$$

$$= [5\cosh^4 3x\, 3\sinh 3x]\sinh^3 5x + \cosh^5 3x[3\sinh^2 5x\, 5\cosh 5x]$$

$$= 15\cosh^4 3x \sinh^2 5x(\sinh 3x \sinh 5x + \cosh 3x \cosh 5x)$$

$$= 15\cosh^4 3x \sinh^2 5x \cosh 8x.$$

The final simplification uses Example 15.4.6. $\qquad \square$

Example 15.4.9 Find the point of inflection on the curve $y = \operatorname{sech} x$ $(x > 0)$ and determine its nature.

Solution We have $y' = -\operatorname{sech} x \tanh x$ so that, using the product rule,

$$y'' = -[-\operatorname{sech} x \tanh x \tanh x + \operatorname{sech} x \operatorname{sech}^2 x]$$

$$= \operatorname{sech} x[\tanh^2 x - \operatorname{sech}^2 x]$$

$$= 0 \iff \tanh^2 x = \operatorname{sech}^2 x.$$

For $x > 0$, $\sinh x > 0$ so that

$$\tanh^2 x = \operatorname{sech}^2 x \iff \frac{\sinh^2 x}{\cosh^2 x} = \frac{1}{\cosh^2 x} \iff \sinh^2 x = 1$$

$$\iff \sinh x = 1$$

Arguing as in the solution of Example 15.4.5,

$$\sinh x = 1 \iff e^x - e^{-x} = 2 \iff e^{2x} - 2e^x - 1 = 0$$

$$\iff e^x = 1 \pm \sqrt{2} \iff e^x = 1 + \sqrt{2} \quad (\text{since } 1 - \sqrt{2} < 0)$$

$$\iff x = \log(1 + \sqrt{2}),$$

A table of signs for y'' is shown below.

x	$\to$	$\log(1 + \sqrt{2})$	$\to$
$\operatorname{sech} x$	$+$	$+$	$+$
$\tanh^2 x - \operatorname{sech}^2 x$	$-$	0	$+$
y''	$-$	0	$+$
y'		$-$	

By Example 15.4.4, $y = 1/\sqrt{2}$ when $x = \log(1 + \sqrt{2})$. Thus the curve $y = \operatorname{sech} x$ ($x > 0$) has a decreasing down-up point of inflection at $(\log(1 + \sqrt{2}), 1/\sqrt{2})$ and this is the only point of inflection on the curve.   □

Note 15.4.10 We have the following *standard integrals*.

$$\int \cosh x \, dx = \sinh x + C \qquad \text{and} \qquad \int \sinh x \, dx = \cosh x + C.$$

Example 15.4.11 Find $\displaystyle\int \tanh x \, dx$.

Solution Using the fact that $\dfrac{d}{dx} \cosh x = \sinh x$, we have

$$\int \tanh x \, dx = \int \frac{\sinh x}{\cosh x} \, dx = \log|\cosh x| + C = \log(\cosh x) + C. \qquad \square$$

15.5   Inverse Hyperbolic Functions

The inverse hyperbolic functions are defined using, where necessary, restrictions of the hyperbolic functions.

Theorem 15.5.1 *Define*

$$f : [0, \infty) \to [1, \infty) \ \ by \ f(x) = \cosh x,$$
$$g : \mathbb{R} \to \mathbb{R} \ \ by \ g(x) = \sinh x,$$
$$h : \mathbb{R} \to (-1, 1) \ \ by \ h(x) = \tanh x.$$

Then f, g and h are bijective.

[Note that $g = \sinh$.]

Proof We shall prove that f is bijective. The proofs for g and h are similar and are left to the reader.

Since $f'(x) = \sinh x > 0$ for all $x \in (0, \infty)$, it follows from the monotonicity theorem and Note 9.4.2 that f is strictly increasing on $[0, \infty)$. Hence, by Theorem 2.6.3, f is injective.

To see that f is surjective, first, observe that $f(0) = 1$. Then let $t \in (1, \infty)$. Since $f(x) = \frac{1}{2}[e^x + e^{-x}] \to \infty$ as $x \to \infty$, there exists $u \in (0, \infty)$ such that $t < f(u)$. Hence $f(0) < t < f(u)$ and the intermediate value theorem applies to give $f(s) = t$ for some $s \in (0, u)$.   □

Definitions 15.5.2 The *inverse hyperbolic cosine function* is the inverse of the bijection $f : [0, \infty) \to [1, \infty)$ defined by $f(x) = \cosh x$. Thus,

$$\cosh^{-1} : [1, \infty) \to [0, \infty)$$

is defined by
$$\cosh^{-1} x = y \quad \text{where} \quad y \in [0, \infty) \text{ and } \cosh y = x.$$

The *inverse hyperbolic sine function* is the inverse of the bijection $g : \mathbb{R} \to \mathbb{R}$ defined by $g(x) = \sinh x$. Thus,

$$\sinh^{-1} : \mathbb{R} \to \mathbb{R}$$

is defined by
$$\sinh^{-1} x = y \quad \text{where} \quad \sinh y = x.$$

The *inverse hyperbolic tangent function* is the inverse of the bijection $h : \mathbb{R} \to (-1, 1)$ defined by $h(x) = \tanh x$. Thus,

$$\tanh^{-1} : (-1, 1) \to \mathbb{R}$$

is defined by
$$\tanh^{-1} x = y \quad \text{where} \quad \tanh y = x.$$

 Hazard The functions $\cosh^{-1}$ and $\tanh^{-1}$ are not inverses of cosh and tanh, but of *restrictions* of these functions (*cf.* the inverse trigonometric functions). However, the function $\sinh^{-1}$ *is* the inverse of sinh. We did not have to restrict $\sinh : \mathbb{R} \to \mathbb{R}$ to obtain a bijection.

The graphs of the inverse hyperbolic functions can be deduced using the mirror property for inverse functions, Note 2.5.8. They are sketched in Figure 15.5.1.

Since the functions cosh, sinh and tanh can be expresses in terms of the exponential function, it is perhaps not surprising that the functions $\cosh^{-1}$, $\sinh^{-1}$ and $\tanh^{-1}$ can be expressed as logarithms.

Theorem 15.5.3 *We have*

(1) *for* $x \in [1, \infty)$, $\cosh^{-1} x = \log(x + \sqrt{x^2 - 1})$,

(2) *for* $x \in \mathbb{R}$, $\sinh^{-1} x = \log(x + \sqrt{x^2 + 1})$,

(3) *for* $x \in (-1, 1)$, $\tanh^{-1} x = \log \sqrt{(1 + x)/(1 - x)}$.

Proof (1) Observe first that, when $x > 1$,
$$-\sqrt{x^2 - 1} < 0 < x - 1 = \sqrt{x - 1}\sqrt{x - 1} < \sqrt{x - 1}\sqrt{x + 1} = \sqrt{x^2 - 1}$$

and, in particular,
$$x - 1 < \sqrt{x^2 - 1} \quad \text{and} \quad -\sqrt{x^2 - 1} < x - 1.$$

So, when $x > 1$,
$$x - \sqrt{x^2 - 1} < 1 < x + \sqrt{x^2 - 1}.$$

Also, when $x = 1$,
$$x - \sqrt{x^2 - 1} = 1 = x + \sqrt{x^2 - 1}.$$

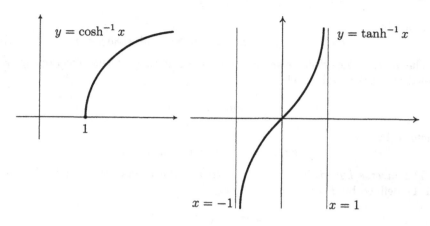

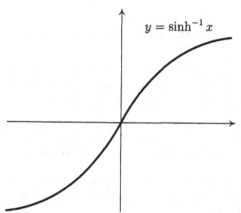

Figure 15.5.1 : Inverse Hyperbolic Functions

Hence, for $x \in [1, \infty)$,

$$\cosh^{-1} x = y \iff \cosh y = x \text{ and } y \geq 0$$
$$\iff e^y + e^{-y} = 2x \text{ and } y \geq 0 \iff e^{2y} - 2xe^y + 1 = 0 \text{ and } y \geq 0$$
$$\iff e^y = x \pm \sqrt{x^2 - 1} \text{ and } e^y \geq 1 \iff e^y = x + \sqrt{x^2 - 1}$$
$$\iff y = \log(x + \sqrt{x^2 - 1}).$$

(2) This is similar to (1) and is left to the reader.

(3) For $x \in (-1, 1)$,

$$\tanh^{-1} x = y \iff \tanh y = x \iff e^y - e^{-y} = x(e^y + e^{-y})$$
$$\iff e^y(1 - x) = e^{-y}(1 + x) \iff e^{2y} = (1 + x)/(1 - x)$$
$$\iff 2y = \log[(1 + x)/(1 - x)] \iff y = \tfrac{1}{2}\log[(1 + x)/(1 - x)]$$
$$\iff y = \log\sqrt{(1 + x)/(1 - x)}.$$

$\square$

Theorem 15.5.4 *The functions* $\cosh^{-1}$, $\sinh^{-1}$ *and* $\tanh^{-1}$ *are differentiable on* $(1, \infty)$, $\mathbb{R}$ *and* $(-1, 1)$, *respectively, with*

$$\frac{d}{dx} \cosh^{-1} x = \frac{1}{\sqrt{x^2 - 1}}, \quad \frac{d}{dx} \sinh^{-1} x = \frac{1}{\sqrt{x^2 + 1}}, \quad \frac{d}{dx} \tanh^{-1} x = \frac{1}{1 - x^2}.$$

Proof Use the logarithmic expressions from Theorem 15.5.3. For example, the differentiability of $\cosh^{-1}$ on $(1, \infty)$ follows from the differentiability of $\log$ on $(0, \infty)$ and, for $x \in (1, \infty)$,

$$\frac{d}{dx} \cosh^{-1} x = \frac{d}{dx} \log(x + \sqrt{x^2 - 1}) = \frac{1}{x + \sqrt{x^2 - 1}} \left(1 + \frac{x}{\sqrt{x^2 - 1}}\right)$$

$$= \frac{1}{x + \sqrt{x^2 - 1}} \cdot \frac{\sqrt{x^2 - 1} + x}{\sqrt{x^2 - 1}} = \frac{1}{\sqrt{x^2 - 1}}. \qquad \square$$

Example 15.5.5 Let $a > 0$. Find $\dfrac{d}{dx}\left(\dfrac{1}{a} \tanh^{-1} \dfrac{x}{a}\right)$.

Solution Using the chain rule, for $x \in (-a, a)$,

$$\frac{d}{dx}\left(\frac{1}{a} \tanh^{-1} \frac{x}{a}\right) = \frac{1}{a}\left(\frac{1}{1 - (x/a)^2} \cdot \frac{1}{a}\right) = \frac{1}{a^2 - x^2}. \qquad \square$$

15.X   Exercises

1. Differentiate, with respect to x,

     (a) $\log(7x - 4)$,   (b) $\log(4 - 7x)$,

     (c) $\log|7x - 4|$,   (d) $\log[(7x - 4)^2]$,

     (e) $\log(x^2 - 2x + 3)$,   (f) $\log[(x^2 - 2x + 3)^5]$,

     (g) $[\log(x^2 - 2x + 3)]^5$,   (h) $\log\sqrt{1 + \sin^2 x}$,

     (i) $\log(\cos x)$,   (j) $\cos(\log x)$,

     (k) $\log(\log x)$,   (l) $x \log x - x$.

     (m) $\log\left(\dfrac{x + 1}{x - 1}\right)$,   (n) $\log\sqrt{\dfrac{x + 1}{x - 1}}$,

2. Use logarithmic differentiation to find $f'(x)$ when $f(x)$ is

     (a) $\dfrac{(x^2 + 2)^{1/2}(x^8 + 8)^{1/8}}{(x^4 + 4)^{1/4}(x^6 + 6)^{1/6}}$,   (b) $\sqrt{\dfrac{\sin^{-1} x}{x^3(2 + \sin x)}}$.

3. Find

(a) $\displaystyle\int \frac{dx}{3x}$,　　　　(b) $\displaystyle\int \frac{dx}{3x+4}$,　　　　(c) $\displaystyle\int \frac{2x}{x^2+4}\,dx$,

(d) $\displaystyle\int \frac{x^4}{x^5+6}\,dx$,　　(e) $\displaystyle\int \frac{dx}{(x+1)\log(x+1)}$,　　(f) $\displaystyle\int \frac{2x-1}{x^2+4}\,dx$,

(g) $\displaystyle\int \tan 7x\,dx$,　　(h) $\displaystyle\int \cot(4x-3)\,dx$,　　(i) $\displaystyle\int \sec 5x\,dx$,

(j) $\displaystyle\int \frac{dx}{\sqrt{3-x^2}}$.　　(k) $\displaystyle\int \frac{dx}{\sqrt{x^2+3}}$,　　(l) $\displaystyle\int \frac{dx}{\sqrt{x^2-3}}$,

4. Evaluate

(a) $\displaystyle\int_{\pi/8}^{\pi/4} \operatorname{cosec} 2x\,dx$,　　(b) $\displaystyle\int_0^3 \frac{2x+1}{2x+3}\,dx$,

(c) $\displaystyle\int_0^2 \frac{dx}{\sqrt{x^2-2x+4}}$,　　(d) $\displaystyle\int_1^9 \frac{dx}{\sqrt{x}(\sqrt{x}+1)}$.

5. For each of the following, establish the inequalities by applying the mean value theorem to the function f.

(a) $\dfrac{5}{4} < \exp(\tfrac{1}{4}) < \dfrac{4}{3}$,　　$f : [0, \tfrac{1}{4}] \to \mathbb{R}$, defined by $f(x) = \exp x$,

(b) $\dfrac{1}{4} < \log(\tfrac{4}{3}) < \dfrac{1}{3}$,　　$f : [1, \tfrac{4}{3}] \to \mathbb{R}$, defined by $f(x) = \log x$,

(c) $\dfrac{2}{7} < \log(\tfrac{4}{3}) < \dfrac{7}{24}$,　　$f : [0, \tfrac{1}{7}] \to \mathbb{R}$, defined by $f(x) = \log\left(\dfrac{1+x}{1-x}\right)$.

6. Show that $x^2/(3x^2 - 2x + 1) \to 1/3$ as $x \to \infty$. Deduce that

$$\frac{\log x}{3x^2 - 2x + 1} \to 0 \text{ as } x \to \infty.$$

What happens to $(3x^2 - 2x + 1)\log x$ as $x \to 0^+$?

7. By considering $1 + \log x = \log(ex)$, show that

$$\frac{1+\log x}{x} \to \begin{cases} 0 & \text{as } x \to \infty, \\ -\infty & \text{as } x \to 0^+. \end{cases}$$

8. Sketch the curves　(a) $y = \dfrac{1+\log x}{x}$,　　(b) $y = \sqrt{x}\log x$.

9. Differentiate, with respect to x,

(a) e^{5x},　　　　　　(b) $e^{(3-2x)}$,　　　　(c) $e^{\sin x}$,

(d) x^e,　　　　　　　(e) 3^{4x},　　　　　　(f) π^{1-x},

(g) $(2x^3+4)e^{-5x}$,　　(h) $(x^2+5)^{\tan x}$,　　(i) $(\cos x)^{(x\sin x)}$.

10. Find

$$\text{(a)} \int e^{5x} \, dx, \qquad \text{(b)} \int e^{(3-2x)} \, dx, \qquad \text{(c)} \int x^e \, dx,$$

$$\text{(d)} \int 3^{4x} \, dx, \qquad \text{(e)} \int \pi^{(1-x)} \, dx, \qquad \text{(f)} \int 2x e^{(x^2-9)} \, dx.$$

11. Determine the behaviour, as $x \to \infty$, of

$$\text{(a)} \ \sqrt{x} e^{-x}, \qquad \text{(b)} \ x^3 e^{-2x}, \qquad \text{(c)} \ x^{-3} e^{5x}.$$

12. Sketch the curves (a) $y = x^2 e^{2x}$, (b) $y = (x^2 + 4x + 5)e^{-x}$.

13. Use L'Hôpital's rule to evaluate

$$\text{(a)} \ \lim_{x \to 0} \frac{\log(7x+1)}{5x}, \qquad \text{(b)} \ \lim_{x \to 3} \frac{\sqrt{3x} - 3}{\log(3x - 8)},$$

$$\text{(c)} \ \lim_{x \to 0} \frac{e^{2x} - 1}{\sin x}, \qquad \text{(d)} \ \lim_{x \to 0} \frac{5^x - 2^x}{x}.$$

14. The temperature of a metal rod drops from 150 degrees to 60 degrees in 1 hour. The temperature of the surrounding air remains constant at 15 degrees. Use Newton's law of cooling (see Example 15.2.11) to determine how much longer it will take for the temperature of the rod to drop to 30 degrees?

15. A radioactive substance is subject to the *law of natural decay* which states that $\dfrac{dV}{dt} = -kV$, where V is the volume of the substance at time t and k is a positive constant. The *half-life* of the substance is the time it takes for half the substance to disappear. Calculate the half-life if 20% of the substance disappears in 15 years.

16. Verify that cosh and sech are even functions, and that sinh, cosech, tanh and coth are odd functions.

17. Solve the equations

$$\text{(a)} \ \ 3\cosh x - \sinh x = 3, \qquad \text{(b)} \ \ 5\sinh x - 3\cosh x = \sqrt{2}.$$

18. Show that, for all $x \in \mathbb{R}$,

$$\text{(a)} \ \ \cosh 2x = \cosh^2 x + \sinh^2 x, \qquad \text{(b)} \ \ \sinh 2x = 2\sinh x \cosh x.$$

19. Differentiate, with respect to x,

$$\text{(a)} \ \ \sinh(x/3), \qquad \qquad \text{(b)} \ \ \sinh(x^2 - x),$$

$$\text{(c)} \ \ \tanh \sqrt{x - 1}, \qquad \qquad \text{(d)} \ \ \log(\cosh 4x),$$

20. Differentiate $\cosh^3 2x \cosh^2 3x$ with respect to x and use Example 15.4.6 to simplify your answer.

21. Find/evaluate (a) $\displaystyle\int \sinh(5x+4)\, dx$, (b) $\displaystyle\int_0^{\log 2} \tanh 3x\, dx$.

22. Let $a > 0$. Differentiate, with respect to x,

$$\text{(a)}\quad \cosh^{-1}\frac{x}{a}, \qquad \text{(b)}\quad \sinh^{-1}\frac{x}{a}.$$

For what ranges of values of x are your answers valid?

23. Determine which of the following improper integrals converge and evaluate those that do.

$$\text{(a)}\quad \int_0^\infty \frac{x}{x^2+1}\, dx, \qquad \text{(b)}\quad \int_0^\infty e^{ax}\, dx \quad (a \neq 0),$$

$$\text{(c)}\quad \int_{-\infty}^\infty \operatorname{sech}^2 x\, dx, \qquad \text{(d)}\quad \int_0^3 \frac{dx}{9-x^2}\, dx.$$

For (d), use Example 15.5.5 and consider the graph of $\tanh^{-1}$.

Chapter 16

Integration—Methods and Applications

In Chapter 14, we introduced indefinite integrals and in Chapter 15, we extended the list of functions we can integrate. At this point, we can only integrate those functions in Appendix E and linear combinations of them.

In the first four sections of this chapter, we introduce methods which extend further the list of functions we can integrate. The first method, and probably the most useful, is *substitution*. It is based on the chain rule for differentiation. In Sections 16.2 and 16.3 we show how the method allows us to integrate rational functions and certain trigonometric functions. The discussion of *t*-formulae is of considerable theoretical interest. They show that certain trigonometric integrals can be transformed into rational integrals. Then, as indicated in Section 16.2, there is a systematic way of evaluating them. Section 16.4 covers the second method, called *integration by parts*. This is based on the product rule for differentiation. It allows us to integrate certain functions which cannot be integrated by earlier methods, but the reader should note that, unless there is a particular reason for using integration by parts, other methods should normally be investigated first.

The remaining sections show how integration can be used to evaluate the lengths of curves, and the surface areas and volumes of solids of revolution. In Section 14.2, we defined the definite integral as the limit of Riemann sums. Here, we reverse this process. If a quantity can be defined as the limit of a sum, and this limit may be interpreted as a Riemann sum, then the quantity is equal to a definite integral. If we can evaluate this definite integral (using the fundamental theorem of calculus), then we can find the required value.

16.1   Substitution

Let f be a continuous function with antiderivative F, and let g be a differentiable function such that g' is continuous and $F \circ g$ is defined. Writing $F \circ g = H$ and $h = H'$, the chain rule gives

$$h(x) = H'(x) = (F \circ g)'(x) = F'(g(x))g'(x) = f(g(x))g'(x).$$

So $\int h(x)\,dx = H(x) + C$ can be written

$$\int f(g(x))g'(x)\,dx = F(g(x)) + C \tag{16.1.1}$$

and we also have

$$\int f(u)\,du = F(u) + C. \tag{16.1.2}$$

We may transfer between equations (16.1.1) and (16.1.2) by substituting

$$g(x) = u, \qquad g'(x)\,dx = du.$$

We are not implying that $g'(x)\,dx$ or du mean anything on their own, merely that one may be replaced by the other. We shall refer to $g(x) = u$ as the *basic substitution*.

As an example, consider

$$\mathcal{I} = \int (1 + \sin x)^3 \cos x\,dx.$$

Substituting

$$1 + \sin x = u, \qquad \cos x\,dx = du$$

we get

$$\mathcal{I} = \int u^3\,du = \frac{u^4}{4} + C = \frac{1}{4}(1 + \sin x)^4 + C.$$

Definite integrals may also be found by substitution. With the functions f, F and g as above, and assuming that g is defined on $[a,b]$ and f on $g([a,b])$, we have

$$\int_a^b f(g(x))g'(x)\,dx = \left[F(g(x))\right]_a^b = F(g(b)) - F(g(a)) \tag{16.1.3}$$

and

$$\int_{g(a)}^{g(b)} f(u)\,du = \left[F(u)\right]_{g(a)}^{g(b)} = F(g(b)) - F(g(a)). \tag{16.1.4}$$

We may transfer between equations 16.1.2 and 16.1.4 by substituting

$$g(x) = u, \qquad g'(x)\,dx = du,$$

x	a	b
u	$g(a)$	$g(b)$

where, as before, $g(x) = u$ is the *basic substitution*.

For example, to evaluate

$$\mathcal{J} = \int_0^{\pi/2} (1 + \sin x)^3 \cos x\,dx$$

we may substitute

$$1 + \sin x = u, \qquad \cos x\,dx = du,$$

x	0	$\pi/2$
u	1	2

to get

$$\mathcal{J} = \int_1^2 u^3\,du = \left[\frac{u^4}{4}\right]_1^2 = \frac{16}{4} - \frac{1}{4} = \frac{15}{4}.$$

Hazard When substituting in an indefinite integral, remember to return to the original variable after integrating.

When substituting in a definite integral, remember to substitute the limits of integration. Alternatively, return to the original variable and use the original limits. Thus, in the above example,

$$\mathcal{J} = \left[\frac{u^4}{4}\right]_{x=0}^{x=\pi/2} = \left[\frac{1}{4}(1+\sin x)^4\right]_0^{\pi/2}.$$

In general, the process of substitution is straightforward. What can be somewhat harder, is deciding which substitution to use. The next example should provide the reader with some ideas.

Example 16.1.1 Find/evaluate

(a) $\displaystyle\int \frac{dx}{x\log x},$ (b) $\displaystyle\int x^2\sqrt{x^3+5}\,dx,$

(c) $\displaystyle\int_0^{\sqrt{3}} \frac{x^2}{x^6+9}\,dx,$ (d) $\displaystyle\int_0^{\pi/2} \sin^5 x\sqrt{\cos x}\,dx,$

Solution (a) Here we substitute

$$\log x = u, \qquad \frac{1}{x}\,dx = du$$

to get

$$\int \frac{dx}{x\log x} = \int \left(\frac{1}{\log x}\right)\frac{1}{x}\,dx = \int \frac{du}{u} = \log|u| + C = \log|\log x| + C.$$

(b) Here we substitute

$$x^3 + 5 = u, \qquad 3x^2\,dx = du$$

to get

$$\int x^2\sqrt{x^3+5}\,dx = \frac{1}{3}\int \sqrt{x^3+5}\,(3x^2)\,dx = \frac{1}{3}\int \sqrt{u}\,du$$

$$= \frac{1}{3}\frac{u^{3/2}}{(3/2)} + C = \frac{2}{9}(x^3+5)^{3/2} + C.$$

(c) Here we substitute

$$x^3 = u, \qquad 3x^2\,dx = du,$$

x	0	$\sqrt{3}$
u	0	$3\sqrt{3}$

to get

$$\int_0^{\sqrt{3}} \frac{x^2}{x^6+9}\,dx = \frac{1}{3}\int_0^{\sqrt{3}} \frac{1}{x^6+9}\,3x^2\,dx = \int_0^{3\sqrt{3}} \frac{1}{u^2+9}\,du$$

$$= \frac{1}{3}\left[\frac{1}{3}\tan^{-1}\frac{u}{3}\right]_0^{3\sqrt{3}} = \frac{1}{9}\tan^{-1}\sqrt{3} = \frac{\pi}{27}.$$

(d) The method used here is generalised later in Section 16.3. We substitute

$$\cos x = u, \qquad -\sin x \, dx = du,$$

x	0	$\pi/2$
u	1	0

to get

$$\int_0^{\pi/2} \sin^5 x \sqrt{\cos x} \, dx$$

$$= -\int_0^{\pi/2} \sin^4 x \sqrt{\cos x} \, (-\sin x) \, dx$$

$$= -\int_1^0 (1 - u^2)^2 \sqrt{u} \, du \quad (\text{using } \sin^2 x = 1 - \cos^2 x = 1 - u^2)$$

$$= \int_0^1 (u^{1/2} - 2u^{5/2} + u^{9/2}) \, du$$

$$= \left[\frac{2}{3} u^{3/2} - \frac{4}{7} u^{7/2} + \frac{2}{11} u^{11/2} \right]_0^1 = \frac{64}{231}. \qquad \square$$

The process of substitution can be generalised as follows. Let f be a continuous function with antiderivative F, and let g and h be differentiable functions such that g' and h' are continuous and $F \circ g$ and $F \circ h$ are defined. Then, arguing as before, we may transfer between the equations

$$\int f(g(x)) g'(x) \, dx = F(g(x)) + C$$

and

$$\int f(h(u)) h'(u) \, du = F(h(u)) + C$$

by substituting

$$g(x) = h(u), \qquad g'(x) \, dx = h'(u) \, du,$$

with appropriate changes of limits for definite integrals. Here, $g(x) = h(u)$ is the *basic substitution*.

Example 16.1.2 Find/evaluate

$$\text{(a)} \int \frac{dx}{\sqrt{2x+1}}, \qquad \text{(b)} \int_1^4 \frac{dx}{(x^2 - 2x + 10)^2}.$$

Solution (a) Here we substitute

$$2x = u^2 \ (u \geq 0), \qquad 2 \, dx = 2u \, du$$

to get

$$\int \frac{dx}{\sqrt{2x+1}} = \int \frac{u \, du}{u+1} = \int \left(1 - \frac{1}{u+1} \right) du$$

$$= u - \log |u+1| + C = \sqrt{2x} - \log |\sqrt{2x} + 1| + C.$$

(b) The method used here is generalised in Section 16.2. We complete the square of the quadric,

$$x^2 - 2x + 10 = (x-1)^2 + 9,$$

and substitute

$$x - 1 = 3\tan\theta \ (-\pi/2 < \theta < \pi/2), \qquad dx = 3\sec^2\theta\, d\theta,$$

x	1	4
θ	0	$\pi/4$

to get

$$\int_1^4 \frac{dx}{(x^2 - 2x + 10)^2} = \int_1^4 \frac{dx}{[(x-1)^2 + 9]^2} = \int_0^{\pi/4} \frac{3\sec^2\theta\, d\theta}{81\sec^4\theta}$$

$$(\text{using } (x-1)^2 + 9 = 9(\tan^2\theta + 1) = 9\sec^2\theta)$$

$$= \frac{1}{27}\int_0^{\pi/4} \cos^2\theta\, d\theta = \frac{1}{54}\int_0^{\pi/4} (1 + \cos 2\theta)\, d\theta$$

$$= \frac{1}{54}\left[\theta + \frac{\sin 2\theta}{2}\right]_0^{\pi/4} = \frac{\pi + 2}{216}. \qquad \square$$

Remarks These refer to the solution of the above example.

(1) The basic substitutions give implicit relationships between x and u, and between x and θ. We could have been explicit by writing (a) $\sqrt{2x} = u$ and (b) $\tan^{-1}[(x-1)/3] = u$, but the implicit forms are easier to use.

(2) In (a) we would normally omit the condition $u \geq 0$ from the substitution, but have it in mind when returning to the original variable after integrating. In fact the condition $u \leq 0$ would do just as well.

Similarly, in (b) we would normally omit the condition $-\pi/2 < \theta < \pi/2$. Some care is then needed in choosing the limits for θ to ensure that the integrand is defined on the range of integration.

Example 16.1.3 Find the area of the finite region bounded by the ellipse

$$\frac{x^2}{a^2} + \frac{y^2}{b^2} = 1,$$

where $a, b > 0$.

Solution For a discussion of the ellipse see Section 10.5. By Theorem 10.5.3, it has parametric equations

$$x = a\cos t, \quad y = b\sin t \quad (0 \leq t < 2\pi).$$

Since the curve is symmetric about both the x- and y-axes, taking $y > 0$, the required area is

$$A = 4\int_0^a y\, dx.$$

See Figure 16.1.1. Then substituting

$$x = a\cos t, \qquad dx = -a\sin t\, dt$$

x	0	a
t	$\pi/2$	0

and $y = b\sin t$

gives

$$A = 4 \int_{\pi/2}^{0} (b \sin t)(-a \sin t)\, dt = 4ab \int_{0}^{\pi/2} \sin^2 t\, dt$$

$$= 2ab \int_{0}^{\pi/2} (1 - \cos 2t)\, dt = 2ab\left[t - \frac{\sin 2t}{2}\right]_{0}^{\pi/2} = \pi ab. \qquad \square$$

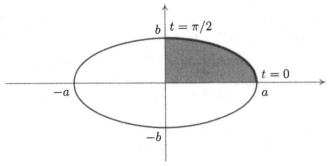

Figure 16.1.1

The next result is used in Sections 16.5, 16.6 and 16.7.

Lemma 16.1.4 *Let f be a real function, continuous on $[a, b]$, and let $x : [\alpha, \beta] \to [a, b]$ be a bijection with a continuous derivative on $[\alpha, \beta]$. Then*

$$\int_{a}^{b} f(x)\, dx = \int_{\alpha}^{\beta} f(x(t))|\dot{x}(t)|\, dt.$$

Proof By Theorem 7.3.18, x is strictly monotonic on $[\alpha, \beta]$.

Case 1. Let x be strictly increasing on $[\alpha, \beta]$. Then

$$\alpha < t < \beta \implies x(\alpha) < x(t) < x(\beta)$$

and it follows, since x is surjective, that $a = x(\alpha)$ and $b = x(\beta)$. Also, for all $t \in [\alpha, \beta]$, $\dot{x}(t) \geq 0$. Otherwise, for some $s \in [\alpha, \beta]$, $\dot{x}(s) < 0$. Then Theorem 7.3.9 (3) and the monotonicity theorem give $\dot{x}$ positive and strictly decreasing on a neighbourhood of s (one-sided if $s = \alpha$ or $s = \beta$). Hence, substituting

$$x = x(t), \qquad dx = \dot{x}(t)\, dt \qquad \begin{array}{|c|c c|} \hline x & a & b \\ \hline t & \alpha & \beta \\ \hline \end{array}$$

gives

$$\int_{a}^{b} f(x)\, dx = \int_{\alpha}^{\beta} f(x(t))\dot{x}(t)\, dt = \int_{\alpha}^{\beta} f(x(t))|\dot{x}(t)|\, dt.$$

Case 2. Let x be strictly decreasing on $[\alpha, \beta]$. Arguing as in Case 1, we get $a = x(\beta)$, $b = x(\alpha)$ and, for all $t \in [\alpha, \beta]$, $\dot{x}(t) \leq 0$.

Hence, substituting

$$x = x(t), \qquad dx = \dot{x}(t)\, dt$$

x	a	b
t	β	α

gives

$$\int_a^b f(x)\, dx = \int_\beta^\alpha f(x(t))\dot{x}(t)\, dt = \int_\alpha^\beta f(x(t))(-\dot{x}(t))\, dt$$

$$= \int_\alpha^\beta f(x(t))|\dot{x}(t)|\, dt. \qquad \square$$

16.2 Rational Integrals

Here we consider integrals of rational functions. In Section 3.4 we saw that for a rational function h the partial fraction decomposition of $h(x)$ is the sum of a polynomial in x together with terms of the form

$$\frac{A}{(x-\alpha)^m} \qquad \text{and} \qquad \frac{Bx+D}{(x^2+\beta x+\gamma)^n}.$$

where $m, n \in \mathbb{N}$. We can easily integrate polynomial functions, and standard integrals give

$$\int \frac{A}{x-\alpha}\, dx = A\log|x-\alpha| + C$$

and, for $m \geq 2$,

$$\int \frac{A}{(x-\alpha)^m}\, dx = \frac{-A}{(m-1)(x-\alpha)^{m-1}} + C.$$

It therefore remains to consider integrals of the form

$$\mathcal{I} = \int \frac{Bx+D}{(x^2+\beta x+\gamma)^n}\, dx$$

where $x^2 + \beta x + \gamma$ is irreducible. First we choose constants P and Q such that

$$Bx + D = P\frac{d}{dx}(x^2+\beta x+\gamma) + Q,$$

the constant P being chosen first, to get the coefficient of x correct. Then we write the integral in the form

$$\int \frac{P(2x+\beta)}{(x^2+\beta x+\gamma)^n}\, dx + \int \frac{Q}{(x^2+\beta x+\gamma)^n}\, dx = \mathcal{I}_1 + \mathcal{I}_2 \quad \text{(respectively)}.$$

For $\mathcal{I}_1$ we substitute

$$x^2 + \beta x + \gamma = u, \qquad (2x+\beta)\, dx = du.$$

For $\mathcal{I}_2$ we complete the square of the quadratic to get

$$x^2 + \beta x + \gamma = (x+\xi)^2 + \eta^2$$

(where $\eta > 0$) then substitute

$$x + \xi = \eta \tan \theta, \qquad dx = \eta \sec^2 \theta \, d\theta$$

and

$$(x + \xi)^2 + \eta^2 = \eta^2 \tan^2 \theta + \eta^2 = \eta^2 \sec^2 \theta.$$

Remarks (1) When $n = 1$ substitution is unnecessary. Standard integrals apply. Thus,

$$\int \frac{P(2x + \beta)}{x^2 + \beta x + \gamma} \, dx = P \log |x^2 + \beta x + \gamma| + C$$

and

$$\int \frac{Q}{(x + \xi)^2 + \eta^2} \, dx = \frac{Q}{\eta} \tan^{-1}\left(\frac{x + \xi}{\eta}\right) + C.$$

(2) The technique extends to integrands of the form (linear)(quadratic)^r for certain non-integer values of r. See Exercises 16.X.4 and 16.X.5.

Example 16.2.1 Find/evaluate

(a) $\mathcal{I} = \displaystyle\int \frac{x^4 + 3x^3 - 1}{x^4 + 3x^3 + 4x^2 + 3x + 1} \, dx,$

(b) $\mathcal{J} = \displaystyle\int_2^5 \frac{x + 7}{(x^2 - 4x + 7)^3} \, dx,$ (c) $\mathcal{K} = \displaystyle\int_0^5 \frac{25 - x^2}{x^4 + 6x^2 + 25} \, dx.$

Solution (a) The partial fraction decomposition of the integrand was found in Example 3.4.6. Hence,

$$\mathcal{I} = \int \left(1 + \frac{2}{x + 1} - \frac{3}{(x + 1)^2} - \frac{2x + 1}{x^2 + x + 1}\right) dx$$

$$= x + 2\log |x + 1| + \frac{3}{x + 1} - \log |x^2 + x + 1| + C.$$

(b) Since

$$\frac{d}{dx}(x^2 - 4x + 7) = 2x - 4 \quad \text{and} \quad x + 7 = \tfrac{1}{2}(2x - 4) + 9,$$

we have

$$\mathcal{J} = \int_2^5 \frac{\tfrac{1}{2}(2x - 4)}{(x^2 - 4x + 7)^3} \, dx + \int_2^5 \frac{9}{(x^2 - 4x + 7)^3} \, dx$$

$$= \mathcal{J}_1 + \mathcal{J}_2 \quad \text{(respectively)}.$$

For $\mathcal{J}_1$, substitute

$$x^2 - 4x + 7 = u, \qquad (2x - 4) \, dx = du,$$

x	2	5
u	3	12

to get

$$\mathcal{J}_1 = \frac{1}{2} \int_3^{12} \frac{du}{u^3} = \frac{1}{2} \left[-\frac{1}{2u^2} \right]_3^{12} = \frac{5}{192}.$$

For $\mathcal{J}_2$, complete the square of the quadratic,

$$x^2 - 4x + 7 = (x-2)^2 + 3,$$

and substitute

$$x - 2 = \sqrt{3} \tan \theta, \qquad dx = \sqrt{3} \sec^2 \theta \, d\theta,$$

x	2	5
θ	0	$\pi/3$

to get

$$\mathcal{J}_2 = 9 \int_2^5 \frac{dx}{[(x-2)^2 + 3]^3} = 9 \int_0^{\pi/3} \frac{\sqrt{3} \sec^2 \theta \, d\theta}{(3 \tan^2 \theta + 3)^3}$$

$$= 9 \int_0^{\pi/3} \frac{\sqrt{3} \sec^2 \theta}{(3 \sec^2 \theta)^3} \, d\theta \quad (\text{using } \tan^2 \theta + 1 = \sec^2 \theta)$$

$$= \frac{1}{\sqrt{3}} \int_0^{\pi/3} \cos^4 \theta \, d\theta = \frac{1}{\sqrt{3}} \int_0^{\pi/3} \left(\frac{1}{8} \cos 4\theta + \frac{1}{2} \cos 2\theta + \frac{3}{8} \right) d\theta$$

(using the methods of Section 6.3)

$$= \frac{1}{\sqrt{3}} \left[\frac{1}{32} \sin 4\theta + \frac{1}{4} \sin 2\theta + \frac{3\theta}{8} \right]_0^{\pi/3}$$

$$= \frac{1}{\sqrt{3}} \left(\frac{1}{32} \left(-\frac{\sqrt{3}}{2} \right) + \frac{1}{4} \frac{\sqrt{3}}{2} + \frac{3}{8} \frac{\pi}{3} \right) = \frac{7}{64} + \frac{\pi}{8\sqrt{3}}.$$

Hence,

$$\mathcal{J} = \frac{5}{192} + \frac{7}{64} + \frac{\pi}{8\sqrt{3}} = \frac{13}{96} + \frac{\pi}{8\sqrt{3}}.$$

For an alternative method of evaluating $\int_0^{\pi/3} \cos^4 \theta \, d\theta$, see Example 16.4.7.

(c) The methods of Section 3.4 give the partial fraction decomposition of the integrand. Hence

$$\mathcal{K} = \int_0^5 \frac{1}{2} \left(\frac{3x+5}{x^2 + 2x + 5} \right) dx + \int_0^5 \frac{1}{2} \left(\frac{-3x+5}{x^2 - 2x + 5} \right) dx$$

$$= \mathcal{K}_1 + \mathcal{K}_2 \quad (\text{respectively}).$$

Then $\dfrac{d}{dx}(x^2 + 2x + 5) = 2x + 2$ and $\dfrac{1}{2}(3x + 5) = \dfrac{3}{4}(2x+2) + 1.$ So

$$\mathcal{K}_1 = \frac{3}{4} \int_0^5 \frac{2x+2}{x^2 + 2x + 5} \, dx + \int_0^5 \frac{dx}{(x+1)^2 + 4}$$

$$= \frac{3}{4} \left[\log |x^2 + 2x + 5| \right]_0^5 + \left[\frac{1}{2} \tan^{-1} \frac{(x+1)}{2} \right]_0^5$$

$$= \frac{3}{4} \log 8 + \frac{1}{2} \tan^{-1} 3 - \frac{1}{2} \tan^{-1} \left(\frac{1}{2} \right).$$

Similarly,

$$K_2 = -\frac{3}{4}\int_0^5 \frac{2x-2}{x^2-2x+5}\,dx + \int_0^5 \frac{dx}{(x-1)^2+4}$$

$$= -\frac{3}{4}\Big[\log|x^2-2x+5|\Big]_0^5 + \Big[\frac{1}{2}\tan^{-1}\frac{(x-1)}{2}\Big]_0^5$$

$$= -\frac{3}{4}\log 4 + \frac{1}{2}\tan^{-1}2 - \frac{1}{2}\tan^{-1}\Big(-\frac{1}{2}\Big).$$

Then, since $\tan^{-1}(-1/2) = -\tan^{-1}(1/2)$, we have

$$K = \frac{3}{4}(\log 8 - \log 4) + \frac{1}{2}(\tan^{-1}3 + \tan^{-1}2)$$

$$= \frac{3}{4}\log 2 + \frac{1}{2}\Big(\frac{3\pi}{4}\Big) = \frac{3}{4}\Big(\log 2 + \frac{\pi}{2}\Big).$$

For the evaluation of $\tan^{-1}3 + \tan^{-1}2$ see Example 5.5.5 (b).   □

16.3   Trigonometric Integrals

In this section we gather together some techniques for dealing with integrals involving trigonometric functions.

Products

Simple products of sines and cosines can be converted to sums of sines and cosines. We used this method in the solution of Example 16.2.1 (b).

Example 16.3.1 Evaluate/find

$$\text{(a)} \int_0^{\pi/2} \sin 7x \cos 5x\,dx, \qquad \text{(b)} \int \sin^6 x\,dx.$$

Solution   (a)   Using Corollary 5.2.5, we have

$$\int_0^{\pi/2} \sin 7x \cos 5x\,dx = \frac{1}{2}\int_0^{\pi/2}(\sin 12x + \sin 2x)\,dx$$

$$= \frac{1}{2}\Big[-\frac{1}{12}\cos 12x - \frac{1}{2}\cos 2x\Big]_0^{\pi/2} = \frac{1}{2}.$$

(b)   By Example 6.3.8,

$$\sin^6 x = -\frac{1}{32}\cos 6x + \frac{3}{16}\cos 4x - \frac{15}{32}\cos 2x + \frac{5}{16}.$$

Hence,

$$\int \sin^6 x\,dx = -\frac{1}{192}\sin 6x + \frac{3}{64}\sin 4x - \frac{15}{64}\sin 2x + \frac{5x}{16} + C.$$ □

Certain trigonometric products are amenable to substitution. Consider the integral

$$\mathcal{I} = \int \sin^2 x \cos^3 x \, dx.$$

We can write

$$\mathcal{I} = \int \sin^2 x \cos^2 x \cos x \, dx$$

so that substituting

$$\sin x = u, \qquad \cos x \, dx = du \qquad \text{and} \qquad \cos^2 x = 1 - \sin^2 x = 1 - u^2$$

gives

$$\mathcal{I} = \int u^2 (1 - u^2) \, du = \int (u^2 - u^4) \, du$$

which is easily found.

In the table, $k \in \mathbb{Z}$ and $r \in \mathbb{R}$, but see Remark (2), below.

Integrand	k	Substitution
$(\sin x)^r (\cos x)^k$	odd	$\sin x = u$
$(\sin x)^k (\cos x)^r$	odd	$\cos x = u$
$(\tan x)^r (\sec x)^k$	even	$\tan x = u$
$(\tan x)^k (\sec x)^r$	odd	$\sec x = u$

Remarks (1)   We have written $(\sin x)^r$ instead of $\sin^r x$ to avoid confusion when $r = -1$, and similarly for $(\cos x)^k$, *etc.*

(2)   The substitutions $\sin x = u$, $\cos x = u$ and $\sec x = u$ work when either (a) $k \in \mathbb{N}$ and $r \in \mathbb{R}$, or (b) $k, r \in \mathbb{Z}$. We met case (a) in Example 16.1.1 (d).
The substitution $\tan x = u$ also works in case (a) and usually, but not always, in case (b). For example, substituting $\tan x = u$ in the integral $\int (\sec x)^{-4} \, dx$ becomes the rational integral $\int (u^2 + 1)^{-2} \, du$. However, the standard substitution for this is $u = \tan \theta$ which leads to $\int \cos^4 \theta \, d\theta$, *i.e.* $\int (\sec \theta)^{-4} \, d\theta$.

(3)   In some cases there is a choice of substitution. For example, if the integrand is $\sin^3 x \cos^5 x$, we could substitute either $\cos x = u$ or $\sin x = u$.

(4)   One or other of the trigonometric identities

$$\sin^2 x + \cos^2 x = 1 \qquad \text{and} \qquad \tan^2 x + 1 = \sec^2 x$$

usually plays a part in these substitutions.

Example 16.3.2 Find/evaluate

(a) $\displaystyle \int \frac{\sin^3 x}{\cos^2 x} \, dx,$ (b) $\displaystyle \int_0^{\pi/4} \sec^6 x \, dx,$ (c) $\displaystyle \int_0^{\pi/3} \tan^3 x (\sec x)^{1/3} \, dx.$

Solution   (a)   Here we substitute

$$\cos x = u, \qquad -\sin x \, dx = du \qquad \text{and} \qquad \sin^2 x = 1 - u^2$$

to get

$$\int \frac{\sin^3 x}{\cos^2 x}\, dx = \int \frac{-\sin^2 x}{\cos^2 x}(-\sin x)\, dx = \int \frac{u^2 - 1}{u^2}\, du = \int (1 - u^{-2})\, du$$

$$= u + u^{-1} + C = \cos x + \sec x + C.$$

(b) Here we substitute

$$\tan x = u, \qquad \sec^2 x\, dx = du, \qquad \begin{array}{c|cc} x & 0 & \pi/4 \\ \hline u & 0 & 1 \end{array} \qquad \text{and} \qquad \sec^2 x = u^2 + 1$$

to get

$$\int_0^{\pi/4} \sec^6 x\, dx = \int_0^{\pi/4} \sec^4 x \sec^2 x\, dx = \int_0^1 (u^2 + 1)^2\, du$$

$$= \int_0^1 (u^4 + 2u^2 + 1)\, du$$

$$= \left[\frac{u^5}{5} + \frac{2u^3}{3} + u\right]_0^1 = \frac{28}{15}.$$

(c) Here we substitute

$$\sec x = u, \qquad \sec x \tan x\, dx = du \quad \text{and} \quad \begin{array}{c|cc} x & 0 & \pi/3 \\ \hline u & 1 & 2 \end{array} \quad \text{and} \quad \tan^2 x = u^2 - 1$$

to get

$$\int_0^{\pi/3} \tan^3 x (\sec x)^{1/3}\, dx = \int_0^{\pi/3} \tan^2 x (\sec x)^{-2/3} (\sec x \tan x)\, dx$$

$$= \int_1^2 (u^2 - 1) u^{-2/3}\, du = \int_1^2 (u^{4/3} - u^{-2/3})\, du$$

$$= \left[\frac{3}{7}u^{7/3} - 3u^{1/3}\right]_1^2 = \frac{9}{7}(2 - 2^{1/3}). \qquad \square$$

Example 16.3.3 For $n \in \mathbb{N} \cup \{0\}$, define

$$I_n = \int \tan^n x\, dx.$$

Show that, for $n \geq 2$,

$$I_n = \frac{1}{n-1} \tan^{n-1} x - I_{n-2}. \qquad (16.3.1)$$

Hence find I_4.

Solution For $n \geq 2$,

$$I_n = \int \tan^{n-2} x \tan^2 x\, dx = \int \tan^{n-2} x (\sec^2 x - 1)\, dx$$

$$= \int \tan^{n-2} x \sec^2 x\, dx - \int \tan^{n-2} x\, dx = \int \tan^{n-2} x \sec^2 x\, dx - I_{n-2}.$$

Then, substituting $\tan x = u$, $\sec^2 x \, dx = du$,

$$\int \tan^{n-2} x \sec^2 x \, dx = \int u^{n-2} \, du = \frac{1}{n-1} u^{n-1} + C = \frac{1}{n-1} \tan^{n-1} x + C$$

so that

$$\mathcal{I}_n = \frac{1}{n-1} \tan^{n-1} x - \mathcal{I}_{n-2}$$

as required. [Note that the constant of integration has 'disappeared'. We can think of it as included in $\mathcal{I}_n$ or $\mathcal{I}_{n-2}$.]

Putting $n = 4$ and then $n = 2$ in (16.3.1) gives

$$\mathcal{I}_4 = \frac{1}{3} \tan^3 x - \mathcal{I}_2 = \frac{1}{3} \tan^3 x - \left(\frac{1}{1} \tan^1 x - \mathcal{I}_0 \right)$$

$$= \frac{1}{3} \tan^3 x - \tan x + x + C. \qquad \square$$

Note 16.3.4 Equation (16.3.1) is an example of a *reduction formula*, also called a *recurrence relation*.

The formula for $\int \tan^n x \, dx$ is particularly useful for n even, when direct substitution is unhelpful.

Similar formulae for $\int \sin^n x \, dx$ and $\int \cos^n x \, dx$ are found in Examples 16.4.5 and 16.4.7.

The *t*-formulae

In Section 5.4 we found expressions for $\sin \theta$, $\cos \theta$ and $\tan \theta$ in terms of $t = \tan \frac{1}{2}\theta$. These can be used to convert certain trigonometric integrals into integrals of rational functions. In particular, they can be used to find integrals of the form

$$\int \frac{d\theta}{a + b \sin \theta + c \cos \theta},$$

where b and c are not both 0.

The basic substitution is $t = \tan \frac{1}{2}\theta$. Then

$$\frac{dt}{d\theta} = \frac{1}{2} \sec^2 \frac{1}{2}\theta = \frac{1}{2}(1 + \tan^2 \frac{1}{2}\theta) = \frac{1}{2}(1 + t^2).$$

As necessary, we substitute

$$t = \tan \frac{1}{2}\theta, \qquad d\theta = \frac{2}{1+t^2} \, dt$$

and, from Theorem 5.4.1,

$$\sin \theta = \frac{2t}{1+t^2}, \qquad \cos \theta = \frac{1-t^2}{1+t^2}, \qquad \tan \theta = \frac{2t}{1-t^2}.$$

Example 16.3.5 Evaluate

$$\text{(a)} \ \ \mathcal{I} = \int_0^{\pi/2} \frac{d\theta}{2 + \sin \theta}, \qquad \text{(b)} \ \ \mathcal{J} = \int_0^{\pi/2} \frac{dx}{3 + 3\sin x - 2\cos x}.$$

Solution   (a) Here, the basic substitution $t = \tan \frac{1}{2}\theta$, leads to the change of limits shown in the table.

θ	0	$\pi/2$
t	0	1

Hence,

$$\mathcal{I} = \int_0^1 \frac{[2/(1+t^2)]\,dt}{2 + [2t/(1+t^2)]} = \int_0^1 \frac{2dt}{2(1+t^2) + 2t}$$

(multiplying numerator and denominator by $1 + t^2$)

$$= \int_0^1 \frac{dt}{t^2 + t + 1} = \int_0^1 \frac{dt}{(t+1/2)^2 + 3/4} = \left[\frac{2}{\sqrt{3}} \tan^{-1}\left(\frac{t+1/2}{\sqrt{3}/2}\right) \right]_0^1$$

$$= \frac{2}{\sqrt{3}}\left(\tan^{-1}\sqrt{3} - \tan^{-1}\frac{1}{\sqrt{3}} \right) = \frac{\pi}{3\sqrt{3}}.$$

    (b) Here the basic substitution is $t = \tan \frac{1}{2}x$ with the same change of limits as in (a), leading to

$$\mathcal{J} = \int_0^1 \frac{2dt}{3(1+t^2) + 3(2t) - 2(1-t^2)} = \int_0^1 \frac{2dt}{5t^2 + 6t + 1}$$

$$= \int_0^1 \frac{2dt}{(5t+1)(t+1)} = \int_0^1 \left(\frac{5/2}{5t+1} - \frac{1/2}{t+1} \right) dt \quad \text{(see Note 3.4.9)}$$

$$= \frac{1}{2}\Big[\log|5t + 1| - \log|t + 1| \Big]_0^1 = \log\sqrt{3}. \qquad\qquad \square$$

16.4   Integration by Parts

Let f and G be differentiable functions such that fG is defined. Then the product rule gives

$$\frac{d}{dx}[f(x)G(x)] = f'(x)G(x) + f(x)G'(x),$$

i.e.

$$f(x)G'(x) = \frac{d}{dx}[f(x)G(x)] - f'(x)G(x).$$

Writing $G' = g$ and integrating, this becomes

$$\int f(x)g(x)\,dx = f(x)G(x) - \int f'(x)G(x)\,dx. \qquad\qquad (16.4.1)$$

This is the formula for *integration by parts*. For definite integrals it becomes

$$\int_a^b f(x)g(x)\,dx = \big[f(x)G(x)\big]_a^b - \int_a^b f'(x)G(x)\,dx, \qquad\qquad (16.4.2)$$

assuming that f and g are defined on $[a, b]$.

Example 16.4.1 Find $\int x\sqrt{x+1}\,dx$.

Solution Putting $f(x) = x$ and $g(x) = \sqrt{x+1}$ in (16.4.1) gives

$$\int x\sqrt{x+1}\,dx = x\frac{(x+1)^{3/2}}{3/2} - \int 1\frac{(x+1)^{3/2}}{3/2}\,dx$$

$$= \frac{2}{3}x(x+1)^{3/2} - \frac{2}{3}\cdot\frac{2}{5}(x+1)^{5/2} + C$$

$$= \frac{2}{15}(x+1)^{3/2}(3x-2) + C. \qquad \square$$

Example 16.4.2 Evaluate $\int_0^1 (x^2+1)e^{2x}\,dx$.

Solution Here we integrate by parts twice to reduce $x^2 + 1$ to a constant by differentiation. Putting $f(x) = x^2 + 1$ and $g(x) = e^{2x}$ in (16.4.2) gives

$$\int_0^1 (x^2+1)e^{2x}\,dx = \left[(x^2+1)\frac{e^{2x}}{2}\right]_0^1 - \int_0^1 2x\frac{e^{2x}}{2}\,dx$$

$$= e^2 - \frac{1}{2} - \int_0^1 xe^{2x}\,dx$$

$$= e^2 - \frac{1}{2} - \left(\left[x\frac{e^{2x}}{2}\right]_0^1 - \int_0^1 1\frac{e^{2x}}{2}\,dx\right)$$

$$= e^2 - \frac{1}{2} - \frac{e^2}{2} + \left[\frac{e^{2x}}{4}\right]_0^1$$

$$= \frac{e^2}{2} - \frac{1}{2} + \frac{e^2}{4} - \frac{1}{4} = \frac{3}{4}(e^2-1). \qquad \square$$

Example 16.4.3 Evaluate $\mathcal{I} = \int_0^{\pi/2} e^{2x}\sin 3x\,dx$.

Solution Here we use the fact that integrating $\sin$ twice gives $-\sin$. Putting $f(x) = e^{2x}$ and $g(x) = \sin 3x$ in (16.4.2) gives

$$\mathcal{I} = \left[e^{2x}\left(-\frac{\cos 3x}{3}\right)\right]_0^{\pi/2} - \int_0^{\pi/2} 2e^{2x}\left(-\frac{\cos 3x}{3}\right)\,dx$$

$$= \frac{1}{3} + \frac{2}{3}\int_0^{\pi/2} e^{2x}\cos 3x\,dx$$

$$= \frac{1}{3} + \frac{2}{3}\left(\left[e^{2x}\frac{\sin 3x}{3}\right]_0^{\pi/2} - \int_0^{\pi/2} 2e^{2x}\frac{\sin 3x}{3}\,dx\right)$$

$$= \frac{1}{3} - \frac{2e^\pi}{9} - \frac{4}{9}\int_0^{\pi/2} e^{2x}\sin 3x\,dx = \frac{1}{3} - \frac{2e^\pi}{9} - \frac{4}{9}\mathcal{I}.$$

Hence,

$$\left(1+\frac{4}{9}\right)\mathcal{I} = \frac{1}{3} - \frac{2e^\pi}{9}, \quad i.e.\ \mathcal{I} = \frac{9}{13}\left(\frac{1}{3} - \frac{2e^\pi}{9}\right) = \frac{1}{13}(3 - 2e^\pi).$$

Alternatively, we can use the fact that differentiating $\sin$ twice gives $-\sin$. Putting $f(x) = \sin 3x$ and $g(x) = e^{2x}$ in (16.4.2) gives

$$\mathcal{I} = \int_0^{\pi/2} \sin 3x e^{2x}\, dx = \left[\sin 3x \frac{e^{2x}}{2}\right]_0^{\pi/2} - \int_0^{\pi/2} 3\cos 3x \frac{e^{2x}}{2}\, dx$$

$$= -\frac{e^{\pi}}{2} - \frac{3}{2}\int_0^{\pi/2} \cos 3x e^{2x}\, dx$$

$$= -\frac{e^{\pi}}{2} - \frac{3}{2}\left(\left[\cos 3x \frac{e^{2x}}{2}\right]_0^{\pi/2} - \int_0^{\pi/2}(-3\sin 3x)\frac{e^{2x}}{2}\, dx\right)$$

$$= -\frac{e^{\pi}}{2} + \frac{3}{4} - \frac{9}{4}\int_0^{\pi/2} \sin 3x e^{2x}\, dx = \frac{3}{4} - \frac{e^{\pi}}{2} - \frac{9}{4}\mathcal{I}.$$

Hence,

$$\left(1 + \frac{9}{4}\right)\mathcal{I} = \frac{3}{4} - \frac{e^{\pi}}{2}, \quad \textit{i.e. } \mathcal{I} = \frac{4}{13}\left(\frac{3}{4} - \frac{e^{\pi}}{2}\right) = \frac{1}{13}(3 - 2e^{\pi}). \qquad \square$$

Example 16.4.4 Find antiderivatives for (a) $\log$ and (b) $\sin^{-1}$.

Solution (a) The 'trick' here is to think of $\log x$ as $(\log x).1$ so that putting $f(x) = \log x$ and $g(x) = 1$ in (16.4.1) gives

$$\int \log x\, dx = \int (\log x).1\, dx = (\log x)x - \int \frac{1}{x}x\, dx$$

$$= x\log x - x + C.$$

(b) The technique here is similar. We have

$$\int \sin^{-1} x\, dx = \int (\sin^{-1} x).1\, dx = (\sin^{-1} x)x - \int \frac{1}{\sqrt{1 - x^2}}x\, dx$$

$$= x\sin^{-1} x + \int \frac{(-2x)\, dx}{2\sqrt{1 - x^2}}.$$

Then, substituting $1 - x^2 = u$, $-2x\, dx = du$,

$$\int \sin^{-1} x\, dx = x\sin^{-1} x + \int \frac{du}{2\sqrt{u}} = x\sin^{-1} x + \sqrt{u} + C$$

$$= x\sin^{-1} x + \sqrt{1 - x^2} + C. \qquad \square$$

Remark For an alternative approach to finding antiderivatives for $\log$ and $\sin^{-1}$, see Exercises 16.X.12.

In Example 16.3.3 we found a reduction formula for $\int \tan^n x\, dx$. We now use integration by parts to find similar formulae for $\int \sin^n x\, dx$ and $\int \cos^n x\, dx$.

Example 16.4.5 For $n \in \mathbb{N} \cup \{0\}$, define

$$\mathcal{I}_n = \int \sin^n x\, dx \quad \text{and} \quad \mathcal{J}_n = \int_0^{\pi/2} \sin^n x\, dx.$$

Show that, for $n \geq 2$,

$$\mathcal{I}_n = -\frac{1}{n} \sin^{n-1} x \cos x + \frac{n-1}{n} \mathcal{I}_{n-2} \qquad (16.4.3)$$

and

$$\mathcal{J}_n = \frac{n-1}{n} \mathcal{J}_{n-2}. \qquad (16.4.4)$$

Hence evaluate $\mathcal{J}_8$.

Solution For $n \geq 2$, integrating by parts gives

$$\mathcal{I}_n = \int \sin^n x \, dx = \int \sin^{n-1} x \sin x \, dx$$

$$= \sin^{n-1} x(-\cos x) - \int (n-1) \sin^{n-2} x \cos x(-\cos x) \, dx$$

$$= -\sin^{n-1} x \cos x + (n-1) \int \sin^{n-2} x(1 - \sin^2 x) \, dx$$

$$= -\sin^{n-1} x \cos x + (n-1) \int \sin^{n-2} x \, dx - (n-1) \int \sin^n x \, dx$$

$$= -\sin^{n-1} x \cos x + (n-1)\mathcal{I}_{n-2} - (n-1)\mathcal{I}_n.$$

Hence

$$n\mathcal{I}_n = -\sin^{n-1} x \cos x + (n-1)\mathcal{I}_{n-2}$$

and dividing both sides by n gives (16.4.3).

From (16.4.3) it follows that, for $n \geq 2$,

$$\mathcal{J}_n = \left[-\frac{1}{n} \sin^{n-1} x \cos x \right]_0^{\pi/2} + \frac{n-1}{n} \mathcal{J}_{n-2} = \frac{n-1}{n} \mathcal{J}_{n-2},$$

establishing (16.4.4).

Finally, using (16.4.4), we have

$$\mathcal{J}_8 = \frac{7}{8} \mathcal{J}_6 = \frac{7}{8} \frac{5}{6} \mathcal{J}_4 = \frac{7}{8} \frac{5}{6} \frac{3}{4} \mathcal{J}_2 = \frac{7}{8} \frac{5}{6} \frac{3}{4} \frac{1}{2} \mathcal{J}_0 = \frac{7}{8} \frac{5}{6} \frac{3}{4} \frac{1}{2} \frac{\pi}{2} = \frac{35\pi}{256}. \qquad \square$$

Problem 16.4.6 Show that, for all $m \in \mathbb{N}$, $\displaystyle\int_0^{\pi/2} \sin^{2m} x \, dx = \binom{2m}{m} \frac{\pi}{2^{2m+1}}$.

Example 16.4.7 For $n \in \mathbb{N} \cup \{0\}$, define

$$\mathcal{I}_n = \int \cos^n x \, dx \quad \text{and} \quad \mathcal{J}_n = \int_0^{\pi/3} \cos^n x \, dx.$$

Show that, for $n \geq 2$,

$$\mathcal{I}_n = \frac{1}{n} \cos^{n-1} x \sin x + \frac{n-1}{n} \mathcal{I}_{n-2} \qquad (16.4.5)$$

and

$$\mathcal{J}_n = \frac{\sqrt{3}}{n2^n} + \frac{n-1}{n} \mathcal{J}_{n-2}. \qquad (16.4.6)$$

Hence evaluate $\mathcal{J}_4$.

Solution   We leave the reader to establish (16.4.5) by following the solution for (16.4.3).

For $n \geq 2$, integrating by parts gives

$$\mathcal{J}_n = \int_0^{\pi/3} \cos^{n-1} x \cos x \, dx$$

$$= \left[\cos^{n-1} x \sin x\right]_0^{\pi/3} - \int_0^{\pi/3} (n-1) \cos^{n-2} x(-\sin x) \sin x \, dx$$

$$= \left(\frac{1}{2}\right)^{n-1} \frac{\sqrt{3}}{2} + (n-1) \int_0^{\pi/3} \cos^{n-2} x(1 - \cos^2 x) \, dx$$

$$= \frac{\sqrt{3}}{2^n} + (n-1) \int \cos^{n-2} x \, dx - (n-1) \int \cos^n x \, dx$$

$$= \frac{\sqrt{3}}{2^n} + (n-1)\mathcal{J}_{n-2} - (n-1)\mathcal{J}_n.$$

Hence

$$n\mathcal{J}_n = \frac{\sqrt{3}}{2^n} + (n-1)\mathcal{J}_{n-2}$$

and dividing both sides by n gives (16.4.6).

Using (16.4.6), we have

$$\mathcal{J}_4 = \frac{\sqrt{3}}{4.2^4} + \frac{3}{4}\mathcal{J}_2 = \frac{\sqrt{3}}{64} + \frac{3}{4}\left(\frac{\sqrt{3}}{2.2^2} + \frac{1}{2}\mathcal{J}_0\right)$$

$$= \frac{\sqrt{3}}{64} + \frac{3\sqrt{3}}{32} + \frac{3}{8}\frac{\pi}{3} = \frac{7\sqrt{3}}{64} + \frac{\pi}{8}. \qquad \square$$

In the final example of this section we use integration by parts to establish the convergence of an improper integral.

Example 16.4.8   Show that the improper integral $\displaystyle\int_0^\infty x^3 e^{-x} \, dx$ converges and find its value.

Solution   It follows from Theorems 15.2.7 and 15.3.7 (4) that, for $n = 0, 1, 2, 3$,

$$u^n e^{-u} \to 0 \text{ as } u \to \infty.$$

Hence, for $u > 0$, integrating by parts gives

$$\int_0^u x^3 e^{-x} \, dx = \left[x^3(-e^{-x})\right]_0^u - \int_0^u 3x^2(-e^{-x}) \, dx = -u^3 e^{-u} + 3\int_0^u x^2 e^{-x} \, dx$$

$$= -u^3 e^{-u} + 3\left(\left[x^2(-e^{-x})\right]_0^u - \int_0^u 2x(-e^{-x}) \, dx\right)$$

$$= -u^3 e^{-u} - 3u^2 e^{-u} + 6\int_0^u xe^{-x} \, dx$$

$$= -u^3 e^{-u} - 3u^2 e^{-u} + 6\left(\left[x(-e^{-x}) \right]_0^u - \int_0^u (-e^{-x})\, dx \right)$$

$$= -u^3 e^{-u} - 3u^2 e^{-u} - 6u e^{-u} + 6\left[-e^{-x} \right]_0^u$$

$$= -u^3 e^{-u} - 3u^2 e^{-u} - 6u e^{-u} - 6 e^{-u} + 6$$

$$\to 6 \text{ as } u \to \infty.$$

Thus $\displaystyle\int_0^\infty x^3 e^{-x}\, dx$ converges to 6. $\square$

Problem 16.4.9 Prove by induction that, for all $n \in \mathbb{N}$, the improper integral

$$\Gamma(n) = \int_0^\infty x^{n-1} e^{-x}\, dx$$

converges to $(n-1)!$.

16.5   Volumes of Revolution

Let f be a real function, continuous on $[a, b]$, and consider the *solid of revolution* formed by revolving, once about the x-axis, the curve $\mathcal{C}$ given by

$$y = f(x) \quad (a \leq x \leq b).$$

See Figure 16.5.1. We define the volume of such a solid by first approximating with cylinders. Recall the formula $\pi R^2 H$ for the volume of a cylinder with radius R and height H.

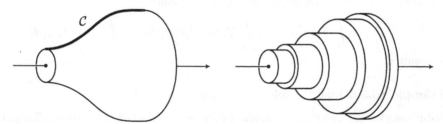

Figure 16.5.1

Let $\mathcal{P} = (x_0, x_1, \ldots, x_n)$ be a partition of $[a, b]$ and, for each $r = 1, \ldots, n$, let $c_r \in [x_{r-1}, x_r]$. Consider the cylinder shown in Figure 16.5.2. It is the solid of revolution formed by revolving, once about the x-axis, the corresponding rectangle, also shown in Figure 16.5.2. There are n such cylinders. The sum of their volumes is

$$S = \sum_{r=1}^n \pi f(c_r)^2 (x_r - x_{r-1}),$$

which is a Riemann sum for the function $g : [a, b] \to \mathbb{R}$ defined by $g(x) = \pi f(x)^2$. Since g is continuous on $[a, b]$, $S \to \int_a^b g(x)\, dx$ as $|\mathcal{P}| \to 0$.

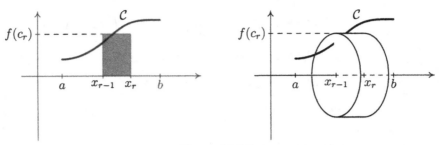

Figure 16.5.2

Definition 16.5.1 Let f be a real function, continuous on $[a, b]$. Then the *volume of revolution V generated* by the curve $y = f(x)$ $(a \leq x \leq b)$ is defined by

$$V = \pi \int_a^b f(x)^2 \, dx = \pi \int_a^b y^2 \, dx. \tag{16.5.1}$$

Theorem 16.5.2 *Let C be the curve given by the parametric equations*

$$x = x(t), \quad y = y(t) \quad (\alpha \leq t \leq \beta),$$

where x is injective with a continuous derivative, and y is continuous, on $[\alpha, \beta]$. Then the volume of revolution generated by C is

$$V = \pi \int_\alpha^\beta y(t)^2 |\dot{x}(t)| \, dt. \tag{16.5.2}$$

Proof Suppose, without loss, that $x : [\alpha, \beta] \to I$ where $I = x([\alpha, \beta])$. Then x is a bijection and, by Example 7.3.16, $I = [a, b]$ (say). Define $f : [a, b] \to \mathbb{R}$ by $f(s) = y(x^{-1}(s))$. By Theorems 7.3.18 and 7.3.9 (3), f is continuous. Since C is the curve $y = f(x)$ $(a \leq x \leq b)$, Definition 16.5.1 and Lemma 16.1.4 give

$$V = \pi \int_a^b f(x)^2 \, dx = \pi \int_\alpha^\beta f(x(t))^2 |\dot{x}(t)| \, dt = \pi \int_\alpha^\beta y(t)^2 |\dot{x}(t)| \, dt$$

as required.   □

Example 16.5.3 Find the volume of a sphere, radius R.

Solution We can think of the required volume V as the volume of revolution generated by the upper half of the circle $x^2 + y^2 = R^2$. Then, using (16.5.1),

$$V = \pi \int_{-R}^R (R^2 - x^2) \, dx = \left[R^2 x - \frac{x^3}{3} \right]_{-R}^R = \frac{4}{3} \pi R^3.$$

Alternatively, expressing the semi-circle in parametric form as

$$x = R \cos t, \quad y = R \sin t \quad (0 \leq t \leq \pi),$$

equation (16.5.2) gives

$$V = \pi \int_0^\pi (R \sin t)^2 |-R \sin t| \, dt = \pi R^3 \int_0^\pi \sin^3 t \, dt = \frac{4}{3} \pi R^3.$$

Verification of the final evaluation is left to the reader.   □

16.6   Arc Lengths

Let f be a real function with a continuous derivative on $[a, b]$, and consider the curve $\mathcal{C}$ given by

$$y = f(x) \quad (a \le x \le b).$$

We define the length or *arc length* of such a curve by first approximating with line segments. Recall the formula

$$\sqrt{(x_2 - x_1)^2 + (y_2 - y_1)^2}$$

for the length of the line segment joining points (x_1, y_1) and (x_2, y_2).

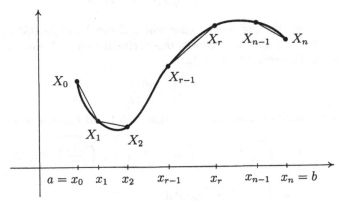

Figure 16.6.1

Let $\mathcal{P} = (x_0, x_1, \ldots, x_n)$ be a partition of $[a, b]$ and for each $r = 0, 1, \ldots, n$, let X_r be the point $(x_r, f(x_r))$ on $\mathcal{C}$. See Figure 16.6.1. The sum of the lengths of the line segments $X_{r-1}X_r$ $(r = 1, \ldots, n)$ is

$$S = \sum_{r=1}^{n} |X_{r-1}X_r| = \sum_{r=1}^{n} \sqrt{(x_r - x_{r-1})^2 + (f(x_r) - f(x_{r-1}))^2}$$

$$= \sum_{r=1}^{n} \sqrt{1 + \left(\frac{f(x_r) - f(x_{r-1})}{x_r - x_{r-1}} \right)^2} \, (x_r - x_{r-1}).$$

For each $r = 1, \ldots, n$, the function f is continuous on $[x_{r-1}, x_r]$ and differentiable on (x_{r-1}, x_r). So, by the mean value theorem, there exist $c_r \in (x_{r-1}, x_r)$ $(r = 1, \ldots, n)$ such that

$$f'(c_r) = \frac{f(x_r) - f(x_{r-1})}{x_r - x_{r-1}}.$$

Hence,

$$S = \sum_{r-1}^{n} \sqrt{1 + f'(c_r)^2} \, (x_r - x_{r-1}),$$

which is a Riemann sum for the function $g : [a, b] \to \mathbb{R}$ defined by $g(x) = \sqrt{1 + f'(x)^2}$. Since g is continuous on $[a, b]$, $S \to \int_a^b g(x) \, dx$ as $|\mathcal{P}| \to 0$.

Definition 16.6.1 Let f be a real function with a continuous derivative on $[a, b]$. Then the *arc length* L of the curve $y = f(x)$ $(a \leq x \leq b)$ is defined by

$$L = \int_a^b \sqrt{1 + f'(x)^2}\, dx = \int_a^b \sqrt{1 + \left(\frac{dy}{dx}\right)^2}\, dx. \qquad (16.6.1)$$

Theorem 16.6.2 *Let C be the curve given by the parametric equations*

$$x = x(t), \quad y = y(t) \quad (\alpha \leq t \leq \beta),$$

where x and y have continuous derivatives on $[\alpha, \beta]$. Then the arc length of C is

$$L = \int_\alpha^\beta \sqrt{\dot{x}(t)^2 + \dot{y}(t)^2}\, dt. \qquad (16.6.2)$$

Proof We prove only the special case where C can be expressed in the form $y = f(x)$ $(a \leq x \leq b)$, f having a continuous derivative on $[a, b]$ and $x : [\alpha, \beta] \to [a, b]$ bijective. From Theorem 10.4.1 we have

$$\frac{dy}{dx} = \frac{\dot{y}(t)}{\dot{x}(t)}.$$

Then under the stated conditions, Definition 16.6.1 and Lemma 16.1.4 give

$$L = \int_a^b \sqrt{1 + \left(\frac{dy}{dx}\right)^2}\, dx = \int_\alpha^\beta \sqrt{1 + \left(\frac{\dot{y}(t)}{\dot{x}(t)}\right)^2}\, |\dot{x}(t)|\, dt$$

$$= \int_\alpha^\beta \sqrt{\dot{x}(t)^2 + \dot{y}(t)^2}\, dt.$$

as required.   □

Corollary 16.6.3 *Let C be the curve given by the polar equation*

$$r = r(\theta) \quad (\alpha \leq \theta \leq \beta),$$

where r has a continuous derivative on $[\alpha, \beta]$. Then the arc length of C is

$$L = \int_\alpha^\beta \sqrt{r^2 + \left(\frac{dr}{d\theta}\right)^2}\, d\theta. \qquad (16.6.3)$$

Proof The curve C has parametric equations

$$x = r \cos \theta, \quad y = r \sin \theta \quad (\alpha \leq \theta \leq \beta).$$

Then

$$\dot{x}(\theta)^2 + \dot{y}(\theta)^2 = \left(\frac{dr}{d\theta} \cos \theta - r \sin \theta\right)^2 + \left(\frac{dr}{d\theta} \sin \theta + r \cos \theta\right)^2$$

$$= r^2 + \left(\frac{dr}{d\theta}\right)^2 \quad (\text{using } \sin^2 \theta + \cos^2 \theta = 1).$$

Hence, by Theorem 16.6.2,

$$L = \int_\alpha^\beta \sqrt{\dot{x}(\theta)^2 + \dot{y}(\theta)^2}\, d\theta = \int_\alpha^\beta \sqrt{r^2 + \left(\frac{dr}{d\theta}\right)^2}\, d\theta,$$

as required.   □

Example 16.6.4 Find the arc lengths of the curves

(a) $y = e^x + \frac{1}{4}e^{-x}$ $(0 \le x \le \log 2)$,

(b) $x = t - \sin t$, $y = 1 - \cos t$ $(0 \le t \le 2\pi)$,

(c) $r = 2 \sec \theta$ $(-\pi/4 \le \theta \le \pi/4)$.

Solution (a) Here,

$$1 + \left(\frac{dy}{dx}\right)^2 = 1 + \left(e^x - \frac{1}{4}e^{-x}\right) = 1 + e^{2x} - \frac{1}{2} + \frac{1}{16}e^{-2x}$$

$$= e^{2x} + \frac{1}{2} + \frac{1}{16}e^{-2x} = \left(e^x + \frac{1}{4}e^{-x}\right)^2.$$

Hence, using (16.6.1), the arc length is

$$\int_0^{\log 2} \left(e^x + \frac{1}{4}e^{-x}\right) dx = \left[e^x - \frac{1}{4}e^{-x}\right]_0^{\log 2} = \frac{9}{8}.$$

(b) Here,

$$\dot{x}(t)^2 + \dot{y}(t)^2 = (1 - \cos t)^2 + (\sin t)^2 = 2(1 - \cos t) = 4\sin^2 \frac{1}{2}t.$$

For $t \in [0, 2\pi]$, $2\sin \frac{1}{2}t \ge 0$. Hence, using (16.6.2), the arc length is

$$\int_0^{2\pi} 2\sin \frac{1}{2}t \, dt = \left[-4\cos \frac{1}{2}t\right]_0^{2\pi} = 8.$$

(c) Here,

$$r^2 + \left(\frac{dr}{d\theta}\right)^2 = (2\sec \theta)^2 + (2\sec \theta \tan \theta)^2 = 4\sec^2 \theta(1 + \tan^2 \theta) = 4\sec^4 \theta.$$

Hence, using (16.6.3), the arc length is

$$\int_{-\pi/4}^{\pi/4} 2\sec^2 \theta \, d\theta = \left[2\tan \theta\right]_{-\pi/4}^{\pi/4} = 4. \qquad \square$$

16.7   Areas of Revolution

Let f be a real function with a continuous derivative on $[a, b]$, and consider the *surface of revolution* formed by revolving, once about the x-axis, the curve

$$y = f(x) \quad (a \le x \le b).$$

See Figure 16.7.1. We define the area of such a surface by first approximating the curve with line segments. Revolving a line segment about the x-axis produces the curved surface of a frustum (a cone cut off parallel to its base), the area of which is given by the formula $\pi(R_1 + R_2)L$, where R_1 and R_2 are the radii, and L is the length of the segment. See Figure 16.7.2.

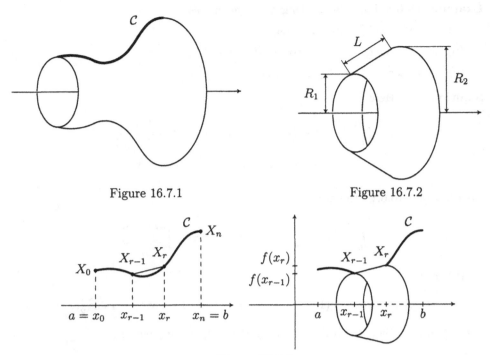

Figure 16.7.1   Figure 16.7.2

Figure 16.7.3

Let $\mathcal{P} = (x_0, x_1, \ldots, x_n)$ be a partition of $[a, b]$ and for each $r = 0, 1, \ldots, n$, let X_r be the point $(x_r, f(x_r))$ on the curve. See Figure 16.7.3. Revolve each line segment $X_{r-1}X_r$ $(r = 1, \ldots, n)$, once about the x-axis, to produce a surface. The sum of the areas of these surfaces is

$$S = \sum_{r=1}^{n} \pi \big(|f(x_{r-1})| + |f(x_r)|\big)|X_{r-1}X_r|.$$

Define $g : [a, b] \to \mathbb{R}$ by $g(x) = 2\pi f(x)\sqrt{1 + f'(x)^2}$. It can be shown that $S \to \int_a^b g(x)\,dx$ as $|\mathcal{P}| \to 0$.

Definition 16.7.1 Let f be a real function with a continuous derivative on $[a, b]$. Then the *area of revolution A generated* by the curve $y = f(x)$ $(a \le x \le b)$ is defined by

$$A = 2\pi \int_a^b |f(x)|\sqrt{1 + f'(x)^2}\,dx = 2\pi \int_a^b |y|\sqrt{1 + \left(\frac{dy}{dx}\right)^2}\,dx. \qquad (16.7.1)$$

Theorem 16.7.2 *Let C be the curve given by the parametric equations*

$$x = x(t), \quad y = y(t) \quad (\alpha \le t \le \beta),$$

where x and y have continuous derivatives on $[\alpha, \beta]$. Then the area of revolution generated by C is

$$A = 2\pi \int_\alpha^\beta |y(t)|\sqrt{\dot{x}(t)^2 + \dot{y}(t)^2}\,dt. \qquad (16.7.2)$$

Proof The proof is omitted. Where C can be expressed in the form $y = f(x)$ ($a \leq x \leq b$), f having a continuous derivative on $[a, b]$ and $x : [\alpha, \beta] \to [a, b]$ bijective, the proof is similar to that of Theorem 16.6.2 under the same restrictions. $\square$

Corollary 16.7.3 *Let C be the curve given by the polar equation*

$$r = r(\theta) \quad (\alpha \leq \theta \leq \beta),$$

where r has a continuous derivative on $[\alpha, \beta]$. Then the area of revolution generated by C is

$$A = 2\pi \int_\alpha^\beta |r \sin\theta| \sqrt{r^2 + \left(\frac{dr}{d\theta}\right)^2}\, d\theta. \qquad (16.7.3)$$

Proof This is left to the reader. See the proof of Corollary 16.6.3. $\square$

Example 16.7.4 Find the areas of revolution generated by the curves

(a) $y = \sqrt{2x + 1}$ $(0 \leq x \leq 3)$,

(b) $x = t - \sin t$, $y = 1 - \cos t$ $(0 \leq t \leq 2\pi)$.

Solution (a) Here

$$1 + \left(\frac{dy}{dx}\right)^2 = 1 + \left(\frac{1}{\sqrt{2x+1}}\right)^2 = \frac{2x+2}{2x+1}.$$

For $x \in [0, 3]$, $(2x + 2)/(2x + 1) \geq 0$. Hence, using (16.7.1), the area of revolution is

$$2\pi \int_0^3 \sqrt{2x+1}\, \sqrt{\frac{2x+2}{2x+1}}\, dx = 2\sqrt{2}\pi \int_0^3 \sqrt{x+1}\, dx$$

$$= 2\sqrt{2}\pi \left[\frac{2}{3}(x+1)^{3/2}\right]_0^3 = \frac{28\sqrt{2}\pi}{3}.$$

(b) We saw in the solution to Example 16.6.4 (b) that, for $t \in [0, 2\pi]$,

$$\sqrt{\dot{x}(t)^2 + \dot{y}(t)^2} = 2\sin\tfrac{1}{2}t.$$

Hence, using (16.7.2), the area of revolution is

$$A = 2\pi \int_0^{2\pi} (2\sin^2\tfrac{1}{2}t)(2\sin\tfrac{1}{2}t)\, dt = 8\pi \int_0^{2\pi} \sin^3\tfrac{1}{2}t\, dt.$$

Substituting $\cos\tfrac{1}{2}t = u$, gives

$$A = -16\pi \int_0^{2\pi} (\sin^2\tfrac{1}{2}t)(-\tfrac{1}{2}\sin t)\, dt = -16\pi \int_1^{-1} (1 - u^2)\, du$$

$$= -16\pi \left[u - \frac{u^3}{3}\right]_1^{-1} = \frac{64\pi}{3}. \qquad \square$$

Example 16.7.5 Find the surface area of a sphere, radius R.

Solution We can think of the required area A as the area of revolution generated by the upper half of the circle $x^2 + y^2 = R^2$ which has the polar equation

$$r = R \quad (0 \le \theta \le \pi).$$

Then, using (16.7.3),

$$A = 2\pi \int_0^\pi |R\sin\theta|\sqrt{R^2 + 0^2}\, d\theta = 2\pi R^2 \int_0^\pi \sin\theta\, d\theta = 4\pi R^2. \qquad \square$$

16.X   Exercises

1. Use standard integrals to find

(a) $\displaystyle\int \frac{dx}{\sqrt{1 - 4x^2}}$,   (b) $\displaystyle\int \frac{dx}{\sqrt{9 - 25x^2}}$,   (c) $\displaystyle\int \frac{dx}{\sqrt{3 + 2x - x^2}}$,

(d) $\displaystyle\int \frac{dx}{9x^2 + 4}$,   (e) $\displaystyle\int \frac{dx}{x^2 + 8x + 18}$,   (f) $\displaystyle\int \frac{dx}{3x^2 + 2x + 2}$,

(g) $\displaystyle\int \frac{dx}{\sqrt{x^2 - 2x - 3}}$,   (h) $\displaystyle\int \frac{dx}{\sqrt{x^2 - 2x}}$,   (i) $\displaystyle\int \frac{dx}{\sqrt{2x^2 + 1}}$.

2. Use suitable substitutions to find/evaluate

(a) $\displaystyle\int x\sqrt{x^2 - 5}\, dx$,   (b) $\displaystyle\int x(3x^2 + 7)^6\, dx$,

(c) $\displaystyle\int \frac{x^2}{18 + x^6}\, dx$,   (d) $\displaystyle\int \frac{\cos 3x}{\sqrt{2 - \sin 3x}}\, dx$,

(e) $\displaystyle\int \sqrt{\frac{\sin^{-1} x}{1 - x^2}}\, dx$,   (f) $\displaystyle\int \frac{x}{x^4 - 2x^2 + 4}\, dx$,

(g) $\displaystyle\int \frac{dx}{x(\log x)^2}$,   (h) $\displaystyle\int_1^2 x^2\sqrt{4 + x^3}\, dx$,

(i) $\displaystyle\int_{1/e}^e \frac{\log x}{x}\, dx$,   (j) $\displaystyle\int_0^{\sqrt{2}} \frac{x}{\sqrt{8 - x^4}}\, dx$,

(k) $\displaystyle\int_0^1 \frac{\sqrt{\tan^{-1} x}}{x^2 + 1}\, dx$,   (l) $\displaystyle\int_{\pi/3}^{\pi/2} \frac{\sin x}{1 - \cos x}\, dx$,

3. Use partial fractions to find/evaluate

(a) $\displaystyle\int \frac{5x - 8}{(x + 4)(x - 3)}\, dx$,   (b) $\displaystyle\int_{-1}^1 \frac{x^2 - 6x + 16}{(x - 2)(x^2 - 4)}\, dx$,

(c) $\displaystyle\int \frac{22x^4}{(x^2 - 1)(2x^2 + 9)}\, dx$,   (d) $\displaystyle\int_0^1 \frac{4x^2}{x^4 + 4}\, dx$.

Hint for (d): $x^4 + 4 = (x^2 + 2)^2 - 4x^2$.

4. Find constants P and Q such that $x + 15 = P(2x - 2) + Q$. Hence evaluate

(a) $\displaystyle \int_1^3 \frac{x + 15}{(x^2 - 2x + 5)^2}\, dx,$ (b) $\displaystyle \int_1^3 \frac{x + 15}{\sqrt{x^2 - 2x + 5}}\, dx.$

5. By differentiating the right-hand side, verify that, for $b \neq 0$,

$$\int \sqrt{x^2 + b}\, dx = \frac{x}{2}\sqrt{x^2 + b} + \frac{b}{2} \log|x + \sqrt{x^2 + b}| + C.$$

Hence evaluate

(a) $\displaystyle \int_1^3 \sqrt{x^2 - 2x + 5}\, dx,$ (b) $\displaystyle \int_1^3 (x + 15)\sqrt{x^2 - 2x + 5}\, dx.$

For (b) use the P and Q from Exercise 16.X.4.

6. Use suitable substitutions to find/evaluate

(a) $\displaystyle \int_0^{\pi/4} \sin^4 x \cos x\, dx,$ (b) $\displaystyle \int_0^{\pi/2} (\cos \theta)^{1/3} \sin^3 \theta\, d\theta,$

(c) $\displaystyle \int_0^{\pi/6} \sec^2 2\theta \tan 2\theta\, d\theta,$ (d) $\displaystyle \int \sin^3 x \cos^4 x\, dx,$

(e) $\displaystyle \int \sin^2 x \cos^5 x\, dx,$ (f) $\displaystyle \int \tan^2 x \sec^4 x\, dx,$

(g) $\displaystyle \int \tan^3 x \sec^3 x\, dx,$ (h) $\displaystyle \int \sinh^2 x \cosh^5 x\, dx,$

(i) $\displaystyle \int \sin^2 3x \cos^3 3x\, dx,$ (j) $\displaystyle \int \tan\left(\frac{x}{2}\right) \sec^4\left(\frac{x}{2}\right) dx.$

7. Use the t-formulae to evaluate

(a) $\displaystyle \int_0^{\pi/2} \frac{d\theta}{2 - \sin \theta},$ (b) $\displaystyle \int_0^{\pi/2} \frac{d\theta}{3 + \sin \theta + 3 \cos \theta},$

(c) $\displaystyle \int_0^{2\pi/3} \frac{dx}{5 + 4 \cos x},$ (d) $\displaystyle \int_0^{\pi/2} \frac{dx}{12 + 13 \sin x}.$

8. Use integration by parts to find/evaluate

(a) $\displaystyle \int x e^{3x}\, dx,$ (b) $\displaystyle \int_{-1}^1 x^2 e^{-x}\, dx,$ (c) $\displaystyle \int (x - 1)^2 e^{2x}\, dx,$

(d) $\displaystyle \int_0^{\pi} e^{-x} \cos x\, dx,$ (e) $\displaystyle \int e^{3x} \sin 2x\, dx,$ (f) $\displaystyle \int e^{3x} \sinh 2x\, dx,$

(g) $\displaystyle \int_0^{\pi/2} x \sin x\, dx,$ (h) $\displaystyle \int_1^e x \log x\, dx,$ (i) $\displaystyle \int_{1/\sqrt{3}}^{\sqrt{3}} x \tan^{-1} x\, dx,$

(j) $\displaystyle \int \cos^{-1} x\, dx,$ (k) $\displaystyle \int \tan^{-1} x\, dx,$ (l) $\displaystyle \int \sqrt{x^2 + 7}\, dx.$

9.   Use any appropriate means to find/evaluate

(a) $\displaystyle\int_0^{\pi/2} \sin^5 x \cos^2 x \, dx,$ (b) $\displaystyle\int_{-2}^{-1/2} (5 - 4x - x^2)^{-1/2} \, dx,$

(c) $\displaystyle\int e^{5x} \cos 2x \, dx,$ (d) $\displaystyle\int e^{5x} \sin^2 x \, dx,$

(e) $\displaystyle\int_1^{\sqrt{3}} \frac{x^3 - x^2 + 1}{x^2(x^2 + 1)} \, dx,$ (f) $\displaystyle\int \frac{x}{2 + x^4} \, dx,$

(g) $\displaystyle\int_0^1 x^5 \sqrt{1 - x^3} \, dx,$ (h) $\displaystyle\int_0^{\pi/2} \frac{dx}{5 + 3\cos x - 2\sin x},$

(i) $\displaystyle\int_{-1}^1 \frac{2x + 1}{x^2 + 2x + 5} \, dx,$ (j) $\displaystyle\int (\log x)^2 \, dx,$

(k) $\displaystyle\int \left(\frac{x - 1}{x + 1}\right)^3 \, dx,$ (l) $\displaystyle\int \frac{dx}{x^{2/3}(x^{2/3} + 1)}.$

10. For $n \in \mathbb{N} \cup \{0\}$, define
$$\mathcal{I}_n = \int \sec^n x \, dx.$$
By writing $\sec^n x = \sec^2 x \sec^{n-2} x$ and integrating by parts, prove that, for $n \geq 2$,
$$\mathcal{I}_n = \frac{1}{n - 1} \sec^{n-2} x \tan x + \frac{n - 2}{n - 1} \mathcal{I}_{n-2}.$$
Use this to evaluate $\displaystyle\int_0^1 (x^2 + 1)^{3/2} \, dx.$

11. For $n \in \mathbb{N} \cup \{0\}$, define
$$\mathcal{J}_n = \int_1^e (\log x)^n \, dx.$$
By writing $(\log x)^n = 1.(\log x)^n$, prove that, for $n \geq 1$,
$$\mathcal{J}_n = e - n\mathcal{J}_{n-1}.$$
Hence evaluate $\mathcal{J}_4$.

12. Let f be a continuous real bijection and let G be an antiderivative for f^{-1}. By differentiating the right-hand side, verify that
$$\int f(x) \, dx = xf(x) - G(f(x)) + C.$$
Hence find

(a) $\displaystyle\int \log x \, dx,$ (b) $\displaystyle\int \sqrt{x} \, dx,$ (c) $\displaystyle\int \sin^{-1} x \, dx.$

13. Find the volume of revolution generated by the curve

$$y = \cos x \quad (-\pi/2 \le x \le \pi/2).$$

14. Find the volume of revolution generated by the curve

$$y = (4 - x^{2/3})^{3/2} \quad (0 \le x \le 8).$$

Find also the arc length of the curve.

15. Find the arc length of the curve

$$y = x^2 + 3x - \frac{1}{8}\log(2x + 3) \quad (1 \le x \le 2).$$

16. Find the arc length of the curve

$$y = \frac{x\sqrt{x+1}}{\sqrt{3}} \quad (0 \le x \le 2).$$

Find also the area of revolution generated by the curve.

17. For each of the following curves given in parametric form, find the volume of revolution, the arc length and the area of revolution.

(a) $x = \sqrt{2}e^t, \quad y = \frac{1}{2}(e^{2t} - t) \quad (0 \le t \le 1),$

(b) $x = 4\sqrt{2}\sin t, \quad y = \sin 2t \quad (0 \le t \le \pi/2).$

18. For each of the following curves given in polar form, find the arc length and, for (a), find also the area of revolution.

(a) $r = \sin \theta \quad (0 \le \theta \le \pi),$

(b) $r = e^\theta \quad (0 \le \theta \le \pi/2),$

(c) $r = 1 + \cos \theta \quad (-\pi \le \theta \le \pi).$

Chapter 17

Ordinary Differential Equations

In applications, we are often faced with the problem of determining a function f given information on some of its derivatives. A typical example would be to find the position of a particle as a function of time, given its velocity as a function of time and position. Such a formulation of a problem is called a *differential equation*. The general form is an equation which involves f and some of its derivatives. The study of such equations is a major area of mathematics and it is not always possible to obtain an exact solution. Here we consider only a few simple types where we can obtain a solution using the techniques on integration developed in Chapters 14 to 16. Our results give a flavour of the subject, and illustrate some important aspects.

The *order* of a differential equation is the order of the highest derivative which actually occurs in the equation—this is analogous to the definition of the degree of a polynomial. In Sections 17.2 to 17.4, we look at three types of first order differential equation. The reader should observe that, in each case, the general solution involves an arbitrary constant. We thus have an infinite family of solutions for each equation. To determine a particular solution, we need one additional piece of information such as a *boundary condition* or an *initial condition*. Typically this will consist of the value of the function at one point. The final section looks at a very special type of second order equation. Now the general solution involves *two* arbitrary constants, so we need two additional conditions to determine a particular solution.

17.1 Introduction

Definitions 17.1.1 An equation involving an *independent* variable x, a *dependent* variable y and derivatives of y with respect to x up to and including the nth derivative, is called an *ordinary differential equation* or ODE (pronounced oh–dee–ee) of *order n*. For example,

$$y\frac{dy}{dx} = 2x(y+1) \qquad \text{and} \qquad \frac{d^2y}{dx^2} - \frac{dy}{dx} - 6y = 0$$

or equivalently,

$$yy' = 2x(y+1) \qquad \text{and} \qquad y'' - y' - 6y = 0,$$

are first and second order ODEs, respectively.

Solving such an equation means expressing y explicitly *or* implicitly as a function of x on an interval. Consider, for example, the second order ODE

$$\frac{d^2y}{dx^2} = 2. \tag{17.1.1}$$

Integrating twice gives

$$y = x^2 + Ax + B \tag{17.1.2}$$

where A and B are constants of integration. We call this the *general solution* since any solution of equation (17.1.1) is of the form (17.1.2) and *vice versa*. Assigning specific values to the *parameters* A and B gives a *particular solution* or *particular integral*.

Remarks (1) The parameters in a general solution may be determined by imposing additional conditions. For example, we may require a solution $y = y(x)$ of equation (17.1.1) for $0 \le x \le 3$, which satisfies the *boundary conditions*

$$y(0) = 5 \quad \text{and} \quad y(3) = 8,$$

i.e.

$$y = 5 \text{ when } x = 0 \quad \text{and} \quad y = 8 \text{ when } x = 3.$$

Substituting these conditions in the general solution (17.1.2) leads to $A = -2$, $B = 5$ and the particular solution

$$y = x^2 - 2x + 5.$$

Alternatively we might require a solution satisfying the *initial conditions*

$$y(0) = 5 \quad \text{and} \quad y'(0) = -2.$$

These also lead to the particular solution

$$y = x^2 - 2x + 5.$$

(2) The general solution of an ODE can appear in different forms. For example, the general solution of equation (17.1.1) can be written

$$y = x^2 + P(x+1) + Q(x-1)$$

where P and Q are parameters. Note, however, that a particular solution determined by additional conditions is independent of the form of the general solution.

(3) Any n^{th} order ODE can be expressed in the form

$$F(x, y, y^{(1)}, y^{(2)}, \ldots, y^{(n)}) = 0.$$

Where F is a real valued function of $n+2$ real variables. Its general solution, when it exists, usually involves n independent parameters and a particular solution will usually be determined by n boundary conditions or initial conditions.

Our aim in this chapter is to study certain specific types of first and second order ODEs.

17.2　First Order Separable Equations

Definition 17.2.1 A first order ODE is called *separable* if it can be written in the form

$$\frac{dy}{dx} = g(x)h(y), \tag{17.2.1}$$

where g and h are real functions.

Method　We assume that g and $1/h$ can be integrated. Then using the substitution $y = y(x)$ and equation (17.2.1) we have

$$\int \frac{1}{h(y)}\, dy = \int \frac{1}{h(y(x))}\frac{dy}{dx}\, dx = \int g(x)\, dx.$$

In effect, we have 'separated the variables' x and y in (17.2.1) so that the left hand side involves the variable y only and the right hand side involves the variable x only. Thus

$$\int \frac{dy}{h(y)} = \int g(x)\, dx$$

is the general solution of equation (17.2.1), valid for x in any interval where $g(x)$ and $h(y)$ are defined and $h(y) \neq 0$.

　　If $k \in \mathbb{R}$ and $h(k) = 0$ then $y = k$ will also be a solution of equation (17.2.1), valid for $x \in \mathbb{R}$.

Example 17.2.2 Find the general solutions of

$$\text{(a)}\ \frac{dy}{dx} = e^{x-2y}, \qquad \text{(b)}\ x\frac{dy}{dx} = \frac{1+x}{2-\sin y}, \qquad \text{(c)}\ \frac{dy}{dx} - py = 0,$$

where $p \in \mathbb{R}$. Find, also, the particular solution of (b) for $x > 0$ which satisfies $y = 2\pi$ when $x = 1$.

Solution　(a) Here,

$$\frac{dy}{dx} = e^{x-2y} = e^x e^{-2y} = e^x \frac{1}{e^{2y}}$$

so that

$$\int e^{2y}\, dy = \int e^x\, dx,$$

i.e.

$$\frac{e^{2y}}{2} = e^x + C,$$

i.e.

$$y = \frac{1}{2}\log 2(e^x + C),$$

where C is a parameter. This is the general solution.

　　(b) Here,

$$\frac{dy}{dx} = \left(\frac{1+x}{x}\right)\frac{1}{2-\sin y}$$

so that

$$\int (2 - \sin y)\, dy = \int \left(\frac{1}{x} + 1 \right) dx,$$

i.e.

$$2y + \cos y = \log |x| + x + C,$$

where C is a parameter. This is the general solution.

For $x > 0$ this becomes

$$2y + \cos y = \log x + x + C.$$

Substituting $x = 1$ and $y = 2\pi$ gives $C = 4\pi$. Hence, the required particular solution is

$$2y + \cos y = \log x + x + 4\pi \quad (x > 0).$$

(c) Here,

$$\frac{dy}{dx} = py$$

so that, for $y \neq 0$,

$$\int \frac{dy}{y} = \int p\, dx,$$

i.e.

$$\log |y| = px + C,$$

i.e.

$$|y| = e^{px+C} = e^{px} e^{C},$$

i.e.

$$|y| = D e^{px},$$

where $D > 0$ is a parameter. This together with $y = 0$ is the general solution. Since $|y| = \pm y$, we can write the general solution in the form

$$y = E e^{px},$$

where E (*any* real number) is a parameter. This includes the solution $y = 0$.   □

Remark The general solution of the ODE in Example 17.2.2 (b) expresses y implicitly as a function of x. This is the best we can do with that solution. When it is reasonable to do so, we shall express y explicitly as a function of x.

Problem 17.2.3 For which values of the parameter C, in the general solution of the ODE in Example 17.2.2 (a), is the solution valid for $x \in \mathbb{R}$?

Example 17.2.4 Find a particular solution of

$$\frac{dy}{dx} = x\sqrt{1 - y^2},$$

valid on an interval containing 0 and satisfying $y(0) = 0$.

Solution For $y^2 \neq 1$, separating the variables gives

$$\int \frac{dy}{\sqrt{1 - y^2}} = \int x \, dx,$$

i.e.

$$\sin^{-1} y = \frac{x^2}{2} + C,$$

where C is a parameter. For $y(0) = 0$ we must take $C = 0$, giving the particular solution

$$\sin^{-1} y = \frac{x^2}{2}.$$

Recalling the definition of $\sin^{-1}$ and remembering that $y^2 \neq 1$, this is equivalent to

$$y = \sin\left(\frac{x^2}{2}\right) \qquad \left(-\frac{\pi}{2} < \frac{x^2}{2} < \frac{\pi}{2}\right),$$

i.e.

$$y = \sin\left(\frac{x^2}{2}\right) \qquad (-\sqrt{\pi} < x < \sqrt{\pi}),$$

which is a suitable particular solution.   □

Remark The ODE in Example 17.2.4 has two further solutions. They are $y = 1$ and $y = -1$. Both are valid for $x \in \mathbb{R}$ but neither satisfies the condition $y(0) = 0$.

Example 17.2.5 Find the general solution of

$$\frac{dy}{dx} = \frac{1}{(x + 4y)^2}.$$

Solution Here we change the variables from (x, y) to (x, u) by making the substitution $u = x + 4y$. This converts the given equation into a separable ODE. We have,

$$\frac{du}{dx} = 1 + 4\frac{dy}{dx} = 1 + \frac{4}{u^2} = \frac{u^2 + 4}{u^2}.$$

Hence,

$$\int \frac{u^2}{u^2 + 4} \, du = \int dx,$$

i.e.

$$\int \left(1 - \frac{4}{u^2 + 4}\right) du = \int dx,$$

i.e.

$$u - \frac{4}{2} \tan^{-1} \frac{u}{2} = x + C,$$

where C is a parameter. Substituting $u = x + 4y$ gives

$$x + 4y - 2 \tan^{-1}\left(\frac{x + 4y}{2}\right) = x + C,$$

i.e.

$$\tan^{-1}\left(\frac{x+4y}{2}\right) = 2y - D,$$

where $D \ (= C/2)$ is a parameter. This is the required general solution. It can be written in the form

$$x + 4y = 2\tan\left(2y - D\right)$$

subject to $-\frac{\pi}{2} < 2y - D < \frac{\pi}{2}$.   □

17.3 First Order Homogeneous Equations

Definition 17.3.1 A first order ODE is called *homogeneous* if it can be written in the form

$$\frac{dy}{dx} = f(y/x) \tag{17.3.1}$$

where f is a real function.

The right-hand side of equation (17.3.1) is a function of y/x but is not always presented in a form which makes this clear. Consider the equation

$$\frac{dy}{dx} = F(x, y) \tag{17.3.2}$$

where F is a real valued function of two real variables. To verify that $F(x, y)$ can be expressed in the form $f(y/x)$, show that, wherever it is defined, $F(x, vx)$ is independent of x, *i.e.* $F(x, vx) = f(v)$. For example,

$$\frac{dy}{dx} = \frac{x^4 + y^4}{xy^3} \qquad \text{and} \qquad \frac{dy}{dx} = \frac{y}{x}(\log y - \log x + 1)$$

are homogeneous since, for $x, v \neq 0$,

$$\frac{x^4 + (vx)^4}{x(vx)^3} = \frac{x^4 + v^4 x^4}{xv^3 x^3} = \frac{x^4(1 + v^4)}{x^4 v^3} = \frac{1 + v^4}{v^3}$$

and, for $x, v > 0$,

$$\frac{vx}{x}(\log(vx) - \log x + 1) = v(\log(vx/x) + 1) = v(\log v + 1).$$

Method To solve equation (17.3.2) we change the variables from (x, y) to (x, v) by making the substitution $y = vx$. This converts the given equation into a separable ODE. The complete substitution is

$$y = vx \qquad \text{and} \qquad \frac{dy}{dx} = v + x\frac{dv}{dx}.$$

Example 17.3.2 Find the general solution of

$$y + \sqrt{x^2 + y^2} - x\frac{dy}{dx} = 0 \qquad (x > 0).$$

Solution Here,

$$\frac{dy}{dx} = \frac{y + \sqrt{x^2 + y^2}}{x}.$$

Substituting $y = vx$ gives

$$v + x\frac{dv}{dx} = \frac{vx + \sqrt{x^2 + v^2 x^2}}{x} = v + \sqrt{1 + v^2}$$

(since $x > 0 \;\Rightarrow\; \sqrt{x^2} = x$). Hence,

$$\frac{dv}{dx} = \frac{1}{x}\sqrt{1 + v^2}.$$

Separating the variables leads to

$$\int \frac{dv}{\sqrt{1 + v^2}} = \int \frac{dx}{x},$$

i.e.

$$\log\left|v + \sqrt{1 + v^2}\right| = \log|x| + C,$$

i.e.

$$\log\left(v + \sqrt{1 + v^2}\right) = \log x + C,$$

where C is a parameter. Replacing v with y/x we get

$$\log\left(\frac{y}{x} + \sqrt{1 + \frac{y^2}{x^2}}\right) = \log x + C.$$

This is the required general solution. We can simplify it as follows. Let $D = e^C$ so that $D > 0$ and $C = \log D$. Then the general solution becomes

$$\log\left(\frac{y}{x} + \sqrt{1 + \frac{y^2}{x^2}}\right) = \log Dx,$$

i.e. (since log is injective)

$$\frac{y}{x} + \sqrt{1 + \frac{y^2}{x^2}} = Dx,$$

i.e. (since $x > 0$)

$$y + \sqrt{x^2 + y^2} = Dx^2.$$

Hence we can write the general solution in the form

$$\sqrt{x^2 + y^2} = Dx^2 - y \qquad (x > 0), \tag{17.3.3}$$

where $D > 0$ is a parameter.   □

Problem 17.3.3 Prove that the general solution (17.3.3) is *equivalent* to the explicit form

$$y = \frac{1}{2}\left(Dx^2 - \frac{1}{D}\right) \qquad (x > 0), \tag{17.3.4}$$

where $D > 0$ is a parameter. You must verify that, when (17.3.4) is satisfied, $Dx^2 - y \geq 0$. Then $\sqrt{(Dx^2 - y)^2} = Dx^2 - y$.

17.4   First Order Linear Equations

Definitions 17.4.1 An ODE is called n^{th} *order linear* if it can be written in the form

$$a_n(x)\frac{d^n y}{dx^n} + a_{n-1}(x)\frac{d^{n-1}y}{dx^{n-1}} + \cdots + a_1(x)\frac{dy}{dx} + a_0(x)y = f(x),$$

where a_n, a_{n-1}, ..., a_1, a_0 and f are real functions and a_n is not the zero function. Thus a first order linear ODE can be written in the form

$$a_1(x)\frac{dy}{dx} + a_0(x)y = f(x),$$

where a_1, a_0 and f are real function, and a_1 is not the zero function. In fact, the *standard form* of such an equation is

$$\frac{dy}{dx} + p(x)y = q(x), \qquad (17.4.1)$$

where p and q are real functions, continuous on an interval I.

Let μ be a real function, differentiable and non-zero on I, such that multiplying the left-hand side of (17.4.1) by $\mu(x)$ gives the derivative of the product $\mu(x)y$. Then $\mu(x)$ is called an *integrating factor*.

Method Let P be *any* antiderivative of p, i.e. $P'(x) = p(x)$ $(x \in I)$, and let μ be a real function, continuous on I and such that

$$|\mu(x)| = e^{P(x)} \quad (x \in I).$$

Since μ is non-zero on I, it follows from the intermediate value theorem that either

$$\mu(x) = e^{P(x)} \ (x \in I) \quad \text{or} \quad \mu(x) = -e^{P(x)} \ (x \in I).$$

In both cases, $\mu'(x) = \mu(x)p(x)$ so that

$$\frac{d}{dx}(\mu(x)y) = \mu(x)\frac{dy}{dx} + \mu'(x)y = \mu(x)\frac{dy}{dx} + \mu(x)p(x)y = \mu(x)\left(\frac{dy}{dx} + p(x)y\right).$$

Hence $\mu(x)$ is an integrating factor for (17.4.1) and

$$\frac{d}{dx}(\mu(x)y) = \mu(x)q(x).$$

Integrating gives

$$\mu(x)y = \int \mu(x)q(x)\,dx,$$

which is the general solution.

Example 17.4.2 Find the general solutions of

(a) $\dfrac{dy}{dx} + 2xy = e^{-x^2}$,   (b) $(9 - x^2)\dfrac{dy}{dx} + 6y = (9 - x^2)^2$,

(c) $\dfrac{dy}{dx} - y\tan x = \sin x \cos x$.

Solution We shall use the above notation.

(a) Here, $p(x) = 2x$ and $q(x) = e^{-x^2}$. Then

$$P(x) = \int p(x)\,dx = \int 2x\,dx = x^2 + C = x^2 \quad \text{(taking } C = 0\text{)}$$

and

$$e^{P(x)} = e^{x^2} = |e^{x^2}|.$$

So we can take $\mu(x) = e^{x^2}$. Hence the required general solution is

$$\mu(x)y = \int \mu(x)q(x)\,dx,$$

i.e.

$$e^{x^2}y = \int e^{x^2}e^{-x^2}\,dx = \int dx = x + C,$$

i.e.

$$y = e^{-x^2}(x + C),$$

where C is a parameter.

(b) Here,

$$\frac{dy}{dx} + \frac{6}{9 - x^2}y = 9 - x^2$$

so that $p(x) = 6/(9 - x^2)$ and $q(x) = 9 - x^2$. Then

$$P(x) = \int p(x)\,dx = \int \frac{6}{9 - x^2}\,dx = \int \left[\frac{1}{3 + x} + \frac{1}{3 - x}\right] dx$$

$$= \log|3 + x| - \log|3 - x| + C$$

$$= \log\left|\frac{3 + x}{3 - x}\right| \quad \text{(taking } C = 0\text{)}$$

and

$$e^{P(x)} = \exp\left(\log\left|\frac{3 + x}{3 - x}\right|\right) = \left|\frac{3 + x}{3 - x}\right|.$$

So we can take

$$\mu(x) = \frac{3 + x}{3 - x}.$$

Hence, the required general solution is

$$\mu(x)y = \int \mu(x)q(x)\,dx,$$

i.e.

$$\left(\frac{3 + x}{3 - x}\right)y = \int \left(\frac{3 + x}{3 - x}\right)(9 - x^2)\,dx$$

$$= \int (3 + x)^2\,dx = \frac{1}{3}(3 + x)^3 + C,$$

i.e.

$$y = \frac{1}{3}(3+x)^2(3-x) + C\left(\frac{3-x}{3+x}\right),$$

where C is a parameter.

(c) Here, $p(x) = -\tan x$ so we can take

$$P(x) = \log|\cos x| \qquad \text{and} \qquad \mu(x) = \cos x.$$

Hence, the required general solution is

$$(\cos x)y \int (\cos x)(\sin x \cos x)\, dx$$

$$\int \cos^2 x \sin x\, dx = -\frac{1}{3}\cos^3 x + C,$$

i.e.

$$y = -\frac{1}{3}\cos^2 x + C \sec x,$$

where C is a parameter. $\qquad\qquad\qquad\qquad\qquad\qquad\qquad\qquad\qquad\qquad\square$

17.5 Second Order Linear Equations

Definitions 17.5.1 The definition of a general n^{th} order linear ODE was given in Section 17.4. Here we restrict our attention to second order linear equations with *constant coefficients*. Such an ODE can be written in the form

$$ay'' + by' + cy = f(x), \tag{17.5.1}$$

where $a, b, c \in \mathbb{R}$ with $a \neq 0$ and f is a real function. In this section only we shall refer to (17.5.1) as the *general equation*.

To solve the general equation we begin by solving the corresponding *homogeneous equation*,

$$ay'' + by' + cy = 0. \tag{17.5.2}$$

The general solution of the homogeneous equation is called the *complementary function*. It depends on the roots of the quadric equation

$$am^2 + bm + c = 0 \tag{17.5.3}$$

which is called the *auxiliary equation*. There are three possibilities: two real roots, one real root or two conjugate complex roots.

Remark In the proof of the next theorem we shall assume that complex numbers and functions behave like real numbers and functions. In the context of the theorem this *is* a valid assumption but we shall not attempt to justify it.

Theorem 17.5.2 (1) *Let p and q be distinct real roots of the auxiliary equation. Then the complementary function is*

$$y = Ae^{px} + Be^{qx}.$$

(2) *Let r be the only root of the auxiliary equation. Then the complementary function is*

$$y = (A + Bx)e^{rx}.$$

(3) *Let $s + it$, $s - it$ $(s, t \in \mathbb{R})$ be distinct complex roots of the auxiliary equation. Then the complementary function is*

$$y = e^{sx}(A \cos tx + B \sin tx).$$

In each case, A and B are parameters.

Proof (1) Since p and q are the roots of the auxiliary equation

$$am^2 + bm + c = 0,$$

it follows that $p + q = -b/a$ and $pq = c/a$ (see Theorem 3.3.3). So the homogeneous equation

$$ay'' + by' + cy = 0$$

is equivalent to

$$y'' - (p + q)y' + pqy = 0,$$

i.e.

$$(y' - qy)' - p(y' - qy) = 0.$$

Let $y' - qy = u$. Then

$$u' - pu = 0.$$

This is a separable equation with general solution $u = De^{px}$, where D is a parameter (see Example 17.2.2 (c)). Hence,

$$y' - qy = De^{px}.$$

This is a linear equation. We can take $\mu(x) = e^{-qx}$ as an integrating factor to get the general solution

$$e^{-qx}y = \int De^{px}e^{-qx}\,dx, \tag{17.5.4}$$

i.e.

$$e^{-qx}y = \int De^{(p-q)x}\,dx = \frac{D}{p-q}e^{(p-q)x} + B,$$

i.e.

$$y = Ae^{px} + Be^{qx},$$

where A and B are parameters.

(2) The argument for (1) applies with $p = q = r$ as far as equation (17.5.4) which becomes

$$e^{-rx}y = \int De^{rx}e^{-rx}\,dx = \int D\,dx = Dx + E,$$

i.e.
$$y = (Dx + E)e^{rx},$$

where D and E are parameters.

(3) The argument for (1) applies with $p = s + it$ and $q = s - it$ giving

$$y = Ae^{(s+it)x} + Be^{(s-it)x} = Ae^{sx}e^{itx} + Be^{sx}e^{-itx},$$

i.e.
$$y = Ae^{sx}(\cos tx + i\sin tx) + Be^{sx}(\cos tx - i\sin tx),$$

i.e.
$$y = e^{sx}\big((A + B)\cos tx + i(A - B)\sin tx\big),$$

i.e.
$$y = e^{sx}(M\cos tx + N\sin tx),$$

where M and N are parameters.   □

Example 17.5.3 Find the general solution of each of the following homogeneous equations.

(a) $y'' + y' - 6y = 0,$ (b) $9y'' - 12y' + 4y = 0,$

(c) $y'' + y' + y = 0,$ (d) $y'' = 7y',$

(e) $y'' - a^2y = 0,$ (f) $y'' + a^2y = 0,$

where $a \in \mathbb{R}$, $a \neq 0$.

Solution (a) The auxiliary equation, $m^2 + m - 6 = 0$, has roots -3 and 2. So the general solution is
$$y = Ae^{-3x} + Be^{2x},$$

where A and B are parameters.

(b) The auxiliary equation, $9m^2 - 12 + 4 = 0$, has root $2/3$ (only). So the general solution is
$$y = (A + Bx)e^{2x/3},$$

where A and B are parameters.

(c) The auxiliary equation, $m^2 + m + 1 = 0$, has roots $-(1/2) \pm i(\sqrt{3}/2)$. So the general solution is
$$y = e^{-x/2}\left(A\cos\frac{\sqrt{3}x}{2} + B\sin\frac{\sqrt{3}x}{2}\right),$$

where A and B are parameters.

(d) The auxiliary equation, $m^2 - 7m = 0$, has roots 0 and 7. So the general solution is
$$y = A + Be^{7x},$$

where A and B are parameters.

(e) The auxiliary equation, $m^2 - a^2 = 0$, has roots $\pm a$. So the general solution is
$$y = Ae^{ax} + Be^{-ax},$$

where A and B are parameters.

(f) The auxiliary equation, $m^2 + a^2 = 0$, has roots $\pm ia$. So the general solution is

$$y = A \cos ax + B \sin ax,$$

where A and B are parameters. $\qquad\qquad\qquad\qquad\qquad\qquad\qquad\qquad$ □

Remark It is sometimes useful to express $A \cos tx + B \sin tx$ in the form $R \cos(tx + \theta)$ or $R \sin(tx + \theta)$ (*cf.* Example 5.3.4). Thus, the general solutions of the equations in Example 17.5.3 (c) and (f) can be written

$$y = Re^{-x/2} \cos\left(\frac{\sqrt{3}x}{2} + \theta\right) \quad \text{and} \quad y = R \sin(ax + \theta),$$

respectively, where R and θ are parameters.

Example 17.5.4 Use the substitution $x = e^t$ to find the general solution of the equation

$$x^2 \frac{d^2y}{dx^2} + x\frac{dy}{dx} - 4y = 0 \qquad (x > 0).$$

Solution Let $x = e^t$. Then

$$\frac{dy}{dt} = \frac{dy}{dx}\frac{dx}{dt} = \frac{dy}{dx}e^t = x\frac{dy}{dx}$$

and

$$\frac{d^2y}{dt^2} = \frac{d}{dt}\left(\frac{dy}{dt}\right) = x\frac{d}{dx}\left(x\frac{dy}{dx}\right) = x\left(x\frac{d^2y}{dx^2} + \frac{dy}{dx}\right) = x^2\frac{d^2y}{dx^2} + x\frac{dy}{dx}.$$

Hence the given ODE becomes

$$\frac{d^2y}{dt^2} - 4y = 0.$$

The auxiliary equation, $m^2 - 4 = 0$, has roots -2 and 2. So the required general solution is

$$y = Ae^{-2t} + Be^{2t}, \qquad\qquad \textit{i.e.} \ y = Ax^{-2} + Bx^2,$$

where A and B are parameters. $\qquad\qquad\qquad\qquad\qquad\qquad\qquad\qquad$ □

Note 17.5.5 If we put $A = 1$, $B = 0$ and then $A = 0$, $B = 1$ in the conclusion of Theorem 17.5.2, in each of the three cases we obtain two particular solutions of the homogeneous equation. Suppose they are $y = y_1 \ [= y_1(x)]$ and $y = y_2 \ [= y_2(x)]$. Then the complementary function is

$$y = Ay_1 + By_2,$$

where A and B are parameters.

Problem 17.5.6 Let $y = u_1 \ [= u_1(x)]$ and $y = u_2 \ [= u_2(x)]$ be solutions of the homogeneous equation

$$ay'' + by' + cy = 0$$

with the property that, if $\alpha, \beta \in \mathbb{R}$ and $\alpha u_1 + \beta u_2$ is the zero function then $\alpha = \beta = 0$. Two functions with this property are said to be *linearly independent*. Prove that the complementary function can be written in the form

$$y = Pu_1 + Qu_2,$$

where P and Q are parameters.

Example 17.5.7 With the notation of Theorem 17.5.2, establish the following by substitution in the homogeneous equation $ay'' + by' + cy = 0$.

(a) In Case (1), $y = e^{px}$ is a solution.

(b) In Case (2), $y = xe^{rx}$ is a solution.

(c) In Case (3), $y = e^{sx}\cos tx$ and $y = e^{sx}\sin tx$ are solutions.

Solution (a) Here, $ap^2 + bp + c = 0$. So

$$a\frac{d^2}{dx^2}(e^{px}) + b\frac{d}{dx}(e^{px}) + ce^{px} = ap^2 e^{px} + bpe^{px} + ce^{px}$$

$$= e^{px}(ap^2 + bp + c) = 0.$$

Hence, $y = e^{px}$ is a solution of the homogeneous equation.

(b) Here, $ar^2 + br + c = 0$ and $r = -b/(2a)$ so that $2ar + b = 0$. So

$$a\frac{d^2}{dx^2}(e^{rx}) + b\frac{d}{dx}(e^{rx}) + ce^{rx} = a(xr^2 e^{rx} + 2re^{rx}) + b(xre^{rx} + e^{rx}) + cxe^{rx}$$

$$= xe^{rx}(ar^2 + br + c) + e^{rx}(2ar + b) = 0.$$

Hence, $y = xe^{rx}$ is a solution of the homogeneous equation.

(c) Here,

$$a(s + it)^2 + b(s + it) + c = 0, \qquad i.e. \ a(s^2 - t^2 + 2ist) + b(s + it) + c = 0,$$

so that, equating real and imaginary parts,

$$a(s^2 - t^2) + bs + c = 0 \qquad \text{and} \qquad 2ast + bt = 0.$$

So

$$a\frac{d^2}{dx^2}(e^{sx}\cos tx) + b\frac{d}{dx}(e^{rx}\cos tx) + ce^{rx}\cos tx$$

$$= ae^{sx}[(s^2 - t^2)\cos tx - 2st\sin tx] + be^{sx}(s\cos tx - t\sin tx) + ce^{sx}\cos tx$$

$$= e^{sx}\cos tx\,[a(s^2 - t^2) + bs + c] - e^{sx}\sin tx\,(2ast + bt) = 0.$$

Hence, $y = e^{sx}\cos tx$ is a solution of the homogeneous equation. The verification for $y = e^{sx}\sin tx$ is similar and is left to the reader. $\qquad\square$

We now return to the general equation

$$ay'' + by' + cy = f(x).$$

As we shall see, the complementary function, *i.e.* the general solution when f is the zero function, plays a part in the general solution of the general equation.

Lemma 17.5.8 *Let* $y = u$ $[= u(x)]$ *be a solution of*

$$ay'' + by' + cy = f(x),$$

let $y = v$ $[= v(x)]$ *be a solution of*

$$ay'' + by' + cy = g(x)$$

and let $\alpha, \beta \in \mathbb{R}$. *Then* $y = \alpha u + \beta v$ *is a solution of*

$$ay'' + by' + cy = \alpha f(x) + \beta g(x).$$

Proof Let $y = \alpha u + \beta v$. Then

$$a(\alpha u + \beta v)'' + b(\alpha u + \beta v)' + c(\alpha u + \beta v)$$
$$= a(\alpha u'' + \beta v'') + b(\alpha u' + \beta v') + c(\alpha u + \beta v)$$
$$= \alpha(au'' + bu' + cu) + \beta(av'' + bv' + cv)$$
$$= \alpha f(x) + \beta g(x)$$

and the proof is complete. □

Theorem 17.5.9 *Let* $y = y_c$ $[= y_c(x)]$ *be the complementary function and let* $y = y_p$ $[= y_p(x)]$ *be a particular solution of the general equation*

$$ay'' + by' + c = f(x).$$

Then the general solution is

$$y = y_c + y_p.$$

Proof Recall (Note 17.5.5) that we can write $y_c = Ay_1 + By_2$, where A and B are parameters. For any values of A and B, $y = y_c$ is a solution of the homogeneous equation. So, by Lemma 17.5.8, $y = y_c + y_p$ is a solution of the general equation. Conversely, let $y = y_q$ be a particular solution of the general equation. Then, using Lemma 17.5.8 again, $y = y_q - y_p$ is a solution of the homogeneous equation. So $y_q - y_p = Ay_1 + By_2$, *i.e.* $y_q = y_c + y_p$, for some values of A and B. This completes the proof. □

Remarks (1) The conclusion of Theorem 17.5.9 is often written

$$GS = CF + PI,$$

where GS is the general solution, CF is the complementary function and PI is a particular integral.

(2) Since the complementary function involves two parameters, the general solution of the general equation also involves two parameters.

We now restrict our attention to solving the general equation

$$ay'' + by' + cy = f(x)$$

for certain types of function f. The problem is to find a particular integral.

Case 1. If $f(x) = he^{kx}$, where $h, k \in \mathbb{R}$, then try, as a particular integral,

$$\begin{cases} He^{kx} & \text{if } k \text{ is not a root of the auxiliary equation,} \\ Hxe^{kx} & \text{if } k \text{ is one of two roots of the auxiliary equation,} \\ Hx^2 e^{kx} & \text{if } k \text{ is the only root of the auxiliary equation.} \end{cases}$$

Case 2. If $f(x) = a_n x^n + a_{n-1} x^{n-1} + \cdots + a_1 x + a_0$, where $a_n, a_{n-1}, \ldots, a_1, a_0 \in \mathbb{R}$, then try, as a particular integral,

$$\begin{cases} P(x) = A_n x^n + A_{n-1} x^{n-1} + \cdots + A_1 x + A_0 \\ \qquad \text{if } 0 \text{ is not a root of the auxiliary equation,} \\ xP(x) & \text{if } 0 \text{ is one of two roots of the auxiliary equation,} \\ x^2 P(x) & \text{if } 0 \text{ is the only root of the auxiliary equation.} \end{cases}$$

Case 3. If $f(x) = u \cos vx + w \sin vx$, where $u, v, w \in \mathbb{R}$, then try, as a particular integral,

$$\begin{cases} Q(x) = U \cos vx + W \sin vx \\ \qquad \text{if } iv \text{ is not a root of the auxiliary equation,} \\ xQ(x) & \text{if } iv \text{ is a root of the auxiliary equation.} \end{cases}$$

The constants H, A_n, A_{n-1}, $\ldots$, A_1, A_0, U and W are found by substituting the proposed particular integral in the general equation.

Remark Multiplying by x or x^2 in the above cases is necessary to avoid solutions of the homogeneous equation.

In the next Example we make use of two results on equating coefficients: Theorems 3.1.7 and 5.2.11. We also use the fact that if $He^u = Ke^u$ for any $u \in \mathbb{R}$ then, since $e^u \neq 0$, $H = K$.

Example 17.5.10 Find the general solution of each of the following equations. Find also the particular solution when additional conditions are given.

(a) $y'' - 6y' + 9y = -e^{5x}$,

(b) $y'' + 5y' + 6y = 3e^{-2x}$,

(c) $y'' + 3y' + 2y = 2x^2 + 1$ $(y(0) = 1, \ h'(0) = 0)$,

(d) $y'' + y' = x$,

(e) $4y'' + y = \cos(x/2) + 3\sin(x/2)$ $(y(0) = 3\pi/4, \ y(\pi) = 0)$.

Solution (a) The auxiliary equation, $m^2 - 6m + 9 = 0$, has root 3 (only). So the complementary function is

$$y = (A + Bx)e^{3x}.$$

For a particular integral we try $y = He^{5x}$. Then $y' = 5He^{5x}$ and $y'' = 25He^{5x}$. Substituting in the given ODE,

$$25He^{5x} - 6(5He^{5x}) + 9He^{5x} = -e^{5x}, \qquad \textit{i.e. } 4He^{5x} = -e^{5x}.$$

Hence $H = -1/4$ and the required general solution is

$$y = (A + Bx)e^{3x} - \frac{e^{5x}}{4},$$

where A and B are parameters.

(b) The auxiliary equation, $m^2 + 5m + 6 = 0$, has roots -3 and -2. So the complementary function is

$$y = Ae^{-3x} + Be^{-2x}.$$

For a particular integral we try $y = Hxe^{-2x}$. Then

$$y' = H(-2xe^{-2x} + e^{-2x}) \qquad \text{and} \qquad y'' = H(4xe^{-2x} - 4e^{-2x}).$$

Substituting in the given ODE,

$$H(4xe^{-2x} - 4e^{-2x}) + 5H(-2xe^{-2x} + e^{-2x}) + 6Hxe^{-2x} = 3e^{-2x},$$

i.e.

$$He^{-2x} = 3e^{-2x}.$$

Hence $H = 3$ and the required general solution is

$$y = Ae^{-3x} + Be^{-2x} + 3xe^{-2x},$$

where A and B are parameters.

(c) The auxiliary equation, $m^2 + 3m + 2 = 0$, has roots -2 and -1, So the complementary function is

$$y = Ae^{-2x} + Be^{-x}.$$

For a particular integral we try

$$y = Px^2 + Qx + R.$$

Then

$$y' = 2Px + Q \qquad \text{and} \qquad y'' = 2P.$$

Substituting in the given ODE,

$$(2P) + 3(2Px + Q) + 2(Px^2 + Qx + R) = 2x^2 + 1,$$

i.e.

$$2Px^2 + (6P + 2Q)x + (2P + 3Q + 2R) = 2x^2 + 1,$$

so that, equating coefficients,

$$2P = 2, \qquad 6P + 2Q = 0, \qquad 2P + 3Q + 2R = 1.$$

Hence $P = 1$, $Q = -3$, $R = 4$ and the required general solution is

$$y = Ae^{-2x} + Be^{-x} + x^2 - 3x + 4,$$

where A and B are parameters.

Differentiating the general solution we have

$$y' = -2Ae^{-2x} - Be^{-x} + 2x - 3.$$

Using the additional conditions, $y(0) = 1$ and $y'(0) = 0$, gives

$$A + B + 4 = 1 \qquad \text{and} \qquad -2A - B - 3 = 0,$$

i.e.

$$A + B = -3 \qquad \text{and} \qquad 2A + B = -3.$$

Hence $A = 0$, $B = -3$ and the required particular solution is

$$y = -3e^{-x} + x^2 - 3x + 4.$$

(d) The auxiliary equation, $m^2 + m = 0$, has roots 0 and -1. So the complementary function is

$$y = A + Be^{-x}.$$

For a particular integral we try

$$y = x(Px + Q) = Px^2 + Qx.$$

Then

$$y' = 2Px + Q \qquad \text{and} \qquad y'' = 2P.$$

Substituting in the given ODE,

$$(2P) + (2Px + Q) = x, \qquad \text{i.e. } 2Px + (2P + Q) = x,$$

so that, equating coefficients, $2P = 1$ and $2P + Q = 0$. Hence $P = 1/2$, $Q = -1$ and the required general solution is

$$y = A + Be^{-x} + \frac{x^2}{2} - x,$$

where A and B are parameters.

(e) The auxiliary equation, $4m^2 + 1 = 0$, has roots $\pm i/2$. So the complementary function is

$$y = A\cos\frac{x}{2} + B\sin\frac{x}{2}.$$

For a particular integral we try

$$y = x\left(U\cos\frac{x}{2} + W\sin\frac{x}{2}\right).$$

Then

$$y' = U \cos \frac{x}{2} + W \sin \frac{x}{2} + x \left(-\frac{U}{2} \sin \frac{x}{2} + \frac{W}{2} \cos \frac{x}{2} \right)$$

and

$$y'' = -U \sin \frac{x}{2} + W \cos \frac{x}{2} + x \left(-\frac{U}{4} \cos \frac{x}{2} - \frac{W}{4} \sin \frac{x}{2} \right).$$

Substituting in the given ODE,

$$4 \left(-U \sin \frac{x}{2} + W \cos \frac{x}{2} \right) + 4x \left(-\frac{U}{4} \cos \frac{x}{2} - \frac{W}{4} \sin \frac{x}{2} \right)$$

$$+ x \left(U \cos \frac{x}{2} + W \sin \frac{x}{2} \right) = \cos \frac{x}{2} + 3 \sin \frac{x}{2},$$

i.e.

$$-4U \sin \frac{x}{2} + 4W \cos \frac{x}{2} = \cos \frac{x}{2} + 3 \sin \frac{x}{2},$$

so that, equating coefficients, $-4U = 3$ and $4W = 1$. Hence $U = -3/4$, $W = 1/4$ and the required general solution is

$$y = A \cos \frac{x}{2} + B \sin \frac{x}{2} + x \left(-\frac{3}{4} \cos \frac{x}{2} + \frac{1}{4} \sin \frac{x}{2} \right),$$

i.e.

$$y = \left(A - \frac{3x}{4} \right) \cos \frac{x}{2} + \left(B + \frac{x}{4} \right) \sin \frac{x}{2},$$

where A and B are parameters.

Using the additional conditions, $y(0) = 3\pi/4$ and $y(\pi) = 0$, gives $A = 3\pi/4$ and $B = -\pi/4$. Hence the required particular solution is

$$y = \frac{3}{4}(\pi - x) \cos \frac{x}{2} + \frac{1}{4}(x - \pi) \sin \frac{x}{2},$$

i.e.

$$y = \frac{1}{4}(\pi - x) \left(3 \cos \frac{x}{2} - \sin \frac{x}{2} \right). \qquad \square$$

Example 17.5.11 Find the general solution of the equation

$$y'' + 4y' + 5y = 8 \sin x + 25x. \tag{17.5.5}$$

Solution The auxiliary equation, $m^2 + 4m + 5 = 0$, has roots $-2 \pm i$. So the complementary function is

$$y = e^{-2x}(A \cos x + B \sin x).$$

We now require a particular integral. To find one, we shall consider separately the two terms $8 \sin x$ and $25x$ (but see Note 17.5.12). If $y = y_1 [= y_1(x)]$ is a particular integral for

$$y'' + 4y' + 5y = 8 \sin x \tag{17.5.6}$$

and $y = y_2 [= y_2(x)]$ is a particular integral for

$$y'' + 4y' + 5y = 25x \tag{17.5.7}$$

then, by Lemma 17.5.8, $y = y_1 + y_2$ is a particular integral for the given ODE (17.5.5).

For y_1 we try
$$y = U \cos x + W \sin x.$$

Then
$$y' = -U \sin x + W \cos x \qquad \text{and} \qquad y'' = -U \cos x - W \sin x.$$

Substituting in equation (17.5.6),

$$(-U \cos x - W \sin x) + 4(-U \sin x + W \cos x)$$
$$+ 5(U \cos x + W \sin x) = 8 \sin x,$$

i.e.

$$(4U + 4W) \cos x + (-4U + 4W) \sin x = 8 \sin x,$$

so that, equating coefficients,

$$U + W = 0 \qquad \text{and} \qquad -U + W = 2.$$

Hence $U = -1$, $W = 1$ and we may take

$$y_1 = -\cos x + \sin x.$$

For y_2 we try $y = Px + Q$. Then $y' = P$ and $y'' = 0$. Substituting in equation (17.5.7) gives

$$0 + 4P + 5(Px + Q) = 25x, \qquad \text{i.e. } 5Px + (4P + 5Q) = 25x,$$

so that, equating coefficients, $5P = 25$ and $4P + 5Q = 0$. Hence $P = 5$, $Q = -4$ and we may take
$$y_2 = 5x - 4.$$

The general solution of the original ODE (17.5.5) is therefore

$$y = e^{-2x}(A \cos x + B \sin x) - \cos x + \sin x + 5x - 4,$$

where A and B are parameters. $\square$

Note 17.5.12 To find a particular integral for the differential equation in Example 17.5.11 we could combine the two stages in the solution and try

$$y = U \cos x + W \sin x + Px + Q.$$

Then substituting in the equation and equating coefficients leads directly to $y_1 + y_2$.

17.X Exercises

1. Find the general solution of each of the following separable equations. Find also the particular solution when additional conditions are given.

(a) $\dfrac{dy}{dx} = 3x^2y^2$ ($y = -1/3$ when $x = 0$),

(b) $\dfrac{dy}{dx} = e^{2x}(4 + y^2)$,

(c) $\dfrac{dy}{dx} = \operatorname{sech} y$ ($y = 1$ when $x = e$),

(d) $\dfrac{dy}{dx} = e^{-x}\tan y$,

(e) $\dfrac{dy}{dx} = \dfrac{y + 1/y}{x + 1/x}$ ($y = 7$ when $x = 3$).

2. Find the general solution of each of the following homogeneous first order equations. Find also the particular solution when additional conditions are given.

(a) $\dfrac{dy}{dx} = \dfrac{x^4 + y^4}{xy^3}$,

(b) $\dfrac{dy}{dx} = \dfrac{y}{x} + \sec\dfrac{y}{x}$ ($y = \pi/2$ when $x = 1$),

(c) $x^2\dfrac{dy}{dx} = x^2 + xy + y^2$,

(d) $x\dfrac{dy}{dx} = y + \sqrt{3x^2 + y^2}$ ($y = 0$ when $x = -\sqrt{3}$),

(e) $y^2\dfrac{dy}{dx} = x^2 - xy + y^2$.

3. Find the general solution of each of the following first order linear equations. Find also the particular solution of (e) which satisfies the given additional conditions.

(a) $\dfrac{dy}{dx} + \dfrac{3}{x}y = 5 + x^2$,

(b) $\dfrac{dy}{dx} - 3y = e^{5x}$,

(c) $\dfrac{dy}{dx} + 2xy + x = \exp(-x^2)$,

(d) $\dfrac{dy}{dx} = y\tan x + \sin x$,

(e) $(x^4 + 3)\dfrac{dy}{dx} + 4x^3y = x^3e^{-x}$ ($y = -2$ when $x = 0$).

4. Find the general solution of each of the following first order equations.

(a) $\dfrac{dy}{dx} = \dfrac{3x - y}{x - 3y}$,

(b) $x\dfrac{dy}{dx} + (x + 2)y - 1 = 0$,

(c) $\dfrac{dy}{dx} = x^2y^2 + x^2 + y^2 + 1$,

(d) $\dfrac{dy}{dx} = \dfrac{y}{x}(1 + \log y - \log x)$,

(e) $\dfrac{dy}{dx} - y\cos x = e^{\sin x} - y$.

5. Find the general solution of each of the following second order equations. Find also the particular solution when additional conditions are given.

(a) $y'' - 3y' + 2y = 12e^{-x}$,

(b) $y'' + 2y' + 5y = 80e^{3x}$,

(c) $y'' - 6y' + 9y = 18x^2 + 3x - 5$,

(d) $y'' - y = x^2 + 1$,

(e) $y'' - 3y' - 10y = 14e^{-2x}$,

(f) $y'' + 4y' + 5y = 32\sin x$,

(g) $4y'' + 12y' + 9y = 144e^{-3x/2}$,

(h) $y'' + y = 2\sin x$,

(i) $y'' + 4y' + 4y = e^{-3x}$ $(y(0) = y'(0) = 0)$,

(j) $y'' - y' - 6y = 6x^2 - 4x + 3$ $(y(0) = 0, \ y'(0) = 9)$,

(k) $y'' + 8y' + 15y = 48\cos x - 16\sin x$ $(y(0) = 3, \ y(\pi/2) = 0)$,

(l) $y'' + 6y' + 9y = 32e^x + 100\cos x$ $(y(0) = 10, \ y(\pi) = -8)$.

Chapter 18

Sequences and Series

A *sequence* is a list of *terms* $x_1, x_2, x_3, \ldots$. Consider the sequence with $x_n = 1/2^n$, *i.e.*

$$\frac{1}{2}, \frac{1}{4}, \frac{1}{8}, \frac{1}{16}, \frac{1}{32}, \ldots$$

As n increases, the terms approach 0, so we say that 0 is the *limit* of the sequence. We can form a new sequence from this with terms s_n equal to the sum of the first n terms,

$$\frac{1}{2}, \frac{3}{4}, \frac{7}{8}, \frac{15}{16}, \frac{31}{32}, \ldots$$

The limit of this new sequence is 1 and we say that the *series*

$$\frac{1}{2} + \frac{1}{4} + \frac{1}{8} + \frac{1}{16} + \frac{1}{32} + \cdots$$

has *sum* 1.

In the first two sections, we consider *real sequences*, *i.e.* sequences whose terms are real numbers. A familiarity with sequences is essential when we discuss iterative processes in Chapter 19.

As the above example indicates, the sum of a series can be defined in terms of a sequence (of sums of the first n terms). In the third section we study series. In particular we establish several theorems which allow us to decide when a series has a sum.

In the last two sections we consider *power series*. Such a series is of the form $a_0 + a_1 x + a_2 x^2 + \cdots$. The value of the sum, and indeed its existence, will depend on the choice of x. The sum is clearly a function of x (whose domain is the set of x for which the sum exists). Taylor's theorem discusses the question of whether a given function can be expressed as a power series. Such series can be used to find good numerical approximations to function values.

18.1   Real Sequences

Definitions 18.1.1 Let A be a non-empty set. A function $x : \mathbb{N} \to A$ is called a *sequence*. When $A \subseteq \mathbb{R}$, x is a *real* sequence. In this chapter, all sequences will be real. For $n \in \mathbb{N}$, we call $x(n)$ the nth *term* of the sequence and usually denote it by x_n.

We can think of a sequence as a list of real numbers, *indexed* by the set $\mathbb{N}$. For example, define $a : \mathbb{N} \to \mathbb{R}$ by

$$a(n) = a_n = \frac{n-1}{n} \quad (n \in \mathbb{N}),$$

so that

$$a_1 = 0, \quad a_2 = \frac{1}{2}, \quad a_3 = \frac{2}{3}, \quad a_4 = \frac{3}{4}, \quad a_5 = \frac{4}{5}, \quad \dots .$$

We can write this sequence in various ways:

$$\{a_1, a_2, a_3, \dots\} \quad \text{or} \quad \left\{0, \frac{1}{2}, \frac{2}{3}, \dots\right\} \quad \text{or} \quad \{a_n\}_{n \in \mathbb{N}} \quad \text{or} \quad \left\{\frac{n-1}{n}\right\}_{n \in \mathbb{N}},$$

or simply

$$\{a_n\} \quad \text{or} \quad \left\{\frac{n-1}{n}\right\}.$$

 Hazard The above use of set notation for sequences is ambiguous. A sequence $\{x_n\}_{n \in \mathbb{N}}$ is not the same as the set of terms $\{x_n : n \in \mathbb{N}\}$. For example, define

$$b_n = (-1)^n \quad (n \in \mathbb{N}).$$

Then the sequence $\{b_n\}_{n \in \mathbb{N}} = \{-1, 1, -1, 1, -1, \dots\}$ has infinitely many terms while the set $\{b_n : n \in \mathbb{N}\} = \{-1, 1\}$ has just two elements.

In the same vein, care is needed when more than one variable is present. For example,

$$\left\{\frac{m}{n}\right\}_{n \in \mathbb{N}} = \left\{m, \frac{m}{2}, \frac{m}{3}, \dots\right\} \quad \text{while} \quad \left\{\frac{m}{n}\right\}_{m \in \mathbb{N}} = \left\{\frac{1}{n}, \frac{2}{n}, \frac{3}{n}, \dots\right\}.$$

We can represent a sequence diagrammatically as points on the real line. The sequences $\{a_n\}$ and $\{b_n\}$, defined above, are shown in Figure 18.1.1. Alternatively, since a sequence is a function, it has a graph. The graphs of $\{a_n\}$ and $\{b_n\}$ are drawn in Figure 18.1.2.

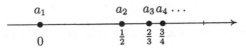

Figure 18.1.1

Note 18.1.2 A sequence variable does not have to 'start' at 1. It is sometimes convenient to extend the definition of sequence to allow functions of the form $x : \mathbb{N} \cup \{0\} \to \mathbb{R}$. Then the sequence x can be written $\{x_0, x_1, x_2, \dots\}$. For example, defining

$$c_r = \frac{r}{r+1} \quad (r \in \mathbb{N} \cup \{0\})$$

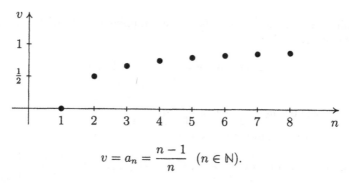

$$v = a_n = \frac{n-1}{n} \quad (n \in \mathbb{N}).$$

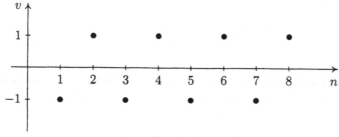

$$v = b_n = (-1)^n \quad (n \in \mathbb{N}).$$

Figure 18.1.2

gives the sequence

$$\{c_0, c_1, c_2, \ldots\} = \left\{0, \frac{1}{2}, \frac{2}{3}, \ldots\right\}.$$

When no domain is specified, assume it to be $\mathbb{N}$.

From Definition 2.4.3, a sequence $\{x_n\}$ is bounded if the set of terms $\{x_n : n \in \mathbb{N}\}$ is bounded, *i.e.* if there exist $A, B \in \mathbb{R}$ such that

$$A \leq x_n \leq B \quad (n \in \mathbb{N})$$

or, equivalently, if there exists $K \geq 0$ such that $|x_n| \leq K \ (n \in \mathbb{N})$.

Example 18.1.3 Find lower and upper bounds for the sequence $\{x_n\}$ defined by

$$x_n = \frac{3n^2 - 4}{2n^2 - n}.$$

Solution We have, $x_1 = -1$ and, for $n \geq 2$, both $3n^2 - 4$ and $2n^2 - n$ are positive so that $x_n > 0$. Hence, -1 is a lower bound for the sequence.
    For $n \in \mathbb{N}$,

$$x_n = \frac{3n^2 - 4}{2n^2 - n} < \frac{3n^2}{2n^2 - n^2} = 3.$$

Hence, 3 is an upper bound for the sequence.   □

Problem 18.1.4 Show that $71/45$ is the least possible upper bound for the sequence $\{x_n\}$ of Example 18.1.3.

From Definition 2.6.1, a sequence $\{x_n\}$ is increasing if, for $m, n \in \mathbb{N}$,

$$m < n \implies x_m \leq x_n.$$

To prove that $\{x_n\}$ is increasing it is enough to show that

$$x_n \leq x_{n+1} \quad (n \in \mathbb{N}),$$

since then, for $m, n \in \mathbb{N}$,

$$m < n \implies x_m \leq x_{m+1} \leq \cdots \leq x_n.$$

Similarly, to prove that $\{x_n\}$ is decreasing it is enough to show that

$$x_n \geq x_{n+1} \quad (n \in \mathbb{N}).$$

Again using Definition 2.6.1 and arguing as above, $\{x_n\}$ is strictly increasing if and only if

$$x_n < x_{n+1} \quad (n \in \mathbb{N})$$

and $\{x_n\}$ is strictly decreasing if and only if

$$x_n > x_{n+1} \quad (n \in \mathbb{N}).$$

Example 18.1.5 Show that the sequence $\{x_n\}$, defined by

$$x_n = \frac{n^2 - 2n}{4n - 3} \quad (n \in \mathbb{N}),$$

is strictly increasing.

Solution For $n \in \mathbb{N}$,

$$x_{n+1} - x_n = \frac{(n+1)^2 - 2(n+1)}{4(n+1) - 3} - \frac{n^2 - 2n}{4n - 3} = \frac{n^2 - 1}{4n + 1} - \frac{n^2 - 2n}{4n - 3}$$

$$= \frac{4n^2 - 2n + 3}{(4n + 1)(4n - 3)} = \frac{4(n - 1/4)^2 + 11/4}{(4n + 1)(4n - 3)} > 0.$$

Hence $x_n < x_{n+1}$ $(n \in \mathbb{N})$ so that $\{x_n\}$ is strictly increasing. $\quad\square$

18.2   Sequence Limits

As with function limits in Chapter 7, our approach to sequence limits in this chapter is informal. We therefore confine ourselves to stating the basic properties and using them to develop further results. A formal definition of sequence limit together with a selection of rigorous proofs can be found in Appendix C, Section C.4.

Definitions 18.2.1 Let $\{x_n\}$ be a sequence and let $L \in \mathbb{R}$. The *limit of* $\{x_n\}$ exists and equals L, and we write

$$\lim_{n \to \infty} x_n = L \quad \text{or} \quad x_n \to L \text{ as } n \to \infty$$

or simply $x_n \to L$, if x_n approaches L as n increases. By this we mean that the distance from x_n to L can be made as small as we wish by choosing n sufficiently large. A sequence with limit L is said to *converge (to L)*.

Note that the distance from x_n to L is $|x_n - L|$, but for $\{x_n\}$ to converge to L the terms $|x_n - L|$ do not have to form a decreasing sequence (see Example 18.2.2).

Example 18.2.2 Show that the sequences $\{a_n\}$ and $\{a_n'\}$, defined by

$$a_n = \frac{n-1}{n} \ (n \in \mathbb{N}), \qquad a_n' = \begin{cases} a_n & \text{if } n \text{ is odd,} \\ 1 & \text{if } n \text{ is even,} \end{cases}$$

both converge to 1.

Solution The graphs of the two sequences are shown in Figure 18.2.1. In each case the line $v = 1$ indicates the limit of the sequence.

The sequence $\{a_n\}$ converges to 1 since

$$|a_n - 1| = \left| \frac{n-1}{n} - 1 \right| = \frac{1}{n}$$

and this can be made as small as we wish by choosing n sufficiently large.

The sequence $\{a_n'\}$ converges to 1 since

$$|a_n' - 1| = \begin{cases} 1/n & \text{if } n \text{ is odd,} \\ 0 & \text{if } n \text{ is even} \end{cases} \leq \frac{1}{n}$$

and hence $|a_n' - 1|$ can be made as small as we wish by choosing n sufficiently large.

Note that, while the terms $|a_n - 1|$ *do* form a decreasing sequence, the terms $|a_n' - 1|$ do *not* (see the end of Definitions 18.2.1). $\qquad \square$

Theorem 18.2.3 *Let* $\{x_n\}$ *be a sequence. Then if* $\lim_{n \to \infty} x_n$ *exists, it is unique.*

Proof A proof is given in Appendix C, Theorem C.4.4. $\qquad \square$

Theorem 18.2.4 *Let* $K \in \mathbb{R}$ *and define*

$$x_n = K, \qquad y_n = \frac{1}{n} \qquad (n \in \mathbb{N}).$$

Then

$$x_n \to K \quad \text{and} \quad y_n \to 0 \quad \text{as } n \to \infty.$$

Proof A proof is given in Appendix C, Theorem C.4.5. $\qquad \square$

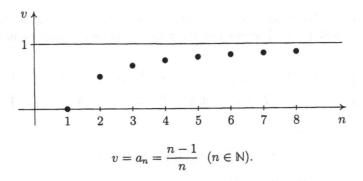

$$v = a_n = \frac{n-1}{n} \quad (n \in \mathbb{N}).$$

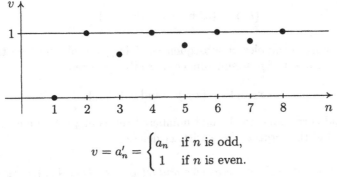

$$v = a'_n = \begin{cases} a_n & \text{if } n \text{ is odd,} \\ 1 & \text{if } n \text{ is even.} \end{cases}$$

Figure 18.2.1

Notes 18.2.5 (1)   The 'first few' terms of a sequence have no bearing on whether or not it converges. Nor do they affect the limit when it exists. Thus

$$x_n \to L \text{ and } x_n = y_n \ (n \geq N) \ \Rightarrow \ y_n \to L.$$

(2)   We say that a property P_n (which depends on a positive integer n) holds *eventually* if, for some $N \in \mathbb{N}$, P_n holds for all $n \geq N$. This might be written

$$P_n \text{ holds } (n \geq N)$$

without specifying N. Thus (1), above, could be stated: if two sequences eventually agree and one converges to a limit then the other converges to the same limit.

(3)   We do not regard $\pm\infty$ as possible limits of a sequence. However, it is convenient to write $x_n \to \infty$ as $n \to \infty$ to mean that x_n is eventually positive and $1/x_n \to 0$ as $n \to \infty$. Similarly, we write $x_n \to -\infty$ as $n \to \infty$ when x_n is eventually negative and $1/x_n \to 0$ as $n \to \infty$. In both cases we gloss over the possibility of $x_n = 0$.

Of course, not every sequence converges. To show that a given sequence does not converge it is often useful to use subsequences.

Definition 18.2.6 Let $\{x_n\}$ and $\{y_n\}$ be sequences. We call $\{y_n\}$ a *subsequence* of $\{x_n\}$ if there is a strictly increasing sequence $\{k_n\}$ of positive integers such that $y_n = x_{k_n} \ (n \in \mathbb{N})$.

Notes 18.2.7 (1) Informally, $\{y_n\}$ is a subsequence of $\{x_n\}$ if it can be formed by deleting terms from $\{x_n\}$. The order of the non-deleted terms must be maintained. For example,

$$\{x_n\} = \{4,\ 7,\ 2,\ 3,\ 9,\ 4,\ 5,\ 8,\ 1,\ 2,\ 0,\ 4,\ 6,\ \dots\ \},$$
$$\{y_n\} = \{\qquad 2,\qquad\quad 4,\qquad 8,\ 1,\qquad\quad 4,\qquad \dots\ \}.$$

In this case,

$$y_1 = x_3,\quad y_2 = x_6,\quad y_3 = x_8,\quad y_4 = x_9,\quad y_5 = x_{12},\quad \dots\ .$$

So, in the notation of Definition 18.2.6,

$$\{k_n\} = \{3,\ 6,\ 8,\ 9,\ 12,\ \dots\ \}.$$

(2) Three simple examples of subsequences of $\{x_n\}$ are obtained by taking k_n to be $2n - 1$, $2n$ and $n + 1$. These subsequences are then written

$$\{x_{2n-1}\},\quad \{x_{2n}\}\quad \text{and}\quad \{x_{n+1}\},$$

respectively, and correspond to the odd numbered terms of $\{x_n\}$, the even numbered terms of $\{x_n\}$, and the terms of $\{x_n\}$ with x_1 deleted.

Theorem 18.2.8 *Let $\{x_n\}$ be a sequence and let $L \in \mathbb{R}$. Then $\{x_n\}$ converges to L if and only if every subsequence of $\{x_n\}$ converges to L.*

Proof A proof is given in Appendix C, Theorem C.4.9.   □

Example 18.2.9 Let $b_n = (-1)^n$ $(n \in \mathbb{N})$. Show that the sequence $\{b_n\}$ does not converge.

Solution Suppose that $b_n \to L$. Then, by Theorem 18.2.8, the sequences $\{b_{2n-1}\}$ and $\{b_{2n}\}$ also converge to L. However,

$$b_{2n-1} = (-1)^{2n-1} = -1 \quad \text{and} \quad b_{2n} = (-1)^{2n} = 1 \quad (n \in \mathbb{N}).$$

So, by Theorem 18.2.4, $b_{2n-1} \to -1$ and $b_{2n} \to 1$. By the uniqueness of sequence limits (Theorem 18.2.3), this implies that $L = -1$ and $L = 1$ which is impossible. Hence, $\{b_n\}$ does not converge.   □

In the next theorem we gather together some basic properties of sequence limits.

Theorem 18.2.10 (Properties of Sequence Limits) *Let $\{x_n\}$, $\{y_n\}$ and $\{z_n\}$ be sequences and let $L, M \in \mathbb{R}$.*

(1) *Let $x_n \to L \neq 0$. Then x_n has the same sign as L $(n \geq N)$.*

(2) *Let $x_n \to L$ and $y_n \to M$. Then*

(a) *for $\lambda \in \mathbb{R}$, $\lambda x_n \to \lambda L$,*

(b) $x_n + y_n \to L + M$,

(c) $x_n y_n \to LM$

and, provided $M \neq 0$,

(d) $\dfrac{x_n}{y_n} \to \dfrac{L}{M}$.

(3) *Let $x_n \to L$ and let f be a real function, continuous at L. Then $f(x_n) \to f(L)$.*

In particular, $|x_n| \to |L|$.

(4) *Let $x_n \leq y_n$ ($n \geq N$), and let $x_n \to L$ and $y_n \to M$. Then $L \leq M$.*

(5) *(The Sandwich Principle) Let $x_n \leq y_n \leq z_n$ ($n \geq N$), and let $x_n \to L$ and $z_n \to L$. Then $y_n \to L$.*

In particular, if $0 \leq y_n \leq z_n$ ($n \geq N$) and $z_n \to 0$ then $y_n \to 0$.

(6) *We have*

$$x_n \to L \iff x_n - L \to 0 \iff |x_n - L| \to 0.$$

In particular, $x_n \to 0 \iff |x_n| \to 0$.

(7) *Let $r > 0$. If $|x_n - L| \geq r$ ($n \geq N$) then $x_n \nrightarrow L$. In particular, if $|x_n| \geq r$ ($n \geq N$) then $x_n \nrightarrow 0$.*

Proof Proofs of (2)(a), (2)(b), (3) and (5) can be found in Appendix C, Theorems C.4.6 and C.4.7.

Observe that, by (1), the condition $M \neq 0$ for (2)(d) implies that $y_n \neq 0$ ($n \geq N$), and we gloss over the possibility that y_n may be 0 for $n < N$.

Here we prove (6) and (7).

(6) ($\Rightarrow$) Let $x_n \to L$. By Theorem 18.2.4 and (2), $x_n - L \to L - L = 0$. Then (3) gives $|x_n - L| \to 0$.

(6) ($\Leftarrow$) Let $|x_n - L| \to 0$. By (2), $-|x_n - L| \to 0$. Since

$$-|x_n - L| \leq x_n - L \leq |x_n - L| \quad (n \in \mathbb{N}),$$

(5) gives $x_n - L \to 0$. Finally, Theorem 18.2.4 and (2) give $x_n = (x_n - L) + L \to 0 + L = L$, as required.

(7) Suppose $|x_n - L| \geq r$ ($n \geq N$) and $x_n \to L$. For $n \in \mathbb{N}$, let $a_n = r$ and $b_n = |x_n - L|$. Then $a_n \leq b_n$ ($n \geq N$) and, by Theorem 18.2.4 and (6), $a_n \to r$ and $b_n \to 0$. Hence, by (4), $r \leq 0$ which is a contradiction. $\square$

Problem 18.2.11 Deduce the general case in Theorem 18.2.10 (5) from the special case where $0 \leq y_n \leq z_n$ ($n \geq N$).

Example 18.2.12 Find $\lim\limits_{n \to \infty} x_n$ when x_n is defined, for $n \in \mathbb{N}$, by

(a) $\dfrac{3n^2 + 2n - 1}{n^2 - n + 1}$,   (b) $\sin\left(\dfrac{n\pi}{2n + 1}\right)$,   (c) $\dfrac{1}{2^n}$,   (d) $\dfrac{2^{2n+1}}{2^{2n} + 1}$.

Solution   (a) We first divide the numerator and denominator by n^2. For $n \in \mathbb{N}$,

$$x_n = \frac{3n^2 + 2n - 1}{n^2 - n + 1} = \frac{3 + (2/n) - (1/n^2)}{1 - (1/n) + (1/n^2)} = \frac{3 + 2(1/n) - (1/n)^2}{1 - (1/n) + (1/n)^2}.$$

Then, using Theorem 18.2.10 (2) and Theorem 18.2.4,

$$x_n \to \frac{3 + 2(0) - 0^2}{1 - 0 + 0^2} = 3 \quad \text{as} \quad n \to \infty.$$

(b)   Arguing as in (a), above,

$$\frac{n\pi}{2n + 1} = \frac{\pi}{2 + (1/n)} \to \frac{\pi}{2} \quad \text{as} \quad n \to \infty.$$

Then, since the sine function is continuous at $\pi/2$, Theorem 18.2.10 (3) applies to give

$$\sin\left(\frac{n\pi}{2n + 1}\right) \to \sin\left(\frac{\pi}{2}\right) = 1 \quad \text{as} \quad n \to \infty.$$

(c)   For $n \in \mathbb{N}$, $2^n = (1 + 1)^n = \sum_{r=0}^{n} \binom{n}{r} \geq \binom{n}{1} = n$ (see Example 4.3.6). Hence

$$0 \leq \frac{1}{2^n} \leq \frac{1}{n} \quad (n \in \mathbb{N}).$$

By Theorem 18.2.4, $1/n \to 0$, and the sandwich principle (Theorem 18.2.10 (5)) gives $1/2^n \to 0$.

(d)   Here we first divide the numerator and denominator by 2^{2n}, then use Theorems 18.2.4 and 18.2.10, and (c), above. For $n \in \mathbb{N}$,

$$\frac{2^{2n+1}}{2^{2n} + 1} = \frac{2^{2n+1}/2^{2n}}{2^{2n}/2^{2n} + 1/2^{2n}} = \frac{2}{1 + (1/2^n)^2} \to \frac{2}{1 + 0^2} = 2. \qquad \square$$

The next two theorems, the first of which generalises Example 18.2.12 (c), will be used later to prove results about series.

Theorem 18.2.13   *Let $x \in \mathbb{R}$, with $|x| < 1$. Then the sequence $\{x^n\}$ converges to 0.*

Proof   By Theorem 18.2.10 (3), since $|x^n| = |x|^n$, $x^n \to 0$ if and only if $|x|^n \to 0$. Thus we may assume, without loss, that $x \geq 0$. Then $0 \leq x < 1$ so that $y = 1 - x > 0$. Hence, for $n \in \mathbb{N}$,

$$1 = (x + y)^n = \sum_{r=0}^{n} \binom{n}{r} x^{n-r} y^r \geq \binom{n}{1} x^{n-1} y = nx^{n-1}y.$$

Thus, $x^{n-1} \leq \dfrac{1}{yn}$ so that

$$0 \leq x^n \leq \frac{A}{n} \quad (n \in \mathbb{N}),$$

where $A = x/y = x/(1 - x)$. By Theorems 18.2.4 and 18.2.10 (2), $A/n \to 0$. Hence, by the sandwich principle (Theorem 18.2.10 (5)), $x^n \to 0$ as required. $\qquad \square$

Theorem 18.2.14 *Let $x \in \mathbb{R}$. Then the sequence $\{x^n/n!\}$ converges to 0.*

Proof The strategy is the same as that in the proof of Theorem 18.2.13. Let $x \geq 0$. Choose $N \in \mathbb{N}$ with $N > x$. Then, for $k \geq N$, $x/k < 1$. Hence, for $n \geq N$,

$$\frac{x^n}{n!} = \frac{x}{n} \cdot \frac{x}{n-1} \cdots \frac{x}{N} \cdot \frac{x^{N-1}}{(N-1)!} \leq \frac{x}{n} \cdot \frac{x^{N-1}}{(N-1)!}.$$

Thus

$$0 \leq \frac{x^n}{n!} \leq \frac{A}{n} \quad (n \geq N),$$

where $A = x^N/(N-1)!$. The sandwich principle gives the result. $\qquad \square$

The last theorem in this section deals with the relationship between convergence and boundedness.

Theorem 18.2.15 (Including the Monotone Convergence Theorem) *Let $\{x_n\}$ be a sequence.*

(1) *If $\{x_n\}$ converges then $\{x_n\}$ is bounded.*

(2) *If $\{x_n\}$ is bounded above and $x_n \leq x_{n+1}$ $(n \geq N)$ then $\{x_n\}$ converges.*

(3) *If $\{x_n\}$ is bounded below and $x_n \geq x_{n+1}$ $(n \geq N)$ then $\{x_n\}$ converges.*

Proof A proof of (1) is given in Appendix C, Theorem C.4.8. The proofs of (2) and (3) are omitted. $\qquad \square$

Notes 18.2.16 (1) A bounded sequence need not converge. To see this, let $b_n = (-1)^n$ $(n \in \mathbb{N})$. Then $\{b_n\}$ is bounded since $-1 \leq b_n \leq 1$ $(n \in \mathbb{N})$ but $\{b_n\}$ does not converge. See Example 18.2.9.

(2) Parts (2) and (3) of Theorem 18.2.15 are together referred to as the *monotone convergence theorem*. It tells us that a bounded sequence which is eventually monotonic converges. The limit is not specified. However, if $x_n \to L$ and A and B are lower and upper bounds, respectively, for $\{x_n\}$, then it follows from Theorems 18.2.4 and 18.2.10 (4) that $A \leq L \leq B$.

Example 18.2.17 The sequence $\{x_n\}$ is defined, recursively, by

$$x_1 = 0, \qquad x_{n+1} = \frac{1}{2}(x_n + 1) \quad (n \in \mathbb{N}).$$

Use induction to show that $x_n \leq 1$ $(n \in \mathbb{N})$. Deduce that $\{x_n\}$ converges to 1.

Solution The details of showing that $x_n \leq 1$ $(n \in \mathbb{N})$ are left to the reader. The induction step follows from

$$x_k \leq 1 \;\Rightarrow\; x_k + 1 \leq 2 \;\Rightarrow\; \leq \frac{1}{2}(x_k + 1) \leq 1 \;\Rightarrow\; x_{k+1} \leq 1.$$

Thus the sequence $\{x_n\}$ is bounded above (by 1) and it follows that $x_n \leq x_{n+1}$ $(n \in \mathbb{N})$ since

$$x_{n+1} = \frac{1}{2}(x_n + 1) \geq \frac{1}{2}(x_n + x_n) = x_n \quad (n \in \mathbb{N}).$$

Hence, by Theorem 18.2.15, $x_n \to L$ (say). By Theorem 18.2.8 (1), $x_{n+1} \to L$, and from Theorems 18.2.4 and 18.2.10,

$$x_{n+1} = \frac{1}{2}(x_n + 1) \to \frac{1}{2}(L + 1).$$

The uniqueness of sequence limits (Theorem 18.2.3) then gives $L = (L+1)/2$ from which $L = 1$ as required. $\square$

Problem 18.2.18 Let $\{x_n\}$ be the sequence of Example 18.2.17. Obtain a formula for x_n in closed form.

18.3   Series

Here we consider the problem of adding up the terms of a sequence.

Definitions 18.3.1 Let $\{a_r\}$ be a sequence and let $\{s_n\}$ be the sequence defined by

$$s_n = \sum_{r=1}^{n} a_r = a_1 + a_2 + \cdots + a_n \quad (n \in \mathbb{N}),$$

so that $s_1 = a_1$, $s_2 = a_1 + a_2$, $s_3 = a_1 + a_2 + a_3$, etc. We call $\{s_n\}$ the sequence of *partial sums* of the *series*

$$\sum_{r=1}^{\infty} a_r = a_1 + a_2 + a_3 + \cdots.$$

The notation $\sum_{r=1}^{\infty} a_r$ is also used. Strictly speaking, the series *is* the sequence of partial sums, but by the rth *term of the series* we mean a_r, the rth term of the original sequence. If the sequence $\{s_n\}$ converges to S, we say that the series $\sum_{r=1}^{\infty} a_r$ *converges*. Then we call S the *sum* of the series and write $\sum_{r=1}^{\infty} a_r = S$.

Any series which does not converge is said to *diverge*.

Hazard We use $\sum_{r=1}^{\infty} a_r$ ambiguously for both the series, which always exists, and the sum, which sometimes exists.

Example 18.3.2 Show that the series

$$\sum_{r=1}^{\infty} \frac{1}{r(r+1)} = \frac{1}{1.2} + \frac{1}{2.3} + \frac{1}{3.4} + \cdots$$

converges and find its sum.

Solution The rth term of the series is

$$\frac{1}{r(r+1)} = \frac{1}{r} - \frac{1}{r+1}.$$

So the nth partial sum is

$$\left(\frac{1}{1} - \frac{1}{2}\right) + \left(\frac{1}{2} - \frac{1}{3}\right) + \left(\frac{1}{3} - \frac{1}{4}\right) + \cdots + \left(\frac{1}{n-1} - \frac{1}{n}\right) + \left(\frac{1}{n} - \frac{1}{n+1}\right)$$

$$= 1 + \left(-\frac{1}{2} + \frac{1}{2}\right) + \left(-\frac{1}{3} + \frac{1}{3}\right) + \cdots + \left(-\frac{1}{n} + \frac{1}{n}\right) - \frac{1}{n+1}$$

$$= 1 - \frac{1}{n+1} = \frac{n}{n+1} = \frac{1}{1+1/n} \to \frac{1}{1+0} = 1.$$

Thus $\displaystyle\sum_{r=1}^{\infty} \frac{1}{r(r+1)}$ converges and has sum 1.   □

Theorem 18.3.3 *Let* $\displaystyle\sum_{r=1}^{\infty} a_r$ *converge. Then* $a_r \to 0$.

Proof Let $\{s_n\}$ be the sequence of partial sums and let S be the sum of the series. Then, for $r \geq 2$,

$$a_r = s_r - s_{r-1} \to S - S = 0.$$

The fact that $s_{r-1} \to S$ is intuitive, but we shall not prove it here. It follows directly from the formal definition of sequence limit in Appendix C, Definition C.4.1.   □

The converse of Theorem 18.3.3 is false as the next example shows.

Example 18.3.4 Show that the *harmonic* series

$$\sum_{r=1}^{\infty} \frac{1}{r} = \frac{1}{1} + \frac{1}{2} + \frac{1}{3} + \cdots$$

diverges.

Solution We compare the nth partial sum

$$s_n = \frac{1}{1} + \frac{1}{2} + \frac{1}{3} + \cdots + \frac{1}{n}$$

with $\displaystyle\int_1^{n+1} \frac{1}{x}\, dx$. The situation is illustrated in Figure 18.3.1. The term $1/r$ in the partial sum is the area of the rectangle on $[r, r+1]$ and this is greater than the corresponding area under the curve $y = 1/x$. Hence,

$$s_n > \int_1^{n+1} \frac{1}{x}\, dx = \big[\log|x|\big]_1^{n+1} = \log(n+1).$$

It follows that $\{s_n\}$ is not bounded (above) and so, by Theorem 18.2.15, $\{s_n\}$ does not converge.   □

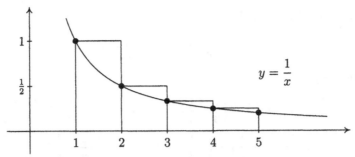

Figure 18.3.1

Example 18.3.5 Let $x \in \mathbb{R}$. Show that the *geometric* series

$$\sum_{r=0}^{\infty} x^r = 1 + x + x^2 + x^3 + x^4 + \cdots$$

converges when $|x| < 1$ and diverges when $|x| \geq 1$. Find the sum of the series when $|x| < 1$.

Solution The series starts with $r = 0$. So the rth term of the series is x^{r-1} and the nth partial sum is

$$s_n = 1 + x + x^2 + \cdots + x^{n-1}.$$

Then, for $n \in \mathbb{N}$,

$$s_n - x s_n = (1 + x + x^2 + \cdots + x^{n-1}) - (x + x^2 + x^3 + \cdots + x^n) = 1 - x^n,$$

so that $(1 - x)s_n = 1 - x^n$. Hence, using Theorem 18.2.13, when $|x| < 1$,

$$s_n = \frac{1 - x^n}{1 - x} \to \frac{1 - 0}{1 - x} = \frac{1}{1 - x}.$$

Thus $\displaystyle\sum_{r=0}^{\infty} x^r$ converges and has sum $\dfrac{1}{1 - x}$ when $|x| < 1$.

When $|x| \geq 1$, $|x^{r-1}| \geq 1$ and so, by Theorem 18.2.10 (7), the rth term of the series $x^{r-1} \not\to 0$. Hence, by Theorem 18.3.3, the series diverges. ☐

Theorem 18.3.6 Let $\displaystyle\sum_{r=1}^{\infty} a_r$ and $\displaystyle\sum_{r=1}^{\infty} b_r$ converge with sums S and T, respectively. Then

   (1) for $\lambda \in \mathbb{R}$, $\displaystyle\sum_{r=1}^{\infty} \lambda a_r$ converges with sum λS,

   (2) $\displaystyle\sum_{r=1}^{\infty} (a_r + b_r)$ converges with sum $S + T$.

Proof Let $\{s_n\}$ and $\{t_n\}$ be the sequences of partial sums for $\sum_{r=1}^{\infty} a_r$ and $\sum_{r=1}^{\infty} b_r$, respectively. Then the nth partial sums for $\sum_{r=1}^{\infty} \lambda a_r$ and $\sum_{r=1}^{\infty} (a_r + b_r)$ are

$$\lambda a_1 + \lambda a_2 + \lambda a_3 + \cdots + \lambda a_n = \lambda(a_1 + a_2 + a_3 + \cdots + a_n) = \lambda s_n$$

and

$$(a_1 + b_1) + (a_2 + b_2) + (a_3 + b_3) + \cdots + (a_n + b_n)$$
$$= (a_1 + a_2 + a_3 + \cdots + a_n) + (b_1 + b_2 + b_3 + \cdots + b_n) = s_n + t_n,$$

respectively. Since $s_n \to S$ and $t_n \to T$, it follows from Theorem 18.2.10 (2) that $\lambda s_n \to \lambda S$ and $s_n + t_n \to S + T$, completing the proof. $\qquad\square$

Corollary 18.3.7 *Let* $\displaystyle\sum_{r=1}^{\infty} a_r$ *diverge and let* $\mu \in \mathbb{R}$, $\mu \neq 0$. *Then* $\displaystyle\sum_{r=1}^{\infty} \mu a_r$ *diverges.*

Proof If $\displaystyle\sum_{r=1}^{\infty} \mu a_r$ converges then, by Theorem 18.3.6 (1), $\displaystyle\sum_{r=1}^{\infty} a_r = \sum_{r=1}^{\infty} \left(\frac{1}{\mu}\right) \mu a_r$ converges, contradicting the given hypotheses. $\qquad\square$

Note 18.3.8 Let $\{s_n\}$ and $\{t_n\}$ be the sequences of partial sums for $\sum_{r=1}^{\infty} a_r$ and $\sum_{r=1}^{\infty} b_r$, respectively. If $a_r = b_r$ $(r \geq N)$ then, for $n \geq N$, $s_n - t_n = s_N - t_N$ is constant so that $\{s_n\}$ converges if and only if $\{t_n\}$ converges, *i.e.* $\sum_{r=1}^{\infty} a_r$ converges if and only if $\sum_{r=1}^{\infty} b_r$ converges.

In other words, the 'first few' terms of a series have no bearing on whether it converges or diverges.

On the other hand, if a series converges, *any* of the terms can affect the sum.

Theorem 18.3.9 (Comparison Test) *Let* $\{a_r\}$ *and* $\{b_r\}$ *be sequences such that* $0 \leq a_r \leq b_r$ $(r \geq N)$.

(1) *If* $\displaystyle\sum_{r=1}^{\infty} b_r$ *converges then* $\displaystyle\sum_{r=1}^{\infty} a_r$ *converges.*

(2) *If* $\displaystyle\sum_{r=1}^{\infty} a_r$ *diverges then* $\displaystyle\sum_{r=1}^{\infty} b_r$ *diverges.*

Proof We prove the case where $N = 1$.

(1) Let $\sum_{r=1}^{\infty} b_r$ converge and let $\{s_n\}$ and $\{t_n\}$ be the sequences of partial sums for $\sum_{r=1}^{\infty} a_r$ and $\sum_{r=1}^{\infty} b_r$, respectively. Then $\{t_n\}$ converges and so, by Theorem 18.2.15 (1), $\{t_n\}$ is bounded above by B (say). Since $0 \leq a_r \leq b_r$ $(r \in \mathbb{N})$,

$$s_n \leq t_n \leq B \quad \text{and} \quad s_n \leq s_n + a_{n+1} = s_{n+1} \qquad (n \in \mathbb{N}).$$

Hence, by Theorem 18.2.15 (2), $\{s_n\}$ converges and the proof is complete.

(2) This is an immediate consequence of (1). $\qquad\square$

Corollary 18.3.10 (Comparison Test—Limit Version) *Let* $\{a_r\}$ *and* $\{b_r\}$ *be sequences of eventually positive terms such that* $a_r/b_r \to L$ *as* $r \to \infty$.

(1) *If* $L \in (0, \infty)$ *then* $\displaystyle\sum_{r=1}^{\infty} a_r$ *converges* $\Leftrightarrow$ $\displaystyle\sum_{r=1}^{\infty} b_r$ *converges.*

(2) *If $L = 0$ then* $\displaystyle\sum_{r=1}^{\infty} b_r$ *converges* $\Rightarrow$ $\displaystyle\sum_{r=1}^{\infty} a_r$ *converges.*

(3) *If $L = \infty$ then* $\displaystyle\sum_{r=1}^{\infty} a_r$ *converges* $\Rightarrow$ $\displaystyle\sum_{r=1}^{\infty} b_r$ *converges.*

Proof We prove the case where $a_r, b_r > 0$ $(r \in \mathbb{N})$

(A) Let $L \in [0, \infty)$. Then $\{a_r/b_r\}$ converges and so is bounded above by M (say). Hence, for $r \in \mathbb{N}$,

$$0 < \frac{a_r}{b_r} \le M \qquad \text{so that} \qquad 0 < a_r \le Mb_r.$$

Then, if $\sum_{r=1}^{\infty} b_r$ converges, so does $\sum_{r=1}^{\infty} Mb_r$ and hence, by Theorem 18.3.9, $\sum_{r=1}^{\infty} a_r$ converges.

(B) Let $L \in (0, \infty)$ or $L = \infty$. Then $b_r/a_r \to 1/L \in (0, \infty)$ or $b_r/a_r \to 0$, *i.e.* $b_r/a_r \to K \in [0, \infty)$. Applying (A), if $\sum_{r=1}^{\infty} a_r$ converges then $\sum_{r=1}^{\infty} b_r$ converges. □

Example 18.3.11 Show that $\displaystyle\sum_{r=1}^{\infty} \frac{1}{r^2}$ converges.

Solution For $r \in \mathbb{N}$, $r(r+1) = r^2 + r \le 2r^2$ so that

$$0 \le \frac{1}{r^2} \le \frac{2}{r(r+1)}.$$

From Example 18.3.2, $\displaystyle\sum_{r=1}^{\infty} \frac{1}{r(r+1)}$ converges and so, by Theorem 18.3.6 (1), $\displaystyle\sum_{r=1}^{\infty} \frac{2}{r(r+1)}$ converges. Hence, by the comparison test (Theorem 18.3.9), $\displaystyle\sum_{r=1}^{\infty} \frac{1}{r^2}$ converges, as required.

Alternatively. Let $a_r = \dfrac{1}{r^2}$, $b_r = \dfrac{1}{r(r+1)}$ $(r \in \mathbb{N})$. Then, for $r \in \mathbb{N}$, $a_r, b_r > 0$ and

$$\frac{a_r}{b_r} = \frac{1}{r^2} \cdot \frac{r(r+1)}{1} = \frac{r+1}{r} = 1 + \frac{1}{r} \to 1 \in (0, \infty) \text{ as } r \to \infty.$$

Since $\displaystyle\sum_{r=1}^{\infty} \frac{1}{r(r+1)}$ converges by Example 18.3.2, $\displaystyle\sum_{r=1}^{\infty} \frac{1}{r^2}$ converges by the comparison test (limit version). □

Problem 18.3.12 Let $\displaystyle\sum_{r=1}^{\infty} a_r$ diverge. Prove that if $a_r, b_r \ge 0$ $(r \ge N)$ then $\displaystyle\sum_{r=1}^{\infty} (a_r + b_r)$ diverges. If there is no restriction on b_r $(r \in \mathbb{N})$, does $\displaystyle\sum_{r=1}^{\infty} (a_r + b_r)$ necessarily diverge?

Notes 18.3.13 (1) The following series of positive terms are useful in connection with the comparison test.

$$\sum_{r=1}^{\infty} \frac{1}{r^2} \quad \text{converges,}$$

$$\sum_{r=1}^{\infty} \frac{1}{r} \quad \text{diverges,}$$

$$\sum_{r=0}^{\infty} x^r \quad \begin{cases} \text{converges if } 0 < x < 1, \\ \text{diverges if } x \geq 1. \end{cases}$$

These follow from Examples 18.3.11, 18.3.4 and 18.3.5, respectively. In view of Theorem 18.3.6 (1) and Corollary 18.3.7 we can also use positive constant multiples of these series. For example,

$$\sum_{r=1}^{\infty} \frac{7}{r^2} \quad \text{converges,} \quad \sum_{r=1}^{\infty} \frac{2}{3r} \quad \text{diverges,} \quad \sum_{r=1}^{\infty} 5\left(\frac{3}{4}\right)^r \quad \text{converges.}$$

(2) There is a more general result which includes the first two series in (1), above. Let $t \in \mathbb{R}$. Then

$$\sum_{r=1}^{\infty} \frac{1}{r^t} \quad \text{converges if and only if } t > 1.$$

Example 18.3.14 For each of the following series, use the comparison test to determine whether it converges or diverges.

(a) $\displaystyle\sum_{r=1}^{\infty} \frac{2r+5}{3r^3+1}$,   (b) $\displaystyle\sum_{r=1}^{\infty} \frac{3r-1}{r^3-5}$,   (c) $\displaystyle\sum_{r=2}^{\infty} \frac{1}{\log(2r-1)}$,   (d) $\displaystyle\sum_{r=1}^{\infty} \frac{r}{3^r}$.

Solution In each case we compare the given series with a series from Note 18.3.13 or a constant multiple of such a series.

(a) For $r \in \mathbb{N}$,

$$0 \leq \frac{2r+5}{3r^3+1} \leq \frac{2r+5r}{3r^3} = \frac{7}{3r^2}.$$

Since $\displaystyle\sum_{r=1}^{\infty} \frac{7}{3r^2}$ converges, $\displaystyle\sum_{r=1}^{\infty} \frac{2r+5}{3r^3+1}$ by the comparison test.

(b) For $r \in \mathbb{N}$, $r^3 - 5 \geq r^3/2$ provided $r^3 \geq 10$, *i.e.* if $r \geq 3$. Hence, for $r \geq 3$,

$$0 \leq \frac{3r-1}{r^3-5} \leq \frac{3r}{r^3/2} = \frac{6}{r^2}.$$

Since $\displaystyle\sum_{r=1}^{\infty} \frac{6}{r^2}$ converges, $\displaystyle\sum_{r=1}^{\infty} \frac{3r-1}{r^3-5}$ converges by the comparison test.

Alternatively. Let $a_r = \dfrac{3r-1}{r^3-5}$, $b_r = \dfrac{1}{r^2}$ $(r \in \mathbb{N})$. Then, for $r \geq 2$, $a_r, b_r > 0$ and

$$\frac{a_r}{b_r} = \left(\frac{3r-1}{r^3-5}\right)\left(\frac{r^2}{1}\right) = \frac{3r^3-r^2}{r^3-5} = \frac{3-(1/r)}{1-(5/r^3)} \to 3 \in (0,\infty) \text{ as } r \to \infty.$$

Since $\displaystyle\sum_{r=1}^{\infty} \frac{1}{r^2}$ converges, $\displaystyle\sum_{r=1}^{\infty} \frac{3r-1}{r^3-5}$ converges by the comparison test.

(c) For $r \in \mathbb{N}$ with $r \geq 2$, $0 < \log(2r-1) \leq 2r-1 \leq 2r$. Hence,

$$\frac{1}{\log(2r-1)} \geq \frac{1}{2r} \geq 0.$$

Since $\displaystyle\sum_{r=2}^{\infty} \frac{1}{2r}$ diverges, $\displaystyle\sum_{r=2}^{\infty} \frac{1}{\log(2r+1)}$ diverges by the comparison test.

(d) For $r \in \mathbb{N}$, the solution to Example 18.2.12 (c) gives, $r \leq 2^r$. Hence

$$0 \leq \frac{r}{3^r} \leq \frac{2^r}{3^r} = \left(\frac{2}{3}\right)^r.$$

Since, $\displaystyle\sum_{r=1}^{\infty} \left(\frac{2}{3}\right)^r$ converges, $\displaystyle\sum_{r=1}^{\infty} \frac{r}{3^r}$ converges by the comparison test. $\square$

Theorem 18.3.15 (Ratio Test) *Let $\{a_r\}$ be a sequence and let $\lambda \in (0,1)$.*

(1) *If $a_r > 0$ and $\dfrac{a_{r+1}}{a_r} \leq \lambda \ (r \geq N)$ then $\displaystyle\sum_{r=1}^{\infty} a_r$ converges.*

(2) *If $a_r > 0$ and $\dfrac{a_{r+1}}{a_r} \geq 1 \ (r \geq N)$ then $\displaystyle\sum_{r=1}^{\infty} a_r$ diverges.*

Proof We prove the case where $N = 1$.

(1) Let $a_r > 0$ and $a_{r+1}/a_r \leq \lambda \ (r \in \mathbb{N})$. Then, for $r \in \mathbb{N}$,

$$a_{r+1} \leq \lambda a_r \leq \lambda^2 a_{r-1} \leq \cdots \leq \lambda^r a_1.$$

Hence $0 < a_r \leq a_1 \lambda^{r-1}$ and, since $\displaystyle\sum_{r=1}^{\infty} a_1 \lambda^{r-1}$ converges, $\displaystyle\sum_{r=1}^{\infty} a_r$ converges by the comparison test.

(2) Let $a_r > 0$ and $a_{r+1}/a_r \geq 1 \ (r \in \mathbb{N})$. Then, for $r \in \mathbb{N}$,

$$a_{r+1} \geq a_r \geq a_{r-1} \geq \cdots \geq a_1 > 0.$$

So, by Theorem 18.2.10 (7), $a_r \nrightarrow 0$. Hence, by Theorem 18.3.3, $\displaystyle\sum_{r=1}^{\infty} a_r$ diverges. $\square$

Corollary 18.3.16 (Ratio Test—Limit Version) *Let $\{a_r\}$ be a sequence of eventually positive terms such that $a_{r+1}/a_r \to L$ as $r \to \infty$.*

(1) *If $L \in [0,1)$ then $\displaystyle\sum_{r=1}^{\infty} a_r$ converges.*

(2) *If $L \in (1,\infty)$ or $L = \infty$ then $\displaystyle\sum_{r=1}^{\infty} a_r$ diverges.*

Proof We prove (1) in the case where $a_r, b_r > 0$ ($r \in \mathbb{N}$), and omit the proof of (2) which is similar.

Let $L \in [0, 1)$ and let $\lambda \in (L, 1)$. Then $a_{r+1}/a_r - \lambda \to L - \lambda$ and, since $L - \lambda < 0$, it follows from Theorem 18.2.10 (1) that $a_{r+1}/a_r - \lambda < 0$ ($r \geq N$). Hence

$$a_r > 0 \quad \text{and} \quad \frac{a_{r+1}}{a_r} < \lambda \quad (r \geq N).$$

So $\displaystyle\sum_{r=1}^{\infty} a_r$ converges by Theorem 18.3.15. $\square$

Notes 18.3.17 (1) In Theorem 18.3.15 (1), the requirement $a_{r+1}/a_r \leq \lambda < 1$ cannot be replaced with $a_{r+1}/a_r < 1$. This latter condition is not enough to guarantee that $\sum_{r=1}^{\infty} a_r$ converges. Consider $\sum_{r=1}^{\infty} 1/r$.

(2)   With the notation of Theorem 18.3.16, if $L = 1$ then, without further information, no decision can be made about the convergence or divergence of $\sum_{r=1}^{\infty} a_r$. Consider $\sum_{r=1}^{\infty} 1/r$ and $\sum_{r=1}^{\infty} 1/r^2$.

Example 18.3.18 For each of the following series, use the ratio test to determine whether it converges or diverges.

(a) $\displaystyle\sum_{r=1}^{\infty} \frac{r}{5^r}$,   (b) $\displaystyle\sum_{r=1}^{\infty} \frac{2^r}{r\sqrt{r}}$,   (c) $\displaystyle\sum_{r=1}^{\infty} \frac{100^r}{r!}$,   (d) $\displaystyle\sum_{r=1}^{\infty} \frac{4^r}{\binom{2r}{r}}$.

Solution (a) Let $a_r = r/5^r$ ($r \in \mathbb{N}$). Then, for $r \in \mathbb{N}$, $a_r > 0$ and

$$\frac{a_{r+1}}{a_r} = \frac{r+1}{5^{r+1}} \cdot \frac{5^r}{r} = \frac{1}{5}\left(\frac{r+1}{r}\right) = \frac{1}{5}\left(1 + \frac{1}{r}\right) \leq \frac{2}{5} < 1.$$

Hence, by the ratio test, $\displaystyle\sum_{r=1}^{\infty} \frac{r}{5^r}$ converges.

(b) Let $a_r = 2^r/(r\sqrt{r})$ ($r \in \mathbb{N}$). Then, for $r \in \mathbb{N}$, $a_r > 0$ and

$$\frac{a_{r+1}}{a_r} = \frac{2^{r+1}}{(r+1)\sqrt{r+1}} \cdot \frac{r\sqrt{r}}{2^r} = 2\left(\frac{r}{r+1}\right)^{3/2} > 1 \text{ when } r \geq 2.$$

Hence, by the ratio test, $\displaystyle\sum_{r=1}^{\infty} \frac{2^r}{r\sqrt{r}}$ diverges.

Alternatively.   Let $a_r = 2^r/(r\sqrt{r})$ ($r \in \mathbb{N}$). Then, for $r \in \mathbb{N}$, $a_r > 0$ and

$$\frac{a_{r+1}}{a_r} = \frac{2^{r+1}}{(r+1)\sqrt{r+1}} \cdot \frac{r\sqrt{r}}{2^r} = 2\left(\frac{r}{r+1}\right)^{3/2} \to 2 \in (1, \infty) \text{ as } r \to \infty.$$

Hence, by the ratio test, $\displaystyle\sum_{r=1}^{\infty} \frac{2^r}{r\sqrt{r}}$ diverges.

(c) Let $a_r = 100^r/r!$ $(r \in \mathbb{N})$. Then, for $r \in \mathbb{N}$, $a_r > 0$ and

$$\frac{a_{r+1}}{a_r} = \frac{100^{r+1}}{(r+1)!} \cdot \frac{r!}{100^r} = \frac{100}{r+1} \le \frac{1}{2} < 1 \text{ when } r \ge 199.$$

Hence, by the ratio test, $\displaystyle\sum_{r=1}^{\infty} \frac{100^r}{r!}$ converges.

(d) Let $a_r = 4^r/\binom{2r}{r}$ $(r \in \mathbb{N})$. Then, for $r \in \mathbb{N}$, $a_r > 0$ and

$$\frac{a_{r+1}}{a_r} = \frac{4^{r+1}[(r+1)!]^2}{(2r+2)!} \cdot \frac{(2r)!}{4^r(r!)^2} = \frac{4(r+1)^2}{(2r+2)(2r+1)} = \frac{2r+2}{2r+1} > 1.$$

Hence, by the ratio test, $\displaystyle\sum_{r=1}^{\infty} \frac{4^r}{\binom{2r}{r}}$ diverges. $\qquad\square$

Definition 18.3.19 A series $\displaystyle\sum_{r=1}^{\infty} a_r$ *converges absolutely* if the series $\displaystyle\sum_{r=1}^{\infty} |a_r|$ converges.

Example 18.3.20 Show that, for any $\theta \in \mathbb{R}$, the series $\displaystyle\sum_{r=1}^{\infty} \frac{\sin(r\theta)}{r!}$ converges absolutely.

Solution We must show that $\displaystyle\sum_{r=1}^{\infty} \left| \frac{\sin(r\theta)}{r!} \right|$ converges. For $r \in \mathbb{N}$,

$$0 \le \left| \frac{\sin(r\theta)}{r!} \right| = \frac{|\sin(r\theta)|}{r!} \le \frac{1}{r!} \le \frac{2}{r^2},$$

since $1! > \dfrac{1^2}{2}$ and, for $r \ge 2$, $r! \ge r(r-1) \ge r\left(\dfrac{r}{2}\right) = \dfrac{r^2}{2}$. By Note 18.3.13, $\displaystyle\sum_{r=1}^{\infty} \frac{2}{r^2}$

converges. Hence, $\displaystyle\sum_{r=1}^{\infty} \left| \frac{\sin(r\theta)}{r!} \right|$ converges by the comparison test, so that $\displaystyle\sum_{r=1}^{\infty} \frac{\sin(r\theta)}{r!}$ converges absolutely. $\qquad\square$

Theorem 18.3.21 Let $\displaystyle\sum_{r=1}^{\infty} a_r$ *converge absolutely. Then* $\displaystyle\sum_{r=1}^{\infty} a_r$ *converges.*

Proof For $r \in \mathbb{N}$,

$$-|a_r| \le a_r \le |a_r|$$

so that

$$0 \le a_r + |a_r| \le 2|a_r|.$$

Since $\sum_{r=1}^{\infty} a_r$ converges absolutely, $\sum_{r=1}^{\infty} |a_r|$ converges. Then $\sum_{r=1}^{\infty} 2|a_r|$ converges, by Theorem 18.3.6 (1). Hence $\sum_{r=1}^{\infty}(a_r + |a_r|)$ converges, by the comparison test. Finally, since

$$a_r = (a_r + |a_r|) - |a_r| \quad (r \in \mathbb{N}),$$

the series $\sum_{r=1}^{\infty} a_r$ converges, by Theorem 18.3.6. $\qquad\square$

The converse of Theorem 18.3.21 is false. The series

$$\sum_{r=1}^{\infty} \frac{(-1)^{r-1}}{r} = \frac{1}{1} - \frac{1}{2} + \frac{1}{3} - \frac{1}{4} + \cdots$$

converges. This follows from the next theorem (see Example 18.3.23). However, the series does not converge absolutely since

$$\sum_{r=1}^{\infty} \left| \frac{(-1)^{r-1}}{r} \right| = \sum_{r=1}^{\infty} \frac{1}{r}$$

and this series diverges, by Example 18.3.4.

Theorem 18.3.22 (Alternating Series Test or Leibniz's Test) *Let $\{a_r\}$ be a sequence such that*

$$(1) \;\; a_r \geq 0 \;\; (r \geq N), \quad (2) \;\; a_r \geq a_{r+1} \;\; (r \geq N), \quad (3) \;\; a_r \to 0.$$

Then $\displaystyle\sum_{r=1}^{\infty} (-1)^{r-1} a_r$ *converges.*

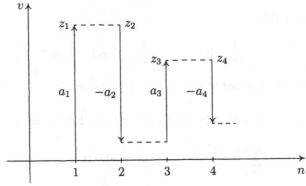

Figure 18.3.2

Proof We prove the case when $N = 1$.

Let $\{s_n\}$ be the sequence of partial sums for the series $\sum_{r=1}^{\infty} (-1)^{r-1} a_r$. For $n \in \mathbb{N}$, define

$$z_n = \begin{cases} s_n & \text{if } n \text{ is odd,} \\ s_{n-1} & \text{if } n \text{ is even.} \end{cases}$$

See Figure 18.3.2. Then

$$z_{n+1} - z_n = \begin{cases} s_n - s_n = 0 & \text{if } n \text{ is odd,} \\ s_{n+1} - s_{n-1} = -a_n + a_{n+1} & \text{if } n \text{ is even.} \end{cases}$$

As $a_n \geq a_{n+1}$ it follows that, for all $n \in \mathbb{N}$,

$$z_{n+1} - z_n \leq 0, \qquad i.e. \; z_n \geq z_{n+1}.$$

Also, from its definition, z_n is the sum of an odd number of terms and so may be written

$$z_n = (a_1 - a_2) + (a_3 - a_4) + \cdots + (a_{m-2} - a_{m-1}) + a_m,$$

where $m = n$ or $n - 1$, according as n is odd or even. Since $a_r \geq a_{r+1}$ and $a_r \geq 0$ ($r \in \mathbb{N}$) it follows that $z_n \geq 0$ ($n \in \mathbb{N}$), i.e. $\{z_n\}$ is bounded below by 0. Hence, by Theorem 18.2.15 (3), $\{z_n\}$ converges to L (say).

Finally, from the definition of z_n,

$$s_n - z_n = \begin{cases} 0 & \text{if } n \text{ is odd}, \\ -a_n & \text{if } n \text{ is even}. \end{cases}$$

Hence, for all $n \in \mathbb{N}$, $-a_n \leq s_n - z_n \leq 0$, i.e.

$$z_n - a_n \leq s_n \leq z_n \quad (n \in \mathbb{N}).$$

Since $z_n \to L$ and $a_n \to 0$, it follows from Theorem 18.2.10 (2) that $z_n - a_n \to L$. Hence, by the sandwich principle, $s_n \to L$, and the series $\sum_{r=1}^{\infty}(-1)^{r-1}a_r$ converges, as required. $\qquad \square$

Example 18.3.23 Show that the series $\displaystyle\sum_{r=1}^{\infty} \frac{(-1)^{r-1}}{r}$ converges.

Solution For $r \in \mathbb{N}$,

$$\frac{1}{r} \geq 0, \quad \frac{1}{r} \geq \frac{1}{r+1} \quad \text{and} \quad \frac{1}{r} \to 0.$$

Hence the alternating series test applies to give the result. $\qquad \square$

Note 18.3.24 Example 18.3.23 can be generalised. Let $t \in \mathbb{R}$. Then

$$\sum_{r=1}^{\infty} \frac{(-1)^{n-1}}{r^t} \quad \text{converges if and only if } t > 0.$$

Compare this with the last series in Note 18.3.13.

Problem 18.3.25 *Dirichlet's test* is a generalisation of Leibniz's test. A special case is stated as follows. Let the partial sums of $\sum_{r=1}^{\infty} x_r$ form a bounded sequence and let $\{y_r\}$ be a decreasing sequence of positive terms converging to 0. Then $\sum_{r=1}^{\infty} x_r y_r$ converges.

Use this test to show that the series

$$\sum_{r=1}^{\infty} \frac{\sin r}{r}.$$

converges. Problem 6.3.10 may be of help.

18.4   Power Series

Definition 18.4.1 A *power series (in x)* is a series of the form

$$\sum_{r=0}^{\infty} a_r x^r = a_0 + a_1 x + a_2 x^2 + \cdots,$$

where $a_0, a_1, a_2, \ldots$ are real constants and x is regarded as a real variable. In general, whether or not the series converges depends on x. For example, the geometric series

$$\sum_{r=0}^{\infty} \left(\frac{x}{2}\right)^r = \sum_{r=0}^{\infty} \frac{1}{2^r} x^r$$

converges for $|x| < 2$ and diverges for $|x| > 2$. This is more or less typical of the general situation.

Lemma 18.4.2 *Let* $S(x) = \sum_{r=0}^{\infty} a_r x^r$ *be a power series and let* $S(y)$ *converge for some* $y \neq 0$. *Then* $S(x)$ *converges absolutely for all* x *with* $|x| < |y|$.

Proof By Theorem 18.3.3, since $S(y)$ converges, $a_r y^r \to 0$. So, by Theorem 18.2.15 (1), there exists $K > 0$ such that $|a_r y^r| \leq K$ $(r \in \mathbb{N} \cup \{0\})$. Let $x \in \mathbb{R}$ with $|x| < |y|$. Then $t = |x|/|y| < 1$ and, for all $r \in \mathbb{N} \cup \{0\}$,

$$0 \leq |a_r x^r| = |a_r y^r| t^r \leq K t^r.$$

Since $\sum_{r=0}^{\infty} t^r$ converges, it follows, by the comparison test, that $S(x)$ converges absolutely. $\qquad\square$

Theorem 18.4.3 *Let* $S(x) = \sum_{r=0}^{\infty} a_r x^r$ *be a power series. Then exactly one of the following holds.*

(1) $S(x)$ *converges only for* $x = 0$.

(2)   *There exists* $R \in (0, \infty)$ *such that* $S(x)$ *converges absolutely for all* $x \in \mathbb{R}$ *with* $|x| < R$ *and diverges for all* $x \in \mathbb{R}$ *with* $|x| > R$.

(3) $S(x)$ *converges absolutely for all* $x \in \mathbb{R}$.

Proof This can be deduced from Lemma 18.4.2 but the proof requires topics not covered by this book and is omitted. $\qquad\square$

Definitions 18.4.4 Let $S(x) = \sum_{r=0}^{\infty} a_r x^r$ be a power series and consider Theorem 18.4.3. In case (2), R is called the *radius of convergence*. In cases (1) and (3), the series is said to have radius of convergence 0 and ∞, respectively. In all three cases, the set of $x \in \mathbb{R}$ for which $S(x)$ converges is called the *interval of convergence*.

The examples given in the table below illustrate some of the possibilities. Note that, in case (2) of Theorem 18.4.3, nothing can be said in general about convergence when $x = \pm R$.

Power Series	Radius of Convergence	Interval of Convergence
$\displaystyle\sum_{r=0}^{\infty} r!x^r$	0	$\{0\}$
$\displaystyle\sum_{r=0}^{\infty} x^r$	1	$(-1,1)$
$\displaystyle\sum_{r=1}^{\infty} \frac{x^r}{r}$	1	$[-1,1)$
$\displaystyle\sum_{r=1}^{\infty} \frac{x^r}{r^2}$	1	$[-1,1]$
$\displaystyle\sum_{r=0}^{\infty} \frac{x^r}{r!}$	∞	$\mathbb{R}$

To verify the entries in the table, first use the next result, Theorem 18.4.5, to establish the radii of convergence. Then, for the second, third and fourth series, determine whether or not the series converges when $x = 1$ and when $x = -1$ by using Examples 18.3.5, 18.3.4, 18.3.23, 18.3.11 and Theorem 18.3.21 in turn. To see some of the techniques in action, look at the solution to Example 18.4.6, below.

Theorem 18.4.5 *Let $\sum_{r=0}^{\infty} a_r x^r$ be a power series with radius of convergence R, and let*

$$\left| \frac{a_{r+1}}{a_r} \right| \to \lambda \quad as \quad r \to \infty.$$

Then

$$R = \begin{cases} 0 & if \ \lambda = \infty, \\ 1/\lambda & if \ 0 < \lambda < \infty, \\ \infty & if \ \lambda = 0. \end{cases}$$

Proof (outline)   Apply the limit version of the comparison test to $\sum_{r=0}^{\infty} |a_r x^r|$. $\square$

Example 18.4.6 For each of the following power series, find R, the radius of convergence, and I, the interval of convergence.

$$\text{(a)} \ \sum_{n=0}^{\infty} \frac{x^r}{(r+4)3^r}, \qquad \text{(b)} \ \sum_{n=0}^{\infty} \frac{(2x)^r}{r^2+1}.$$

Solution   (a) For $r \in \mathbb{N} \cup \{0\}$, let $a_r = \dfrac{1}{(r+4)3^r}$.   Then

$$\left| \frac{a_{r+1}}{a_r} \right| = \frac{1}{(r+5)3^{n+1}} \cdot \frac{(r+4)3^r}{1} = \frac{1}{3}\left(\frac{r+4}{r+5} \right) \to \frac{1}{3} \quad as \ r \to \infty.$$

Hence, by Theorem 18.4.5, $R = 3$.

When $x = 3$, $\displaystyle\sum_{r=0}^{\infty} a_r x^r = \sum_{r=0}^{\infty} \frac{1}{r+4}$ which diverges (by the comparison test), and

when $x = -3$, $\displaystyle\sum_{r=1}^{\infty} a_r x^r = \sum_{r=1}^{\infty} \frac{(-1)^r}{r+4}$ which converges (by the alternating series test).
Thus $I = [-3, 3)$.

(b) For $r \in \mathbb{N} \cup \{0\}$, let $a_r = \dfrac{2^r}{r^2 + 1}$. Then

$$\left| \frac{a_{r+1}}{a_r} \right| = \frac{2^{r+1}}{(r+1)^2 + 1} \cdot \frac{r^2 + 1}{2^r} = 2\left(\frac{r^2 + 1}{r^2 + 2r + 2} \right) \to 2 \quad \text{as } r \to \infty.$$

Hence, by Theorem 18.4.5, $R = 1/2$.

When $x = \pm 1/2$, $\displaystyle\sum_{r=0}^{\infty} |a_r x^r| = \sum_{n=0}^{\infty} \frac{1}{r^2 + 1}$ which converges (by the comparison

test). So $\displaystyle\sum_{r=0}^{\infty} a_r x^r$ is absolutely convergent, and hence convergent, when $x = \pm 1/2$.
Thus $I = [-1/2, 1/2]$. $\square$

Problem 18.4.7 Determine interval of convergence for the power series $\displaystyle\sum_{r=0}^{\infty} \frac{r!(2r)!}{(3r)!} x^r$.

On its interval of convergence, a power series defines a real function which behaves, in many ways, like a polynomial function. Some of these similarities are listed in the next theorem.

Theorem 18.4.8 (Properties of Power Series) *Let $\sum_{r=0}^{\infty} a_r x^r$ and $\sum_{r=0}^{\infty} b_r x^r$ converge for $x \in I = (-A, A)$ where $0 < A \le \infty$. Define $f : I \to \mathbb{R}$ and $g : I \to \mathbb{R}$ by*

$$f(x) = \sum_{r=0}^{\infty} a_r x^r \quad and \quad g(x) = \sum_{r=0}^{\infty} b_r x^r.$$

(1) (Arithmetic) *Let $x \in I$. Then*

(a) *for $\lambda \in \mathbb{R}$, $\lambda f(x) = \displaystyle\sum_{r=0}^{\infty} \lambda a_r x^r$.*

(b) $f(x) + g(x) = \displaystyle\sum_{r=0}^{\infty} (a_r + b_r) x^r$,

(c) $f(x)g(x) = \displaystyle\sum_{r=0}^{\infty} c_r x^r$,

where, for $r \in \mathbb{N}$, $c_r = a_0 b_r + a_1 b_{r-1} + \cdots + a_{r-1} b_1 + a_r b_0$.

(2) (Calculus) *On I, f is differentiable (and therefore continuous) and, for $x \in I$,*

$$f'(x) = \sum_{r=1}^{\infty} a_r r x^{r-1} \quad and \quad \int f(x)\, dx = \sum_{r=0}^{\infty} a_r \frac{x^{r+1}}{r+1} + C.$$

In particular these power series have the same radius of convergence as $\sum_{r=0}^{\infty} a_r x^r$.

(3) (Equating Coefficients) *If $f = g$ then $a_r = b_r$ $(r \in \mathbb{N})$.*

Proof (1) Parts (a) and (b) follow from Theorem 18.3.6. The proof of (c) is omitted.

(2) This proof is omitted.

(3) This can be deduced from (2). $\square$

Notes 18.4.9 The properties referred to are those of Theorem 18.4.8.

(1) Property (2) is sometimes expressed by saying that, on I, the power series representing f can be differentiated and integrated *term by term.*

(2) It follows by repeated applications of Property (2) that, for all $n \in \mathbb{N}$, the nth derivative of f exists on I.

(3) It follows from Property (3) that the *power series representation* of f on I is unique. For a stronger version of Property (3), see Theorem 18.5.8.

18.5   Taylor's Theorem

Here we consider the question: given a real function, when can we find a power series which represents it? Our first step is to generalise the mean value theorem.

Theorem 18.5.1 (Taylor's Theorem) *Let $n \in \mathbb{N}$ and let f be a real function such that f, $f^{(1)}$, $f^{(2)}$, ..., $f^{(n-1)}$ are defined and continuous on $[a, b]$ and $f^{(n)}$ is defined on (a, b). Then there exists $c \in (a, b)$ such that*

$$f(b) = f(a) + \frac{f^{(1)}(a)}{1!}(b - a) + \frac{f^{(2)}(a)}{2!}(b - a)^2 + \cdots$$

$$\cdots + \frac{f^{(n-1)}(a)}{(n-1)!}(b - a)^{n-1} + \frac{f^{(n)}(c)}{n!}(b - a)^n.$$

Proof Let Δ be the real number defined by

$$f(b) = f(a) + \frac{f^{(1)}(a)}{1!}(b - a) + \frac{f^{(2)}(a)}{2!}(b - a)^2 + \cdots$$

$$\cdots + \frac{f^{(n-1)}(a)}{(n-1)!}(b - a)^{n-1} + \frac{\Delta}{n!}(b - a)^n.$$

Define $F : [a, b] \rightarrow \mathbb{R}$ by

$$F(x) = f(b) - \left(f(x) + \frac{f^{(1)}(x)}{1!}(b - x) + \frac{f^{(2)}(x)}{2!}(b - x)^2 + \cdots \right.$$

$$\left. \cdots + \frac{f^{(n-1)}(x)}{(n-1)!}(b - x)^{n-1} + \frac{\Delta}{n!}(b - x)^n \right).$$

Then F is continuous on $[a, b]$ and differentiable on (a, b) with $F(a) = F(b) = 0$. Hence, by Rolle's theorem (Theorem 9.3.1), there exists $c \in (a, b)$ such that $F'(c) = 0$.

Differentiating F we get

$$F'(x) = -\left(f^{(1)}(x) - f^{(1)}(x) + \frac{f^{(2)}(x)}{1!}(b-x) \right.$$

$$- \frac{f^{(2)}(x)}{1!}(b-x) + \frac{f^{(3)}(x)}{2!}(b-x)^2$$

$$- \cdots$$

$$- \frac{f^{(n-1)}(x)}{(n-2)!}(b-x)^{n-2} + \frac{f^{(n)}(x)}{(n-1)!}(b-x)^{n-1}$$

$$\left. - \frac{\Delta}{(n-1)!}(b-x)^{n-1} \right)$$

$$= \frac{\Delta - f^{(n)}(x)}{(n-1)!}(b-x)^{n-1}.$$

Since $b - c \neq 0$, putting $x = c$ gives $\Delta = f^{(n)}(c)$ as required. $\qquad\square$

Definition 18.5.2 The term $\dfrac{f^{(n)}(c)}{n!}(b-a)^n$ in the conclusion of Taylor's theorem is called the *Taylor's theorem remainder*. We shall denote it by R_n.

Notes 18.5.3 (1) The case $n = 1$ in Taylor's theorem is the mean value theorem.

(2) The conclusion of Taylor's theorem remains true if $b < a$ provided $[a, b]$ and (a, b) are replaced by $[b, a]$ and (b, a), respectively. We leave this for the reader to verify by applying Taylor's theorem to the real function g defined by $g(x) = f(-x)$, on $[-a, -b]$.

(3) Putting $a = 0$ and $b = x$ in Taylor's theorem, the conclusion becomes

$$f(x) = f(0) + \frac{f^{(1)}(0)}{1!}x + \frac{f^{(2)}(0)}{2!}x^2 + \cdots + \frac{f^{(n-1)}(0)}{(n-1)!}x^{n-1} + R_n,$$

where

$$R_n = \frac{f^{(n)}(c)}{n!}x^n$$

for some c between 0 and x.

Example 18.5.4 Use Taylor's theorem with $n = 3$ to obtain bounds for $e^{1/4}$.

Solution With the notation of Taylor's theorem, let $n = 3$, $f = \exp$, $a = 0$ and $b = 1/4$. Then the conditions of the theorem are satisfied and so

$$e^{1/4} = e^0 + \frac{e^0}{1!}\left(\frac{1}{4} - 0\right) + \frac{e^0}{2!}\left(\frac{1}{4} - 0\right)^2 + \frac{e^c}{3!}\left(\frac{1}{4} - 0\right)^3$$

for some $c \in (0, 1/4)$. This reduces to

$$e^{1/4} = \frac{41}{32} + \frac{e^c}{384}. \tag{18.5.1}$$

Since exp is strictly increasing on $\mathbb{R}$ and $0 < c < 1/4$ it follows that $1 < e^c < e^{1/4}$. Using this in (18.5.1) gives

$$\frac{41}{32} + \frac{1}{384} < e^{1/4} < \frac{41}{32} + \frac{e^{1/4}}{384}.$$

Hence

$$\frac{493}{384} < e^{1/4} \quad \text{and} \quad e^{1/4}\left(1 - \frac{1}{384}\right) < \frac{41}{32}$$

so that

$$\frac{493}{384} < e^{1/4} < \frac{492}{383}. \qquad \square$$

Definitions 18.5.5 Let f be a real function such that, for all $n \in \mathbb{N}$, $f^{(n)}(0)$ is defined. Then

$$f(0) + \frac{f^{(1)}(0)}{1!}x + \frac{f^{(2)}(0)}{2!}x^2 + \cdots = \sum_{r=0}^{\infty} \frac{f^{(r)}(0)}{r!}x^r$$

is the *Maclaurin series for f* (or *for $f(x)$*). The *range of validity* of the series is the set of x for which the series converges to $f(x)$.

A list of *standard Maclaurin series* and their ranges of validity is given in Appendix G.

Theorem 18.5.6 *Let f be a real function such that, for all $n \in \mathbb{N}$, $f^{(n)}$ exists on $[0, x]$ or $[x, 0]$, and the Taylor's theorem remainder $R_n \to 0$ as $n \to \infty$. Then*

$$f(x) = f(0) + \frac{f^{(1)}(0)}{1!}x + \frac{f^{(2)}(0)}{2!}x^2 + \cdots = \sum_{r=0}^{\infty} \frac{f^{(r)}(0)}{r!}x^r.$$

Proof For $n \in \mathbb{N}$, let $s_n = \sum_{r=0}^{n-1} \frac{f^{(r)}(0)}{r!}x^r$ (the nth partial sum).

Then, using Notes 18.5.3,

$$f(x) = f(0) + \frac{f^{(1)}(0)}{1!}x + \frac{f^{(2)}(0)}{2!}x^2 + \cdots + \frac{f^{(n-1)}(0)}{(n-1)!}x^{n-1} + R_n = s_n + R_n.$$

Hence, using Theorem 18.2.10 (2), $s_n = f(x) - R_n \to f(x)$ and the result follows. $\quad \square$

Hazard A function may be defined on $I = (-A, A)$ and have a Maclaurin series which converges on I without the two being equal on I. For example, let f be the real function defined by $f(0) = 0$ and $f(x) = \exp(-1/x^2)$ ($x \in \mathbb{R} - \{0\}$). Then, for all $r \in \mathbb{N} \cup \{0\}$, $f^{(r)}(0) = 0$. In this case, the function is defined and its Maclaurin series converges on $\mathbb{R}$ but the two are not equal except at 0.

Example 18.5.7 Find the Maclaurin series for exp, and its range of validity.

Solution Let $f = \exp$. Then, for $r \in \mathbb{N}$, $f^{(r)} = \exp$ so that $f^{(r)}(0) = e^0 = 1$. Thus the Maclaurin series for f is

$$1 + \frac{x}{1!} + \frac{x^2}{2!} + \cdots = \sum_{r=0}^{\infty} \frac{x^r}{r!}.$$

When $x = 0$ this series converges to $1 = f(0)$. By Taylor's theorem, for $x \neq 0$, the series converges to $f(x)$ provided the Taylor's theorem remainder $R_n \to 0$. We have

$$R_n = \frac{e^c}{n!} x^n$$

where c lies between 0 and x. Hence, $c < |x|$ and, since $\exp$ is strictly increasing on $\mathbb{R}$,

$$0 \leq |R_n| = \frac{e^c}{n!} |x|^n < e^{|x|} \frac{|x|^n}{n!}.$$

By Theorem 18.2.14, $|x|^n/n! \to 0$. Then the sandwich principle gives $|R_n| \to 0$ and Theorem 18.2.10 (6) gives $R_n \to 0$. It follows that the range of validity is $\mathbb{R}$. Thus we have

$$e^x = 1 + \frac{x}{1!} + \frac{x^2}{2!} + \cdots = \sum_{r=0}^{\infty} \frac{x^r}{r!} \quad (x \in \mathbb{R}). \qquad \square$$

Theorem 18.5.8 *Let $\sum_{r=0}^{\infty} a_r x^r$ converge for $x \in I = (-A, A)$ where $0 < A \leq \infty$. Define $f : I \to \mathbb{R}$ by $f(x) = \sum_{r=0}^{\infty} a_r x^r$. Then $\sum_{r=0}^{\infty} a_r x^r$ is the Maclaurin series for f.*

Proof We must show that $a_0 = f(0)$ and, for $r \in \mathbb{N}$, $a_r = \dfrac{f^{(r)}(0)}{r!}$. For $x \in I$,

$$a_0 + a_1 x + a_2 x^2 + \cdots + a_r x^r + a_{r+1} x^{r+1} + a_{r+2} x^{r+2} + \cdots = f(x). \qquad (18.5.2)$$

Putting $x = 0$ in (18.5.2) gives $a_0 = f(0)$. For $r \in \mathbb{N}$, $f^{(r)}$ is defined on I (see Note 18.4.9 (2)). Differentiating (18.5.2) r times gives

$$a_r r! + a_{r+1} \frac{(r+1)!}{1!} x + a_{r+2} \frac{(r+2)!}{2!} x^2 + \cdots = f^{(r)}(x).$$

Then putting $x = 0$ gives $a_r r! = f^{(r)}(0)$, *i.e.* $a_r = \dfrac{f^{(r)}(0)}{r!}$. $\qquad \square$

Example 18.5.9 Find the Maclaurin series for e^{-2x} and its range of validity.

Solution From Example 18.5.7 we have

$$e^x = 1 + \frac{x}{1!} + \frac{x^2}{2!} + \frac{x^3}{3!} + \cdots = \sum_{r=0}^{\infty} \frac{x^r}{r!} \quad (x \in \mathbb{R}).$$

Replacing x with $-2x$ we get

$$e^{-2x} = 1 - \frac{2x}{1!} + \frac{4x^2}{2!} - \frac{8x^3}{3!} + \cdots = \sum_{r=0}^{\infty} \frac{(-2)^r}{r!} x^r.$$

This holds for $-2x \in \mathbb{R}$, *i.e.* for all $x \in \mathbb{R}$, and it follows from Theorem 18.5.8 that we have found the required Maclaurin series. $\qquad \square$

Example 18.5.10 Find the Maclaurin series for $(1-x)^{-1}$. Deduce the Maclaurin series for (a) $(2+x)^{-1}$, (b) $\log(1-x)$ and (c) $x/(1-x)^2$. Use (c) to find the sum of the series $\sum_{r=1}^{\infty} r/3^r$.

Solution We shall make use of Theorems 18.4.8 and 18.5.8 without further reference. By Example 18.3.5,

$$(1-x)^{-1} = 1 + x + x^2 + x^3 + \cdots = \sum_{r=0}^{\infty} x^r \quad (|x| < 1). \qquad (18.5.3)$$

(a) Using (18.5.3), we have

$$(2+x)^{-1} = \frac{1}{2}\left(1 + \frac{x}{2}\right)^{-1} = \frac{1}{2}(1-y)^{-1} \quad (\text{where } y = -x/2)$$

$$= \frac{1}{2}(1 + y + y^2 + y^3 + \cdots) \quad (|y| < 1)$$

$$= \frac{1}{2}\left(1 - \frac{x}{2} + \frac{x^2}{4} - \frac{x^3}{8} + \cdots\right) \quad (|x/2| < 1),$$

i.e.

$$(2+x)^{-1} = \frac{1}{2} - \frac{x}{4} + \frac{x^2}{8} - \frac{x^3}{16} + \cdots = \sum_{r=0}^{\infty} \frac{(-1)^r}{2^{r+1}} x^r \quad (|x| < 2).$$

(b) Integrating (18.5.3) gives

$$-\log|1-x| = x + \frac{x^2}{2} + \frac{x^3}{3} + \frac{x^4}{4} + \cdots + C \quad (|x| < 1)$$

and putting $x = 0$ gives $C = 0$. Hence,

$$\log(1-x) = -x - \frac{x^2}{2} - \frac{x^3}{3} - \frac{x^4}{4} - \cdots = \sum_{r=1}^{\infty} \left(-\frac{1}{r}\right) x^r \quad (|x| < 1).$$

[It may be shown that the range of validity is $[-1, 1)$.]

(c) Differentiating (18.5.3) gives

$$(1-x)^{-2} = 1 + 2x + 3x^2 + 4x^3 + \cdots \quad (|x| < 1),$$

and multiplying by x,

$$x(1-x)^{-2} = x + 2x^2 + 3x^3 + 4x^4 + \cdots \quad (|x| < 1). \qquad (18.5.4)$$

Finally, putting $x = 1/3$ in (18.5.4) gives

$$\frac{3}{4} = \frac{1}{3} + \frac{2}{9} + \frac{3}{27} + \frac{4}{81} + \cdots.$$

Hence, $\displaystyle\sum_{r=1}^{\infty} \frac{r}{3^r}$ converges to $\dfrac{3}{4}$.   □

For 'small' values of x, the 'first few' non-zero terms of the Maclaurin series for $f(x)$ usually give a reasonable approximation to $f(x)$. With this in mind, it is sometimes useful to find the Maclaurin series as far as the term in x^n for some specified n, without attempting to determine the whole series. When doing this, we may ignore terms whose only contribution to the final series is to terms in x^m where $m > n$.

The last two examples in this chapter illustrate some possible techniques for finding *truncated* Maclaurin series. We shall ignore the ranges of validity. The reader may find it instructive to try and work them out.

Example 18.5.11 Find the Maclaurin series for $\tan x$ as far as the term in x^3.

Solution Let $f(x) = \tan x$ so that $f(0) = 0$ and

$$f^{(1)}(x) = \sec^2 x, \qquad f^{(1)}(0) = 1,$$

$$f^{(2)}(x) = 2\sec^2 x \tan x, \qquad f^{(2)}(0) = 0,$$

$$f^{(3)}(x) = 4\sec^2 x \tan^2 x + 2\sec^4 x, \qquad f^{(3)}(0) = 2.$$

Hence, the Maclaurin series for $\tan x$ is

$$0 + \frac{1}{1!}x + \frac{0}{2!}x^2 + \frac{2}{3!}x^3 + \cdots, \qquad i.e. \ x + \frac{x^3}{3} + \cdots.$$

Alternatively, using the Maclaurin series for $\sin x$, $\cos x$ and $(1-x)^{-1}$ (see Appendix G) together with Theorem 18.4.8 (1)(c),

$$\tan x = \frac{\sin x}{\cos x} = \frac{x - \dfrac{x^3}{3!} + \dfrac{x^5}{5!} - \cdots}{1 - \dfrac{x^2}{2!} + \dfrac{x^4}{4!} - \cdots}$$

$$= \left(x - \frac{x^3}{6} + \cdots\right)(1-y)^{-1} \qquad \left(\text{where } y = \frac{x^2}{2} - \frac{x^4}{24} + \cdots\right)$$

$$= \left(x - \frac{x^3}{6} + \cdots\right)(1 + y + \cdots)$$

$$= \left(x - \frac{x^3}{6} + \cdots\right)\left(1 + \left(\frac{x^2}{2} - \cdots\right) + \cdots\right)$$

$$= x + \frac{x^3}{2} - \frac{x^3}{6} + \cdots = x + \frac{1}{3}x^3 + \cdots,$$

as before. □

Example 18.5.12 Find the Maclaurin series for each of the following, as far as the term in x^4.

(a) $\cos 2x - x\sin x,$ (b) $\dfrac{xe^{-x}}{1+x^2},$ (c) $\log(\cos x).$

Solution We make use of standard Maclaurin series (see Appendix G) and Theorem 18.4.8.

(a) We have

$$\cos 2x - x \sin x = \left(1 - \frac{(2x)^2}{2!} + \frac{(2x)^4}{4!} - \cdots\right) - x\left(x - \frac{x^3}{3!} + \cdots\right)$$

$$= \left(1 - 2x^2 + \frac{2}{3}x^4 - \cdots\right) - \left(x^2 - \frac{1}{6}x^4 + \cdots\right)$$

$$= 1 - 3x^2 + \frac{5}{6}x^4 + \cdots .$$

(b) We have

$$\frac{xe^{-x}}{1+x^2} = xe^{-x}(1+x^2)^{-1} = x\left(1 - x + \frac{x^2}{2} - \frac{x^3}{6} + \cdots\right)(1 - x^2 + \cdots)$$

$$= x\left(1 - x^2 + \cdots - x + x^3 - \cdots + \frac{x^2}{2} - \cdots - \frac{x^3}{6} + \cdots\right)$$

$$= x - x^2 - \frac{1}{2}x^3 + \frac{5}{6}x^4 + \cdots .$$

(c) We have

$$\log(\cos x) = \log\left(1 - \frac{x^2}{2} + \frac{x^4}{24} - \cdots\right)$$

$$= \log(1 - y) \qquad \left(\text{where } y = \frac{x^2}{2} - \frac{x^4}{24} + \cdots\right)$$

$$= -y - \frac{y^2}{2} - \cdots$$

$$= -\left(\frac{x^2}{2} - \frac{x^4}{24} + \cdots\right) - \frac{1}{2}\left(\frac{x^2}{2} + \cdots\right)^2 - \cdots$$

$$= -\frac{x^2}{2} + \frac{x^4}{24} - \cdots - \frac{x^4}{8} + \cdots$$

$$= -\frac{1}{2}x^2 - \frac{1}{12}x^4 + \cdots .$$

Alternatively, in Example 18.5.11 we found that

$$\tan x = x + \frac{1}{3}x^3 + \cdots .$$

Integrating gives

$$-\log(\cos x) = \frac{1}{2}x^2 + \frac{1}{12}x^4 + \cdots + C$$

and putting $x = 0$ gives $C = 0$. Hence

$$\log(\cos x) = -\frac{1}{2}x^2 - \frac{1}{12}x^4 - \cdots ,$$

as before.   □

18.X   Exercises

1. Find lower and upper bounds for the sequence $\{x_n\}$ when x_n is

$$\text{(a) } \frac{4n}{3n-2}, \qquad \text{(b) } \frac{n^2-3n}{3n^2+5}, \qquad \text{(c) } \frac{n^2-7}{n^2-5}.$$

2. Let $\{a_n\}$ and $\{b_n\}$ be the sequences defined by

$$a_n = \frac{n+3}{n^2+3}, \qquad b_n = \frac{3n^2-4}{3n-2} \qquad (n \in \mathbb{N}).$$

Show that $\{a_n\}$ is decreasing and $\{b_n\}$ is increasing.

3. Evaluate $\lim\limits_{n\to\infty} x_n$ when x_n is

$$\text{(a) } \frac{6n-5}{3n+4}, \qquad \text{(b) } \frac{3-4n}{6n-1}, \qquad \text{(c) } \frac{2}{n+\pi},$$

$$\text{(d) } \frac{n-1/n}{n+1/n}, \qquad \text{(e) } \frac{2n^2+n+5}{3n^2-n+1}, \qquad \text{(f) } \frac{n^2-3n+2}{1+n-5n^2},$$

$$\text{(g) } \frac{n+2}{n^2+n+1}, \qquad \text{(h) } \frac{n^3+3n}{2n^3-7}, \qquad \text{(i) } \frac{1}{n^3}\left(\frac{n}{n-3}\right),$$

$$\text{(j) } \tan\left(\frac{n\pi}{4n+3}\right), \qquad \text{(k) } \log(n^2+1) - 2\log(n+1).$$

4. Evaluate $\lim\limits_{n\to\infty} (1/3^n)$ and deduce the values of

$$\text{(a) } \lim_{n\to\infty} \frac{3^n+1}{3^n-1}, \qquad \text{(b) } \lim_{n\to\infty} \frac{3^{n+1}}{3^n-1}.$$

5. Find the sum of each of the following series.

$$\text{(a) } \sum_{r=0}^{\infty} \left(\frac{2}{3}\right)^r, \quad \text{(b) } \sum_{r=0}^{\infty} (-1)^r \frac{1}{3^r}, \quad \text{(c) } \sum_{r=0}^{\infty} \frac{2^{r+1}}{3^r}, \quad \text{(d) } \sum_{r=0}^{\infty} \frac{2^r+1}{3^r}.$$

6. Express the infinite decimal $d = 0{\cdot}101010101\ldots$ in the form m/n where $m, n \in \mathbb{N}$. [Hint. Write $d = 0{\cdot}1 + 0{\cdot}001 + 0{\cdot}00001 + \cdots$.]

7. Prove that, for $n \in \mathbb{N}$, $3^n > n^2$. Deduce that $n/3^n \to 0$ as $n \to \infty$. For $n \in \mathbb{N}$, define

$$s_n = \frac{1}{3} + \frac{2}{9} + \frac{3}{27} + \cdots + \frac{n}{3^n}.$$

By considering $3s_n - s_n$, show that

$$s_n = \frac{3}{4} - \frac{3}{4}\cdot\frac{1}{3^n} - \frac{1}{2}\cdot\frac{n}{3^n}.$$

Hence find the sum of the series $\sum_{r=1}^{\infty} r/3^r$. [An alternative method is given in Example 18.5.10.]

8. For each of the following series, find the partial fraction decomposition of the rth term. Hence find the nth partial sum and the sum of the series.

(a) $\displaystyle\sum_{r=1}^{\infty} \frac{1}{(3r-2)(3r+1)}$,   (b) $\displaystyle\sum_{r=1}^{\infty} \frac{1}{r^2 + 2r}$,   (c) $\displaystyle\sum_{r=1}^{\infty} \frac{1}{r(r+1)(r+2)}$.

9. Use the comparison test to determine whether each of the following series converges or diverges.

(a) $\displaystyle\sum_{r=1}^{\infty} \frac{1}{r^2 + 1}$,   (b) $\displaystyle\sum_{r=1}^{\infty} \frac{2r+1}{r^3 + r}$,   (c) $\displaystyle\sum_{r=1}^{\infty} \frac{2r+1}{r^2 + r}$,

(d) $\displaystyle\sum_{r=1}^{\infty} \frac{1}{2r^3 - 1}$,   (e) $\displaystyle\sum_{r=1}^{\infty} \frac{4}{r^2 - 2}$,   (f) $\displaystyle\sum_{r=1}^{\infty} \frac{3^r + 5^r}{2^r + 7^r}$.

10. Use the ratio test to determine whether each of the following series converges or diverges.

(a) $\displaystyle\sum_{r=1}^{\infty} \frac{7^r}{r^3}$,   (b) $\displaystyle\sum_{r=1}^{\infty} \frac{10^r}{(r+3)!}$,   (c) $\displaystyle\sum_{r=1}^{\infty} \binom{2r}{r}\left(\frac{1}{2}\right)^r$,   (d) $\displaystyle\sum_{r=1}^{\infty} \frac{r+1}{3^{2r+1}}$.

11. Use the alternating series test to show that each of the following series converges.

(a) $\displaystyle\sum_{r=1}^{\infty} (-1)^{r-1} \frac{r}{r^2 + 1}$,   (b) $\displaystyle\sum_{r=1}^{\infty} \frac{(-1)^r}{\sqrt{r+1}}$,   (c) $\displaystyle\sum_{r=2}^{\infty} (-1)^{r-1} \frac{\log r}{r}$.

12. Determine whether $\sum_{r=1}^{\infty} a_r$ converges or diverges when a_r is

(a) $\displaystyle\frac{r^2}{2^r}$,   (b) $\displaystyle\frac{r}{\sqrt{r^4 + 1}}$,   (c) $\displaystyle(-1)^{r-1}\frac{r+1}{r^2+1}$,

(d) $\displaystyle\frac{r!}{100^r}$,   (e) $\displaystyle(-1)^{r-1}\frac{3r+1}{4r+1}$,   (f) $\displaystyle\frac{\cos(r\pi/3)}{r^2 + 3}$.

Information in Appendix F may help you decide which test to use in each case.

13. Find the interval of convergence for each of the following power series.

(a) $\displaystyle\sum_{r=0}^{\infty} \frac{(3x)^r}{r^2 + 1}$,   (b) $\displaystyle\sum_{r=0}^{\infty} \frac{2^r x^r}{3^r}$,   (c) $\displaystyle\sum_{r=0}^{\infty} \frac{2^r x^r}{3^{r+1}}$,   (d) $\displaystyle\sum_{r=1}^{\infty} \frac{(-x)^r}{r 5^r}$,   (e) $\displaystyle\sum_{r=1}^{\infty} \frac{x^r}{r^r}$.

14. By applying Taylor's theorem with $n = 2$, show that

(a) $\displaystyle\frac{61}{50} < e^{1/5} < \frac{60}{49}$,   (b) $\displaystyle\frac{49}{50} < \cos(1/5) < \frac{50}{51}$.

15. By applying Taylor's theorem with $n = 6$, show that $\displaystyle\frac{1957}{720} < e < \frac{1956}{719}$.

16. Use Example 8.4.5 and Theorem 18.2.14 to find the Maclaurin series for $\sin$ and show that its range of validity is $\mathbb{R}$. Deduce the Maclaurin series for $\cos$.

17. Find the Maclaurin series for the polynomial

$$p(x) = a_n x^n + a_{n-1} x^{n-1} + \cdots + a_1 x + a_0$$

and state its range of validity.

18. Use the Maclaurin series for $\exp$ to find the Maclaurin series for $\cosh$ and $\sinh$. What are the ranges of validity of these series?

19. Make use of standard Maclaurin series to find the Maclaurin series for each of the following. In each case determine the range of validity.

    (a) $\exp(x^2)$,   (b) $\cos(-2x)$,   (c) $\dfrac{x}{1+x}$,

    (d) $\log\left(1 - \dfrac{x}{3}\right)$,   (e) $\dfrac{1}{x-2}$,   (f) $\cos x + \cosh x$.

20. Use the Maclaurin series for $(1-x)^{-1}$ to find the Maclaurin series for $(1-3x)^{-1}$. What is its range of validity? Deduce, by differentiating, the Maclaurin series for $(1-3x)^{-2}$.

21. Use the Maclaurin series for $(1-x)^{-1}$ to find the Maclaurin series for $(1+x^2)^{-1}$. Deduce the Maclaurin series for $\tan^{-1} x$.

22. Find the Maclaurin series for $\sec x$ as far as the term in x^4, (a) by differentiating $\sec x$ four times, (b) by using the Maclaurin series for $\cos x$.

23. Find the Maclaurin series for each of the following, as far as the term in x^4.

    (a) $\dfrac{1}{1+x+x^2}$,   (b) $\dfrac{\log(1+x)}{1+x}$,   (c) $e^x \cos x$,

    (d) $\exp(\sin x)$,   (e) $\dfrac{1}{1 - \sin 2x}$,   (f) $\log(1 + x \cos x)$.

24. Find the Maclaurin series for $f(x) = \log\left(\dfrac{1+x}{1-x}\right)$. What is its range of validity? Show that, when $x = 1/11$,

$$f(x) - 2\left(x + \frac{x^3}{3}\right) < \frac{2}{5}x^5(1 + x^2 + x^4 + \cdots) < \frac{1}{3 \cdot 10^5}.$$

Hence find an approximate value for $\log(1 \cdot 2)$, correct to 4 decimal places.

25. (Binomial Expansion) Let $a \in \mathbb{R}$. By differentiating, show that the Maclaurin series for $(1+x)^a$ is

$$1 + ax + \frac{a(a-1)}{2!}x^2 + \frac{a(a-1)(a-2)}{3!}x^3 + \cdots$$

$$= \sum_{r=0}^{\infty} \frac{a(a-1)\ldots(a-r+1)}{r!}x^r.$$

The range of validity depends on a. It may be proved that, in all cases, the range of validity contains $(-1, 1)$. What happens when $a \in \mathbb{N}$?

Find the Maclaurin series for $\sqrt{1+x}$ as far as the term in x^3.

Chapter 19

Numerical Methods

In applications we often have to find numerical values. In many cases, it is not possible to obtain *exact* values. In practice it is usually enough to obtain an *approximate* value, together with a bound for the error involved.

In this chapter, we consider three such problems—finding zeros of a real function, evaluating a definite integral and finding the solution of an initial value problem for an ordinary differential equation.

The methods are of two types—*iteration* and *local approximation*. In the former, we obtain a sequence of approximations, each depending on previous values. In the latter, we approximate the function on an interval by a polynomial function (usually constant, linear or quadratic).

Questions of whether the sequence produced by an iterative scheme converges to the correct value, and the estimation of the error involved, are usually difficult to answer. They form the subject matter of *numerical analysis*. Here, we will only discuss these topics in cases where an answer is accessible at this level. Only in the case of the bisection method (for finding zeros) can we provide a simple answer.

In Sections 19.2 and 19.3, we have two methods for locating zeros. In Section 19.4, we have three methods for evaluating definite integrals. The reader should note that the more complicated methods tend to be more efficient in that they produce a more accurate answer with an equivalent amount of work. The search for more efficient methods is another aspect of numerical analysis.

19.1 Errors

When using numerical methods it is important to be aware of the types of inaccuracy that may be present. In all the methods we will consider both the exact solution c and the approximation $\widehat{c}$, obtained by the numerical method, will be real numbers. The difference $c - \widehat{c}$ is called the (*absolute*) *error*.

Even when the exact solution is unknown, one may be able to prove that there is a limitation on the error. Let $\varepsilon > 0$ and suppose that

$$|c - \widehat{c}| \leq \varepsilon.$$

Then ε is called an *error bound*. This inequality is conventionally written as $c = \widehat{c} \pm \varepsilon$, and we adopt this practice here. Such error bounds are obtained by analysis of the numerical method.

If we use a iterative method to obtain a sequence $\{c_n\}$ of approximations to c then we say that these approximations *converge* to c if $c_n \to c$ as $n \to \infty$ (see Section 18.2). If ε_n is an error bound for the nth approximation c_n, then

$$0 \leq |c - c_n| \leq \varepsilon_n,$$

and if the error bound $\varepsilon_n \to 0$ as $n \to \infty$ then, by the sandwich principle, the sequence of approximations will converge to c.

Two main types of error are encountered. There are *round-off errors* which come about because the computer or calculator we use works to a finite accuracy. For example, on a (primitive) calculator working to three significant figures we might evaluate

$$\frac{13}{14} - \frac{12}{13} \approx 0\cdot929 - 0\cdot923$$
$$= 0\cdot006.$$

However, we know that

$$\frac{13}{14} - \frac{12}{13} = \frac{13^2 - 12.14}{13.14} = \frac{1}{182}$$
$$\approx 0\cdot00549 \quad \text{(correct to 3 significant figures)}.$$

It is the subtraction of numbers that are 'almost' equal (according to the computer) that leads to an inaccuracy in this very simple example.

The second type of error, which is the one of most mathematical interest, is *truncation error*. In iterative methods such errors usually occur because we stop the process when the successive approximations agree to a certain accuracy or when an error bound is known to be sufficiently small. In what follows we will only discuss truncation errors. Also, when adding terms to obtain the approximation, we will write the results as rational numbers, or as rational multiples of irrational numbers, so that rounding errors are eliminated.

19.2   The Bisection Method

Let f be a real function which is continuous on $[a, b]$, and such that $f(a)$ and $f(b)$ have different signs. Then, by the intermediate value theorem, $f(c) = 0$ for some $c \in (a, b)$. The idea of the bisection method is to use this result repeatedly to determine smaller and smaller intervals which contains a zero. The process is halted when the interval is so small that, by taking the mid-point of this interval, we approximate the zero to whatever accuracy is required.

Now let $c \in (a, b)$ be a zero of f. The distance from c to $\widehat{c} = (a+b)/2$, the mid-point of (a, b), is at most half the length of the interval. Hence

$$|c - \widehat{c}| < \frac{b - a}{2}, \tag{19.2.1}$$

and so half of the length of the interval is an error bound.

To illustrate the method, consider the real function f, defined by

$$f(x) = x^3 + x - 23,$$

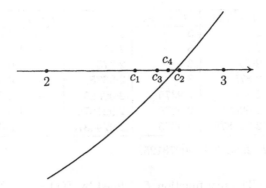

Figure 19.2.1

which is continuous on any interval and has exactly one real zero. We find that $f(2) < 0 < f(3)$, and hence $f(x)$ has a zero $c \in (2,3)$. Figure 19.2.1 illustrates the process we will describe. The first approximation to c is $c_1 = (2+3)/2 = 5/2$, and from (19.2.1) an error bound is $\varepsilon_1 = 1/2$. Further, we find that $f(5/2) < 0 < f(3)$ and so $c \in (5/2, 3)$. Hence we take $c_2 = (5/2 + 3)/2 = 11/4$ as a second approximation to c with $\varepsilon_2 = 1/4$ as error bound. If we carry out this process n times we obtain an nth approximation c_n with an error bound of $\varepsilon_n = 1/2^n$. Since $1/2^n \to 0$ as $n \to \infty$, this enables us, by taking n sufficiently large, to determine the zero c to any desired accuracy. Note that if the function has two zeros in the initial interval only one will be approximated in this process.

 Hazard A good approximation to a zero is a value of x which is 'close' to the zero c. This is not necessarily the same as x being such that $f(x)$ is 'small'.

It is convenient to lay out the iterations in a table. We show four iterations for the example discussed above.

n	$f < 0$ at	$f > 0$ at	c_n	$f(c_n)$	ε_n
1	2	3	2·5	−	0·5
2	2·5	3	2·75	+	0·25
3	2·5	2·75	2·625	−	0·125
4	2·625	2·75	2·6875		0·0625

This calculation shows that $c = 2{\cdot}6875 \pm 0{\cdot}0625$.

Example 19.2.1 Use the bisection method to find $\sqrt[3]{25}$ to within $0{\cdot}01$.

Solution We take $f(x) = x^3 - 25$. Since $2^3 - 25 < 0 < 3^3 - 25$ we first consider the interval $[2, 3]$ and proceed until the error bound is less than $0{\cdot}01$.

n	$f < 0$ at	$f > 0$ at	c_n	$f(c_n)$	ε_n
1	2	3	2·5	−	0·5
2	2·5	3	2·75	−	0·25
3	2·75	3	2·875	−	0·125
4	2·875	3	2·9375	+	0·0625
5	2·875	2·9375	2·90625	−	0·03125
6	2·90625	2·9375	2·921875	−	0·015625
7	2·921875	2·9375	2·9296875		0·0078125

Thus $\sqrt[3]{25} = 2 \cdot 9296875 \pm 0 \cdot 0078125$.   □

Notes 19.2.2 (1) The real function f, defined by $f(x) = x^3 - 25$, is increasing since $f'(x) = 3x^2 \geq 0$ and so, by Theorems 9.4.1 and 2.6.3, has at most one zero. Hence the zero approximated in Example 19.2.1 *is* the cube root of 25.

(2) The convergence of the bisection method is very slow. Since we reduce the error bound by a factor of two with each iteration, and $10 > 2^3$, it requires, on average, more than three iterations to achieve each decimal place accurately. In Example 19.2.1 we have $c_6 \approx 2 \cdot 92$ and $c_7 \approx 2 \cdot 93$ (to two decimal places) and so after seven iterations we are only sure of the first decimal place.

19.3   Newton's Method

Let f be a real function that is differentiable, with f' continuous on some non-trivial interval. Suppose that f has a *simple* zero c in this interval, *i.e.* $f(c) = 0$ but $f'(c) \neq 0$. Since f' is continuous there is a neighbourhood N of c in which $f'(x) \neq 0$. We choose some $c_0 \in N$, with c_0 'sufficiently close' to c in a sense discussed later, and, as illustrated in Figure 19.3.1, approximate the graph of the function by its tangent at $(c_0, f(c_0))$. An approximation to the zero of f is the zero of the linear function determined by this tangent. Since $f'(c_0) \neq 0$, the tangent is not parallel to the x-axis, and so intersects it at some point $(c_1, 0)$. The x-coordinate c_1 provides an approximation to c.

Expressing this description mathematically, the tangent to the graph of f at $(c_0, f(c_0))$ has equation
$$y - f(c_0) = f'(c_0)(x - c_0),$$
and so, since $(c_1, 0)$ lies on this line, $0 - f(c_0) = f'(c_0)(c_1 - c_0)$, *i.e.*

$$c_1 = c_0 - \frac{f(c_0)}{f'(c_0)}.$$

If $c_1 \in N$ we may then repeat this process by considering the tangent to f at $(c_1, f(c_1))$ to obtain a second point $(c_2, 0)$, giving c_2 as a further approximation to c. Continuing in this way, we get a sequence $\{c_n\}$ of approximations defined by

$$c_n = c_{n-1} - \frac{f(c_{n-1})}{f'(c_{n-1})}, \tag{19.3.1}$$

which, we hope, converges to c. This means of finding approximations to zeros is called *Newton's method* or the *Newton-Raphson method*.

To obtain the initial point in the sequence c_0 we may use the intermediate value theorem to find the best integer approximation to the zero.

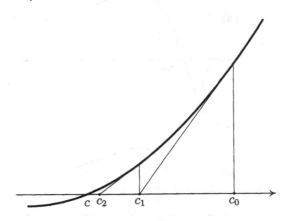

Figure 19.3.1

Example 19.3.1 Show that

$$x^3 - 2x^2 + 3x - 5 = 0 \tag{19.3.2}$$

has exactly one real root. Find the best integer approximation to this root and apply Newton's method to obtain one further approximation.

Solution Let f be the real function defined by

$$f(x) = x^3 - 2x^2 + 3x - 5.$$

Then f is continuous on $\mathbb{R}$ and, since $f(1) < 0 < f(2)$, the intermediate value theorem gives $f(c) = 0$ for some $c \in (1, 2)$. So f has at least one real zero. Since $f(3/2) < 0$ it follows that $c \in (3/2, 2)$ and so 2 is the best integer approximation to c.

Also, for $x \in \mathbb{R}$,

$$f'(x) = 3x^2 - 4x + 3 = 3(x - 2/3)^2 + 5/3 > 0$$

and so, by Theorem 9.4.1, f is a strictly increasing function. By Theorem 2.6.3, f is injective and thus (19.3.2) has exactly one real root.

By Newton's method, using (19.3.1), we get

$$c_1 = c_0 - \frac{f(c_0)}{f'(c_0)} = c_0 - \frac{c_0^3 - 2c_0^2 + 3c_0 - 5}{3c_0^2 - 4c_0 + 3},$$

and, for $c_0 = 2$, we get

$$c_1 = \frac{13}{7}. \qquad \square$$

Remark Unlike the bisection method, there is no *a priori* guarantee that successive approximations c_n will approach the root c or even converge. However, we will show that if a sequence of approximations does converge then the limit must be a root. As

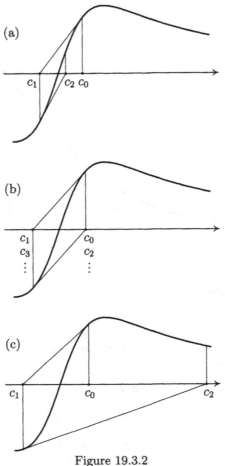

Figure 19.3.2

shown in Figure 19.3.2, the convergence, or otherwise, is dependent on the choice of the starting value c_0.

In (a) c_0 is sufficiently close to c that the method converges, in (b) the approximations alternate between c_0 and c_1 and in (c) successive approximations move further from c.

This uncertainty about convergence is compensated for by the greater efficiency of Newton's method, when it does converge.

Theorem 19.3.2 *Let f be a real function that is differentiable with a continuous non-zero derivative on an interval I. Let $\{c_n\}$, the sequence defined by (19.3.1), lie in the interval I and let $c_n \to c \in I$. Then $f(c) = 0$.*

Proof From (19.3.1) we have

$$c_n = c_{n-1} - \frac{f(c_{n-1})}{f'(c_{n-1})},$$

the left hand side of which converges to c. Now, by Theorem 18.2.8, $c_{n-1} \to c$ since $c_n \to c$. Further, by Theorem 18.2.10 (3), since f and f' are continuous on I, $f(c_{n-1}) \to f(c)$ and $f'(c_{n-1}) \to f'(c)$. Thus, the right hand side converges to $c - \dfrac{f(c)}{f'(c)}$ giving

$$c = c - \frac{f(c)}{f'(c)},$$

and hence $f(c) = 0$. □

Example 19.3.3 Using Newton's method, find successive approximations to $\sqrt[3]{25}$, taking 3 as starting value and continuing until the iterations agree to 9 significant figures.

Solution We consider the real function f, defined by $f(x) = x^3 - 25$. By Note 19.2.2 (1), the unique real zero of $f(x)$ is the cube root of 25.

Using (19.3.1), successive approximations to the root are related by

$$c_n = c_{n-1} - \frac{c_{n-1}^3 - 25}{3c_{n-1}^2} = \frac{2c_{n-1}^3 + 25}{3c_{n-1}^2} \quad (n \in \mathbb{N}), \tag{19.3.3}$$

where $c_0 = 3$.

The iterations are shown in the following table.

n	c_n
0	3·00000000
1	2·92592593
2	2·92401898
3	2·92401774
4	2·92401774

giving $\sqrt[3]{25} \approx 2 \cdot 92401774$. □

Remarks (1) While the fact that c_3 and c_4 in Example 19.3.3 agreed to 9 significant figures *suggests* that these approximate $\sqrt[3]{25}$ to this accuracy. It is difficult to prove this. A result like the one that follows is required.

(2) To illustrate the accuracy of the successive approximations in Example 19.3.3, we show an approximate truncation error, the difference between c_n and the actual value of $\sqrt[3]{25}$, at each stage.

n	Error
0	8×10^{-2}
1	2×10^{-3}
2	1×10^{-6}
3	5×10^{-13}
4	1×10^{-23}

Observe how the number of correct decimal places more or less double each time; this is very much faster than the bisection method.

In this example the value of $\sqrt[3]{25}$ may readily be found, to any desired accuracy, using any calculator or computer, but this is not always the case. In situations where the precise value can *only* be approximated one has to rely on results like the one that follows to have confidence that the approximations converge to the actual value, and to have a measure of how accurate the approximations are.

Theorem 19.3.4 *Let f be a real function with a continuous second derivative on $[a, b]$ such that $f'(x) > 0$ and $f''(x) > 0$ for $x \in [a, b]$. Let $c \in (a, b)$ be such that $f(c) = 0$ and let $\{c_n\}$ be the sequence obtained by Newton's method with $c_0 = b$. Then*

    (1) $c < c_n < c_{n-1} \leq b$ for $n \in \mathbb{N}$,

    (2) $c_n \to c$,

    (3) $|c - c_n| \leq \dfrac{A}{2f'(a)}|c - c_{n-1}|^2$, for $n \in \mathbb{N}$,

where A is an upper bound for $\{f''(x) : a \leq x \leq b\}$.

Proof (1) (By induction) From (19.3.1), for $n = 1$,

$$c_1 - c_0 = -\frac{f(c_0)}{f'(c_0)}. \tag{19.3.4}$$

Since $f'(x) > 0$ for $x \in [a, b]$, $f'(c_0) = f'(b) > 0$ and since f is strictly increasing on $[a, b]$, $f(c_0) = f(b) > f(c) = 0$. Therefore $c_1 - c_0 < 0$, *i.e.* $c_1 < c_0$.

    By Taylor's theorem (Theorem 18.5.1) on $[c, c_0]$, we have

$$0 = f(c) = f(c_0) + (c - c_0)f'(c_0) + \frac{1}{2}(c - c_0)^2 f''(\xi)$$

for some $\xi \in (c, c_0)$. Dividing by $f'(c_0)$ ($\neq 0$) and rearranging we get

$$c - \left(c_0 - \frac{f(c_0)}{f'(c_0)}\right) + \frac{1}{2}(c - c_0)^2 \frac{f''(\xi)}{f'(c_0)} = 0,$$

then using (19.3.4), this gives

$$c - c_1 = -\frac{1}{2}(c - c_0)^2 \frac{f''(\xi)}{f'(c_0)}.$$

Since $f''(\xi) > 0$ and $f'(c_0) > 0$, it follows that $c - c_1 < 0$, *i.e.* $c < c_1$. So (1) holds for $n = 1$.

    Now suppose that (1) holds for $n = k - 1$ for some $k \geq 2$. From (19.3.1) for $n = k$,

$$c_k - c_{k-1} = -\frac{f(c_{k-1})}{f'(c_{k-1})}. \tag{19.3.5}$$

By the induction hypothesis, $c < c_{k-1} \leq b$ and then by Taylor's theorem on $[c, c_{k-1}]$,

$$c - c_k = -\frac{1}{2}(c - c_{k-1})^2 \frac{f''(\xi)}{f'(c_{k-1})}, \tag{19.3.6}$$

where $\xi \in (c, c_{k-1})$. Since f is strictly increasing on $[a, b]$, $f(c_{k-1}) > f(c) = 0$ and we are given that $f'(c_{k-1}) > 0$ and $f''(\xi) > 0$. Thus by (19.3.5), $c_k - c_{k-1} < 0$ and, by

(19.3.6), $c - c_k < 0$, that is $c < c_k < c_{k-1} \leq b$. Hence, by induction, (1) holds for $n \in \mathbb{N}$.

(2) By (1),
$$c < \cdots < c_k < c_{k-1} < \cdots < c_1 < c_0 = b,$$

so that $\{c_n\}$ is decreasing and bounded below. By Theorem 18.2.15 (3), this sequence has a limit d, say, where $c \leq d$.

On the interval $[c, c_0]$, f' is continuous and non-zero and $\{c_n\} \subset [c, c_0]$ converges to d. Using Theorem 19.3.2, $f(d) = 0$ and since $f' > 0$ on $[c, c_0]$, f has at most one zero. It follows that $d = c$ and $c_n \to c$ as required.

(3) Using Taylor's theorem on $[c, c_{n-1}]$,

$$c - c_n = -\frac{1}{2}(c - c_{n-1})^2 \frac{f''(\xi)}{f'(c_{n-1})},$$

for some $\xi \in (c, c_{n-1})$. Since $f'' > 0$ on $[a, b]$, f' is strictly increasing, and as $a < c < c_{n-1} < b$, $f'(a) < f'(c_{n-1})$. Also, since f'' is continuous on $[a, b]$, it has an upper bound A. Since $f''(x) > 0$ for $x \in [a, b]$, $A > 0$. Using these estimates (3) follows immediately. $\qquad \square$

Example 19.3.5 Use Theorem 19.3.4 on the interval $[5/2, 3]$ to prove that the approximations to $\sqrt[3]{25}$ found by Newton's method in Example 19.3.3 converge. Further, find an error bound for each of the first three iterations.

Solution Let $f(x) = x^3 - 25$, then $f'(x) = 3x^2$ and $f''(x) = 6x$. Clearly, f'' is continuous on $[5/2, 3]$ and $f'(x) > 0$ and $f''(x) > 0$ for all $x \in [5/2, 3]$. Finally, by Example 19.2.1, f has a zero in $(5/2, 3)$ and so, by Theorem 19.3.4 (2), the sequence $\{c_n\}$ of approximations obtained by Newton's method with $c_0 = 3$ converges.

The error bound in Theorem 19.3.4 (3) uses $f'(5/2) = 75/4$ and an upper bound A for $f''(x)$ for $x \in [5/2, 3]$. Since f'' is strictly increasing, the obvious choice for A is $f''(3) = 18$. Hence

$$|c - c_n| \leq \frac{12}{25}|c - c_{n-1}|^2, \tag{19.3.7}$$

where $c = \sqrt[3]{25}$.

Since $c \in (5/2, 3)$ and $c_0 = 3$, by direct calculation we have

$$|c - c_0| < \frac{1}{2} = 0\cdot 5.$$

To use (19.3.7) to determine error bounds for each iteration it is convenient to write the fraction $12/25$ as $48/10^2$. Then

$$|c - c_1| \le \frac{48}{10^2}|c - c_0|^2 < \frac{48}{10^2}\frac{1}{2^2} = \frac{12}{10^2}$$

$$= 0{\cdot}12,$$

$$|c - c_2| \le \frac{48}{10^2}|c - c_1|^2 < \frac{48}{10^2}\frac{12^2}{10^4} = \frac{6912}{10^6}$$

$$= 0{\cdot}006912,$$

$$|c - c_3| \le \frac{48}{10^2}|c - c_2|^2 < \frac{48}{10^2}\frac{6912^2}{10^{12}} = \frac{2293235712}{10^{14}}$$

$$= 0{\cdot}00002293235712,$$

provide the required error bounds.   □

Problem 19.3.6 Use Theorem 19.3.4, for a subinterval of $[5/2, 3]$, to show that the third iteration obtained in Example 19.3.3 gives an approximation to $\sqrt[3]{25}$ in which the first seven decimal places are correct.

19.4   Definite Integrals

We will consider three methods of approximating definite integrals by replacing the integrand with easily integrated expressions. The first method involves a special case of the Riemann sum used to define the definite integral.

     Let f be a real function, continuous on $[a, b]$. Recall from Chapter 14 that we form a partition $\mathcal{P}$ of $[a, b]$ into n subintervals of equal length. We will introduce some shorthand notation since throughout this section we refer to the points that define this partition, as well as intervening points, and the values of the integrand at these points.

Notation 19.4.1 Let $a, b \in \mathbb{R}$ with $a < b$, let $n \in \mathbb{N}$ and define $h = (b - a)/n$. For each $c \in \mathbb{R}$ we write x_c for $a + ch$. In particular, the points in the partition are $x_0 = a + 0h = a$, $x_1 = a + 1h, \dots, x_n = a + nh = b$, in agreement with the notation used in Chapter 14. Also, for example, $x_{1/2} = a + (1/2)h$ is the mid-point of $[x_0, x_1]$.

     Further, if the integrand is $f(x)$, we will use f_c to denote $f(x_c)$ so that, in particular, $f_i = f(x_i)$ for $i = 0, 1, \dots, n$.

Rectangle rule   Consider any subinterval $[\alpha, \beta] \subseteq [a, b]$ and define the constant function $P \colon [\alpha, \beta] \to \mathbb{R}$ by

$$P(x) = f\left(\frac{\alpha + \beta}{2}\right).$$

The function P has the value of f at the mid-point of the interval on which P is defined. We obtain an approximation to the integral of f on this interval,

$$\int_\alpha^\beta f(x)\,dx \approx \int_\alpha^\beta P(x)\,dx = (\beta - \alpha)\,f\left(\frac{\alpha + \beta}{2}\right),$$

i.e. the area under the curve $y = f(x)$ between $x = \alpha$ and $x = \beta$ is approximated by the area of the rectangle with base $[\alpha, \beta]$ and height determined by the value of the function at the mid-point of the interval, as shown in Figure 19.4.1.

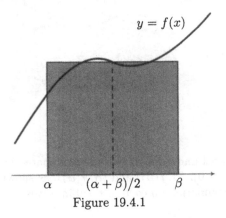

Figure 19.4.1

Taking such an approximation on each of the subintervals $[x_{i-1}, x_i]$ of length h we get

$$\int_a^b f(x)\,dx = \int_{x_0}^{x_1} f(x)\,dx + \int_{x_1}^{x_2} f(x)\,dx + \cdots + \int_{x_{n-1}}^{x_n} f(x)\,dx$$
$$\approx hf(a + (1/2)h) + hf(a + h + (1/2)h) +$$
$$\cdots + hf(a + (n-1)h + (1/2)h)$$

and so we obtain the *rectangle rule*

$$\int_a^b f(x)\,dx \approx h\big[f_{1/2} + f_{3/2} + \cdots + f_{(2n-1)/2}\big]. \tag{19.4.1}$$

Trapezoidal rule   In a similar way, we may take a straight line approximation to the integrand and define the function $P : [\alpha, \beta] \to \mathbb{R}$ by

$$P(x) = qx + r,$$

where q and r are such that P and f agree at the end-points of the interval, *i.e.* $q\alpha + r = f(\alpha)$ and $q\beta + r = f(\beta)$. The constant q may be zero so that P is either linear or constant. This gives the approximation

$$\int_\alpha^\beta f(x)\,dx \approx \int_\alpha^\beta P(x)\,dx = \left[\frac{q}{2}x^2 + rx\right]_\alpha^\beta$$
$$= \frac{q}{2}(\beta^2 - \alpha^2) + r(\beta - \alpha)$$
$$= \frac{\beta - \alpha}{2}[(\alpha + \beta)q + 2r] = \frac{\beta - \alpha}{2}[(q\alpha + r) + (q\beta + r)]$$
$$= \frac{\beta - \alpha}{2}[f(\alpha) + f(\beta)].$$

Note that we do not need to determine the constants q and r explicitly.

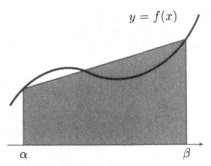

Figure 19.4.2

In this case, the area under the curve is approximated by the area of the trapezium with vertices $(\alpha, 0)$, $(\beta, 0)$, $(\beta, f(\beta))$ and $(\alpha, f(\alpha))$, as shown in Figure 19.4.2. Taking this approximation on each of n subintervals of length $h = (b - a)/n$,

$$\int_a^b f(x)\,dx \approx \frac{h}{2}\big[(f_0 + f_1) + (f_1 + f_2) + \cdots + (f_{n-1} + f_n)\big]$$

and this gives the *trapezoidal rule*

$$\int_a^b f(x)\,dx \approx \frac{h}{2}\big[f_0 + 2f_1 + \cdots + 2f_{n-1} + f_n\big]. \tag{19.4.2}$$

Remark For a particular type of integrand, a combination of the rectangle and trapezoidal rules may be used to obtain an error bound. To prove the necessary theorem we will need to use the fact that a concave-up curve lies between a chord and a tangent.

Lemma 19.4.2 *Let f be a real function with f' continuous on $[a, b]$ and f'' positive on (a, b). Let $c \in [a, b]$ and let the tangent to the graph of f at $(c, f(c))$ and the chord through $(a, f(a))$ and $(b, f(b))$ have equations $y = t(x)$ and $y = h(x)$ respectively. Then*

$$t(x) \le f(x) \le h(x) \qquad (x \in [a, b])$$

and strict inequality holds except for $t(c) = f(c)$, $f(a) = h(a)$ and $f(b) = h(b)$.

Proof The tangent has equation $y - f(c) = f'(c)(x - c)$. Hence, for $x \in [a, b]$,

$$t(x) = f(c) + f'(c)(x - c). \tag{19.4.3}$$

The chord has equation $y - f(a) = \dfrac{f(b) - f(a)}{b - a}(x - a)$. Hence, for $x \in [a, b]$,

$$h(x) = f(a) + \frac{f(b) - f(a)}{b - a}(x - a) = \Big(\frac{b - x}{b - a}\Big)f(a) + \Big(\frac{x - a}{b - a}\Big)f(b). \tag{19.4.4}$$

For any $s, t \in [a, b]$ with $s \ne t$ by Taylor's theorem on $[s, t]$, there exists $u \in (s, t)$ such that

$$f(t) = f(s) + f'(s)(t - s) + \frac{f''(u)}{2}(t - s)^2.$$

Hence, since $f''(u) > 0$,

$$f(t) > f(s) + f'(s)(t - s). \tag{19.4.5}$$

Putting $s = c$, $t = x$ in (19.4.5) and using (19.4.3) gives

$$f(x) > t(x) \qquad (x \in [a, b] - \{c\}). \tag{19.4.6}$$

Putting $s = x$, $t = a$ then $s = x$, $t = b$ in (19.4.5) gives,

$$f(a) > f(x) - f'(x)(x - a) \qquad (x \in (a, b]), \tag{19.4.7}$$

$$f(b) > f(x) + f'(x)(b - x) \qquad (x \in [a, b)). \tag{19.4.8}$$

Adding $(b - x) \times (19.4.7)$ and $(x - a) \times (19.4.8)$ gives, for $x \in (a, b)$,

$$(b - x)f(a) + (x - b)f(b) > (b - x)f(x) + (x - a)f(x) = (b - a)f(x).$$

Hence, using (19.4.4),

$$h(x) > f(x) \qquad (x \in (a, b)). \tag{19.4.9}$$

Combining (19.4.6) and (19.4.9), and observing that $t(c) = f(c)$, $f(a) = h(a)$ and $f(b) = h(b)$, completes the proof. $\qquad\qquad\qquad\qquad\qquad\qquad\qquad\qquad\qquad\square$

Theorem 19.4.3 *Let f be a real function, with f' continuous on $[a, b]$ and f'' positive on (a, b). Let I_R and I_T be the approximations to $\int_a^b f(x)\, dx$, obtained using the rectangle and trapezoidal rules, respectively, for the simplest case, $n = 1$. Then*

$$I_R < \int_a^b f(x)\, dx < I_T. \tag{19.4.10}$$

Proof Construct two trapezia with base $[a, b]$ and parallel sides perpendicular to the x-axis and through $(a, 0)$ and $(b, 0)$, respectively. First, taking the slant side to be the chord joining $(a, f(a))$ and $(b, f(b))$, one obtains the trapezium used in the trapezoidal rule, of area I_T. To obtain the second, take the tangent to the curve at the mid-point $((a + b)/2, f((a + b)/2))$ as the slant side (see Figure 19.4.3).

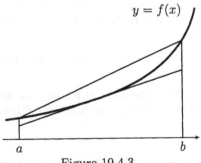

Figure 19.4.3

The area of the second trapezium is the average of the lengths of the parallel sides, which equals the height of the trapezium at the mid-point of the interval, multiplied by the length of the interval. Hence the area is $f((a+b)/2)(b-a)$ which is exactly the approximation obtained by the rectangle rule, I_R.

By Lemma 19.4.2, the chord lies above the curve and the tangent lies below it. Thus

$$I_R < I < I_T.$$ □

Example 19.4.4 Use the rectangle and trapezoidal rules to obtain approximations to

$$I = \int_0^1 1 - \cos x^2 \, dx,$$

by considering four subintervals. Obtain a further approximation and an error bound.

Solution Let $f(x) = 1 - \cos(x^2)$. For the trapezoidal rule we need the value of the integrand at the end-points, and for the rectangle rule, the values at the mid-point of each subinterval. Overall we will need the values of the integrand f_c at the nine values given in the following table

c	x_c	$f_c = f(x_c)$
0	0	0·00000000
1/2	0·125	0·00012207
1	0·25	0·00195249
3/2	0·325	0·00987141
2	0·5	0·03108758
5/2	0·625	0·07532874
3	0·75	0·15407550
7/2	0·825	0·27905062
4	1	0·45969769

Since there are four subintervals of $[0, 1]$, $h = 1/4$ and, from (19.4.1) and (19.4.2),

$$I_R = \frac{1}{4}\left(f_{1/2} + f_{3/2} + f_{5/2} + f_{7/2}\right) = 0·091093209$$

$$I_T = \frac{1}{8}\left(f_0 + 2f_1 + 2f_2 + 2f_3 + f_4\right) = 0·104241104$$

Now $f'(x) = 2x \sin(x^2)$ and $f''(x) = 2\sin(x^2) + 4x^2\cos(x^2) > 0$ for $x \in [0, 1]$. Thus, by Theorem 19.4.3, we have $0·091093209 < I < 0·104241104$. Taking the mid-point as the approximation we get

$$I = 0·09766716 \pm 0·00657395$$ □

Problem 19.4.5 Show how Theorem 19.4.3 may be adapted to be used with real functions f such that f' is continuous on $[a, b]$ and $f''(x) < 0$ (rather that $f''(x) > 0$) for $x \in (a, b)$.

Problem 19.4.6 Let I_R and I_T be approximations to an integral obtained using the rectangle rule and trapezoidal rule, respectively, on n subintervals. If I_T' is the trapezoidal approximation, on $2n$ subintervals, to the same integral, show that

$$I_T' = \frac{1}{2}(I_R + I_T).$$

Using this, and the results of Example 19.4.4, obtain a further approximation to

$$\int_0^1 1 - \cos x^2 \, dx.$$

Simpson's rule Taking another approximation for the integrand, we consider the function $P \colon [\alpha, \beta] \to \mathbb{R}$ defined by

$$P(x) = px^2 + qx + r,$$

where any of the constants p, q and r may be zero so that P is a quadratic, linear or constant function. The function P is to agree with f at *three* points: the end-points and mid-point of $[\alpha, \beta]$. Thus

$$p\alpha^2 + q\alpha + r = f(\alpha), \qquad (19.4.11)$$
$$p\beta^2 + q\beta + r = f(\beta) \qquad (19.4.12)$$

and

$$p\left(\frac{\alpha + \beta}{2}\right)^2 + q\left(\frac{\alpha + \beta}{2}\right) + r = f\left(\frac{\alpha + \beta}{2}\right). \qquad (19.4.13)$$

The approximation to the integral determined by P is

$$\int_\alpha^\beta f(x) \, dx \approx \int_\alpha^\beta P(x) \, dx$$

$$= \left[\frac{1}{3}px^3 + \frac{1}{2}qx^2 + rx\right]_\alpha^\beta = \frac{1}{3}p(\beta^3 - \alpha^3) + \frac{1}{2}(\beta^2 - \alpha^2) + r(\beta - \alpha)$$

$$= \frac{\beta - \alpha}{6}\left[2p(\alpha^2 + \alpha\beta + \beta^2) + 3q(\alpha + \beta) + 6r\right]$$

$$= \frac{\beta - \alpha}{6}\left[(p\alpha^2 + q\alpha + r) + (p(\alpha + \beta)^2 + 2q(\alpha + \beta) + 4r)\right.$$
$$\left. + (p\beta^2 + q\beta + r)\right]$$

$$= \frac{\beta - \alpha}{6}\left[f(\alpha) + 4f\left(\frac{\alpha + \beta}{2}\right) + f(\beta)\right],$$

using (19.4.11)–(19.4.13). Once more, it is unnecessary to calculate the constants p, q and r determining the approximation.

Applying this result on each of n subintervals of $[a, b]$

$$\int_a^b f(x) \, dx \approx \frac{h}{6}\left[(f_0 + 4f_{1/2} + f_1) + (f_1 + 4f_{3/2} + f_2) + \cdots + (f_{n-1} + 4f_{(2n-1)/2} + f_n)\right],$$

and so we get *Simpson's rule*

$$\int_a^b f(x) \, dx \approx \frac{h}{6}\left[f_0 + 4f_{1/2} + 2f_1 + 4f_{3/2} + 2f_2 + \cdots + 2f_{n-1} + 4f_{(2n-1)/2} + f_n\right].$$

This is usually called *Simpson's rule with $2n + 1$ ordinates* as the function has to be evaluated at $2n + 1$ points. The cases we will consider are Simpson's rule with three ordinates (one subinterval of length $h = b - a$), with the approximation

$$\int_a^b f(x) \, dx \approx \frac{b - a}{6}\left[f_0 + 4f_{1/2} + f_1\right], \qquad (19.4.14)$$

with five ordinates (two subintervals of length $h = (b-a)/2$),

$$\int_a^b f(x)\,dx \approx \frac{b-a}{12}\left[f_0 + 4f_{1/2} + 2f_1 + 4f_{3/2} + f_2\right], \tag{19.4.15}$$

and with seven ordinates (three subintervals of length $h = (b-a)/3$),

$$\int_a^b f(x)\,dx \approx \frac{b-a}{18}\left[f_0 + 4f_{1/2} + 2f_1 + 4f_{3/2} + 2f_2 + 4f_{5/2} + f_3\right]. \tag{19.4.16}$$

Example 19.4.7 Compare the exact value of

$$\int_0^1 \frac{dx}{1+x^2}$$

with the approximation obtained by using Simpson's rule with three ordinates.

Solution Let $f(x) = 1/(1+x^2)$. Then

$$\int_0^1 f(x)\,dx = \left[\tan^{-1} x\right]_0^1 = \frac{\pi}{4} \quad (\approx 0{\cdot}78540)$$

and, from (19.4.14),

$$\int_0^1 f(x)\,dx \approx \frac{1-0}{6}\left[f(0) + 4f(1/2) + f(1)\right]$$

$$= \frac{1}{6}\left[1 + \frac{16}{5} + \frac{1}{2}\right] = \frac{47}{60} \quad (\approx 0{\cdot}78333) \qquad \square$$

Note 19.4.8 For a function f for which $f^{(4)}$ exists and is continuous on $[\alpha, \beta]$ it may be shown that the error in Simpson's rule (on one subinterval), namely

$$E = \int_\alpha^\beta f(x)\,dx - \frac{\beta-\alpha}{6}\left[f(\alpha) + 4f\left(\frac{\alpha+\beta}{2}\right) + f(\beta)\right],$$

is given by

$$E = -\frac{(\beta-\alpha)^5}{180}\frac{f^{(4)}(\xi)}{2^4},$$

for some $\xi \in (\alpha, \beta)$. It follows that if the fourth derivative of f is identically zero, *i.e.* if $f(x)$ is a polynomial of degree less than or equal to three, then Simpson's rule is *exact*. This is no surprise if f is constant, linear or quadratic since in such cases $P(x) = f(x)$. In the case that f is cubic, f and P are not identically equal but the two finite regions between the curves (see Figure 19.4.4) have equal areas and so exactly cancel.

Problem 19.4.9 Show directly that if $f(x) = x^3$ then

$$\int_a^b f(x)\,dx = \frac{b-a}{6}\left[f(a) + 4f\left(\frac{a+b}{2}\right) + f(b)\right].$$

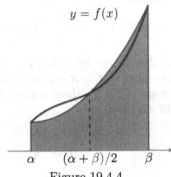

Figure 19.4.4

Example 19.4.10 Use Simpson's rule with five and seven ordinates to obtain approximate values for

$$I = \int_0^4 \frac{dx}{x^2 - 3x + 4}.$$

Solution For Simpson's rule with five ordinates we have $h = (4 - 0)/2 = 2$,

c	x_c	f_c
0	0	1/4
1/2	1	1/2
1	2	1/2
3/2	3	1/4
2	4	1/8

where $f_c = 1/(x_c^2 - 3x_c + 4)$, and (19.4.15) gives

$$I \approx I_5 = \frac{4-0}{12}\left[\frac{1}{4} + 4\left(\frac{1}{2}\right) + 2\left(\frac{1}{2}\right) + 4\left(\frac{1}{4}\right) + \frac{1}{8}\right] = \frac{35}{24}.$$

For Simpson's rule with seven ordinates we have $h = (4 - 0)/3 = 4/3$,

c	x_c	f_c
0	0	1/4
1/2	2/3	9/22
1	4/3	9/16
3/2	2	1/2
2	8/3	9/28
5/2	10/3	9/46
3	4	1/8

and (19.4.16) gives

$$I \approx I_7 = \frac{4-0}{18}\left[\frac{1}{4} + 4\left(\frac{9}{22}\right) + 2\left(\frac{9}{16}\right) + 4\left(\frac{1}{2}\right) + 2\left(\frac{9}{28}\right) + 4\left(\frac{9}{46}\right) + \frac{1}{8}\right]$$
$$= \frac{23242}{15939}.$$

To five decimal places we have

$$I_5 \approx 1.45833 \quad \text{and} \quad I_7 \approx 1.45818. \qquad \square$$

Problem 19.4.11 Let E_n denote the error in Simpson's rule using n subintervals, *i.e.* $2n + 1$ ordinates. Suppose that f is a real function such that $f^{(4)}$ exists and is continuous on $[a, b]$. By Theorem 7.3.11, there exists $M \in \mathbb{R}$ such that $|f^{(4)}(x)| \le M$ for all $x \in \mathbb{R}$. From Note 19.4.8 we have

$$|E_1| \le \frac{(b-a)^5}{180} \frac{M}{2^4}.$$

Deduce that

$$|E_2| \le \frac{(b-a)^5}{180} \frac{M}{4^4}$$

and obtain a bound for E_n.

19.5 Euler's Method

Suppose we have a first order ODE

$$y' = F(x, y) \tag{19.5.1}$$

with a solution $y = y(x)$ in $[a, b]$ satisfying $y(a) = y_0$. We wish to find an approximate value for $y(b)$. Euler's method consists of approximating the solution curve $y = y(x)$ on $[a, b]$, by its tangent at (a, y_0), $y = T(x)$. Then $T(b)$ gives an approximation to $y(b)$. This process is illustrated in Figure 19.5.1.

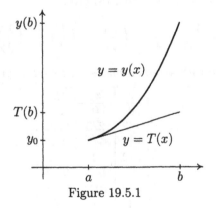

Figure 19.5.1

The tangent to $y = y(x)$ at (a, y_0) has equation $y - y_0 = y'(a)(x - a)$, and, since $y(x)$ satisfies (19.5.1), $y'(a) = F(a, y(a))$. So the tangent is

$$y = y_0 + F(a, y_0)(x - a),$$

and

$$y(b) \approx y_0 + (b - a)F(a, y_0). \tag{19.5.2}$$

As is evident for the solution curve shown in Figure 19.5.1, this approximation is rather crude. A better approximation is obtained by taking a partition of $[a, b]$ into n equal subintervals of length $h = (b - a)/n$, by defining points $x_i = a + ih$ for $i = 0, 1, \ldots, n$, and making similar approximations on each subinterval. In this context h is called the *step length*.

On the first subinterval $[x_0, x_1]$ we approximate the solution curve by the tangent joining (x_0, y_0) to (x_1, y_1) where $y_1 = y_0 + hF(x_0, y_0)$ is the approximation to $y(x_1)$ given by (19.5.2). In the $(i + 1)^{\text{th}}$ interval we approximate the solution curve by the line through (x_i, y_i) with gradient $F(x_i, y(x_i))$. In this way we get points (x_i, y_i) for $i = 0, 1, \ldots, n$, where

$$y_{i+1} = y_i + hF(x_i, y_i), \tag{19.5.3}$$

giving y_n as the approximation to $y(b)$. This is called *Euler's method with step length h*. Note also that the lines joining (x_0, y_0) to (x_1, y_1) to ... to (x_n, y_n) give an approximation to the solution curve $y = y(x)$ on $[a, b]$. In Figure 19.5.2, these approximations are shown for $n = 3$, $n = 6$ and $n = 12$.

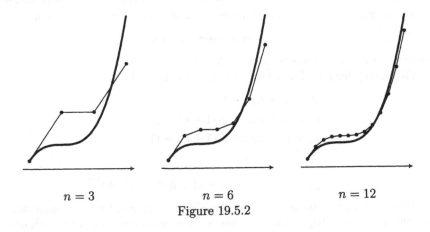

$$n = 3 \qquad\qquad\qquad n = 6 \qquad\qquad\qquad n = 12$$

Figure 19.5.2

Example 19.5.1 Apply Euler's method with step lengths $h = 1$, $h = 0{\cdot}5$ and $h = 0{\cdot}2$ to find approximate values for $y(1)$ given that

$$y' = x + y \quad \text{and} \quad y(0) = 1.$$

Solution   With $h = 1$,

$$y_1 = y_0 + h(x_0 + y_0) = 1 + 1(0 + 1) = 2,$$

giving $y(1) \approx 2$.

With $h = 0{\cdot}5$,

$$y_1 = 1 + 0{\cdot}5(0 + 1) = 1{\cdot}5,$$
$$y_2 = 1{\cdot}5 + 0{\cdot}5(0{\cdot}5 + 1{\cdot}5) = 2{\cdot}5,$$

giving $y(1) \approx 2{\cdot}5$.

With $h = 0{\cdot}2$,

$$y_1 = 1 + 0{\cdot}2(0 + 1) = 1{\cdot}2,$$
$$y_2 = 1{\cdot}2 + 0{\cdot}2(0{\cdot}2 + 1{\cdot}2) = 1{\cdot}48,$$
$$y_3 = 1{\cdot}48 + 0{\cdot}2(0{\cdot}4 + 1{\cdot}48) = 1{\cdot}856,$$
$$y_4 = 1{\cdot}856 + 0{\cdot}2(0{\cdot}6 + 1{\cdot}856) = 2{\cdot}3472,$$
$$y_5 = 2{\cdot}3472 + 0{\cdot}2(0{\cdot}8 + 2{\cdot}3472) = 2{\cdot}97664,$$

giving $y(1) \approx 2{\cdot}97664$.   □

Problem 19.5.2 Find the particular solution of the first order ODE, $y' = x + y$, which satisfies $y(0) = 1$. Deduce that the exact value of $y(1)$ is $2(e - 1) \approx 3{\cdot}43656$.

Remark Reducing the step length in Euler's method does not guarantee a better approximation and even when approximations do converge to the exact answer they often do so very slowly. Typically the error will be of the same order of magnitude as the step length. As an example that we can investigate explicitly consider

$$y' = y \quad \text{and} \quad y(0) = 1.$$

The exact solution is $y = \exp(x)$ so that $y(1) = e$.

With a step length of $h = 1/n$ we get, from (19.5.3),

$$y_1 = y_0 + hy_0 = 1 + h,$$
$$y_2 = y_1 + hy_1 = y_1(1 + h) = (1 + h)^2,$$
$$y_3 = y_2 + hy_2 = y_2(1 + h) = (1 + h)^3,$$
$$\vdots$$
$$y_n = y_{n-1} + hy_{n-1} = (1 + h)^n = (1 + h)^{1/h}.$$

The value of e correct to five decimal places is $2{\cdot}71828$ and below we show the approximations y_n, correct to five decimal places, that we obtain for various choices of step length, and the error $\varepsilon_n = e - y_n$

n	h	y_n	ε_n
1	$1{\cdot}0$	$2{\cdot}0$	$0{\cdot}71828$
10	$0{\cdot}1$	$2{\cdot}59374$	$0{\cdot}12454$
100	$0{\cdot}01$	$2{\cdot}70481$	$0{\cdot}01347$
1000	$0{\cdot}001$	$2{\cdot}71692$	$0{\cdot}00136$
10000	$0{\cdot}0001$	$2{\cdot}71815$	$0{\cdot}00013$

It is possible to prove that the sequence of approximations $\{y_n\}$ tends to e as $n \to \infty$, or equivalently, as $h \to 0$. Consider

$$\log y_n = \log(1 + h)^{1/h} = \frac{\log(1 + h)}{h}.$$

Then, using l'Hôpital's rule, Theorem 9.6.1,

$$\lim_{n \to \infty} \log y_n = \lim_{h \to 0} \frac{\log(1 + h)}{h} = 1.$$

Thus $\log y_n \to 1$, and hence $y_n \to e$, as $n \to \infty$.

19.X   Exercises

1. Use the bisection method to find an approximation to the real root of each of the
    following equations to within $0 \cdot 1$.

    (a) $x^3 - 2 = 0$,   (b) $x^3 + 8x - 7 = 0$,

    (c) $x^5 + 16x^3 + 1 = 0$,   (d) $3^x - 2 = 0$.

2. Let f be the real function defined by $f(x) = x^3 - x^2 - 2x + 1$. Show that f has a
    zero in each of the intervals $(-2, -1)$, $(0, 1)$ and $(1, 2)$. Use the bisection method
    to find an approximation to each of these zeros to within $0 \cdot 05$.

3. Show that each of the following equations has exactly one real root in the specified
    range. Find the best integer approximation to the root and apply Newton's
    method once to obtain a further approximation.

    (a) $x^7 + 10x - 13 = 0$ $(x \in \mathbb{R})$,

    (b) $x^5 + x^2 - 3 = 0$ $(x > 0)$,

    (c) $2x^3 + 12x^2 - 59 = 0$ $(x > 0)$,

    (d) $x^3 + 6x^2 + 16x - 121 = 0$ $(x \in \mathbb{R})$.

4. Show that the equation $x^3 + x + 1 = 0$ has exactly one real root. Find the best
    integer approximation to the root and use Newton's method twice to find two
    further approximations.

5. Let f be the real function defined by $f(x) = x^3 - x - 1$. By differentiating $f(x)/x$
    with respect to x, show that f has exactly one positive zero. Find the best
    integer approximation to this zero and apply Newton's method twice to obtain
    two further approximations.

6. Find the best integer approximation to the cube root of 14. Use Newton's method
    to find two further approximations.

7. Use the rectangle rule with three subintervals to find an approximate value for

    $$\int_0^2 \frac{dx}{x^2 + 1}.$$

8. Use the trapezoidal rule with two subintervals to find an approximate value for

    $$\int_0^1 \frac{dx}{1 + \cos^{-1} x}.$$

9. Use the rectangle and trapezoidal rules with three subintervals to find approximate values for

$$\int_1^4 \frac{e^x}{x}\,dx,$$

correct to four decimal places. Obtain a further approximation and a bound for the error.

10. Use Simpson's rule with three ordinates to find an approximate value for each of the following integrals.

$$\text{(a)}\ \int_0^1 \sqrt{1 + 48x^4}\,dx, \qquad \text{(b)}\ \int_1^2 \frac{dx}{x}.$$

11. Use Simpson's rule with five ordinates to find an approximate value for each of the following integrals.

$$\text{(a)}\ \int_1^3 \frac{dx}{2x-1}, \qquad \text{(b)}\ \int_0^2 \frac{dx}{1+x^3}.$$

12. Use Simpson's rule with five ordinates to show that

$$\text{(a)}\ \int_0^\pi \frac{x}{1 - \sin x \cos x}\,dx \approx \frac{\pi^2}{2}, \qquad \text{(b)}\ \int_0^1 \frac{dx}{\pi - \sin^{-1}\sqrt{x}} \approx \frac{247}{180\pi}.$$

13. Use Euler's method with step lengths $h = 1$, $h = 1/2$ and $h = 1/3$ to find approximate values for $y(b)$ in each of the following cases.

$$\text{(a)}\ \ y' = 2x - y, \qquad y(0) = 1, \qquad b = 1,$$
$$\text{(b)}\ \ y' = x^2 - y, \qquad y(0) = 2, \qquad b = 1,$$
$$\text{(c)}\ \ y' = y^2 + 3, \qquad y(0) = 0, \qquad b = 1,$$
$$\text{(d)}\ \ y' = x - \frac{y}{x}, \qquad y(1) = 0, \qquad b = 2.$$

14. Use Euler's method with step length $h = 0{\cdot}1$ to find an approximate value for $y(0{\cdot}5)$ given $y' = x - y$ and $y(0) = 1$.

Appendix A

Answers to Exercises

Chapter 1

1. $S_1 = (-2,6)$, $S_2 = [-9,-1]$, $S_3 = [2/3, 4/3]$,
    $S_4 = (0, \infty)$, $S_5 = (-\infty, 3]$, $S_6 = (-\infty, 3)$.

2. Use $y = x + |x| = \begin{cases} 2x & \text{if } x \geq 0, \\ 0 & \text{if } x < 0. \end{cases}$

3. $|x - y| = 2 \Leftrightarrow (y = x + 2 \text{ or } y = x - 2)$. $(x, y) = (3, 1)$ or $(1, 3)$.

6. (a) $x = 9$, (b) $x = 1$, (c) $x = 5$, (d) $x = -19/17$, (e) $x = 3/4$.

7. 3, 1, 1/3, 8, 1/5, $\sqrt{2}$.

8. x^6, x^2, $x^{3/2}$, $2x^2$, $x^{1/2}y^{3/2}$.

9. $\dfrac{-2}{\sqrt{x^2 - 1}}$.

11. $S_1 = (-\infty, -1] \cup (1, 3]$, $S_2 = (-\infty, -7] \cup (-2, \infty)$.
    $S_3 = (-\infty, -1) \cup (1, 3)$, $S_4 = (-\infty, -2) \cup (-1, 3) \cup (11/3, \infty)$,

14. For $x > 1$, $\dfrac{x^3}{x^3 + 1} < \dfrac{x^4}{x^4 + 1}$.

16. The factorisation is $x^2 - y^2 + x + y = (x + y)(x - y + 1)$.

Chapter 2

1. Graphs: (a) $\{(1, -4), (2, -1), (3, 2)\}$, (b) $\{(-1, 11), (0, 7), (1, 5)\}$,
    (c) $\{(1, 5), (2, 5), (3, 7)\}$.
    Images: (a) $\{-4, -1, 2\}$, (b) $\{5, 7, 11\}$, (b) $\{5, 7\}$.

2. Images: (a) $[1/2, 1]$, (b) $[1, \infty)$, (c) $(-\infty, 2]$, (d) $\mathbb{R}$.

3. $f \circ f : [-1, 1] \to \mathbb{R}$, $(f \circ f)(x) = |x|$.

4. (a) $\mathbb{R} - \{-1/2\}$, (b) $[1/4, \infty)$, (c) $\mathbb{R} - \{(1 - \sqrt{11})/2, (1 + \sqrt{11})/2\}$,
    (d) $\mathbb{R} - (-2, 1)$, (e) $\mathbb{R} - [-2, 1]$, (f) $\mathbb{R} - \{-2, 1\}$.

5. (a) $g \circ f : \mathbb{R} \to \mathbb{R}$, $(g \circ f)(x) = 12x^2 + 4x + 9$,
        $f \circ g : \mathbb{R} \to \mathbb{R}$, $(f \circ g)(x) = 6x^2 + 16x + 27$,

    (b) $g \circ f : \mathbb{R} - \{-2\} \to \mathbb{R}$, $(g \circ f)(x) = -(x + 1)$,
        $f \circ g : \mathbb{R} - \{1/3, 1\} \to \mathbb{R}$, $(f \circ g)(x) = (x + 1)/(3x - 1)$,

    (c) $g \circ f : \mathbb{R} \to \mathbb{R}$, $(g \circ f)(x) = 32x + 31$,
        $f \circ g : \mathbb{R} \to \mathbb{R}$, $(f \circ g)(x) = 32x + 31$,

    (d) $g \circ f : \mathbb{R} \to \mathbb{R}$, $(g \circ f)(x) = \begin{cases} x - 3 & \text{if } x < 0, \\ x - 1 & \text{if } x \geq 0. \end{cases}$'

        $f \circ g : \mathbb{R} \to \mathbb{R}$, $(f \circ g)(x) = \begin{cases} x - 3 & \text{if } x < 2, \\ x - 1 & \text{if } x \geq 2. \end{cases}$'

    Required sets are:
    (a) $\{-1, 3\}$, (b) $\{-1, 0\}$, (c) $\mathbb{R}$, (d) $(-\infty, 0) \cup [2, \infty)$.

7. Bounded below: p, q, u.   Bounded above: p.

8. (a) Restriction: $f_1 : \{1, 2, 3\} \to \{-4, -1, 2\}$.
        Graph of f_1: $\{(-4, 1), (-1, 2), (2, 3)\}$.

    (b) Restriction: $g_1 : \{-1, 0, 1\} \to \{5, 7, 11\}$.
        Graph of f_1: $\{(5, 1), (7, 0), (11, -1)\}$.

    (c) Restriction: $h_1 : \{1, 3\} \to \{5, 7\}$.
        Graph of h_1: $\{(5, 1), (7, 3)\}$.

        Restriction: $h_2 : \{2, 3\} \to \{5, 7\}$.
        Graph of h_2: $\{(5, 2), (7, 3)\}$.

9. (a) $f^{-1} : \mathbb{R} \to \mathbb{R}$, $f^{-1}(x) = (x - 5)/3$,
    (b) $g^{-1} : [1, 2] \to [0, 3]$, $g^{-1}(x) = 4 - x^2$,
    (c) $h^{-1} : [1, 2] \to [2, 3]$, $h^{-1}(x) = 1 + (2/x)$,
    (d) $u^{-1} : (1, \infty) \to (0, 1)$, $u^{-1}(x) = (x - 1)/(x + 1)$,
    (e) $v^{-1} : [-1, 3] \to [-1, 1]$, $v^{-1}(x) = 1 - \sqrt{1 + x}$.

Chapter 3

1. (a) $q(x) = 2x^2 + x + 3$, $r(x) = x - 7$,

(b) $q(x) = x^2 - x - 5$, $r(x) = -x - 2$,

(c) $q(x) = 2x^3 + x^2 - 5x - 1$, $r(x) = 0$,

(d) $q(x) = x^2 - x + 3$, $r(x) = -2x - 2$,

(e) $q(x) = \frac{1}{2}x$, $r(x) = \frac{3}{2}x + 7$.

2. $a = 4$, $b = 8$, $c = 12$, $d = -7$.

3. $111/32$ (put $x = -1/2$ in the polynomial).

4. (a) $2, 2 \pm \sqrt{5}$, (b) -1, (d) $\pm 1, 5/4, 4/3$, (c) ± 2, (e) none.

5. (a) $(x^2 + x + 2)(x^2 - x + 2)$ (b) $(x^2 + 2x + 2)(x^2 - 2x + 2)$

6. $p(x) = (x^2 + x + 2)(2x^2 - 3x + 4)$.

7. 7, at $(-2, 3)$ (by completing squares).

8. $p(x) = (x - 1)^2(x^2 + 2x + (3 - a))$. For the second factor $\Delta = 4(a - 2)$.

9. 1.

10. (a) $(2, 3)$, (b) $[-3, 1]$, (c) $\mathbb{R}$, (d) $\{5/3\}$, (e) $(3, \infty)$.

11. $\Delta = 1 + 2k - 3k^2 < 0$ for $k \notin [-1/3, 1]$. The zeros are $(k + 1 \pm \sqrt{\Delta})/2$. The required length is the distance between the zeros, *i.e.* $= \sqrt{\Delta}$.

12. (a) $\dfrac{1}{2}\left[\dfrac{3}{x-3} - \dfrac{1}{x+1}\right]$,

   (b) $\dfrac{4}{x} - \dfrac{3}{x+1} - \dfrac{3}{(x+1)^2}$,

   (c) $\dfrac{1}{x+3} - \dfrac{6}{(x+3)^2} + \dfrac{9}{(x+3)^3}$,

   (d) $2 + \dfrac{1}{x} + \dfrac{3}{x+2} - \dfrac{1}{x-2}$,

   (e) $\dfrac{1}{3}\left[\dfrac{1}{x+1} - \dfrac{x-2}{x^2-x+1}\right]$,

   (f) $11 - \dfrac{81}{2x^2+9} - \dfrac{1}{x+1} + \dfrac{1}{x-1}$,

   (g) $\dfrac{2x-3}{x^2+4} + \dfrac{3}{x-2}$,

   (h) $\dfrac{18}{x-2} - \dfrac{18x-14}{x^2+1} + \dfrac{40x+30}{(x^2+1)^2}$.

Chapter 4

2. $\displaystyle\sum_{r=1}^{n}(2r - 1)^3 = n^2(2n^2 - 1)$.

5. $n = 3$.

6. (a) $n = 10$, (b) $r = 4$ or $r = 6$.

9. $10.9.8.7.6 = 30240$.

10. (a) $7.6.5.4.3 = 2520$,   (b) $7^5 = 16807$.

    For even numbers choose the final even digit first.
    (a) $3.6.5.4.3 = 1080$,   (b) $3.7^4 = 7203$.

11. (a) $a^8 - 4a^6b^3 + 6a^4b^6 - 4a^2b^9 + b^{12}$,

    (b) $x^{10} - 15x^7 + 90x^4 - 270x + 405x^{-2} - 243x^{-5}$,

    (c) $z^6 + 6z^3 + 15 + 20z^{-3} + 15z^{-6} + 6z^{-9} + z^{-12}$.

12. $105/2$.

13. 84, 0.

14. $\binom{7}{2}\binom{5}{3} = 210$.

15. 15.

16. $\dfrac{3^n - 1}{2}$.

Chapter 5

1. $\sin\theta = -5/13$, $\cos\theta = -12/13$, $\cot\theta = 12/5$, $\sec\theta = -13/12$,
    $\operatorname{cosec}\theta = -13/5$.

3. (a) $\mathbb{R}$,

    (b) $\cdots \cup [-\frac{5\pi}{2}, -\frac{3\pi}{2}] \cup [-\frac{\pi}{2}, \frac{\pi}{2}] \cup [\frac{3\pi}{2}, \frac{5\pi}{2}] \cup \cdots = \bigcup_{k\in\mathbb{Z}} [(4k-1)\frac{\pi}{2}, (4k+1)\frac{\pi}{2}]$.

4. $g \circ f : \mathbb{R} \to \mathbb{R}$, $(g \circ f)(x) = |\sin x|$.

6. The image of P is the circle, centre the origin, radius 1.

8. $\cos\dfrac{\pi}{12} = \dfrac{1}{\sqrt{2}(\sqrt{3}-1)} = \dfrac{\sqrt{3}+1}{2\sqrt{2}}$, $\sin\dfrac{\pi}{12} = \dfrac{\sqrt{3}-1}{2\sqrt{2}}$.

10. (a) $x = -\dfrac{\pi}{8} + k\pi$ or $x = \dfrac{5\pi}{8} + k\pi$ $(k \in \mathbb{Z})$,

    (b) $x = \pm\dfrac{\pi}{3} + 2k\pi$ or $x = \pm\dfrac{\pi}{6} + \dfrac{2k\pi}{3}$ $(k \in \mathbb{Z})$,

    (c) $x = \dfrac{\pi}{2} + 2k\pi$ or $x = \dfrac{7\pi}{6} + 2k\pi$ $(k \in \mathbb{Z})$.

11. (a) $\pi/2$,   (b) $2\pi/3$,   (c) $\pi/6$,   (d) $\pi/4$,   (e) $2\pi/3$,   (f) $12/5$,
    (g) $-1/7$,   (h) $(a^2 - 1)/(2a)$,   (i) $\sqrt{2/3}$,   (j) $\pi/4$.

12. (a) $(-1, 1)$,   (b) $(-\infty, -2] \cup [2, \infty)$,

14. $x = 12/5$.

Chapter 6

1. (a) n divisible by 4, *i.e.* n of the form $4k$ $(k \in \mathbb{N})$,
   (b) n of the form $4k - 2$ $(k \in \mathbb{N})$,
   (c) n of the form $4k - 3$ $(k \in \mathbb{N})$,
   (d) n of the form $4k - 1$ $(k \in \mathbb{N})$.

2. $1 + 5i$, $5 + 3i$, $-10 - 5i$, $-\dfrac{2}{5} - \dfrac{11}{5}i$, $-6 + 2i$.

3. (a) $\dfrac{1}{5} - \dfrac{2}{5}i$,   (b) $\dfrac{2}{5} + \dfrac{1}{5}i$,   (c) $\dfrac{1}{2} - \dfrac{3}{2}i$,

   (d) $\dfrac{17}{25} - \dfrac{19}{25}i$,   (e) $-\dfrac{4}{29} - \dfrac{19}{29}i$,   (f) $\dfrac{1}{2} - \dfrac{1}{10}i$.

5. (a) $\dfrac{1 - 7i}{2}$,   (b) $-3 + i$.

6. -1, $1 + 2i$, $1 - 2i$.

10. $\sqrt{2}e^{i\pi/4}$, $\sqrt{2}e^{-i\pi/4}$, $2\sqrt{2}e^{i3\pi/4}$, $3e^{i\pi}$, $2e^{-i2\pi/3}$, $4e^{-i\pi/2}$.

11. 5, α, $-\alpha$, $\pi - \alpha$, $\alpha - \pi$, $\alpha - \dfrac{\pi}{2}$.

12. 1.

13. $\dfrac{\cos\theta - 2\sin\theta}{5 - 4\sin 2\theta}$, $\dfrac{2\cos\theta - \sin\theta}{5 - 4\sin 2\theta}$.

14. $\dfrac{1}{4}\left(\sqrt{3} + 1\right) + \dfrac{1}{4}\left(\sqrt{3} - 1\right)i$, $1 + i = \sqrt{2}e^{i\pi/4}$, $\sqrt{3} + i = 2e^{i\pi/6}$.

15. (a) 512, $512\sqrt{3}$, (b) $-\sqrt{3}/256$, $-1/256$.
    (c) n of the form $6k$ $(k \in \mathbb{Z})$, (d) n of the form $6k + 3$ $(k \in \mathbb{Z})$.

16. $16\sin^5\theta - 20\sin^3\theta + 5\sin\theta$,
    $1 - 18\sin^2\theta + 48\sin^4\theta - 32\sin^6\theta$.

17. (a) $\dfrac{1}{16}(\sin 5\theta - 3\sin 5\theta + 10\sin\theta)$,

    (b) $\dfrac{1}{64}(-\sin 7\theta - \sin 5\theta + 3\sin 3\theta + 3\sin\theta)$,

    (c) $\dfrac{1}{32}(\cos 6\theta + 6\cos 4\theta + 15\cos 2\theta + 10)$.

18. (a) $x = \pm i\sqrt{6}$, (b) $x = -3 \pm 5i$.

19. $-1 \pm 4i$, $(-1 \pm 3i)/2$.

20. (a) 1, $e^{i2\pi/3}$, $e^{i4\pi/3}$,   (b) $e^{i\pi/3}$, -1, $e^{i5\pi/3}$,
    (c) $e^{i\pi/6}$, $e^{i5\pi/6}$, $-i$,   (d) i, $e^{i7\pi/6}$, $e^{i11\pi/6}$,

(e) $\sqrt{2}e^{i\pi/4}$, $\sqrt{2}e^{i11\pi/12}$, $\sqrt{2}e^{i19\pi/12}$.

(f) $2e^{i\pi/3}$, $2e^{i5\pi/6}$, $2e^{i4\pi/3}$, $2e^{i11\pi/6}$,   (g) $e^{i\pi/4}$, $e^{i5\pi/4}$.

21. $(z+1)(z^2 - 2\cos(\pi/5) + 1)(z^2 - 2\cos(3\pi/5) + 1)$.

Chapter 7

1.   (a) $1/2$, (b) 2, (c) $7/5$, (d) $2/(3a)$, (e) $-1/(2\sqrt{2})$, (f) $-1/x^2$,
     (g) 1, (h) $5/3$, (i) $5/3$, (j) $5/3$, (k) $1/2$, (l) 2.

2.   Continuous at $-2, 0$ and $\pi/2$, discontinuous at -1.

3. $f(0) = 3$, $f(2) = -1$.

5.   (a) -1, (b) 3, (c) $-1, 1, 2$, (d) 2.

8.   (a) $-3/4$, (b) 3, (c) $1/5$, (d) 0, (e) 2, (f) 2π.

9.   (a) $\to -\infty$, $\to \infty$, (b) $\to \infty$, $\to \infty$.

10. (a) $-\pi/2$, (b) $\pi/2$.

11. (a) $\to 1/5$, (b) $\to \infty$, $\to -\infty$, (c) $\to \infty$, (d) $\to \infty$, $\to -\infty$.

Chapter 8

1.   (a) $5x^4$, (b) $3x^2 + 2$, (c) $6x + 2$, (d) $-\dfrac{1}{x^2}$, (e) $\dfrac{-3}{(3x-1)^2}$,

     (f) $\dfrac{5}{(2-5x)^2}$, (g) $\dfrac{1}{2\sqrt{x+2}}$, (h) $\dfrac{-1}{2\sqrt{1-x}}$. (i) $\dfrac{x}{\sqrt{x^2+1}}$, (j) $-\dfrac{3}{x^4}$,

     (k) $\dfrac{-2(x+1)}{(x^2+2x)^2}$, (l) $\dfrac{1}{2(1-x)^{3/2}}$, (m) $\dfrac{2}{3x^{1/3}}$, (n) $2\cos(2x+1)$,

     (o) $-\sin x$.

3.   (a) $y = x + 3$, $x + y + 1 = 0$, $(5, 8)$,

     (b) $y = 2x - 1$, $x + 2y = 8$, $(1, 1)$, $(-3, -7)$.

     (c) $y = -2x + 3$, $2y = x + 11$, $(\sqrt{2}, 3 - 2\sqrt{2})$, $(-\sqrt{2}, 3 + 2\sqrt{2})$,

     (d) $y = -x - 2$, $y = x - 6$, no other point, $(1, -5)$, $(3, -3)$.

4. $(-1, 2)$, $(1, 4)$, $(3/2, 37/16)$.

5.   (a) $10(2x - 3)^4$, (b) $12x(x^2 - 4)^5$, (c) $3(2x^2 + 5x - 1)^2(4x + 5)$,

     (d) $\dfrac{3}{2\sqrt{3x - 5}}$ (e) $\dfrac{-x}{\sqrt{1 - x^2}}$,

(f) $\dfrac{3x^2}{2(x^3+5)^{1/2}}$, (g) $x(10x^3+15x-28)$,

(h) $(x-5)^2(5x^2-10x+27)$, (i) $2(2x+1)^3(3x-4)(18x-13)$,

(j) $(4-x)^2(1+2x)^4(37-16x)$,

(k) & (l) $\dfrac{12x+7}{2(2x+3)^{1/2}(3x-1)^{1/2}} = \dfrac{12x+7}{2(6x^2+7x-3)^{1/2}}$,

(m) $\dfrac{-2x}{(1+x^2)^2}$, (n) $\dfrac{31}{(3x+1)^2}$, (o) $\dfrac{-17}{(3+x)^2}$, (p) $\dfrac{5x^2+4x+1}{(x^2+3x+1)^2}$,

(q) $\dfrac{2(2-x)}{(x+3)^3}$, (r) $\dfrac{-183(9x+7)^2}{(x-6)^4}$, (s) $\dfrac{-19}{2(2x+7)^{1/2}(3x+1)^{3/2}}$,

(t) $\dfrac{4}{(x-4)^{1/2}(x+4)^{3/2}}$, (u) $\dfrac{x}{(1-x^2)^{3/2}}$, (v) $\dfrac{\sin x \cos x}{\sqrt{2+\sin^2 x}}$,

(w) $\dfrac{x\cos\sqrt{2+x^2}}{\sqrt{2+x^2}}$, (x) $5\sin^4 x\sin 6x$, (y) 0,

(z) $\cos(\sin(\sin x))\cos(\sin x)\cos x$.

6. (a) $\dfrac{4-x}{y-3}$ $(y\neq 3)$, (b) $\dfrac{5y-3x^2}{3y^2-5x}$ $(3y^2\neq 5x)$,

(c) $\dfrac{2xy-y^2+\cos(x+y)}{2xy-x^2-\cos(x+y)}$ $(2xy-x^2-\cos(x+y)\neq 0)$,

(d) $\dfrac{4(x-y)^3+y\cos(xy)}{4(x-y)^3-x\cos(xy)}$ $(4(x-y)^3\neq x\cos(xy))$.

7. $(21/\sqrt{13}, 12/\sqrt{13})$, tangents have gradients $-9/7$ and $7/9$.

8. (a) $3x(x^2+2)^{1/2}$, (b) $\dfrac{3(6x+5)}{4(3x^2+5x-1)^{1/4}}$, (c) $\dfrac{-39}{5(2x+1)^{2/5}(7x-3)^{8/5}}$,

(d) $\dfrac{-2(18x+11)}{3(7+3x)^{1/3}(1-4x)^{2/3}}$, (e) $\dfrac{2\cos(x^{2/3})}{3x^{1/3}}$, (f) $\dfrac{2\cos x}{3(\sin x)^{1/3}}$,

(g) $\dfrac{-2x\sin\sqrt{2x^2+3}}{\sqrt{2x^2+3}}$, (h) $\dfrac{-2\sin x\cos x}{\sqrt{2\cos^2 x+3}}$, (i) $-2x\sin(\sin(x^2))\cos(x^2)$,

(j) $-\cos(\cos(\tan x))\sin(\tan x)\sec^2 x$, (k) $3\tan^2 x\sec^2 x$,

(l) $\sec x(\sec^2 x+\tan^2 x)$, (m) $\dfrac{-\operatorname{cosec}^2\sqrt{x}\cot\sqrt{x}}{\sqrt{x}}$,

(n) $12\sin^2 4x\cos 6x\cos 10x$,

(o) $(2/3)(\sin 5x)^{1/3}(10\cos 5x\sin 2x+3\sin 5x\cos 2x)$,

(p) $\dfrac{-2\sin 10x}{(\cos 4x)^{1/2}(\cos 6x)^{2/3}}$, (q) $7\cos^6 x\cos 8x$, (r) $\dfrac{-7\cos^6 x\cos 6x}{\sin^2 7x}$,

(s) $\dfrac{-(2\sin 4x\sin 5x+5\cos 4x\cos 5x)}{(\cos 4x)^{1/2}\sin^2 5x}$, (t) $\dfrac{\sec x\left((1+x^2)\tan x-2x\right)}{(1+x^2)^2}$,

(u) $\dfrac{(4x+1)\cos 2}{\cos^2(2x^2+x-1)}$, (v) $2\sec x(\tan x+\sec x)^2$,

(w) $\dfrac{\sqrt{2}}{\sqrt{1-2x}\sqrt{4x-1}}$, (x) $\dfrac{-\sqrt{2}x}{|x|\sqrt{1-x^2}(1+x^2)}$,

(y) $\dfrac{-1}{\sqrt{x}|x+1|} = \dfrac{-1}{\sqrt{x}(x+1)}$ (since $x \geq 0$), (z) $-\tan^{-1}\sqrt{x} + \dfrac{1-x}{2\sqrt{x}(1+x)}$.

Chapter 9

1. (a) $(-1, 12)$ is a maximum turning point,
    $(2, -15)$ is a minimum turning point,

   (b) $(2/3, 17/9)$ is a down-up horizontal point of inflection,

   (c) $(-2, -21)$ and $(2, -21)$ are minimum turning points,
    $(0, -5)$ is a maximum turning point,

   (d) $(4, -1/4)$ is a maximum turning point,

   (e) $(1/2, 2)$ is a maximum turning point,
    $(3/2, 10)$ is a minimum turning point,

   (f) $(-2, 0)$ and $(2, 0)$ are down-up horizontal points of inflection.

2. (a) $(\pi/6, 3\sqrt{3}/8)$ is a maximum turning point,
    $(\pi/2, 0)$ is an up-down horizontal point of inflection,
    $(5\pi/6, -3\sqrt{3}/8)$ is a minimum turning point,

   (b) $(-\pi/4, -2)$ is a minimum turning point,
    $(\pi/4, 2)$ is a maximum turning point,

   (c) $(-2\pi/3, 3\sqrt{3}/2)$ is a maximum turning point,
    $(0, 0)$ is an up-down horizontal point of inflection,
    $(2\pi/3, -3\sqrt{3}/2)$ is a minimum turning point,

3. (a) -24, 25, (b) -17, $41/27$, (c) $\dfrac{\pi}{4} - 1$, $\dfrac{7\pi}{4} + 1$.

4. $30^2 45^3 = 82,012,500.$

5. $\sqrt{3}/9 = 1/(3\sqrt{3}).$

7. Radius $= 0.5$ metres, height (length) $= 2$ metres.

14 & 15.

function	strictly increasing on	strictly decreasing on
14. (a)	$[-1/2, 0]$ and $[1/2, \infty)$	$(-\infty, -1/2]$ and $[0, 1/2]$
14. (b)	$(-\infty, -1]$ and $[3, \infty)$	$[-1, 1)$ and $(1, 3]$
14. (c)	$[-\sqrt{3}, \sqrt{3}]$	$(-\infty, -\sqrt{3}]$ and $[\sqrt{3}, \infty)$
15.	$[0, \pi/6]$ and $[\pi/2, 5\pi/6]$	$[\pi/6, \pi/2]$ and $[5\pi/6, \pi]$

16. Velocity $= 254$, acceleration $= 210$.

17. When $t = 1/3$, position $= 196/27$, acceleration $= -2$.
    When $t = 1/2$, position $= 29/4$, acceleration $= 2$.

18. 0·8 square metres per minute.

19. $5\sqrt{2}$ knots.

20. $2/\pi$ metres per minute.

21. (a) 5/3, (b) 3/16, (c) 0, (d) 2/3, (e) −5, (f) 0.

22. (a) $\to -\infty$, (b) $\to 2$.

Chapter 10

2. (a) $(2,3)$ is a minimum turning point,
$x = 0$ and $y = x$ are asymptotes,

(b) $(0,0)$ is a minimum turning point,
$(2^{1/3}, 2^{2/3}/3)$ is a maximum turning point,
$x = -1$ and $y = 0$ are asymptotes,

(c) $(2,0)$ is a maximum turning point,
$(6, 8/9)$ is a minimum turning point,
$x = 0$, $x = 3$ and $y = 1$ are asymptotes,

(d) $(0, -2)$ and $(-2, -2/3)$ are minimum turning points,
$(-1/3, -63/32)$ is a maximum turning point,
$x = -1$, $x = 1$ and $y = 0$ are asymptotes.

3. (a) $(0,0)$ is an increasing up-down point of inflection,

(b) $(-2, -16)$ is an increasing up-down point of inflection,
$(0,0)$ is a horizontal down-up point of inflection,

(c) No point of inflection,

(d) $(-5^{1/4}, -5^{1/4}/8)$ is a decreasing down-up point of inflection,
$(0,0)$ is an increasing up-down point of inflection,
$(5^{1/4}, 5^{1/4}/8)$ is a decreasing down-up point of inflection,

(e) $(-\pi/2, -\pi)$ is an increasing up-down point of inflection,
$(\pi/2, \pi)$ is an increasing down-up point of inflection,

(f) $(0,0)$ is an increasing up-down point of inflection,
(π, π) is a decreasing down-up point of inflection,

(g) $(0,0)$ is an increasing up-down point of inflection.

4. (a) $(-1/\sqrt{3}, -1/(2\sqrt{3}))$ is a minimum turning point,
       $(1/\sqrt{3}, 1/(2\sqrt{3}))$ is a maximum turning point,
       $x = -1$ and $y = 0$ are asymptotes,
       $(0,0)$ is a increasing up-down point of inflection,
       $(\pm 1, \pm 1/4)$ are decreasing down-up points of inflection.

   (b) $(1, 1/4)$ is a maximum turning point,
       $y = 0$ is an asymptote,
       $(2, 2/9)$ is a decreasing down-up point of inflection.

   (c) $(-3, -9/2)$ is a maximum turning point,
       $(3, 9/2)$ is a minimum turning point,
       $x = -\sqrt{3}$, $x = \sqrt{3}$ and $y = x$ are asymptotes,
       $(0,0)$ is a horizontal up-down point of inflection,

   (d) $(-1, -1/4)$ is a minimum turning point,
       $x = 3$ and $y = 0$ are asymptotes,
       $(-2, -128/625)$ is a decreasing down-up point of inflection,

   (e) $(0,0)$ is a minimum turning point,
       $(4, -8)$ is a maximum turning point,
       $x = 2$ and $y = -x - 2$ are asymptotes,
       no point of inflection.

5. $(-1, -2)$ is a maximum turning point, $(3, 6)$ is a minimum turning point.

6. $(1, 0)$ is a minimum turning point, $(5, 2/27)$ is a maximum turning point,
   $(5 - 2\sqrt{3}, (3 - \sqrt{3})/72$ is an increasing up-down point of inflection,
   $(5 + 2\sqrt{3}, (3 + \sqrt{3})/72$ is a decreasing down-up point of inflection.

7. (a) all,   (b) none,   (c) origin only,   (d) x-axis only.

9. $(-3, 6)$, $(3, -6)$.

10. $15y - 3x = 4$, $t = -1/4$.

Chapter 11

1. (b) $AX = \begin{bmatrix} -9 & 6 & 9 \\ -18 & 15 & 24 \end{bmatrix}$, $AY = \begin{bmatrix} 22 & 28 \\ 49 & 64 \end{bmatrix}$,

$BX = \begin{bmatrix} -3 & 3 & 5 \\ 2 & 1 & 5 \end{bmatrix}$, $BY = \begin{bmatrix} 9 & 12 \\ 0 & 2 \end{bmatrix}$,

$CX = \begin{bmatrix} 3 & 3 & 0 \\ -6 & 3 & 4 \end{bmatrix}$, $CY = \begin{bmatrix} -2 & -2 \\ 13 & 16 \end{bmatrix}$,

$X^2 = \begin{bmatrix} -7 & 4 & 8 \\ -4 & 1 & -4 \\ -8 & -1 & 4 \end{bmatrix}$, $XY = \begin{bmatrix} 17 & 22 \\ 5 & 4 \\ 9 & 10 \end{bmatrix}$,

$YA = \begin{bmatrix} 9 & 12 & 15 \\ 19 & 26 & 33 \\ 29 & 40 & 51 \end{bmatrix}$, $YB = \begin{bmatrix} 5 & 3 & -1 \\ 11 & 7 & -1 \\ 17 & 11 & -1 \end{bmatrix}$,

$YC = \begin{bmatrix} 1 & 1 & 4 \\ 3 & 1 & 8 \\ 5 & 1 & 12 \end{bmatrix}$, $YZ = \begin{bmatrix} 4 & 3 \\ 10 & 5 \\ 16 & 7 \end{bmatrix}$,

$ZA = \begin{bmatrix} -2 & -1 & 0 \\ 9 & 12 & 15 \end{bmatrix}$, $ZB = \begin{bmatrix} 0 & 1 & 3 \\ 5 & 3 & -1 \end{bmatrix}$,

$ZC = \begin{bmatrix} 2 & -3 & -2 \\ 1 & 1 & 4 \end{bmatrix}$, $Z^2 = \begin{bmatrix} 3 & -4 \\ 4 & 3 \end{bmatrix}$.

(d) $\begin{bmatrix} 9 & 0 & -4 \\ 0 & 9 & 4 \\ -4 & 4 & 9 \end{bmatrix}$, $\begin{bmatrix} 9 & 0 & 4 \\ 0 & 9 & 4 \\ 4 & 4 & 9 \end{bmatrix}$.

2. (a) $\left\{ \begin{bmatrix} 2 & 0 \\ 0 & 1 \end{bmatrix}, \begin{bmatrix} 2 & 0 \\ 0 & -1 \end{bmatrix}, \begin{bmatrix} -2 & 0 \\ 0 & 1 \end{bmatrix}, \begin{bmatrix} -2 & 0 \\ 0 & -1 \end{bmatrix} \right\}$, (b) $\emptyset$.

3. (b) $a = 6$, $b = 1$.

10. (a) $\begin{bmatrix} 1 & 0 & 0 & -3 \\ 0 & 1 & 0 & -5 \\ 0 & 0 & 1 & 4 \end{bmatrix}$, (b) $\begin{bmatrix} 1 & 2 & 0 & 0 \\ 0 & 0 & 1 & 0 \\ 0 & 0 & 0 & 1 \end{bmatrix}$, (c) $\begin{bmatrix} 1 & 0 & -1 & 0 \\ 0 & 1 & 6 & 0 \\ 0 & 0 & 0 & 1 \end{bmatrix}$.

11. (a) $x = 1 + z + t$, $y = -2z - 2t$, (b) $x = 1/3$, $y = 4/3$, $z = -2/3$,
    (c) no solution, (d) $x = 5 - t$, $y = 3 - 2t$, $z = t$.

12. $2a + b = c$, $x = 2/3 - 2z/3$, $y = 1/3 - z/3 + t$.

19. (a) $B^{-1} = B^2 - 3B + 3I$, (b) $(B + 3I)^{-1} = (1/64)(B^2 - 6B + 21I)$.

21. (a) True, (b) False, $A = \begin{bmatrix} 1 & 0 \\ 0 & 0 \end{bmatrix}$, (c) False, $A = \begin{bmatrix} 0 & 1 \\ 0 & 0 \end{bmatrix}$,

    (d) True, (e) False, $A = \begin{bmatrix} 0 & 1 \\ 0 & 0 \end{bmatrix}$, (f) False, $A = \begin{bmatrix} 0 & 1 \\ -1 & 0 \end{bmatrix}$.

22. (a) $\begin{bmatrix} 1/7 & -4/7 \\ 1/7 & 3/7 \end{bmatrix}$, (b) $\begin{bmatrix} 1 & -2 & 0 \\ -3 & 8 & -3 \\ 2 & -5 & 2 \end{bmatrix}$, (c) non-invertible,

(d) $\begin{bmatrix} -1/2 & 1/2 & 1/2 \\ 7/6 & -5/6 & -1/2 \\ 5/6 & -1/6 & -1/2 \end{bmatrix}$, (e) $\begin{bmatrix} -2/3 & 1/3 & 1/3 \\ 7/3 & -5/3 & 1/3 \\ 8/3 & -7/3 & 2/3 \end{bmatrix}$,

(f) $\begin{bmatrix} 0 & 2 & -1 & -1 \\ -2 & 0 & -1 & 2 \\ 1 & 1 & 0 & -1 \\ 1 & -2 & 1 & 0 \end{bmatrix}$.

23. $a = 6$, $\begin{bmatrix} -21/25 & 2/5 & 13/25 \\ 8/25 & -1/5 & 1/25 \\ 2/5 & 0 & -1/5 \end{bmatrix}$.

Chapter 12

1. $(0, -2, 11)$, $(-7, 9, -4)$, $(-11, 9, -13)$.

2. $\alpha = 0$, $\beta = 1$, $\gamma = 1$.

3. -2, $-4/3$.

4. $\alpha = 2$, $\beta = -1$ (for example), no.

5. $\mathbf{b} - \mathbf{a}$, $-\mathbf{a}$, $-\mathbf{b}$, $\mathbf{a} - \mathbf{b}$.

6. (a) collinear, (b) not collinear.

7. $3 : 2$.

8. $\frac{1}{2}(\mathbf{b} + \mathbf{c})$, $\frac{1}{2}(\mathbf{c} + \mathbf{a})$, $\frac{1}{2}(\mathbf{a} + \mathbf{b})$,

$\frac{1}{6}(4\mathbf{a} + \mathbf{b} + \mathbf{c})$, $\frac{1}{6}(\mathbf{a} + 4\mathbf{b} + \mathbf{c})$, $\frac{1}{6}(\mathbf{a} + \mathbf{b} + 4\mathbf{c})$.

9. $\frac{1}{2}(\mathbf{b} + \mathbf{c})$, $\frac{1}{2}(\mathbf{c} + \mathbf{d})$, $\frac{1}{5}(\mathbf{a} + \mathbf{b} + 2\mathbf{c} + \mathbf{d})$, $4 : 1$.

10. -3, $2 : 3$.

11. (a) $\frac{1}{\lambda + 1}(\lambda\mathbf{a} + \mathbf{b})$, $\frac{1}{\lambda + 1}(\lambda\mathbf{a} + \mathbf{c})$,

(b) $\frac{1}{(\alpha + 1)(\lambda + 1)}(\alpha\lambda\mathbf{a} + (\lambda + 1)\mathbf{b} + \alpha\mathbf{c})$,

$\frac{1}{(\beta + 1)(\lambda + 1)}(\beta\lambda\mathbf{a} + \beta\mathbf{b} + (\lambda + 1)\mathbf{c})$,

(c) $\alpha = \beta = \lambda + 1$.

13. $|x|$, $\sqrt{y^2 + z^2}$.

14. (a) $\pm\dfrac{1}{3}(1, 2, -2)$,

   (b) $(x, y, z) = (-2, -5, 6) + t(1, 2, -2)$,
   $x = -2 + t$, $y = -5 + 2t$, $z = 6 - 2t$,
   $$\dfrac{x + 2}{1} = \dfrac{y + 5}{2} = \dfrac{z - 6}{-2},$$

   (c) $(-4, -9, 10)$, $(0, -1, 2)$,   (d) $(3, 5, -4)$, $(-7/5, -19/5, 24/5)$.

15. $(1, 1, 2)$, $(2, -1, 5)$.

16. 9, $(2, 4, -3)$.

17. $\dfrac{x}{2} = \dfrac{y}{-4} = \dfrac{z + 3}{3}$ (for example).

Chapter 13

1. $\sqrt{7}$.

2. $\pi/6$.

3. -1, 3.

5. (a) -3,   (b) $-1/\sqrt{14}$,   (c) -2.

6. $(0, 0, -1)$, $(8/9, 4/9, -1/9)$.

7. $\sqrt{2/3}$.

9. $(-1, -4, -10)$, $\sqrt{\dfrac{13}{14}}$, $\pm\dfrac{1}{3\sqrt{13}}(1, 4, 10)$.

10. (c) $-2 : 3$.

11. $2(\mathbf{b} \times \mathbf{a})$.

12. $(-2, -3, 1)$, 3, $(4, 3, 4)$, 6, $(0, -1, 2)$, $2x + 2y + z = 0$.

13. 2.

14. $x^2 + y^2 + z^2 - 6x + 6y - 6z + 13 = 0$.

15. $x^2 + y^2 + z^2 - 12z = 0$, $z = 12$.

16. $(6, 6, 3)$, 21, right-handed.

18. (a) $(5, -2, -1)$,   (c) $\pi/3$,   (d) $(9, -4, 1)$.

19. (a) $3x + 2y + 4z = -2$, (b) $3x - 2y + 2z = 0$, (c) $2x - 3y + z = 16$, (d) $3x - y = 1$.

20. $x + 2y - z = -3$, $(-3, 1, 2)$.

21. $\dfrac{x}{1} = \dfrac{y - 4}{-10} = \dfrac{z - 6}{-4}$.

22. $4x - 3y - z = 3$, $\dfrac{x - 13}{7} = \dfrac{y - 18}{11} = \dfrac{z + 5}{-5}$.

23. $x + 20y - 28z = -42$.

24. (a) $x + 4y + 7z = -3$, (b) $\dfrac{x - 2}{1} = \dfrac{y + 3}{4} = \dfrac{z - 1}{7}$.

25. $(1, 1, 1)$, $(-1, 2, 1)$, $\mathcal{L}_1$.

26. $(3, -1, 3)$.

27. $\dfrac{x - 2}{2} = \dfrac{y}{1} = \dfrac{z + 1}{-1}$.

Chapter 14

1. (a) $\dfrac{x^4}{4} + C$, (b) $-\dfrac{1}{x} + C$, (c) $\dfrac{2x^{3/2}}{3} + C$, (d) $2\sqrt{x} + C$,

    (e) $\dfrac{(2x + 7)^6}{12} + C$, (f) $\dfrac{-(5x - 4)^{-2}}{10} + C$, (g) $\dfrac{-3(3 - 2x)^{1/3}}{2} + C$,

    (h) $x^4 + x^3 + x^2 + x + C$, (i) $\dfrac{x^2}{2} - x - \dfrac{3}{2(2x + 1)} + C$,

    (j) $-\dfrac{\cos 3x}{3} + C$, (k) $\dfrac{\sin(2x + 1)}{2} + C$, (l) $-\tan(2 - x) + C$,

    (m) $\dfrac{x^3}{3} - \tan x + C$, (n) $\dfrac{\tan 3x + \sec 3x}{3} + C$, (o) $\sin^{-1}\dfrac{x}{3} + C$,

    (p) $\dfrac{1}{2}\sin^{-1}\dfrac{2x}{3} + C$, (q) $\dfrac{1}{\sqrt{3}}\tan^{-1}\dfrac{x}{\sqrt{3}} + C$, (r) $\dfrac{1}{2\sqrt{3}}\tan^{-1}\dfrac{2x}{\sqrt{3}} + C$,

    (s) & (t) $\dfrac{1}{3}\tan^{-1}\left(\dfrac{x - 1}{3}\right) + C$.

2. $2x\sqrt{1 + x^4}$.

4. (a) $-1/2$, (b) $8/3$, (c) -6.

5. (a) 8, (b) 4, (c) $-28/3$, (d) 18, (e) 0, (f) 1,

    (g) $\dfrac{5^{3/2} - 3^{3/2}}{3}$, (h) $\sqrt{5} - \sqrt{3}$, (i) $\dfrac{3}{2}$, (j) $\dfrac{\pi}{12} + \dfrac{\sqrt{3}}{8}$,

    (k) $1 - \pi/4$, (l) $5/2$, (m) 5, (n) 2, (o) $\pi/4$, (p) $\pi/8$.

6. (a) 0, (b) 2/3, (c) 0.

8. $(22/7) - \pi$.

9. (a) $(-1, 1)$, $(3, -7)$, 32, (b) $(0, 0)$, $(1, 1)$, $(2, 2)$, 1/2,
   (c) $(3, -1)$, $(3, 3)$, 32/3.

10. (a) $\dfrac{2\pi}{3} - \dfrac{\sqrt{3}}{2}$, (b) $\dfrac{\pi}{2} + 1$.

11. (a), (c) and (e) do not converge.
    (b) converges to 2. (d) converges to $3^{4/3}$. (f) converges to π.

Chapter 15

1. (a), (b) & (c) $\dfrac{7}{7x - 4}$, (d) $\dfrac{14}{7x - 4}$, (e) $\dfrac{2(x - 1)}{x^2 - 2x + 3}$,

   (f) $\dfrac{10(x - 1)}{x^2 - 2x + 3}$, (g) $10[\log(x^2 - 2x + 3)]^4 \left(\dfrac{x - 1}{x^2 - 2x + 3} \right)$,

   (h) $\dfrac{\sin x \cos x}{1 + \sin^2 x}$, (i) $-\tan x$, (j) $-\dfrac{1}{x} \sin(\log x)$, (k) $\dfrac{1}{x \log x}$,

   (l) $\log x$, (m) $\dfrac{2}{1 - x^2}$, (n) $\dfrac{1}{1 - x^2}$.

2. (a) $\dfrac{(x^2 + 2)^{1/2}(x^8 + 8)^{1/8}}{(x^4 + 4)^{1/4}(x^6 + 6)^{1/6}} \left[\dfrac{x}{x^2 + 2} - \dfrac{x^3}{x^4 + 4} - \dfrac{x^5}{x^6 + 6} + \dfrac{x^7}{x^8 + 8} \right]$,

   (b) $\dfrac{1}{2} \sqrt{\dfrac{\sin^{-1} x}{x^3(2 + \sin x)}} \left[\dfrac{1}{\sqrt{1 - x^2} \sin^{-1} x} - \dfrac{3}{x} - \dfrac{\cos x}{2 + \sin x} \right]$.

3. (a) $\dfrac{1}{3} \log|3x| + C$ or $\dfrac{1}{3} \log|x| + C$, (b) $\dfrac{1}{3} \log|3x + 4| + C$,

   (c) $\log|x^2 + 4| + C$, (d) $\dfrac{1}{5} \log|x^5 + 6| + C$, (e) $\log|\log(x + 1)| + C$,

   (f) $\log|x^2 + 4| - \dfrac{1}{2} \tan^{-1} \dfrac{x}{2} + C$, (g) $-\dfrac{1}{7} \log|\cos 7x| + C$,

   (h) $\dfrac{1}{4} \log|\sin(4x - 3)| + C$, (i) $\dfrac{1}{5} \log|\sec 5x + \tan 5x| + C$,

   (j) $\sin^{-1}(x/\sqrt{3}) + C$, (k) $\log|x + \sqrt{x^2 + 3}| + C$,

   (l) $\log|x + \sqrt{x^2 - 3}| + C$.

4. (a) $\dfrac{1}{2} \log(1 + \sqrt{2})$, (b) $3 - \log 3$, (c) $\log 3$, (d) $2 \log 2$.

6. $(3x^2 - 2x + 1) \log x \to -\infty$ as $x \to 0^+$.

9. (a) $5e^{5x}$, (b) $-2e^{3-2x}$, (c) $(\cos x)e^{\sin x}$, (d) $e x^{e-1}$,
   (e) $(4 \log 3)3^{4x}$, (f) $-(\log \pi)\pi^{1-x}$, (g) $-2e^{-5x}(5x^3 - 3x^2 + 10)$,

(h) $(x^2 + 5)^{\tan x}\left[\sec^2 x \log(x^2 + 5) + \dfrac{2x \tan x}{x^2 + 5}\right]$,

(i) $(\cos x)^{x \sin x}\left[\sin x \log(\cos x) + x \cos x \log(\cos x) - \dfrac{x \sin^2 x}{\cos x}\right]$.

10. (a) $\dfrac{e^{5x}}{5} + C$,   (b) $-\dfrac{e^{3-2x}}{2} + C$,   (c) $\dfrac{x^{e+1}}{e+1} + C$,

(d) $\dfrac{3^{4x}}{4 \log 3} + C$,   (e) $-\dfrac{\pi^{1-x}}{\log \pi} + C$,   (f) $e^{x^2 - 9} + C$.

11. (a) $\to 0$,   (b) $\to 0$,   (c) $\to \infty$.

13. (a) $7/5$,   (b) $1/6$,   (c) 2,   (d) $\log(5/2)$.

14. 1 hour longer.

15. $\dfrac{15 \log 2}{\log(5/4)}$ years,   *i.e.* approximately 46·594 years.

17. (a) $0, \log 2$,   (b) $\log(2\sqrt{2})$.

19. (a) $\dfrac{1}{3} \cosh \dfrac{x}{3}$,   (b) $(2x - 1)\cosh(x^2 - x)$,   (c) $\dfrac{\text{sech}^2 \sqrt{x - 1}}{2\sqrt{x - 1}}$,

(d) $4 \tanh 4x$.

20. $6 \cosh^2 2x \cosh 3x \sinh 5x$.

21. (a) $\dfrac{1}{5} \cosh(5x + 4) + C$,   (b) $\dfrac{1}{3} \log \dfrac{65}{16}$.

22. (a) $\dfrac{1}{\sqrt{x^2 - a^2}}$ $(x > a)$,   (b) $\dfrac{1}{\sqrt{x^2 + a^2}}$ $(x \in \mathbb{R})$.

23. (a) and (d) do not converge.

(b) does not converge when $a > 0$, converges to $-1/a$ when $a < 0$.

(c) converges to 2.

Chapter 16

1. (a) $\dfrac{1}{2} \sin^{-1} 2x + C$,   (b) $\dfrac{1}{5} \sin^{-1} \dfrac{5x}{3} + C$,   (c) $\sin^{-1} \dfrac{x-1}{2} + C$,

(d) $\dfrac{1}{6} \tan^{-1} \dfrac{3x}{2} + C$,   (e) $\dfrac{1}{\sqrt{2}} \tan^{-1} \dfrac{x+4}{\sqrt{2}} + C$,   (f) $\dfrac{1}{\sqrt{5}} \tan^{-1} \dfrac{3x+1}{\sqrt{5}} + C$,

(g) $\log|x - 1 + \sqrt{x^2 - 2x - 3}| + C$,   (h) $\log|x - 1 + \sqrt{x^2 - 2x}| + C$,

(i) $\dfrac{1}{\sqrt{2}} \log\left|x + \sqrt{x^2 + \dfrac{1}{2}}\right| + C$ or $\dfrac{1}{\sqrt{2}} \log|\sqrt{2}x + \sqrt{2x^2 + 1}| + C$.

2. (a) $\dfrac{(x^2-5)^{3/2}}{3}+C$, (b) $\dfrac{(3x^2+7)^7}{42}+C$, (c) $\dfrac{1}{9\sqrt{2}}\tan^{-1}\dfrac{x^3}{3\sqrt{2}}+C$,

(d) $-\dfrac{2}{3}\sqrt{2-\sin 3x}+C$, (e) $\dfrac{2}{3}(\sin^{-1}x)^{3/2}+C$, (f) $\dfrac{1}{2\sqrt{3}}\tan^{-1}\dfrac{x^2-1}{\sqrt{3}}+C$,

(g) $\dfrac{-1}{\log x}+C$, (h) $\dfrac{16\sqrt{3}}{3}-\dfrac{10\sqrt{5}}{9}$, (i) 0, (j) $\dfrac{\pi}{8}$, (k) $\dfrac{\pi^{3/2}}{12}$,

(l) $\log 2$.

3. (a) $4\log|x+4|+\log|x-3|+C$, (b) $3\log 3+(4/3)$,

(c) $11x+\log\left|\dfrac{x-1}{x+1}\right|-\dfrac{27}{\sqrt{2}}\tan^{-1}\dfrac{\sqrt{2}x}{3}+C$, (d) $\tan^{-1}2-\dfrac{1}{2}\log 5$.

4. $P=1/2$, $Q=16$, (a) $9/16+\pi/4$, (b) $2(\sqrt{2}-1)+16\log(\sqrt{2}+1)$.

5. (a) $2\sqrt{2}+2\log(1+\sqrt{2})$, (b) $\dfrac{8}{3}(14\sqrt{2}-1)+32\log(1+\sqrt{2})$.

6. (a) $1/(20\sqrt{2})$, (b) $9/20$, (c) $3/4$, (d) $\dfrac{1}{7}\cos^7 x-\dfrac{1}{5}\cos^5 x+C$,

(e) $\dfrac{1}{3}\sin^3 x-\dfrac{2}{5}\sin^5 x+\dfrac{1}{7}\sin^7 x+C$, (f) $\dfrac{1}{3}\tan^3 x+\dfrac{1}{5}\tan^5 x+C$,

(g) $\dfrac{1}{5}\sec^5 x-\dfrac{1}{3}\sec^3 x+C$, (h) $\dfrac{1}{3}\sinh^3 x+\dfrac{2}{5}\sinh^5 x+\dfrac{1}{7}\sinh^7 x+C$,

(i) $\dfrac{1}{9}\sin^3 3x-\dfrac{1}{15}\sin^5 3x+C$,

(j) $\tan^2\dfrac{x}{2}+\dfrac{1}{2}\tan^4\dfrac{x}{2}+C$ or $\dfrac{1}{2}\sec^4\dfrac{x}{2}+C$.

7. (a) $2\pi/(3\sqrt{3})$, (b) $\log(4/3)$, (c) $\pi/9$, (d) $(1/5)\log(3/2)$.

8. (a) $e^{3x}(3x-1)/9+C$, (b) $e-5/e$, (c) $e^{2x}(2x^2-6x+5)/4+C$,

(d) $(e^{-\pi}+1)/2$, (e) $e^{3x}(3\sin 2x-2\cos 2x)/13+C$,

(f) $(e^{3x}/5)(3\sinh 2x-2\cosh 2x)+C$, (g) 1, (h) $(e^2+1)/4$,

(i) $5\pi/9-1/\sqrt{3}$, (j) $x\cos^{-1}x-\sqrt{1-x^2}+C$,

(k) $x\tan^{-1}x-\dfrac{1}{2}\log|x^2+1|+C$,

(l) $\dfrac{1}{2}\left(x\sqrt{x^2+7}+7\log|x+\sqrt{x^2+7}|\right)+C$.

9. (a) $\dfrac{8}{105}$, (b) $\dfrac{\pi}{6}$, (c) $\dfrac{e^{5x}}{29}(5\cos 2x+2\sin 2x)+C$,

(d) $\dfrac{e^{5x}}{290}(29-25\cos 2x-10\sin 2x)+C$, (e) $1-\dfrac{1}{\sqrt{3}}+\dfrac{1}{2}\log 2-\dfrac{\pi}{6}$,

(f) $\dfrac{1}{2\sqrt{2}}\tan^{-1}\dfrac{x^2}{\sqrt{2}}+C$, (g) $\dfrac{4}{45}$, (h) $\dfrac{\pi}{6\sqrt{3}}$, (i) $\log 2-\dfrac{\pi}{8}$,

(j) $x\left((\log x)^2-2\log x+2\right)+C$,

(k) $x+\dfrac{4}{(x+1)^2}-\dfrac{12}{x+1}-6\log|x+1|+C$, (l) $3\tan^{-1}x^{1/3}$.

10. $\frac{1}{8}(7\sqrt{2} + 3\log(1 + \sqrt{2}))$.

11. $9e - 24$.

12. (a) $x\log x - x + C$,   (b) $\frac{2}{3}x^{3/2} + C$,   (c) $x\sin^{-1}x + \sqrt{1 - x^2} + C$.

13. $\pi^2/2$.

14. $2^{13}\pi/105$,   12.

15. $6 + \frac{1}{8}\log\frac{7}{5}$.

16. $4 - \frac{2}{\sqrt{3}}$, $\frac{16\pi}{3}$.

17. (a) $\frac{\sqrt{2}\pi}{4}\left(\frac{e^5}{5} - \frac{4e^3}{9} + e - \frac{109}{45}\right)$, $\frac{e^2}{2}$, $\frac{\pi}{4}(e^4 - 4)$.
    (b) $32\sqrt{2}\pi/15$, 2π, 8π.

18. (a) π, π^2,   (b) $\sqrt{2}(e^{\pi/2} - 1)$,   (c) 8.

Chapter 17

1. (a) $y = -1/(x^3 + C)$ or $y = 0$, $C = 3$.
    (b) $\tan^{-1}(y/2) = e^{2x} + K$,   (c) $\sinh y = x + C$, $C = -\cosh 1$.
    (d) $\log|\sin y| = -e^{-x} + C$ or $y = k\pi$ $(k \in \mathbb{Z})$.
    (e) $y^2 + 1 = D(x^2 + 1)$ $(D > 0)$, $D = 5$.

2. (a) $(y/x)^4 = 4\log|x| + K$,   (b) $\sin(y/x) = \log|x| + C$, $C = 1$.
    (c) $\tan^{-1}(y/x) = \log|x| + C$,
    (d) $\sin^{-1}(y/x) = \begin{cases} \log|x| + C & (x > 0), \\ -\log|x| + K & (x < 0) \end{cases}$ or $y = -x$ or $y = x$,
    $K = \log 2 - \pi/6$.
    (e) $\tan^{-1}(y/x) = -\log(D|x - y|\sqrt{x^2 + y^2})$ $(D > 0)$ or $y = -x$.

3. (a) $y = 5x/4 + x^2/6 + C/x^3$,   (b) $y = e^{5x}/2 + Ce^{3x}$,
    (c) $y = xe^{-x^2} - 1/2 + Ce^{-x^2}$,   (d) $y = (\sin x \tan x)/2 + C\sec x$,
    (e) $y(x^4 + 3) = -e^{-x}(x^3 + 3x^2 + 6x + 6) + C$, $C = 0$.

4. (a) $3y^2 - 2xy + 3x^2 = D$ $(D > 0)$,   (b) $y = 1/x - 1/x^2 + Ce^{-x}/x^2$,
    (c) $\tan^{-1}y = x^3/3 + x + C$,   (d) $y = xe^{Dx}$ $(x > 0, D > 0)$,
    (e) $y = e^{\sin x}(1 + Ce^{-x})$.

5. (a) $y = Ae^x + Be^{2x} + 2e^{-x}$, (b) $y = e^{-x}(A\cos 2x + B\sin 2x) + 4e^{3x}$,
    (c) $y = (A + Bx)e^{3x} + 2x^2 + 3x + 1$, (d) $y = Ae^{-x} + Be^x - x^2 - 3$,
    (e) $y = Ae^{5x} + Be^{-2x} - 2xe^{-2x}$,
    (f) $y = e^{-2x}(A\cos x + B\sin x) - 4\cos x + 4\sin x$,
    (g) $y = (A + Bx)e^{-3x/2} + 18x^2 e^{-3x/2}$,
    (h) $y = A\cos x + B\sin x - x\cos x$,
    (i) $y = (A + Bx)e^{-2x} + e^{-3x}$, $A = -1$, $B = 1$,
    (j) $y = Ae^{3x} + Be^{-2x} - x^2 + x - 1$, $A = 2$, $B = -1$,
    (k) $y = Ae^{-5x} + Be^{-3x} + 2\cos x$, $A = e^\pi/(e^\pi - 1)$, $B = -1/(e^\pi - 1)$,
    (l) $y = (A + Bx)e^{-3x} + 2e^x + 8\cos x + 6\sin x$, $A = 0$, $B = -2e^{4\pi}/\pi$.

Chapter 18

1. (a) 4/3, 4, (b) −1/4, 1/3, (c) 1/2, 3. These are all best possible.

3. (a) 2, (b) −2/3, (c) 0, (d) 1, (e) 2/3, (f) −1/5, (g) 0, (h) 1/2,
    (i) 1/6, (j) 1, (k) 0.

4. 0, (a) 1, (b) 3.

5. (a) 3, (b) 3/4, (c) 6, (d) 9/2.

6. 10/99.

7. 3/4.

8. (a) $\dfrac{n}{3n+1}$, $\dfrac{1}{3}$, (b) $\dfrac{1}{2}\left(\dfrac{3}{2} - \dfrac{1}{n+1} - \dfrac{1}{n+2}\right)$, $\dfrac{3}{4}$,
    (c) $\dfrac{1}{2}\left(\dfrac{1}{2} - \dfrac{1}{n+1} + \dfrac{1}{n+2}\right)$, $\dfrac{1}{4}$.

9. (a), (b), (d), (e) and (f) converge, (c) diverges.

10. (b) and (d) converge, (a) and (c) diverge.

12. (a), (c) and (f) converge, (b), (d) and (e) diverge.

13. (a) $[-1/3, 1/3]$, (b) $(-3/2, 3/2)$, (c) $(-3/2, 3/2)$, (d) $(-5, 5]$, (e) $\mathbb{R}$.

16. See Appendix G.

17. $a_0 + a_1 x + \cdots + a_{n-1}x^{n-1} + a_n x^n$, $\mathbb{R}$.

18. See Appendix G

19. (a) $\displaystyle\sum_{r=0}^{\infty} \frac{x^{2r}}{r!}$, $\mathbb{R}$,   (b) $\displaystyle\sum_{m=0}^{\infty} \frac{(-1)^m 2^{2m} x^{2m}}{(2m)!}$, $\mathbb{R}$,

   (c) $\displaystyle\sum_{r=1}^{\infty} (-1)^{r-1} x^r$, $(-1, 1)$,   (d) $\displaystyle\sum_{r=1}^{\infty} \left(-\frac{1}{r3^r}\right) x^r$, $[-3, 3)$,

   (e) $\displaystyle\sum_{r=0}^{\infty} \left(-\frac{1}{2^{r+1}}\right) x^r$, $(-2, 2)$,   (f) $\displaystyle\sum_{k=0}^{\infty} \frac{2x^{4k}}{(4k)!}$, $\mathbb{R}$.

20. $\displaystyle\sum_{r=0}^{\infty} 3^r x^r$, $(-1/3, 1/3)$, $\displaystyle\sum_{r=1}^{\infty} r3^{r-1} x^{r-1} = \sum_{s=0}^{\infty} (s+1) 3^s x^s$.

21. $\displaystyle\sum_{m=0}^{\infty} (-1)^m x^{2m}$, $\displaystyle\sum_{m=0}^{\infty} \frac{(-1)^m x^{2m+1}}{2m+1}$.

22. $1 + \dfrac{1}{2} x^2 + \dfrac{5}{24} x^4 + \cdots$.

23. (a) $1 - x + x^3 - x^4 + \cdots$,   (b) $x - \dfrac{3}{2} x^2 + \dfrac{11}{6} x^3 - \dfrac{25}{12} x^4 + \cdots$,

   (c) $1 + x - \dfrac{1}{3} x^3 - \dfrac{1}{6} x^4 + \cdots$,   (d) $1 + x + \dfrac{1}{2} x^2 - \dfrac{1}{8} x^4 + \cdots$,

   (e) $1 + 2x + 4x^2 + \dfrac{20}{3} x^3 + \dfrac{32}{3} x^4 + \cdots$,   (f) $x - \dfrac{1}{2} x^2 - \dfrac{1}{6} x^3 + \dfrac{1}{4} x^4 + \cdots$.

24. $2x + \dfrac{2}{3} x^3 + \dfrac{2}{5} x^5 + \dfrac{2}{7} x^7 + \cdots = \displaystyle\sum_{m=0}^{\infty} \frac{2x^{2m+1}}{2m+1}$, $(-1, 1)$, $0{\cdot}1823$.

25. For $n \in \mathbb{N}$, $(1 + x)^n = \displaystyle\sum_{r=0}^{n} \binom{n}{r} x^r$ $(x \in \mathbb{R})$.

   $1 + \dfrac{1}{2} x - \dfrac{1}{8} x^2 + \dfrac{1}{16} x^3 + \cdots$.

Chapter 19

1. (a) $21/16 = 1{\cdot}3125$,   (b) $13/16 = 0{\cdot}8125$,   (c) $-7/16 = -0{\cdot}4375$,
   (d) $11/16 = 0{\cdot}6875$.   All $\pm 0{\cdot}0625$.

2. $-37/32 = -1{\cdot}15625$, $15/32 = 0{\cdot}46875$, $57/32 = 1{\cdot}78125$.
   All $\pm 0{\cdot}03125$.

3. (a) $1, 19/17$,   (b) $1, 8/7$,   (c) $2, 139/72$,   (d) $3, 229/79$.

4. $-1, -3/4, -59/86$.

5. $1, 3/2, 31/23$.

6. 2, 5/2, 181/75.

7. 283/255.

8. $\dfrac{\pi^2 + 13\pi + 24}{4(2 + \pi)(3 + \pi)}$.

9. 17·3223, 18·5736, 17·9480 ± 0·6256.

10. (a) 8/3, (b) 25/36.

11. (a) 73/90, (b) 691/630.

13. (a) 0, 3/4, 8/9, (b) 0, 5/8, 62/81, (c) 3, 33/8, 139/27,
    (d) 1, 13/12, 10/9.

14. 0·68098.

Appendix B

Solutions to Problems

Problem 1.1.2 The set $\mathbb{D}$ is not a number system since $1, 3 \in \mathbb{D}$ but $1 + 3 = 4 \notin \mathbb{D}$.

The set $\mathbb{E}$ *is* a number system. The elements of $\mathbb{E}$ are of the form $2n$ where $n \in \mathbb{N}$. Let $m, n \in \mathbb{N}$. Then $m + n \in \mathbb{N}$ so that $2m + 2n = 2(m + n) \in \mathbb{E}$, and $2mn \in \mathbb{N}$ so that $(2m)(2n) = 2(2mn) \in \mathbb{E}$.

Define $\mathbb{F} = \{k/2^n : k \in \mathbb{Z}, n \in \mathbb{N}\}$. Let $h, k \in \mathbb{Z}$ and $m, n \in \mathbb{N}$. Then $\mathbb{F}$ is a number system since

$$\frac{h}{2^m} + \frac{k}{2^n} = \frac{h2^n + k2^m}{2^{m+n}} = \frac{t}{2^s} \quad \text{and} \quad \frac{h}{2^m} \cdot \frac{k}{2^n} = \frac{hk}{2^{m+n}} = \frac{u}{2^s},$$

where $t, u \in \mathbb{Z}$ and $s \in \mathbb{N}$. To see that $\mathbb{Z} \subset \mathbb{F}$ observe that $2k \in \mathbb{Z}$ so that $k = (2k)/2^1 \in \mathbb{F}$. Also, $1/2 \in \mathbb{F}$ but $1/2 \notin \mathbb{Z}$. To see that $\mathbb{F} \subset \mathbb{Q}$ observe that $2^n \in \mathbb{N}$ so that $k/2^n \in \mathbb{Q}$. Also, $1/3 \in \mathbb{Q}$ but $1/3 \notin \mathbb{F}$ since $1/3 = k/2^n$ implies that $2^n = 3k$ which is impossible (the right-hand side is divisible by 3, the left-hand side is not).

Problem 2.1.6 Both $g \circ f$ and $f \circ g$ have domain $\mathbb{R}$ and codomain $\mathbb{R}$. Hence,

$$
\begin{aligned}
g \circ f = f \circ g \;\;\Leftrightarrow\;\; & (g \circ f)(x) = (f \circ g)(x) \quad (x \in \mathbb{R}) \\
\Leftrightarrow\;\; & g(f(x)) = f(g(x)) \quad (x \in \mathbb{R}) \\
\Leftrightarrow\;\; & p(ax + b) + q = a(px + q) + b \quad (x \in \mathbb{R}) \\
\Leftrightarrow\;\; & apx + bp + q = apx + aq + b \quad (x \in \mathbb{R}) \\
\Leftrightarrow\;\; & (a - 1)q = (p - 1)b.
\end{aligned}
$$

Problem 2.2.12 Putting $k = \pm 1$ in (2.2.1) gives (2.2.2).

Now let (2.2.1) hold. Since $x + 0.p = x$, (2.2.1) holds for $k = 0$. Let $k = \pm n$, $n \in \mathbb{N}$, and let $x \in A$. Then successive applications of (2.2.2) give

$$x \pm p \in A, \quad x \pm 2p \in A, \quad \ldots, \quad x \pm np \in A,$$

$$f(x + np) = f(x + np - p) = f(x + np - 2p) = \cdots = f(x),$$

$$f(x) = f(n - p) = f(n - 2p) = \cdots = f(x - np).$$

Thus (2.2.1) holds for all $k \in \mathbb{Z}$.

For a more formal proof, use induction (see Section 4.1).

Problem 2.5.11 Let $A = [-1, 0] \cup (1, \infty)$. Observe that if $x \in A$ then $-x \notin A$. Define $f : A \to [1, \infty)$ by $f(x) = x^2 + 1$. Then f is a restriction of g with the same image as g. To see that f is injective, let $s_1, s_2 \in A$. Then $s_1 \neq -s_2$, so that

$$f(s_1) = f(s_2) \implies s_1^2 + 1 = s_2^2 + 1 \implies s_1^2 = s_2^2 \implies s_1 = s_2.$$

To see that f is surjective, let $t \in [1, \infty)$. If $t \in [1, 2]$, let $s = -\sqrt{t-1}$. Then $s \in [-1, 0] \subset A$ and $f(s) = t$. If $t \in (2, \infty)$, let $s = \sqrt{t-1}$. Then $s \in (1, \infty) \subset A$ and $f(s) = t$.

Problem 2.6.4 Define $f : [0, 1] \to \mathbb{R}$ by $f(x) = x$ $(0 \leq x < 1)$ and $f(1) = -1$. Then f is not increasing since $0 < 1$ and $f(0) > f(1)$, and f is not decreasing since $0 < 1/2$ and $f(0) < f(1/2)$. To see that f is injective, let $s_1, s_2 \in [0, 1]$ with $f(s_1) = f(s_2) = t$. If $t \geq 0$ then $s_1, s_2 \in [0, 1)$ and $s_1 = s_2$ $(= t)$. If $t < 0$ then $t = -1$ and $s_1 = s_2$ $(= 1)$.

Problem 3.3.7 The quadratic $F_k(x)$ has two real zeros provided $\Delta > 0$. In this case, the zeros are $\frac{1}{2}[-(3k+2) \pm \sqrt{\Delta}]$ and they differ by $\sqrt{\Delta}$. We have

$$\Delta > 0 \text{ and } \sqrt{\Delta} = 6 \iff \Delta > 0 \text{ and } \Delta = 36$$
$$\iff k^2 + 12k - 28 > 0 \text{ and } k^2 + 12k - 28 = 36$$
$$\iff (k+14)(k-2) > 0 \text{ and } (k-16)(k+4) = 0$$
$$\iff [k < -14 \text{ or } k > 2] \text{ and } [k = -16 \text{ or } k = 4]$$
$$\iff k = -16 \text{ or } k = 4.$$

Problem 4.1.9 Let $x \in \mathbb{R}$. For $n \in \mathbb{N}$, define x^n (the nth power of x) by

$$x^1 = x \quad \text{and} \quad x^{n+1} = x.x^n \quad (n \in \mathbb{N}).$$

Problem 4.1.12 For $n \in \mathbb{N}$, let $Q(n)$ be the statement:

$$P(1) \text{ and } P(2) \text{ and } \dots \text{ and } P(n).$$

First, $Q(1)$ is $P(1)$ so that, by (1), $Q(1)$ is true. Next, let $k \in \mathbb{N}$ and assume that $Q(k)$ is true. Then $P(1), \dots, P(k)$ are all true so that, by (2), $P(k+1)$ is true. Hence, $P(1), \dots, P(k+1)$ are all true so that $Q(k+1)$ is true. Thus, if $Q(k)$ is true then $Q(k+1)$ is true also. Hence, by induction, $Q(n)$ is true for all $n \in \mathbb{N}$ and, in particular, $P(n)$ is true for all $n \in \mathbb{N}$.

Problem 4.3.9 Using the binomial theorem we have

$$n^n = [k + (n-k)]^n = \sum_{r=0}^{n} \binom{n}{r} k^r (n-k)^{n-r} > \binom{n}{k} k^k (n-k)^{n-k}$$

and, since $k, (n-k) > 0$, the result follows.

Problem 5.1.2 It follows from the definitions of $\tan\theta$ and $\cot\theta$, that $\cot\theta = 1/\tan\theta$ provided both sides are defined. Now $\cot\theta$ is defined if and only if $\sin\theta \neq 0$, but $1/\tan\theta$ is defined if and only if $\tan\theta$ is defined and not zero, *i.e.* if and only if $\cos\theta \neq 0$ and $\sin\theta \neq 0$. Hence, $\cot\theta = 1/\tan\theta$ if and only if $\cos\theta \neq 0$ and $\sin\theta \neq 0$, *i.e.* if and only if $\theta \in \mathbb{R} - \{k\pi/2 : k \in \mathbb{Z}\}$.

Problem 5.2.9 Let a and b be such that $(a+b\sqrt{3})^2 = 2-\sqrt{3}$. Then $a^2+3b^2+2ab\sqrt{3} = 2 - \sqrt{3}$. This may be satisfied by taking

$$a^2 + 3b^2 = 2 \ \text{ and } \ 2ab = -1.$$

Then, eliminating b one obtains

$$4a^4 - 8a^2 + 3 = (2a^2 - 1)(2a^2 - 3) = 0,$$

for which we may choose the solution $a = 1/\sqrt{2}$ and hence $b = -1/\sqrt{2}$. (There are other choices for a and b but all of these lead to the same expressions for $\sin(\pi/12)$ and $\cos(\pi/12)$.)

Thus

$$\sin\frac{\pi}{12} = \frac{\sqrt{2-\sqrt{3}}}{2} = \frac{1}{2}\left|\frac{1-\sqrt{3}}{\sqrt{2}}\right| = \frac{\sqrt{3}-1}{2\sqrt{2}},$$

as required. The result for $\cos(\pi/12)$ is obtained similarly.

Problem 5.2.10 Using Example 5.2.8, we have

$$\tan\frac{\pi}{12} = \frac{\sin(\pi/12)}{\cos(\pi/12)} = \frac{\sqrt{3}-1}{\sqrt{3}+1} = 2 - \sqrt{3}.$$

Putting $\theta = \pi/12$ in Theorem 5.1.6 gives the required bounds for π.

Problem 5.2.12 From the given equations we get

$$(a - p)\cos\theta + (b - q)\sin\theta = 0, \tag{1}$$

$$(a - p)\cos\phi + (b - q)\sin\phi = 0. \tag{2}$$

Then $\sin\theta \times (2) - \sin\phi \times (1)$ and $\cos\phi \times (1) - \cos\theta \times (2)$ give

$$(a - p)\sin(\theta - \phi) = 0, \tag{3}$$

$$(b - q)\sin(\theta - \phi) = 0. \tag{4}$$

Since $\theta - \phi \neq 0$ for any $k \in \mathbb{Z}$, $\sin(\theta - \phi) \neq 0$. Hence, from (3), $a = p$, and from (4), $b = q$.

Problem 5.5.6 [For an alternative approach, see Example 9.4.5.] We have

$$\sin(\sin^{-1} x) = x, \qquad \cos(\cos^{-1} x) = x$$

and, since

$$\sin^{-1} x \in [-\tfrac{\pi}{2}, \tfrac{\pi}{2}], \qquad \cos^{-1} x \in [0, \pi], \qquad (*)$$

we also have

$$\cos(\sin^{-1} x) = \sqrt{1 - x^2}, \qquad \sin(\cos^{-1} x) = \sqrt{1 - x^2}.$$

Then

$$\begin{aligned}
\sin \theta &= \sin(\sin^{-1} x + \cos^{-1} x) \\
&= \sin(\sin^{-1} x) \cos(\cos^{-1} x) + \cos(\sin^{-1} x) \sin(\cos^{-1} x) \\
&= x.x + \sqrt{1 - x^2}\sqrt{1 - x^2} = x^2 + (1 - x^2) = 1.
\end{aligned}$$

Hence $\theta = \tfrac{\pi}{2} + 2k\pi$ where $k \in \mathbb{Z}$. However, from $(*)$,

$$\theta = \sin^{-1} x + \cos^{-1} x \in [-\tfrac{\pi}{2}, \tfrac{3\pi}{2}]$$

so that $k = 0$ and $\theta = \tfrac{\pi}{2}$.

Problem 6.3.6 Using the binomial theorem and de Moivre's theorem,

$$(1 + e^{2i\theta})^n = \sum_{r=0}^{n} \binom{n}{r} e^{2ir\theta}. \qquad (*)$$

Since $\cos 2\theta = 2\cos^2 \theta - 1$ and $\sin 2\theta = 2\sin \theta \cos \theta$,

$$1 + e^{2i\theta} = 1 + \cos 2\theta + i\sin 2\theta = 2\cos\theta(\cos\theta + i\sin\theta) = 2\cos\theta\, e^{i\theta}.$$

Hence $1 + e^{2i\theta} = 2\cos\theta\, e^{i\theta}$. Applying de Moivre's theorem again, and using $(*)$,

$$2^n \cos^n \theta\, e^{ni\theta} = \sum_{r=0}^{n} \binom{n}{r} e^{2ir\theta}.$$

Equating real and imaginary parts gives

$$2^n \cos^n \theta \cos n\theta = \sum_{r=0}^{n} \binom{n}{r} \cos 2r\theta,$$

$$2^n \cos^n \theta \sin n\theta = \sum_{r=0}^{n} \binom{n}{r} \sin 2r\theta.$$

Finally, taking the ratio of the above equations,

$$\tan n\theta = \frac{\displaystyle\sum_{r=0}^{n} \binom{n}{r} \sin 2r\theta}{\displaystyle\sum_{r=0}^{n} \binom{n}{r} \cos 2r\theta},$$

provided $\cos \theta \neq 0$ and $\cos n\theta \neq 0$, *i.e.* $\theta \neq (2h + 1)\pi/2$ and $\theta \neq (2k + 1)\pi/(2n)$ for any $h, k \in \mathbb{Z}$.

Problem 6.3.10 We have
$$S - zS = (z + z^2 + \cdots + z^n) - (z^2 + z^3 + \cdots + z^{n+1}) = z - z^{n+1}.$$
Hence $(1 - z)S = z(1 - z^n)$ and so, for $z \neq 1$,
$$S = \frac{z(1 - z^n)}{1 - z}.$$
Then using this with $z = e^i$ we get
$$\sin 1 + \sin 2 + \cdots + \sin n = \mathrm{Im}(e^i + e^{i2} + \cdots + e^{in}) = \mathrm{Im}\,\frac{e^i(1 - e^{in})}{1 - e^i}.$$
Hence
$$|\sin 1 + \sin 2 + \cdots + \sin n| \leq \left| \frac{e^i(1 - e^{in})}{1 - e^i} \right| = \frac{|e^i||1 - e^{in}|}{|1 - e^i|}$$
$$\leq \frac{1 + |e^{in}|}{|1 - e^i|} = \frac{2}{|1 - e^i|},$$
as required.

Problem 7.2.11 Since g is bounded on N, there exists $K \geq 0$ such that, for all $x \in N$, $|g(x)| \leq K$. Hence, for all $x \in N$,
$$0 \leq |f(x)g(x)| \leq K|f(x)|.$$
By Note 7.1.3 (3), $\lim_{x \to c} |f(x)| = 0$. Then, by Theorem 7.2.3 (2)(a), $\lim_{x \to c} K|f(x)| = 0$. Hence, by the sandwich principle, $\lim_{x \to c} |f(x)g(x)| = 0$ and a further application of Note 7.1.3 gives $\lim_{x \to c} f(x)g(x) = 0$, as required.

Problem 7.3.10 Since $\lim_{x \to c}(x - c) = 0$ and g is bounded on a neighbourhood of c, Problem 7.2.11 gives $\lim_{x \to c} h(x) = 0$. Since $h(c) = 0$, this implies that h is continuous at c.

Now let g and h be the real functions defined by
$$g(x) = \begin{cases} 1 & \text{if } x \in \mathbb{Q}, \\ 0 & \text{if } x \in \mathbb{R} - \mathbb{Q}, \end{cases} \qquad h(x) = xg(x).$$
Since g is bounded on $\mathbb{R}$ (for $x \in \mathbb{R}$, $0 \leq g(x) \leq 1$), it follows (from above with $c = 0$) that h is continuous at 0. If h is continuous at $a \neq 0$ then
$$\lim_{x \to a} g(x) = \lim_{x \to a} \frac{h(x)}{x} = \frac{h(a)}{a} = g(a)$$
implying that g is continuous at a which is false.

Let g be defined as above. It follows (from above with $c = -2$ and $c = 2$) that $x \mapsto (x + 2)g(x)$ is continuous at -2, and $x \mapsto (x - 2)g(x)$ is continuous at 2. Define $F : \mathbb{R} \to \mathbb{R}$ by $F(x) = (x + 2)(x - 2)g(x)$. Then, since $x \mapsto x + 2$ and $x \mapsto x - 2$ are continuous on $\mathbb{R}$, F is continuous at -2 and at 2, by Theorem 7.3.9 (2). If F is continuous at $a \notin \{-2, 2\}$ then
$$\lim_{x \to a} g(x) = \lim_{x \to a} \frac{F(x)}{(x + 2)(x - 2)} = \frac{F(a)}{(a + 2)(a - 2)} = g(a)$$
implying that g is continuous at a which is false.

Problem 7.3.12 Define $g : (a, b] \to \mathbb{R}$ by $g(x) = 1/(x - a)$. Then g is continuous. Suppose that g is bounded above by M. Then

$$0 < \frac{1}{b - a} \leq \frac{1}{x - a} \leq M \quad (a < x \leq b) \tag{*}$$

and $\dfrac{1}{M} \leq b - a$. Let $x = a + \dfrac{1}{M + 1}$. Then

$$a < x < a + \frac{1}{M} \leq a + (b - a) = b$$

so that $x \in (a, b]$. However, $g(x) = M + 1 > M$, contradicting $(*)$. Hence g is not bounded (above).

[This is essentially the same as the solution to Example 2.4.4 (b).]

Problem 7.3.15 Since $|f|$ is constant on $\mathbb{R}$ there exists k (≥ 0) such that, for all $x \in \mathbb{R}$, $|f(x)| = k$, *i.e.* for all $x \in \mathbb{R}$, $f(x) = k$ or $f(x) = -k$. If $k = 0$ then, for all $x \in \mathbb{R}$, $f(x) = 0$, so that f is constant on $\mathbb{R}$. Suppose f is not constant on $\mathbb{R}$. Then $k \neq 0$ and there exist $a, b \in \mathbb{R}$, with $a < b$, such that $f(a) = -k$ and $f(b) = k$, or *vice versa*. Hence $f(a) < 0 < f(b)$ or $f(a) > 0 > f(b)$. Since f is continuous on $[a, b]$ it follows from the intermediate value theorem that $f(c) = 0$ for some $c \in (a, b)$. But this contradicts $k \neq 0$. Hence f is constant on $\mathbb{R}$, as required.

Problem 7.3.17 Suitable functions $f, g, h : (0, 1) \to \mathbb{R}$ are defined by

$$f(x) = x, \qquad g(x) = 4x(1 - x), \qquad h(x) = \frac{1}{x} - 1.$$

Problem 7.4.8 From Definition 7.4.5,

$f(x) \to -\infty$ as $x \to c$

$\Leftrightarrow$ $f(x) < 0$ for all x in a punctured neighbourhood of c
and $1/f(x) \to 0$ as $x \to c$

$\Leftrightarrow$ $-f(x) > 0$ for all x in a punctured neighbourhood of c
and $1/[-f(x)] \to 0$ as $x \to c$

$\Leftrightarrow$ $-f(x) \to \infty$ as $x \to c$.

Then, using this together with Theorem 7.2.3 (2)(a) and Note 7.4.7 (2),

$f(x) \to -\infty$ and $g(x) \to L$ as $x \to c$

$\Rightarrow$ $-f(x) \to \infty$ and $-g(x) \to -L$ as $x \to c$

$\Rightarrow$ $-[f(x) + g(x)] = [-f(x)] + [-g(x)] \to \infty$ as $x \to c$

$\Rightarrow$ $f(x) + g(x) \to -\infty$ as $x \to c$.

Problem 8.1.5   This is illustrated for the case $n = 4$ in Example 8.1.3 (a).
Now consider $n \in \mathbb{N}$. Let $x \in \mathbb{R}$. Then, for $h \neq 0$,

$$\frac{(x+h)^n - x^n}{h} = \frac{1}{h} \Bigg[\sum_{r=0}^{n} \binom{n}{r} x^{n-r} h^r - x^n \Bigg]$$

$$= \frac{1}{h} \sum_{r=1}^{n} \binom{n}{r} x^{n-r} h^r = \sum_{r=1}^{n} \binom{n}{r} x^{n-r} h^{r-1}$$

$$= nx^{n-1} + h \sum_{r=2}^{n} \binom{n}{r} x^{n-r} h^{r-2}$$

$$\to nx^{n-1} \quad \text{as} \quad h \to 0.$$

Problem 8.4.6   Using Example 8.4.5, for $x \in \mathbb{R}$ and $n \in \mathbb{N}$, we have

$$\frac{d^n}{dx^n} \cos x = \frac{d^n}{dx^n} \Big(\frac{d}{dx} \sin x \Big) = \frac{d^{n+1}}{dx^{n+1}} \sin x = \sin(x + (n+1)\tfrac{\pi}{2})$$

$$= \sin(x + n\tfrac{\pi}{2} + \tfrac{\pi}{2}) = \cos(x + n\tfrac{\pi}{2}).$$

Problem 8.4.8   Differentiating both sides of

$$(fg)^{(1)} = f^{(1)}g^{(0)} + f^{(0)}g^{(1)},$$

using the product rule, gives

$$(fg)^{(2)} = [f^{(2)}g^{(0)} + f^{(1)}g^{(1)}] + [f^{(1)}g^{(1)} + f^{(0)}g^{(2)}]$$

$$= f^{(2)}g^{(0)} + 2f^{(1)}g^{(1)} + f^{(0)}g^{(2)}.$$

Differentiating again gives

$$(fg)^{(3)} = [f^{(3)}g^{(0)} + f^{(2)}g^{(1)}] + 2[f^{(2)}g^{(1)} + f^{(1)}g^{(2)}] + [f^{(1)}g^{(2)} + f^{(0)}g^{(3)}]$$

$$= f^{(3)}g^{(0)} + 3f^{(2)}g^{(1)} + 3f^{(1)}g^{(2)} + f^{(0)}g^{(3)}.$$

And differentiating again gives

$$(fg)^{(4)} = [f^{(4)}g^{(0)} + f^{(3)}g^{(1)}] + 3[f^{(3)}g^{(1)} + f^{(2)}g^{(2)}]$$

$$+ 3[f^{(2)}g^{(2)} + f^{(1)}g^{(3)}] + [f^{(1)}g^{(3)} + f^{(0)}g^{(4)}]$$

$$= f^{(4)}g^{(0)} + 4f^{(3)}g^{(1)} + 6f^{(2)}g^{(2)} + 4f^{(1)}g^{(3)} + f^{(0)}g^{(4)}.$$

Observe that the coefficients on the right-hand-sides of the formulae for $(fg)^{(1)}$, $(fg)^{(2)}$,
$(fg)^{(3)}$ and $(fg)^{(4)}$ are, respectively, 1-1, 1-2-1, 1-3-3-1 and 1-4-6-4-1, which are the first
four rows of Pascal's triangle (see Section 4.3). The formula for $(fg)^{(n)}$ $(n \in \mathbb{N})$ is given
by

$$(fg)^{(n)} = \sum_{r=0}^{n} \binom{n}{r} f^{(n-r)}g^{(r)}.$$

This result, for suitable functions f and g, is known as *Leibniz's theorem*.

Problem 9.4.7 Let $f(x) = L$ $(x > 1)$. Then, using properties of limits,

$$\lim_{x \to \infty} \frac{x+1}{x-1} = 1, \qquad \lim_{x \to \infty} \frac{x-1}{x+1} = 1,$$

and hence

$$L = \lim_{x \to \infty} f(x) = \tan^{-1} 1 + \tan^{-1} 1 = \frac{\pi}{4} + \frac{\pi}{4} = \frac{\pi}{2}.$$

Problem 10.3.4 Let $\mathcal{C}$ be the curve defined by $R(x, y) = 0$, where $R : A \to \mathbb{R}$, $A \subseteq \mathbb{R}^2$. Let $\mathcal{C}$ be symmetric about the x- and y-axes. Then

$$(x, y) \in A \text{ and } R(x, y) = 0 \;\Rightarrow\; (x, -y) \in A \text{ and } R(x, -y) = 0$$

and

$$(x, y) \in A \text{ and } R(x, y) = 0 \;\Rightarrow\; (-x, y) \in A \text{ and } R(-x, y) = 0.$$

Hence,

$$(x, y) \in A \text{ and } R(x, y) = 0 \;\Rightarrow\; (x, -y) \in A \text{ and } R(x, -y) = 0$$
$$\Rightarrow\; (-x, -y) \in A \text{ and } R(-x, -y) = 0$$

so that $\mathcal{C}$ is symmetric about the origin.

Problem 10.5.2 The parabola (10.5.1), with $a > 0$, is symmetric about the x-axis and so it is sufficient to consider

$$\mathcal{P}' : \quad y = \sqrt{ax} \quad (x \geq 0).$$

Zeros: $x = 0$.

Asymptotes: None.

Signs: $y > 0$ for $x > 0$.

Critical Points: $y' = \dfrac{\sqrt{a}}{2\sqrt{x}} > 0$ for all $x > 0$ and so there are no critical points.

Vertical tangents: $y' \to \infty$ as $x \to 0^+$, so there is a vertical tangent at $(0, 0)$.

Points of inflection: $y'' = -\dfrac{\sqrt{a}}{4x\sqrt{x}} < 0$ for all $x > 0$, so there is no point of inflection and $\mathcal{P}'$ is concave down.

The curve $\mathcal{P}$ is obtained using its symmetry about the x-axis.

The hyperbola (10.5.3) is symmetric about the y-axis as well as the x-axis and so we consider

$$\mathcal{H}' : \quad y = \frac{b}{a}\sqrt{x^2 - a^2} \quad (x \geq a).$$

Zeros: $x = a$.

Asymptotes: Since

$$y - \frac{b}{a}x = \frac{b}{a}\left(\sqrt{x^2 - a^2} - x\right)$$

$$= \frac{-ba}{\sqrt{x^2 - a^2} + x}$$

$$\to 0^- \quad \text{as} \quad x \to \infty,$$

$y = (b/a)x$ is a non-vertical asymptote, approached from below as $x \to \infty$.

Signs: $y > 0$ for $x > a$.

Critical Points: $y' = \dfrac{bx}{a\sqrt{x^2 - a^2}} > 0$ for all $x > a$ and so there are no critical points.

Vertical tangents: $y' \to \infty$ as $x \to a^+$, so there is a vertical tangent at $(a, 0)$.

Points of inflection: $y'' = \dfrac{-ba}{(x^2 - a^2)^{3/2}} < 0$ for all $x > a$, so there is no point of inflection and $\mathcal{H}'$ is concave down.

The curve $\mathcal{H}$ is obtained using its symmetry about the x- and y-axes.

Problem 11.4.10   Let A be an $n \times n$ matrix and let $i, j \in \{1, \ldots, n\}$. Then

$$(A^{\mathrm{T}})_{ij} = (A)_{ji} \quad \text{and} \quad (-A)_{ij} = -(A)_{ij}.$$

Hence,

$$A \text{ skew-symmetric} \implies A^{\mathrm{T}} = -A \implies (A)_{ji} = -(A)_{ij}.$$

In particular,

$$A \text{ skew-symmetric} \implies (A)_{ii} = -(A)_{ii} \implies (A)_{ii} = 0,$$

i.e. all the diagonal entries of A are 0.

A skew symmetric matrix is square. All the main diagonal entries are zero. Pairs of entries placed symmetrically with respect to the main diagonal have the same magnitude but opposite signs.

Problem 12.4.11   The centroids of triangles ABC and DEF have position vectors

$$\frac{1}{3}(\mathbf{a} + \mathbf{b} + \mathbf{c}) \quad \text{and} \quad \frac{1}{3}(\mathbf{d} + \mathbf{e} + \mathbf{f}),$$

respectively. Using the section formula,

$$\mathbf{d} = \frac{\beta\mathbf{b} + \alpha\mathbf{c}}{\alpha + \beta}, \quad \mathbf{e} = \frac{\beta\mathbf{c} + \alpha\mathbf{a}}{\alpha + \beta}, \quad \mathbf{f} = \frac{\beta\mathbf{a} + \alpha\mathbf{b}}{\alpha + \beta}.$$

Hence,

$$\frac{1}{3}(\mathbf{d} + \mathbf{e} + \mathbf{f}) = \frac{1}{3} \cdot \frac{1}{(\alpha + \beta)}[(\beta\mathbf{b} + \alpha\mathbf{c}) + (\beta\mathbf{c} + \alpha\mathbf{a}) + (\beta\mathbf{a} + \alpha\mathbf{b})]$$

$$= \frac{1}{3} \cdot \frac{1}{(\alpha + \beta)}[(\alpha + \beta)\mathbf{a} + (\alpha + \beta)\mathbf{b} + (\alpha + \beta)\mathbf{c}] = \frac{1}{3}(\mathbf{a} + \mathbf{b} + \mathbf{c}).$$

Problem 13.2.13 Let $ABCDEF$ be the plane hexagon and let P, Q, R, S, T and U be the mid-points of AB, BC, CD, DE, EF and FA respectively. We suppose that the lines PS and QT intersect at O and then work in terms of the position vectors $\mathbf{a}$, $\mathbf{b}$, ... of A, B, ... relative to this origin. Note that the position vectors of the mid-points are $\mathbf{p} = \frac{1}{2}(\mathbf{a} + \mathbf{b})$ and so on.

Since opposite sides are parallel, we have

$$(\mathbf{a} - \mathbf{b}) \times (\mathbf{d} - \mathbf{e}) = \mathbf{0}, \tag{1}$$

$$(\mathbf{b} - \mathbf{c}) \times (\mathbf{e} - \mathbf{f}) = \mathbf{0}, \tag{2}$$

$$(\mathbf{c} - \mathbf{d}) \times (\mathbf{f} - \mathbf{a}) = \mathbf{0}. \tag{3}$$

Now P, O, S and Q, O, T are collinear, and hence $\mathbf{p}$ is parallel to $\mathbf{s}$ and $\mathbf{q}$ is parallel to $\mathbf{t}$. Thus

$$(\mathbf{a} + \mathbf{b}) \times (\mathbf{d} + \mathbf{e}) = \mathbf{0}, \tag{4}$$

$$(\mathbf{b} + \mathbf{c}) \times (\mathbf{e} + \mathbf{f}) = \mathbf{0}. \tag{5}$$

Also, introduce vector $\mathbf{z}$:

$$(\mathbf{c} + \mathbf{d}) \times (\mathbf{f} + \mathbf{a}) = \mathbf{z}. \tag{6}$$

If we can show that $\mathbf{z} = \mathbf{0}$ then we can conclude that R, O, U are also collinear and the proof is complete.

The sums $(1) + (4)$, $(2) + (5)$ and $(3) + (6)$, respectively, give

$$2(\mathbf{a} \times \mathbf{d} + \mathbf{b} \times \mathbf{e}) = \mathbf{0}, \tag{7}$$

$$2(\mathbf{b} \times \mathbf{e} + \mathbf{c} \times \mathbf{f}) = \mathbf{0}, \tag{8}$$

$$2(\mathbf{c} \times \mathbf{f} + \mathbf{d} \times \mathbf{a}) = \mathbf{z}, \tag{9}$$

and since $\mathbf{d} \times \mathbf{a} = -\mathbf{a} \times \mathbf{d}$, the combination $(7) - (8) + (9)$ gives $\mathbf{z} = \mathbf{0}$, as required.

Problem 13.6.6 Let $Q(\alpha, \beta, \gamma)$ be the projection of P on $\mathcal{P}$. A normal vector to $\mathcal{P}$ is $\mathbf{n} = (l, m, n)$. Thus, since $\mathbf{q} - \mathbf{p}$ is normal to $\mathcal{P}$, for some $\lambda \in \mathbb{R}$,

$$\mathbf{q} - \mathbf{p} = (\alpha - a, \beta - b, \gamma - c) = \lambda(l, m, n). \tag{$*$}$$

Also, $Q \in \mathcal{P}$ and hence, using $(*)$,

$$d = l\alpha + m\beta + n\gamma = l(a + \lambda l) + m(b + \lambda m) + n(c + \lambda n)$$
$$= la + mb + nc + (l^2 + m^2 + n^2)\lambda.$$

Solving for λ and substituting in $(*)$ gives

$$\mathbf{q} - \mathbf{p} = \frac{d - la - mb - nc}{l^2 + m^2 + n^2}(l, m, n),$$

from which we get the required distance

$$|PQ| = \frac{|la + mb + nc - d|}{\sqrt{l^2 + m^2 + n^2}}.$$

Problem 14.2.4 We have

$$\sum_{r=1}^{n} 2\sin r\theta \sin \tfrac{1}{2}\theta = \sum_{r=1}^{n}[\cos(r-\tfrac{1}{2})\theta - \cos(r+\tfrac{1}{2})\theta]$$

$$= [\cos \tfrac{1}{2}\theta - \cos \tfrac{3}{2}\theta] + [\cos \tfrac{3}{2}\theta - \cos \tfrac{5}{2}\theta] + \cdots$$

$$\cdots + [\cos(n-\tfrac{1}{2})\theta - \cos(n+\tfrac{1}{2})\theta]$$

$$= \cos \tfrac{1}{2}\theta - \cos(n+\tfrac{1}{2})\theta.$$

It follows that, when $\sin \tfrac{1}{2}\theta \neq 0$,

$$\sum_{r=1}^{n} \sin r\theta = \frac{\cos \tfrac{1}{2}\theta - \cos(n+\tfrac{1}{2})\theta}{2\sin \tfrac{1}{2}\theta}. \qquad (*)$$

For $n \in \mathbb{N}$, let $\theta_n = \pi/(2n)$ and let $\mathcal{P}_n$ be the partition $(0, \theta_n, 2\theta_n, \ldots, n\theta_n)$ of $(0, \pi/2)$. Then $|\mathcal{P}_n| = \pi/(2n) \to 0$ as $n \to \infty$. Taking right end-points we get the Riemann sum

$$\sum_{r=1}^{n} \theta_n \sin r\theta_n = \theta_n \frac{\cos \tfrac{1}{2}\theta_n - \cos(n+\tfrac{1}{2})\theta_n}{2\sin \tfrac{1}{2}\theta_n} \quad \text{(using } (*))$$

$$= \left(\cos \tfrac{1}{2}\theta_n - \cos(n+\tfrac{1}{2})\theta_n\right) \Big/ \left(\frac{\sin \tfrac{1}{2}\theta_n}{\tfrac{1}{2}\theta_n}\right)$$

$$\to (\cos 0 - \cos \tfrac{\pi}{2})/1 = 1 \text{ as } n \to \infty$$

since $\tfrac{1}{2}\theta_n = \pi/(4n) \to 0$ and $(n+\tfrac{1}{2})\theta_n = \tfrac{\pi}{2} + \tfrac{1}{2}\theta_n \to \tfrac{\pi}{2}$ as $n \to \infty$, and $\sin\theta/\theta \to 1$ as $\theta \to 0$. Hence

$$\int_{1}^{\pi/2} \sin x \, dx = 1.$$

Problem 14.2.6 Define $f : [0, 1] \to [0, 1]$ by $f(x) = 1 - x^2$. Then f is a continuous bijection and $f^{-1} : [0, 1] \to [0, 1]$ is defined by $f^{-1}(x) = \sqrt{1-x}$. It follows from the mirror property for inverse functions that

$$\int_{0}^{1} f^{-1}(x) \, dx = \int_{0}^{1} f(x) \, dx.$$

Convince yourself of this by sketching the curves $y = f(x)$ and $y = f^{-1}(x)$ on the same diagram. Hence, using Example 14.2.3 (a),

$$\int_{0}^{1} \sqrt{1-x} \, dx = \int_{0}^{1} (1-x^2) \, dx = \frac{2}{3}.$$

Problem 14.3.10 The straight line through the points $(a, 0)$ and $(0, c)$ has equation $y = -(c/a)(x-a)$, and the straight line through the points $(b, 0)$ and $(0, c)$ has equation $y = -(c/b)(x-b)$. Hence, the required area is

$$\int_{a}^{0} \left[-\frac{c}{a}(x-a)\right] dx + \int_{0}^{b} \left[-\frac{c}{b}(x-b)\right] dx$$

$$= \left[-\frac{c}{a}\cdot\frac{(x-a)^2}{2}\right]_{a}^{0} + \left[-\frac{c}{b}\cdot\frac{(x-b)^2}{2}\right]_{0}^{b} = -\frac{c}{a}\cdot\frac{a^2}{2} + \frac{c}{b}\cdot\frac{b^2}{2} = \frac{(b-a)c}{2}.$$

Problem 15.1.9 For $x < 0$ we have

$$-x > 0, \quad (-x)^4 > 0, \quad -\tan^{-1} x > 0, \quad x^2 + 1 > 0$$

and hence $-y > 0$. Using Theorem 15.1.5,

$$\log(-y) = \log \frac{(-x)^4(-\tan^{-1} x)}{x^2 + 1} = 4\log(-x) + \log(-\tan^{-1} x) - \log(x^2 + 1)$$

Then differentiating implicitly, with respect to x,

$$\frac{1}{-y}\left(-\frac{dy}{dx}\right) = \frac{4}{-x}(-1) + \frac{1}{(-\tan^{-1} x)}\frac{-1}{(x^2 + 1)} - \frac{1}{x^2 + 1}2x,$$

so that

$$\frac{dy}{dx} = y\left[\frac{4}{x} + \frac{1}{(x^2 + 1)\tan^{-1} x} - \frac{2x}{x^2 + 1}\right].$$

Problem 16.4.9 For $n \in \mathbb{N}$, let $P(n)$ be the statement: $\Gamma(n) = (n-1)!$. First, for $u > 0$,

$$\int_0^u e^{-x}\, dx = 1 - e^{-u} \to 1 \quad \text{as} \quad n \to \infty$$

so that $\Gamma(1) = \int_0^\infty e^{-x}\, dx = 1 = (1-1)!$. Thus $P(1)$ is true. Next, Let $k \in \mathbb{N}$ and assume that $P(k)$ is true, *i.e.* that $\Gamma(k) = (k-1)!$. Then, for $u > 0$, integrating by parts gives

$$\int_0^u x^k e^{-x}\, dx = [x^k(-e^{-x})]_0^u - \int_0^u kx^{k-1}(-e^{-x})\, dx$$

$$= -u^k e^{-u} + k\int_0^u x^{k-1}e^{-x}\, dx$$

$$\to 0 + k\Gamma(k) \quad \text{as} \quad u \to \infty$$

so that

$$\Gamma(k+1) = \int_0^\infty x^k e^{-x}\, dx = k\Gamma(k) = k(k-1)! = k! = ((k+1) - 1)!.$$

Thus, if $P(k)$ is true then $P(k+1)$ is true also. Hence, by induction, $P(n)$ is true for all $n \in \mathbb{N}$.

Problem 17.2.3 For $C \geq 0$ and all $x \in \mathbb{R}$, $e^x + C > 0$ and $y = \frac{1}{2}\log 2(e^x + C)$ is defined and satisfies the given ODE. For $C < 0$, $x = \log(-C)$ gives $e^x + C = 0$ so that $y = \frac{1}{2}\log 2(e^x + C)$ is not defined. Hence, the general solution is valid for $x \in \mathbb{R}$ if and only if $C \geq 0$.

Problem 17.3.3 Let $x, D > 0$. Then

$$x^2 + y^2 = (Dx^2 - y)^2 \iff x^2 + y^2 = D^2x^4 - 2Dx^2y + y^2$$

$$\iff 2Dx^2y = x^2(D^2x^2 - 1)$$

$$\iff y = \frac{1}{2}\left(Dx^2 - \frac{1}{D}\right)$$

and

$$y = \frac{1}{2}\left(Dx^2 - \frac{1}{D}\right) \quad \Rightarrow \quad Dx^2 - y = Dx^2 - \frac{1}{2}\left(Dx^2 - \frac{1}{D}\right)$$
$$= \frac{1}{2}\left(Dx^2 + \frac{1}{D}\right) > 0.$$

Hence,

$$\sqrt{x^2 + y^2} = Dx^2 - y \quad \Leftrightarrow \quad x^2 + y^2 = (Dx^2 - y)^2 \quad \Leftrightarrow \quad y = \frac{1}{2}\left(Dx^2 - \frac{1}{D}\right).$$

Problem 17.5.6 The complementary function is the general solution of the homogeneous equation $ay'' + by' + cy = 0$. For all $P, Q \in \mathbb{R}$, $y = Pu_1 + Qu_2$ is a solution. Verify this directly, by substituting in the equation, or use Lemma 17.5.8 with $f(x) = g(x) = 0$. We prove that every solution can be written in this form. Let $y = Ay_1 + By_2$ be the complementary function as described in Note 17.5.5. Then there exist $A_1, B_1, A_2, B_2 \in \mathbb{R}$ such that

$$u_1 = A_1y_1 + B_1y_2 \quad \text{and} \quad u_2 = A_2y_1 + B_2y_2.$$

Hence, writing $D = A_1B_2 - A_2B_1$,

$$B_2u_1 - B_1u_2 = Dy_1 \quad \text{and} \quad -A_2u_1 + A_1u_2 = Dy_2.$$

If $D = 0$ then, since u_1 and u_2 are linearly independent, $A_1 = B_1 = A_2 = B_2 = 0$ so that both u_1 and u_2 are the zero function. This contradicts linear independence (consider, for example, $1.u_1 + 0.u_2$). Thus $D \neq 0$ and

$$y = Ay_1 + By_2 = \frac{AB_2 - A_2B}{D}u_1 + \frac{A_1B - AB_1}{D}u_2 = Pu_1 + Qu_2,$$

where $P, Q \in \mathbb{R}$.

Problem 18.1.4 For all $n \in \mathbb{N}$,

$$\frac{71}{45} - x_n = \frac{71}{45} - \frac{3n^2 - 4}{2n^2 - n} = \frac{7n^2 - 71n + 180}{45(2n^2 - n)} = \frac{(n-5)(7n-36)}{45(2n^2 - n)} \geq 0$$

(consider separately $n \leq 4$, $n = 5$, $n \geq 6$), so that $x_n \leq 71/45$. Thus $71/45$ is an upper bound for $\{x_n\}$. Since $x_5 = 71/45$ it follows that $71/45$ is the least possible upper bound.

Problem 18.2.11 Let $x_n \leq y_n \leq z_n$ $(n \geq N)$, and let $x_n \to L$ and $z_n \to L$. Then

$$0 \leq y_n = x_n \leq z_n - x_n \quad (n \geq N)$$

and, by Theorem 18.2.10 (2), $z_n - x_n \to L - L = 0$. Hence, by the special case, $y_n - x_n \to 0$ so that a further application of Theorem 18.2.10 (2) gives

$$y_n = x_n + (y_n - x_n) \to L + 0 = L.$$

Problem 18.2.18 For $n \in \mathbb{N}$, $2x_{n+1} = x_n + 1$. Hence,

$$2x_2 = x_1 + 1, \qquad \text{so} \quad 2^1 x_2 - 2^0 x_1 = 2^0,$$

$$2x_3 = x_2 + 1, \qquad \text{so} \quad 2^2 x_3 - 2^1 x_2 = 2^1,$$

$$2x_4 = x_3 + 1, \qquad \text{so} \quad 2^3 x_4 - 2^2 x_3 = 2^2,$$

$$\vdots$$

$$2x_n = x_{n-1} + 1, \qquad \text{so} \quad 2^{n-1} x_n - 2^{n-2} x_{n-1} = 2^{n-2} \quad (n \geq 2).$$

Adding the equations on the right, cancelling and using $x_1 = 0$, gives,

$$2^{n-1} x_n = 2^0 + 2^1 + 2^2 + \cdots + 2^{n-2} = 2^{n-1} - 1. \qquad i.e. \qquad x_n = \frac{2^{n-1} - 1}{2^{n-1}},$$

and this also holds for $n = 1$.

Problem 18.3.12 Since $0 \leq a_r \leq a_r + b_r$ $(r \geq N)$ and $\sum_{r=1}^{\infty} a_r$ diverges, it follows from the comparison test that $\sum_{r=1}^{\infty} (a_r + b_r)$ diverges.

If there is no restriction on b_r $(r \in \mathbb{N})$, then $\sum_{r=1}^{\infty} (a_r + b_r)$ may not diverge. For example, let $b_r = -a_r$ $(r \in \mathbb{N})$.

Problem 18.3.25 For $n \in \mathbb{N}$, let $x_r = \sin(r)$ and $y_r = 1/r$. Let $\{s_n\}$ be the sequence of partial sums of $\sum_{r=1}^{\infty} x_r$. From the solution to Problem 6.3.10, for $n \in \mathbb{N}$,

$$|s_n| \leq \frac{2}{|1 - e^i|}.$$

Hence $\{s_n\}$ is bounded. Then, since $\{y_r\}$ is a decreasing sequence of positive terms converging to 0, it follows from the given test that $\sum_{r=1}^{\infty} x_r y_r$ converges, $i.e.$ $\sum_{r=1}^{\infty} \dfrac{\sin r}{r}$ converges.

Problem 18.4.7 For $n \in \mathbb{N} \cup \{0\}$, let $a_r = \dfrac{r!(2r)!}{(3r)!}$. Then

$$\left| \frac{a_{r+1}}{a_r} \right| = \frac{(r+1)!(2r+2)!}{(3r+3)!} \cdot \frac{(3r)!}{r!(2r)!} = \frac{(r+1)(2r+2)(2r+1)}{(3r+3)(3r+2)(3r+1)}$$

$$= \frac{2(r+1)(2r+1)}{3(3r+2)(3r+1)} \to \frac{4}{27} \qquad \text{as } r \to \infty.$$

Hence the radius of convergence is $27/4$.

When $|x| = 27/4$,

$$\left| \frac{a_{r+1} x^{r+1}}{a_r x^r} \right| = \frac{27}{4} \left| \frac{a_{r+1}}{a_r} \right| = \frac{18r^2 + 27r + 9}{18r^2 + 18r + 4} > 1 \qquad (r \in \mathbb{N} \cup \{0\})$$

giving

$$|a_{r+1} x^{r+1}| > |a_r x^r| > \cdots > |a_0| = 1.$$

So $a_r x^r \not\to 0$ (see Theorem 18.2.10 (7)) and $\sum a_r x^r$ diverges. Hence the interval of convergence is $(-27/4, 27/4)$.

Problem 19.3.6 Let $f(x) = x^3 - 25$. From Example 19.2.1, the root $c = \sqrt[3]{25}$ lies in the interval $(11/4, 3)$. Hence we repeat the solution to Example 19.3.5 for the interval $[11/4, 3]$ rather than $[5/2, 3]$.

In this case we need $f'(11/4) = 363/16$, $A = f''(3) = 18$ and $|c - c_0| < 1/4$. So,

$$|c - c_1| \le \frac{48}{121}|c - c_0|^2 < \frac{3}{121} = 0.024793388,$$

$$|c - c_2| \le \frac{48}{121}|c - c_1|^2 < \frac{432}{1771561} = 0.000243853,$$

$$|c - c_3| \le \frac{48}{121}|c - c_2|^2 < \frac{8957952}{379749833583241} = 0.000000024.$$

Hence the first seven decimal places in the approximation c_3 are guaranteed to be correct.

Problem 19.4.5 Let f be a real function such that f' is continuous on $[a, b]$ and $f''(x) < 0$ for all $x \in [a, b]$. Define $g = -f$ then g' is continuous on $[a, b]$ and $g''(x) = -f''(x) > 0$ for all $x \in [a, b]$. It follows from (19.4.10) that

$$I_R(g) < \int_a^b g(x)\, dx < I_T(g),$$

where $I_R(g)$ and $I_T(g)$ are the rectangle and trapezoidal approximations to $\int_a^b g(x)\, dx$. It is clear from (19.4.1) and (19.4.2) that $I_R(f) = -I_R(g)$ and $I_T(f) = -I_T(g)$ and hence

$$-I_R(f) < -\int_a^b f(x)\, dx < -I_T(f),$$

i.e.

$$I_T(f) < \int_a^b f(x)\, dx < I_R(f).$$

Problem 19.4.6 From (19.4.1) and (19.4.2),

$$I_R = h\left[f_{1/2} + f_{3/2} + \cdots + f_{n-1/2}\right],$$

and

$$I_T = \frac{h}{2}\left[f_0 + 2f_1 + \cdots + 2f_{n-1} + f_n\right],$$

where h is the length of the interval divided by n. If we take $2n$ subintervals, of length $h/2$, then

$$I_T' = \frac{h}{4}\left[f_0 + 2f_{1/2} + 2f_1 + \cdots + 2f_{n-1} + 2f_{n-1/2} + f_n\right],$$

and it follows that $I_T' = (1/2)(I_R + I_T)$.

The further approximation obtained from Example 19.4.4 is the trapezoidal rule for eight subintervals,

$$\int_0^1 [1 - \cos(x^2)]\, dx \approx 0.097667160.$$

Problem 19.4.9   We have

$$\int_a^b x^3 \, dx = \left[\frac{x^4}{4}\right]_a^b = \frac{b-a}{4}\left[b^3 + b^2 a + ba^2 + a^3\right].$$

Also,

$$a^3 + 4\left(\frac{a+b}{2}\right)^3 + b^3 = \frac{3}{2}\left(a^3 + a^2 b + ab^2 + b^3\right),$$

and so the result follows.

Problem 19.4.11   The error in Simpson's rule on n intervals is bounded by n times the error on each of the subintervals. In particular, for $n = 2$, each subinterval has length $(b-a)/2$ and so the error on each is bounded by

$$\frac{[(b-a)/2]^5}{180}\frac{M}{2^4} = \frac{(b-a)^5}{180}\frac{M}{2^5 2^4}.$$

Hence

$$|E_2| \le 2\frac{(b-a)^5}{180}\frac{M}{2^5 2^4} = \frac{(b-a)^5}{180}\frac{M}{4^4}.$$

For the general case,

$$|E_n| \le n\frac{[(b-a)/n]^5}{180}\frac{M}{2^4} = \frac{(b-a)^5}{180}\frac{M}{(2n)^4}.$$

Problem 19.5.2   An integrating factor for the ODE is e^{-x} and so

$$e^{-x}y = \int xe^{-x} \, dx = -(1+x)e^{-x} + C,$$

and hence the general solution is

$$y = -1 - x + Ce^x.$$

If $y(0) = 1$ then $C = 2$ and hence $y(1) = -2 + 2e = 2(e-1)$ as required.

Appendix C

Limits and Continuity—A Rigorous Approach

In this appendix we introduce the formal definitions of function limit, continuity and sequence limit. Their application is illustrated with examples and a selection of theorems. A proof of L'Hôpital's rule is also included.

C.1 Function Limits

Definition C.1.1 Let f be a real function whose domain contains a punctured neighbourhood of c and let $L \in \mathbb{R}$. Then the *limit of f at c exists and equals L*, and we write

$$f \to L \quad \text{as} \quad x \to c \qquad \text{or} \qquad \lim_{x \to c} f(x) = L,$$

if and only if, for each $\varepsilon > 0$, there exists $\delta > 0$ such that

$$0 < |x - c| < \delta \implies |f(x) - L| < \varepsilon.$$

Notes C.1.2 (1) In general, δ will depend on ε.

(2) If $\delta > 0$ 'works' for a particular $\varepsilon > 0$ then the same δ will 'work' for any larger ε. Thus we are concerned, essentially, with 'small' positive ε.

(3) If $\delta > 0$ 'works' for a particular $\varepsilon > 0$ then any smaller $\delta > 0$ will 'work' for the same ε. Choosing $\delta > 0$ 'sufficiently small' guarantees that $x \in \mathbb{R}$, satisfying $0 < |x - c| < \delta$, lies in the domain of f.

(4) The function f need not be defined at c and even when it is, $f(c)$ has no bearing on the existence or value of $\lim_{x \to c} f(x)$.

In Figure C.1.1, $f(x) = 3/(x + 1)$, $c = 2$ and $L = 1$. When $\varepsilon = 0{\cdot}2$ we can take $\delta = 0{\cdot}5$ (or any smaller positive number). Geometrically, for each $\varepsilon > 0$ there must exist $\delta > 0$ such that, for x in the ranges $2 - \delta < x < 2$ and $2 < x < 2 + \delta$, the graph of f lies between the lines $y = 1 - \varepsilon$ and $y = 1 + \varepsilon$.

Example C.1.3 Show that

(a) $\lim_{x \to 1} (3x + 1) = 4,$ (b) $\lim_{x \to 3} x^2 = 9,$ (c) $\lim_{x \to 2} \dfrac{3}{x + 1} = 1.$

526

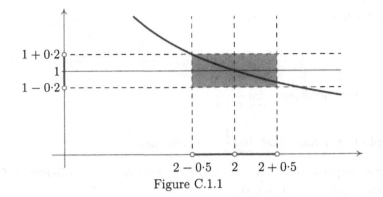

Figure C.1.1

Solution (a) Let $\varepsilon > 0$. Then, for $x \in \mathbb{R}$,

$$|(3x + 1) - 4| = |3x - 3| = 3|x - 1|$$

$$< \varepsilon \quad \text{if} \quad |x - 1| < \varepsilon/3.$$

Choose $\delta = \varepsilon/3$. Then $\delta > 0$ and

$$0 < |x - 1| < \delta \;\Rightarrow\; |(3x + 1) - 4| < \varepsilon.$$

Since $\varepsilon > 0$ was arbitrary, $\lim_{x \to 1}(3x + 1) = 4$.

(b) Let $\varepsilon > 0$. Then, for $x \in \mathbb{R}$,

$$|x - 3| < 1 \;\Rightarrow\; 2 < x < 4 \;\Rightarrow\; 5 < x + 3 < 7$$

and hence

$$|x^2 - 9| = |(x + 3)(x - 3)| = |x + 3||x - 3|$$

$$< 7|x - 3| \quad \text{if} \quad |x - 3| < 1$$

$$< \varepsilon \quad \text{if} \quad |x - 3| < \varepsilon/7.$$

Choose $\delta = \min\{1, \varepsilon/7\}$. Then $\delta > 0$ and

$$0 < |x - 3| < \delta \;\Rightarrow\; |x^2 - 9| < \varepsilon.$$

Since $\varepsilon > 0$ was arbitrary, $\lim_{x \to 3} x^2 = 9$.

(c) Let $\varepsilon > 0$. Then, for $x \in \mathbb{R}$,

$$|x - 2| < 1 \;\Rightarrow\; 1 < x < 3 \;\Rightarrow\; 2 < x + 1 < 4$$

and hence

$$\left|\frac{3}{x + 1} - 1\right| = \left|\frac{2 - x}{x + 1}\right| = \frac{|x - 2|}{|x + 1|}$$

$$< \frac{|x - 2|}{2} \quad \text{if} \quad |x - 2| < 1$$

$$< \varepsilon \quad \text{if} \quad |x - 2| < 2\varepsilon.$$

Choose $\delta = \min\{1, 2\varepsilon\}$. Then $\delta > 0$ and

$$0 < |x - 2| < \delta \ \Rightarrow \ \left| \frac{3}{x+1} - 1 \right| < \varepsilon.$$

Since $\varepsilon > 0$ was arbitrary, $\lim\limits_{x \to 2} \dfrac{3}{x+1} = 1$. □

Example C.1.4 Show that $\lim\limits_{x \to 0} \dfrac{1}{x}$ does not exist.

Solution Suppose that $\lim\limits_{x \to 0} (1/x) = L \in \mathbb{R}$. We take $\varepsilon = 1$ in Definition C.1.1. Since $1 > 0$, there exists $\delta > 0$ such that

$$0 < |x - 0| < \delta \ \Rightarrow \ \left| \frac{1}{x} - L \right| < 1$$

so that

$$0 < |x| < \delta \ \Rightarrow \ \left| \frac{1}{x} \right| < |L| + 1 \ \Rightarrow \ |x| > \frac{1}{|L| + 1}.$$

But this is false when $x = \min\{\delta/2, 1/(|L| + 1)\}$. The result follows. □

The next three results were stated in Section 7.1. Throughout, f and g are real functions whose domains contain a punctured neighbourhood of c.

Theorem C.1.5 *Let* $\lim\limits_{x \to c} f(x)$ *exist. Then this limit is unique.*

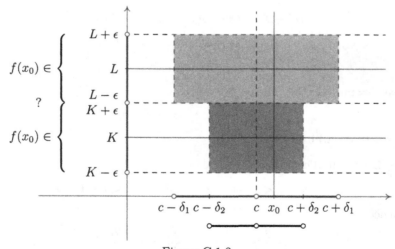

Figure C.1.2

Proof The proof is illustrated in Figure C.1.2.

Suppose that $\lim_{x \to c} f(x) = L$ and $\lim_{x \to c} f(x) = K$ where $L \neq K$. Let $\varepsilon = |L - K|/2$. Then $\varepsilon > 0$. Since $\lim_{x \to c} f(x) = L$, there exists $\delta_1 > 0$ such that

$$0 < |x - c| < \delta_1 \ \Rightarrow \ |f(x) - L| < \varepsilon$$

and, since $\lim_{x \to c} f(x) = K$, there exists $\delta_2 > 0$ such that

$$0 < |x - c| < \delta_2 \ \Rightarrow \ |f(x) - K| < \varepsilon.$$

Choose (and fix) x_0 with $0 < |x_0 - c| < \min\{\delta_1, \delta_2\}$. Then

$$2\varepsilon = |L - K| = |[L - f(x_0)] + [f(x_0) - K]|$$
$$\leq |f(x_0) - L| + |f(x_0) - K| < \varepsilon + \varepsilon = 2\varepsilon$$

which is impossible. The result follows. $\qquad\square$

Theorem C.1.6 *Let* $\lim_{x \to c} f(x) = L > 0$. *Then there exists* $\delta > 0$ *such that*

$$0 < |x - c| < \delta \ \Rightarrow \ f(x) > 0.$$

Proof Since $L > 0$ and $\lim_{x \to c} f(x) = L$, there exists $\delta > 0$ such that

$$0 < |x - c| < \delta \ \Rightarrow \ |f(x) - L| < L.$$

Hence,

$$0 < |x - c| < \delta \ \Rightarrow \ -L < f(x) - L < L \ \Rightarrow \ f(x) > 0,$$

as required. $\qquad\square$

Theorem C.1.7 *Let* $\lim_{x \to c} f(x) = L$ *and* $\lim_{x \to c} g(x) = K$. *Then*

$$\lim_{x \to c} [f(x) + g(x)] = L + K.$$

Proof Let $\varepsilon > 0$. Then $\varepsilon/2 > 0$. Since $\lim_{x \to c} f(x) = L$, there exists $\delta_1 > 0$ such that

$$0 < |x - c| < \delta_1 \ \Rightarrow \ |f(x) - L| < \varepsilon/2$$

and, since $\lim_{x \to c} g(x) = K$, there exists $\delta_2 > 0$ such that

$$0 < |x - c| < \delta_2 \ \Rightarrow \ |g(x) - K| < \varepsilon/2.$$

Let $\delta = \min\{\delta_1, \delta_2\}$. Then $\delta > 0$ and

$$0 < |x - c| < \delta \ \Rightarrow \ |[f(x) + g(x)] - [L + K]|$$
$$= |[f(x) - L] + [g(x) - K]|$$
$$\leq |f(x) - L| + |g(x) - K|$$
$$< \varepsilon/2 + \varepsilon/2 = \varepsilon.$$

Since $\varepsilon > 0$ was arbitrary, $\lim_{x \to c} [f(x) + g(x)] = L + K$ as required. $\qquad\square$

Notes C.1.8 (1) Left and right function limits are defined along the same lines as the (two sided) function limit. For example, let f be a real function whose domain contains a punctured right neighbourhood of c and let $L \in \mathbb{R}$. Then the *right limit of f at c* exists and equals L, and we write

$$f(x) \to L \quad \text{as} \quad x \to c^+ \qquad \text{or} \qquad \lim_{x \to c^+} f(x) = L,$$

if and only, for each $\varepsilon > 0$, there exists $\delta > 0$ such that

$$c < x < c + \delta \;\Rightarrow\; |f(x) - L| < \varepsilon.$$

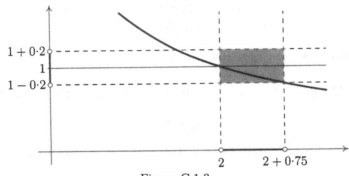

Figure C.1.3

In Figure C.1.3, $f(x) = 3/(x+1)$, $c = 2$ and $L = 1$. When $\varepsilon = 0{\cdot}2$ we can take $\delta = 0{\cdot}75$ (or any smaller positive number). Compare this with the situation in Figure C.1.1.

(2) When the dom f contains a punctured neighbourhood of c, $\lim_{x \to c} f(x) = L$ is equivalent to $\lim_{x \to c^-} f(x) = \lim_{x \to c^+} f(x) = L$.

C.2   Continuity

Definitions C.2.1 Let f be a real function whose domain contains a neighbourhood of c. Then f is *continuous at c* if and only if , for each $\varepsilon > 0$, there exists $\delta > 0$ such that

$$|x - c| < \delta \;\Rightarrow\; |f(x) - f(c)| < \varepsilon.$$

If f is continuous at each point of an open interval I then f is *continuous on I*.

Notes C.2.2 (1) In general, δ will depend on ε.

(2) Choosing $\delta > 0$ 'sufficiently small' guarantees that $x \in \mathbb{R}$, satisfying $|x - c| < \delta$, lies in the domain of f.

(3) The continuity of f at c is equivalent to $\lim_{x \to c} f(x) = f(c)$.

Example C.2.3 Show that the real function f, defined by $f(x) = \sqrt{x}$, is continuous on $(0, \infty)$. See also Example C.2.7.

Solution Let $c \in (0, \infty)$ and let $\varepsilon > 0$. Then, for $x \in (0, \infty)$,

$$|\sqrt{x} - \sqrt{c}| = \left|\frac{(\sqrt{x} - \sqrt{c})(\sqrt{x} + \sqrt{c})}{\sqrt{x} + \sqrt{c}}\right| = \frac{|x - c|}{\sqrt{x} + \sqrt{c}}$$

$$< \frac{|x - c|}{\sqrt{c}} < \varepsilon \quad \text{if} \quad |x - c| < \sqrt{c}\varepsilon.$$

Choose $\delta = \sqrt{c}\varepsilon$. Then $\delta > 0$ and

$$|x - c| < \delta \quad \Rightarrow \quad |\sqrt{x} - \sqrt{c}| < \varepsilon.$$

Since $\varepsilon > 0$ was arbitrary, f is continuous at c. Since $c \in (0, \infty)$ was arbitrary, f is continuous on $(0, \infty)$. □

Example C.2.4 Prove that the sine function is continuous on $\mathbb{R}$.

Solution Let $c \in \mathbb{R}$ and let $\varepsilon > 0$. Then, for $x \in \mathbb{R}$,

$$|\sin x - \sin c| = \left|2\cos\frac{x + c}{2}\sin\frac{x - c}{2}\right|$$

$$= 2\left|\cos\frac{x + c}{2}\right|\left|\sin\frac{x - c}{2}\right|$$

$$\leq 2(1)\left|\frac{x - c}{2}\right| = |x - c|$$

$$< \varepsilon \quad \text{if} \quad |x - c| < \varepsilon.$$

Choose $\delta = \varepsilon$. Then $\delta > o$ and

$$|x - c| < \delta \quad \Rightarrow \quad |\sin x - \sin c| < \varepsilon.$$

Since $\varepsilon > 0$ was arbitrary, sin is continuous at c. Since $c \in \mathbb{R}$ was arbitrary, sin is continuous on $\mathbb{R}$. □

The next result was stated in Section 7.3.

Theorem C.2.5 *Let f and g be real functions with f continuous at c and g continuous at $f(c)$. Then $g \circ f$ is continuous at c.*

Proof Let $\varepsilon > 0$. Since g is continuous at $f(c)$, there exists $\delta' > 0$ such that

$$|y - f(c)| < \delta' \quad \Rightarrow \quad |g(y) - g(f(c))| < \varepsilon$$

and, since f is continuous at c, there exists $\delta > 0$ such that

$$|x - c| < \delta \quad \Rightarrow \quad |f(x) - f(c)| < \delta'.$$

Hence $g \circ f$ is defined on, $(c - \delta, c + \delta)$, a neighbourhood of c, and

$$|x - c| < \delta \quad \Rightarrow \quad |g(f(x)) - g(f(c))| < \varepsilon,$$

i.e.

$$|x - c| < \delta \quad \Rightarrow \quad |(g \circ f)(x) - (g \circ f)(c)| < \varepsilon.$$

Since $\varepsilon > 0$ was arbitrary, $g \circ f$ is continuous at c as required. □

Notes C.2.6 (1) Left and right continuity is defined along the same lines as (two sided) continuity. For example, let f be a real function whose domain contains a punctured right neighbourhood of c. Then f is *right continuous at c* if and only if, for each $\varepsilon > 0$, there exists $\delta > 0$ such that

$$c \le x < c + \delta \quad \Rightarrow \quad |f(x) - f(c)| < \varepsilon.$$

This is equivalent to $\lim_{x \to c^+} f(x) = f(c)$.

(2) When the domain of f contains a neighbourhood of c, f continuous at c is equivalent to f left continuous *and* right continuous at c.

(3) When the domain of f contains a closed interval $[a, b]$, we say that f is continuous on $[a, b]$ if and only if it is continuous on (a, b), right continuous at a and left continuous at b.

Example C.2.7 Show that the real function f, defined by $f(x) = \sqrt{x}$, is right continuous at 0.

Solution Let $\varepsilon > 0$. Choose $\delta = \varepsilon^2$. Then $\delta > 0$ and

$$0 \le x < \delta \quad \Rightarrow \quad \sqrt{x} < \sqrt{\delta} = \varepsilon,$$

i.e.

$$0 \le x < 0 + \delta \Rightarrow |f(x) - f(0)| < \varepsilon.$$

Since $\varepsilon > 0$ was arbitrary, f is right continuous at 0. □

C.3   L'Hôpital's Rule

L'Hôpital's rule was introduced, without proof, in Section 9.6. Here we provide a proof based on the definition of limit given in Section C.1 of this appendix. First we prove a generalisation of the mean value theorem.

Theorem C.3.1 (Cauchy's Mean Value Theorem) *Let f and g be real functions, continuous on $[a, b]$ and differentiable on (a, b), with $g'(x) \ne 0$ $(x \in (a, b))$. Then there exists $c \in (a, b)$ such that*

$$\frac{f'(c)}{g'(c)} = \frac{f(b) - f(a)}{g(b) - g(a)}.$$

Proof By Rolle's theorem, $g(b) - g(a) \ne 0$. Define $h : [a, b] \to \mathbb{R}$ by $h(x) = f(x) - kg(x)$, where k is chosen so that $h(a) = h(b)$, *i.e.*

$$k = \frac{f(b) - f(a)}{g(b) - g(a)}. \tag{$*$}$$

Then h is continuous on $[a, b]$ and differentiable on (a, b). So, by Rolle's theorem, there exists $c \in (a, b)$ such that $h'(c) = 0$,   *i.e.* $f'(c) = kg'(c)$. Using (*) the result follows.   □

Theorem C.3.2 (L'Hôpital's Rule) *Let f and g be real functions, continuous on a neighbourhood N of c and differentiable on $N - \{c\}$, with $f(c) = g(c) = 0$ and $g'(x) \neq 0$ $(x \in \mathbb{N} - \{c\})$. Let $L \in \mathbb{R}$. Then*

$$\lim_{x \to c} \frac{f'(x)}{g'(x)} = L \quad \Rightarrow \quad \lim_{x \to c} \frac{f(x)}{g(x)} = L.$$

Proof Let $\displaystyle\lim_{x \to c} \frac{f'(x)}{g'(x)} = L$ and let $\varepsilon > 0$. Then there exists $\delta > 0$ such that $M = [c - \delta, c + \delta] \subseteq N$ and

$$0 < |x - c| < \delta \quad \Rightarrow \quad \left| \frac{f'(x)}{g'(x)} - L \right| < \varepsilon.$$

By Cauchy's mean value theorem, for each $x \in (c, c + \delta)$, there exists $c_x \in (c, x)$ such that

$$\frac{f'(c_x)}{g'(c_x)} = \frac{f(x) - f(c)}{g(x) - g(c)} = \frac{f(x)}{g(x)},$$

and similarly, for each $x \in (c - \delta, c)$, there exists $c_x \in (x, c)$ with the same property. Hence, $f(x)/g(x)$ is defined for $x \in M - \{c\}$ and

$$0 < |x - c| < \delta \quad \Rightarrow \quad 0 < |c_x - c| < \delta \quad \Rightarrow \quad \left| \frac{f'(c_x)}{g'(c_x)} - L \right| < \varepsilon$$

$$\Rightarrow \quad \left| \frac{f(x)}{g(x)} - L \right| < \varepsilon.$$

Since $\varepsilon > 0$ was arbitrary, $\displaystyle\lim_{x \to c} f(x)/g(x) = L$.   □

Note C.3.3 Since the existence of $\displaystyle\lim_{x \to c}[f'(x)/g'(x)]$ tacitly implies that $g'(x) \neq 0$ for all x in a punctured neighbourhood of c, this latter condition might be omitted from the statement of Theorem C.3.2.

C.4   Sequence Limits

Recall that a real sequence $\{x_n\}$ is a function $x : \mathbb{N} \to \mathbb{R}$ where, for each $n \in \mathbb{N}$, $x(n) = x_n$. It is sometimes convenient to regard $\{x_n\}$ as an infinite list of real numbers, indexed by $\mathbb{N}$, where, for each $n \in \mathbb{N}$, the nth term is x_n.

Definition C.4.1 Let $\{x_n\}$ be a real sequence and let $L \in \mathbb{R}$. We say that $\{x_n\}$ *converges* to L or has *limit* L, and write

$$x_n \to L \text{ as } x \to \infty \quad \text{or} \quad \lim_{n \to \infty} x_n = L,$$

if and only if, for each $\varepsilon > 0$, there exists $n_0 \in \mathbb{N}$ such that

$$n > n_0 \quad \Rightarrow \quad |x_n - L| < \varepsilon.$$

Notes C.4.2 (1) In general, n_0 will depend on ε.

(2) If $n_0 \in \mathbb{N}$ 'works' for a particular $\varepsilon > 0$ then the same n_0 will 'work' for any larger ε. Thus we are concerned, essentially, with 'small' positive ε.

(3) If $n_0 \in \mathbb{N}$ 'works' for a particular $\varepsilon > 0$ then any larger n_0 will 'work' for the same ε.

In Figure C.4.1, $x_n = (n-1)/n$ $(n \in \mathbb{N})$ and $L = 1$. When $\varepsilon = 0{\cdot}4$ we can take $n_0 = 2$ (or any larger integer) and when $\varepsilon = 0{\cdot}2$ we can take $n_0 = 5$. Geometrically, for each $\varepsilon > 0$ there must exist $n_0 \in \mathbb{N}$ such that, to the right of the vertical line $n = n_0$, the graph of the sequence lies between the horizontal lines $v = 1 - \varepsilon$ and $v = 1 + \varepsilon$.

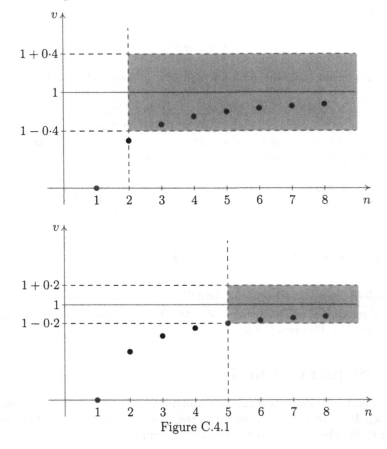

Figure C.4.1

Example C.4.3 Show that

(a) $\displaystyle\lim_{n\to\infty} \frac{n-1}{n} = 1,$ (b) $\displaystyle\lim_{n\to\infty} \frac{2n^2 + 3n + 4}{3n^2 + 2n + 1} = \frac{2}{3},$ (c) $\displaystyle\lim_{n\to\infty} \frac{2n-4}{2n-5} = 1.$

Solution (a) Let $\varepsilon > 0$. Then

$$\left|\frac{n-1}{n} - 1\right| = \frac{1}{n} < \varepsilon \quad \text{if} \quad n > \frac{1}{\varepsilon}.$$

Choose $n_0 \in \mathbb{N}$ with $n_0 \geq 1/\varepsilon$. Then

$$n > n_0 \quad \Rightarrow \quad \left| \frac{n-1}{n} - 1 \right| < \varepsilon.$$

Since $\varepsilon > 0$ was arbitrary, $\displaystyle \lim_{n \to \infty} \frac{n-1}{n} = 1$.

(b) Let $\varepsilon > 0$. Then

$$\left| \frac{2n^2 + 3n + 4}{3n^2 + 2n + 1} - \frac{2}{3} \right| = \frac{5n + 10}{3(3n^2 + 2n + 1)} < \frac{5n + 10n}{3(3n^2)} = \frac{5}{3n}$$

$$< \varepsilon \quad \text{if} \quad n > \frac{5}{3\varepsilon}.$$

Choose $n_0 \in \mathbb{N}$ with $n_0 \geq \dfrac{5}{3\varepsilon}$. Then

$$n > n_0 \quad \Rightarrow \quad \left| \frac{2n^2 + 3n + 4}{3n^2 + 2n + 2} - \frac{2}{3} \right| < \varepsilon.$$

Since $\varepsilon > 0$ was arbitrary, $\displaystyle \lim_{n \to \infty} \frac{2n^2 + 3n + 4}{3n^2 + 2n + 1} = \frac{2}{3}$.

(c) Let $\varepsilon > 0$. Then

$$\left| \frac{2n-4}{2n-5} - 1 \right| = \left| \frac{1}{2n-5} \right| = \frac{1}{2n-5} \quad \text{if} \quad n > 2$$

$$< \frac{1}{2n-n} = \frac{1}{n} \quad \text{if} \quad n > 5$$

$$< \varepsilon \quad \text{if} \quad n > \frac{1}{\varepsilon}.$$

Choose $n_0 \in \mathbb{N}$ with $n_0 \geq \max\{5, 1/\varepsilon\}$. Then

$$n > n_0 \quad \Rightarrow \quad \left| \frac{2n-4}{2n-5} - 1 \right| < \varepsilon.$$

Since $\varepsilon > 0$ was arbitrary, $\displaystyle \lim_{n \to \infty} \frac{2n-4}{2n-5} = 1$. $\qquad \square$

We next prove a selection of results which were stated in Section 18.2. Throughout, $\{x_n\}$, $\{y_n\}$ and $\{z_n\}$ are real sequences.

Theorem C.4.4 *Let $\{x_n\}$ converge. Then its limit is unique.*

Proof Suppose that $\{x_n\}$ converges to L and $\{x_n\}$ converges to K where $L \neq K$. Let $\varepsilon = |L - K|/2$. Then $\varepsilon > 0$. Since $\{x_n\}$ converges to L, there exists $n_1 \in \mathbb{N}$ such that

$$n > n_1 \quad \Rightarrow \quad |x_n - L| < \varepsilon$$

and, since $\{x_n\}$ converges to K, there exists $n_2 \in \mathbb{N}$ such that

$$n > n_2 \implies |x_n - K| < \varepsilon.$$

Choose (and fix) $m \in \mathbb{N}$ with $m > \max\{n_1, n_2\}$. Then

$$2\varepsilon = |L - K| = |(L - x_m) + (x_m - K)|$$
$$\leq |L - x_m| + |x_m - K| < \varepsilon + \varepsilon = 2\varepsilon$$

which is impossible. The result follows.   □

Theorem C.4.5 *Let $K \in \mathbb{R}$ and define*

$$x_n = K, \qquad y_n = \frac{1}{n} \qquad (n \in \mathbb{N}).$$

Then $\{x_n\}$ converges to K and $\{y_n\}$ converges to 0.

Proof Let $\varepsilon > 0$. Then, for $n \in \mathbb{N}$,

$$|x_n - K| = 0 < \varepsilon \quad \text{for all } n \in \mathbb{N}$$

and
$$|y_n - 0| = 1/n < \varepsilon \quad \text{if } n > 1/\varepsilon.$$

Choose $n_1 = 1$ and $n_2 \in \mathbb{N}$ with $n_2 > 1/\varepsilon$. Then

$$n > n_1 \implies |x_n - K| < \varepsilon \qquad \text{and} \qquad n > n_2 \implies |y_n - 0| < \varepsilon.$$

Since $\varepsilon > 0$ was arbitrary, $\{x_n\}$ converges to K and $\{y_n\}$ converges to 0.   □

Theorem C.4.6 *Let $\{x_n\}$ and $\{y_n\}$ converge to L and K, respectively, let $\lambda \in \mathbb{R}$ and let f be a real function continuous at L. Then*

(1) *$\{\lambda x_n\}$ converges to λL,*

(2) *$\{x_n + y_n\}$ converges to $L + K$,*

(3) *$\{f(x_n)\}$ converges to $f(L)$,*

Proof (1) If $\lambda = 0$ then the result follows from Theorem C.4.5. Suppose $\lambda \neq 0$. Let $\varepsilon > 0$. Then $\varepsilon/|\lambda| > 0$. Since $\{x_n\}$ converges to L, there exists $n_0 \in \mathbb{N}$ such that

$$n > n_0 \implies |x_n - L| < \varepsilon/|\lambda|$$

Then
$$n > n_0 \implies |\lambda x_n - \lambda L| = |\lambda||x_n - L| < \varepsilon.$$

Since $\varepsilon > 0$ was arbitrary, $\{\lambda x_n\}$ converges to λL.

(2) Let $\varepsilon > 0$. Then $\varepsilon/2 > 0$. Since $\{x_n\}$ converges to L, there exists $n_1 \in \mathbb{N}$ such that

$$n > n_1 \implies |x_n - L| < \varepsilon/2$$

and, since $\{y_n\}$ converges to K, there exists $n_2 \in \mathbb{N}$ such that

$$n > n_2 \Rightarrow |y_n - K| < \varepsilon/2.$$

Let $n_0 = \max\{n_1, n_2\}$. Then $n_0 \in \mathbb{N}$ and

$$
\begin{aligned}
n > n_0 \Rightarrow |(x_n + y_n) - (L + K)| &= |(x_n - L) + (y_n - K)| \\
&\leq |x_n - L| + |y_n - K| \\
&< \varepsilon/2 + \varepsilon/2 = \varepsilon.
\end{aligned}
$$

Since $\varepsilon > 0$ was arbitrary, $\{x_n + y_n\}$ converges to $L + K$.

(3) Let $\varepsilon > 0$. Since f is continuous at L, there exists $\delta > 0$ such that

$$|y - L| < \delta \Rightarrow |f(y) - f(L)| < \varepsilon.$$

Since $\{x_n\}$ converges to L, there exists $n_0 \in \mathbb{N}$ such that

$$n > n_0 \Rightarrow |x_n - L| < \delta.$$

Then

$$n > n_0 \Rightarrow |x_n - L| < \delta \Rightarrow |f(x_n) - f(L)| < \varepsilon.$$

Since $\varepsilon > 0$ was arbitrary, $\{f(x_n)\}$ converges to $f(L)$. $\qquad\square$

Theorem C.4.7 (Sandwich Principle) *Let $\{x_n\}$, $\{y_n\}$ and $\{z_n\}$ be such that $x_n \leq y_n \leq z_n$ $(n \geq N)$ and both $\{x_n\}$ and $\{z_n\}$ converge to L. Then $\{y_n\}$ converges to L.*

Proof Let $\varepsilon > 0$. Then there exist $n_1, n_2 \in \mathbb{N}$ such that

$$n > n_1 \Rightarrow |x_n - L| < \varepsilon \Rightarrow L - \varepsilon < x_n < L + \varepsilon$$

and

$$n > n_2 \Rightarrow |z_n - L| < \varepsilon \Rightarrow L - \varepsilon < z_n < L + \varepsilon.$$

Hence

$$n > n_0 = \max\{n_1, n_2\} \Rightarrow L - \varepsilon < x_n \leq y_n \leq z_n < L + \varepsilon$$

so that

$$n > n_0 \Rightarrow |y_n - L| < \varepsilon.$$

Since $\varepsilon > 0$ was arbitrary, $\{y_n\}$ converges to L, $\qquad\square$

Theorem C.4.8 *Let $\{x_n\}$ converge. Then $\{x_n\}$ is bounded.*

Proof We show that there exists $K \geq 0$ such that $|x_n| \leq K$ $(n \in \mathbb{N})$. Let $\{x_n\}$ converge to L. Since $1 > 0$, there exists $n_0 \in \mathbb{N}$ such that

$$n > n_0 \Rightarrow |x_n - L| < 1.$$

Hence,

$$n > n_0 \Rightarrow |x_n| - |L| < 1 \Rightarrow |x_n| < |L| + 1.$$

Let

$$K = \max\{|x_1|, |x_2|, \ldots, |x_{n_0}|, |L| + 1\}.$$

Then $K \geq 0$ and, for all $n \in \mathbb{N}$, $|x_n| \leq K$. $\qquad\square$

Theorem C.4.9 *Let $L \in \mathbb{R}$. Then $\{x_n\}$ converges to L if and only if every subsequence of $\{x_n\}$ converges to L.*

Proof ($\Rightarrow$) Let $\{x_n\}$ converge to L and let $y_n = x_{k_n}$ ($n \in \mathbb{N}$) where $\{k_n\}$ is a strictly increasing sequence in $\mathbb{N}$. Let $\varepsilon > 0$. Then there exists $n_0 \in \mathbb{N}$ such that

$$n > n_0 \quad \Rightarrow \quad |x_n - L| < \varepsilon.$$

A simple induction argument gives $k_n \geq n$ ($n \in \mathbb{N}$). Hence

$$n > n_0 \quad \Rightarrow \quad k_n > n_0 \quad \Rightarrow \quad |x_{k_n} - L| < \varepsilon \quad \Rightarrow \quad |y_n - L| < \varepsilon.$$

Since $\varepsilon > 0$ was arbitrary, $\{y_n\}$ converges to L.

($\Leftarrow$) This is immediate since $\{x_n\}$ is a subsequence of itself. $\qquad \square$

Example C.4.10 For $n \in \mathbb{N}$, let

$$x_n = (-1)^n \left(\frac{n-1}{n} \right).$$

Show that $\{x_n\}$ does not converge.

Solution For $n \in \mathbb{N}$,

$$x_{2n-1} = (-1) \left(\frac{2n-2}{2n-1} \right) = -\left(\frac{2 - 2(1/n)}{2 - 1/n} \right) \to -1$$

and

$$x_{2n} = (+1) \left(\frac{2n-1}{2n} \right) = \frac{2 - 1/n}{2} \to 1$$

as $n \to \infty$. Thus the subsequences $\{x_{2n-1}\}$ and $\{x_{2n}\}$ converge but to different limits. Hence $\{x_n\}$ does not converge. $\qquad \square$

Remark The definition and many of the proofs in this section apply equally to *complex* sequences, *i.e.* functions $z : \mathbb{N} \to \mathbb{C}$, provided $|\cdot|$ is taken to mean the complex modulus.

Appendix D

Properties of Trigonometric Functions

Throughout, θ and ϕ are real numbers.

1. $-1 \le \sin\theta \le 1, \qquad -1 \le \cos\theta \le 1.$

2. $\sin^2\theta + \cos^2\theta = 1, \qquad \tan^2\theta + 1 = \sec^2\theta, \qquad 1 + \cot^2\theta = \operatorname{cosec}^2\theta.$

3. For $k \in \mathbb{Z}$, $\quad \sin(\theta + 2k\pi) = \sin\theta \quad$ (period 2π),
    $\cos(\theta + 2k\pi) = \cos\theta \quad$ (period 2π),
    $\tan(\theta + k\pi) \ = \tan\theta \quad$ (period π).

4. $\sin(-\theta) = -\sin\theta \quad$ (odd function),
    $\cos(-\theta) = \cos\theta \quad$ (even function),
    $\tan(-\theta) = -\tan\theta \quad$ (odd function).

5. $\sin\left(\frac{\pi}{2} \pm \theta\right) = \cos\theta, \qquad \cos\left(\frac{\pi}{2} \pm \theta\right) = \mp\sin\theta.$

6. The diagram shows in which quadrants lie the angles for which sin, cos and tan are *positive*.

	sin	all
	tan	cos

7.

θ	0	$\pi/6$	$\pi/4$	$\pi/3$	$\pi/2$
$\sin\theta$	0	$1/2$	$1/\sqrt{2}$	$\sqrt{3}/2$	1
$\cos\theta$	1	$\sqrt{3}/2$	$1/\sqrt{2}$	$1/2$	0
$\tan\theta$	0	$1/\sqrt{3}$	1	$\sqrt{3}$	?

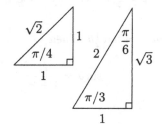

8. Solving for x when α is a given real number:

$$\sin x = \sin \alpha \iff x = \alpha + 2k\pi \text{ or } x = -\alpha + (2k+1)\pi \text{ for some } k \in \mathbb{Z},$$
$$\cos x = \cos \alpha \iff x = \alpha + 2k\pi \text{ or } x = -\alpha + 2k\pi \text{ for some } k \in \mathbb{Z},$$
$$\tan x = \tan \alpha \iff x = \alpha + k\pi \text{ for some } k \in \mathbb{Z}.$$

9. $\sin(\theta + \phi) = \sin\theta\cos\phi + \cos\theta\sin\phi,$
$\sin(\theta - \phi) = \sin\theta\cos\phi - \cos\theta\sin\phi,$
$\cos(\theta + \phi) = \cos\theta\cos\phi - \sin\theta\sin\phi,$
$\cos(\theta - \phi) = \cos\theta\cos\phi + \sin\theta\sin\phi,$
$$\tan(\theta + \phi) = \frac{\tan\theta + \tan\phi}{1 - \tan\theta\tan\phi}, \qquad \tan(\theta - \phi) = \frac{\tan\theta - \tan\phi}{1 + \tan\theta\tan\phi}.$$

10. $\sin\theta\cos\phi = \frac{1}{2}[\sin(\theta + \phi) + \sin(\theta - \phi)],$
$\cos\theta\sin\phi = \frac{1}{2}[\sin(\theta + \phi) - \sin(\theta - \phi)],$
$\cos\theta\cos\phi = \frac{1}{2}[\cos(\theta + \phi) + \cos(\theta - \phi)],$
$\sin\theta\sin\phi = \frac{1}{2}[\cos(\theta - \phi) - \cos(\theta + \phi)]$

11. $\sin\theta + \sin\phi = 2\sin\left(\dfrac{\theta + \phi}{2}\right)\cos\left(\dfrac{\theta - \phi}{2}\right),$
$\sin\theta - \sin\phi = 2\cos\left(\dfrac{\theta + \phi}{2}\right)\sin\left(\dfrac{\theta - \phi}{2}\right),$
$\cos\theta + \cos\phi = 2\cos\left(\dfrac{\theta + \phi}{2}\right)\cos\left(\dfrac{\theta - \phi}{2}\right),$
$\cos\theta - \cos\phi = -2\sin\left(\dfrac{\theta + \phi}{2}\right)\sin\left(\dfrac{\theta - \phi}{2}\right).$

12. $\sin 2\theta = 2\sin\theta\cos\theta,$
$\cos 2\theta = \cos^2\theta - \sin^2\theta = 1 - 2\sin^2\theta = 2\cos^2\theta - 1,$
$$\tan 2\theta = \frac{2\tan\theta}{1 - \tan^2\theta}.$$

13. $\sin^2\theta = \frac{1}{2}(1 - \cos 2\theta)$ and $\cos^2\theta = \frac{1}{2}(1 + \cos 2\theta),$

14. Let $t = \tan\frac{1}{2}\theta$. Then

$$\sin\theta = \frac{2t}{1 + t^2}, \qquad \cos\theta = \frac{1 - t^2}{1 + t^2}, \qquad \tan\theta = \frac{2t}{1 - t^2}.$$

Appendix E

Table of Integrals

The following table can be used to find derivatives as well as integrals. The real functions f and F are such that

$$f(x) = F'(x), \quad \int f(x)\, dx = F(x) + C.$$

$f(x)$	$F(x)$
$x^r \quad (r \neq -1)$	$\dfrac{x^{r+1}}{r+1}$
$\dfrac{1}{x}$	$\log|x|$
$\log x$	$x \log x - x$
e^x	e^x
$a^x \quad (a > 0)$	$\dfrac{a^x}{\log a}$
$\sin x$	$-\cos x$
$\cos x$	$\sin x$
$\tan x$	$-\log|\cos x|, \quad \log|\sec x|$
$\sec x$	$\log|\sec x + \tan x|$
$\operatorname{cosec} x$	$-\log|\operatorname{cosec} x + \cot x|, \quad \log|\operatorname{cosec} x - \cot x|$
$\cot x$	$\log|\sin x|$
$\sec^2 x$	$\tan x$
$\operatorname{cosec}^2 x$	$-\cot x$
$\sec x \tan x$	$\sec x$
$\operatorname{cosec} x \cot x$	$-\operatorname{cosec} x$
$\sin^{-1} x$	$x \sin^{-1} x + \sqrt{1-x^2}$
$\cos^{-1} x$	$x \cos^{-1} x - \sqrt{1-x^2}$
$\tan^{-1} x$	$x \tan^{-1} x - \dfrac{1}{2}\log(x^2 + 1)$

$f(x)$	$F(x)$
$\dfrac{1}{x^2 + a^2}$ $(a \neq 0)$	$\dfrac{1}{a} \tan^{-1} \dfrac{x}{a}$
$\dfrac{1}{x^2 - a^2}$ $(a \neq 0)$	$\dfrac{1}{2a} \log \left\| \dfrac{x-a}{x+a} \right\|, \quad -\dfrac{1}{a} \tanh^{-1} \dfrac{x}{a}$
$\dfrac{1}{\sqrt{a^2 - x^2}}$ $(a > 0)$	$\sin^{-1} \dfrac{x}{a}, \quad -\cos^{-1} \dfrac{x}{a}$
$\dfrac{1}{\sqrt{x^2 + b}}$ $(b \neq 0)$	$\log \|x + \sqrt{x^2 + b}\|,$ $\sinh^{-1} \dfrac{x}{\sqrt{b}} \quad (b > 0), \quad \cosh^{-1} \dfrac{x}{\sqrt{-b}} \quad (b < 0)$
$\sqrt{a^2 - x^2}$ $(a > 0)$	$\dfrac{x}{2} \sqrt{a^2 - x^2} + \dfrac{a^2}{2} \sin^{-1} \dfrac{x}{a}$
$\sqrt{x^2 + b}$ $(b \neq 0)$	$\dfrac{x}{2} \sqrt{x^2 + b} + \dfrac{b}{2} \log \|x + \sqrt{x^2 + b}\|$
$\cosh x$	$\sinh x$
$\sinh x$	$\cosh x$
$\tanh x$	$\log(\cosh x)$
$\operatorname{sech} x$	$2 \tan^{-1}(e^x), \quad \tan^{-1}(\sinh x)$
$\operatorname{cosech} x$	$\log \left\| \tanh \dfrac{x}{2} \right\|$
$\coth x$	$\log \| \sinh x \|$
$\operatorname{sech}^2 x$	$\tanh x$
$\operatorname{cosech}^2 x$	$-\coth x$
$\operatorname{sech} x \tanh x$	$-\operatorname{sech} x$
$\operatorname{cosech} x \coth x$	$-\operatorname{cosech} x$

Appendix F

Which Test for Convergence?

This 'route map' is designed to help you decide which test for convergence might apply to a given series. But it is only a rough guide, success is not guaranteed!

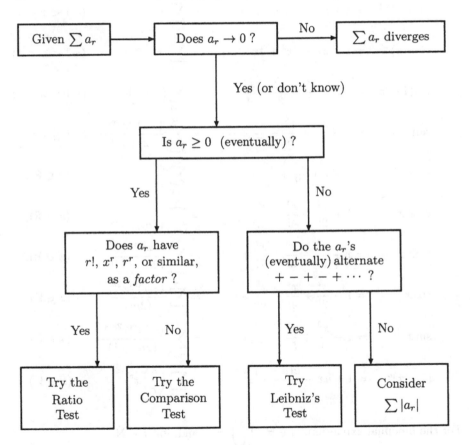

Appendix G

Standard Maclaurin Series

The following is a list of some standard Maclaurin series.

$$(1-x)^{-1} \;=\; 1 + x + x^2 + x^3 + \cdots \qquad = \sum_{r=0}^{\infty} x^r \qquad (-1 < x < 1),$$

$$(1-x)^{-2} \;=\; 1 + 2x + 3x^2 + 4x^3 + \cdots \;=\; \sum_{r=0}^{\infty} (r+1)x^r \qquad (-1 < x < 1),$$

$$\log(1-x) \;=\; -x - \frac{x^2}{2} - \frac{x^3}{3} - \frac{x^4}{4} - \cdots \;=\; \sum_{r=1}^{\infty} \left(-\frac{x^r}{r}\right) \qquad (-1 \le x < 1),$$

$$\tan^{-1} x \;=\; x - \frac{x^3}{3} + \frac{x^5}{5} - \frac{x^7}{7} + \cdots \;=\; \sum_{m=0}^{\infty} \frac{(-1)^m x^{2m+1}}{2m+1} \quad (-1 \le x \le 1),$$

$$e^x \;=\; 1 + x + \frac{x^2}{2!} + \frac{x^3}{3!} + \cdots \;=\; \sum_{r=0}^{\infty} \frac{x^r}{r!} \qquad (x \in \mathbb{R}),$$

$$\cosh x \;=\; 1 + \frac{x^2}{2!} + \frac{x^4}{4!} + \frac{x^6}{6!} + \cdots \;=\; \sum_{m=0}^{\infty} \frac{x^{2m}}{(2m)!} \qquad (x \in \mathbb{R}),$$

$$\sinh x \;=\; x + \frac{x^3}{3!} + \frac{x^5}{5!} + \frac{x^7}{7!} + \cdots \;=\; \sum_{m=0}^{\infty} \frac{x^{2m+1}}{(2m+1)!} \qquad (x \in \mathbb{R}),$$

$$\cos x \;=\; 1 - \frac{x^2}{2!} + \frac{x^4}{4!} - \frac{x^6}{6!} + \cdots \;=\; \sum_{m=0}^{\infty} \frac{(-1)^m x^{2m}}{(2m)!} \qquad (x \in \mathbb{R}),$$

$$\sin x \;=\; x - \frac{x^3}{3!} + \frac{x^5}{5!} - \frac{x^7}{7!} + \cdots \;=\; \sum_{m=0}^{\infty} \frac{(-1)^m x^{2m+1}}{(2m+1)!} \qquad (x \in \mathbb{R}),$$

$$(1+x)^a \;=\; 1 + ax + \frac{a(a-1)}{2!}x^2 + \cdots \;=\; \sum_{r=0}^{\infty} \binom{a}{r} x^r \qquad (x \in V).$$

For the binomial expansion, $a \in \mathbb{R}$, $\binom{a}{0} = 1$ and, for $r \in \mathbb{N}$,

$$\binom{a}{r} = \frac{a(a-1)\ldots(a-r+1)}{r!}.$$

The set V depends on a. When $a \in \mathbb{N}$, $V = \mathbb{R}$. In all cases, $(-1, 1) \subseteq V$.

Index

In this index **bold face** is used to specify page numbers for major references such as definitions or sections on the topic. *Italic* text indicates an example.